Developments in Mathematics

Volume 58

The **Developments in Mathematics** (DEVM) book series is devoted to publishing well-written monographs within the broad spectrum of pure and applied mathematics.

Ideally, each book should be self-contained and fairly comprehensive in treating a particular subject. Topics in the forefront of mathematical research that present new results and/or a unique and engaging approach with a potential relationship to other fields are most welcome. High quality edited volumes conveying current state-of-the-art research will occasionally also be considered for publication. The DEVM series appeals to a variety of audiences including researchers, postdocs, and advanced graduate students.

More information about this series at http://www.springer.com/series/5834

George E. Andrews · Christian Krattenthaler ·
Alan Krinik
Editors

Lattice Path Combinatorics and Applications

Springer

Editors
George E. Andrews
Department of Mathematics
The Pennsylvania State University
University Park, PA, USA

Christian Krattenthaler
Fakultät für Mathematik
Universität Wien
Vienna, Austria

Alan Krinik
Department of Mathematics and Statistics
California State Polytechnic University, Pomona
Pomona, CA, USA

ISSN 1389-2177 ISSN 2197-795X (electronic)
Developments in Mathematics
ISBN 978-3-030-11101-4 ISBN 978-3-030-11102-1 (eBook)
https://doi.org/10.1007/978-3-030-11102-1

Library of Congress Control Number: 2018966396

This Springer imprint is published by the registered company Springer Nature Switzerland AG
The registered company address is: Gewerbestrasse 11, 6330 Cham, Switzerland

Foreword

Lattice Path Conference: A Journey[1]

Everything should be made as simple as possible but not simpler.

Albert Einstein

Out of curiosity, someone may ask: How did the words "Lattice Path" become associated with a series of conferences that encompass a broad range of intellectual activities involving mathematics and applications? Representation of quite a few combinatorial objects by lattice paths, conceptually and visually simple enough without much mathematical training, drew the attention of T. V. Narayana, my supervisor, and me. We started using this name as frequently as possible. It eventually has become part of vocabulary in combinatorics. Here, we may note that H. D. Grossman published a series of papers with the title *Fun with lattice Points* in Scripta Mathematica during the late forties of the last millennium. As time progressed, we often have witnessed that "bijectivity" mapping, a simple idea but sometimes demanding ingenious talent without much mathematical preparation, does the trick in solving challenging problems. Our stories are to be told in a simple manner as far as possible.

Just after almost simultaneous publications in 1979 of two books, *Lattice Path Combinatorics with Statistical Applications* by T. V. Narayana and *Lattice Path Counting and Applications* by me, I realized that there was a substantial growing interest in lattice path combinatorics and applications in the fields of applied probability, statistics, and computer science. I also realized that the distribution of researchers was worldwide. In order to increase the awareness of the subject, my intention to organize a conference to bring eminent and young researchers together and to promote interaction between the theory group and those involved in applications resulted in the first *Conference on Lattice Path Combinatorics and*

[1]This is a revised and updated form of the article *Reminiscing Over* that appeared in Fundamenta Informaticae, vol. 117, 2012.

Applications (in short, LP Conference) that was held at McMaster University, Canada, in 1984. Incidentally, I have been at McMaster University since 1964 and the University was highly supportive of my initiative to organize the conference. Its success prompted quite a few to voice an encore for it. In the meantime, the publication of two books, *Combinatorial Enumeration* by I. P. Goulden and D. Jackson in 1983, and *Enumerative Combinatorics, vol. 1* by R. P. Stanley in 1986 encouraged me to move ahead in organizing another conference.

The second conference was held again at McMaster University in 1990, although some of the enthusiasts wished it to be held earlier. Due to the popularity of the first conference, it attracted more to participate. Also for the first time, an organizing committee of which I was a member was formed to look after the arrangements. A special feature of the second conference was a session dedicated to the memory of T. V. Narayana who passed away in 1987.

Both conferences had international participation and triggered so much interest that participants showed their willingness to organize LP Conferences elsewhere. Thus, subsequent conferences were called "International" and were held at the University of Delhi, India, in 1994; University of Vienna, Austria, in 1998; University of Athens, Greece, in 2002; East Tennessee State University, Johnson City, USA, in 2007. The Seventh International LP Conference at Siena, Italy, in 2010 was a continuation of the same trend. The main persons at local level were Kanwar Sen in India, Walter Böhm and Christian Krattenthaler in Austria, Ch. A. Charalambides in Greece, Anant Godbole in USA, and Renzo Pinzani and Simone Rinaldi in Italy. The true international nature is also reflected by past participation from Australia, Austria, Bangladesh, Canada, China, France, Germany, Greece, India, Italy, Japan, Kazakhstan, Scotland, South Korea, South Africa, Sweden, Taiwan, UK, and USA.

Throughout the years, the topics covered range over wide but related varieties like lattice path and other combinatorial problems, q-calculus, orthogonal polynomials, plane partitions, Stirling numbers, hypergeometric functions, partial orders, spanning surfaces, generating functions, recurrence relations, bijectivity, algebraic geometry, asymptotics, random walks, nonparametric inference, discrete distributions, urn models, queueing theory, quality control, and other fields of applications such as probability, statistics, physics, psychology, management science, and computer science. During the Greece Conference, the title changed to Lattice Path Combinatorics and Discrete Distributions in order to emphasize the "Discrete Distributions" content. Because of the diverse nature of topics, an international scientific committee was formed for guidance and reviewing process for the third conference, and since then, the practice has been continuing.

A new initiative started by dedicating the fourth conference to the memory of Germain Kreweras (1918–1998) and T. V. Narayana (1930–1987), both of whom made a significant contribution to the field. In the same spirit, the fifth conference was organized in the memory of István Vincze (1912–1999) and a special paper on his life and contribution was presented.

The number of participants is small and is remarkably steady to be 60–70. For that reason or otherwise, the format of the conference has been to allot the same duration, usually being 20–25 min, to each paper without discrimination and without any specially invited speakers. However, on occasions, instructional lectures of longer duration, reviewing topics of current interest have been programmed. In the Delhi Conference, the speakers were X. G. Viennot and E. Csáki, with the titles of their presentation being *Gessel–Viennot Methodology* and *Some Aspects of Random Walk*, respectively. In the sixth conference at Johnson City, USA, there were four such lectures given by George E. Andrews on *Partitions, Ferrers Graphs and q-Hypergeometric Functions*, Ira Gessel on *An Introduction to Lattice Path Enumeration*, Adrienne W. Kemp on *Discrete Distributions*, and Walter Böhm on *Lattice Path Counting and the Theory of Queues*.

The conferences are of two and a half day duration, and the organizers have so far been able to arrange some entertainment programs and sometimes after-dinner speakers.

The refereed papers among those presented at each conference have often been published as special issues of the Journal of Statistical Planning and Inference. The past issues are: vol. 14, no. 1 (1986), vol. 34, nos. 1–2 (1993), vol. 54, no. 1 (1996), vol. 101, nos. 1–2 (2002), vol. 135. no. 1 (2005), vol. 140, no. 8 (2010). The refereed papers of the seventh conference appeared in Fundamenta Informaticae, vol. 117 (2012). While I have been the guest editor, I acknowledge, with my sincere appreciation, the joint editorship of W. Böhm and C. Krattenthaler for 2002, Ch. A. Charalambides for 2005 and A. Godbole for 2010. J. N. Srivastava, Editor-in-Chief of special issues deserves my gratitude for being a source of encouragement right from the beginning of the publication of the first issue.

The eighth conference was held in 2015 at California State Polytechnic University, Pomona, California, USA, for four days instead of the usual two and a half days. In addition to all earlier features including social events, it had a special feature of having a 10-min gap between talks to allow a good discussion on the paper. This conference was dedicated to Shreeram Shankar Abhyankar (1930–2013) and Philippe Flajolet (1948–2011) and paid tribute to them in special sessions. In addition, the eighth conference had special sessions recognizing the contributions of George E. Andrews and Lajos Takács. Unfortunately, Lajos Takács was unable to attend this conference and later passed away in December of 2015. We grieve his passing with deep sadness. Two more colleagues, J. N. Srivastava and J. L. Jain, were also memorialized at the eighth conference; both were closely associated and very helpful with previous lattice path conferences.

The non-threatening title "Lattice Path" instead of one with specialized mathematical jargon had its well-expected impact when two bright high school students came forward to present papers. Another encouraging sign was the significant participation from younger people.

We thank Alan Krinik and the organizing committee who worked hard for this successful conference.

Remarkably, the egalitarian structure of the conference has proved to be successful without sacrificing the quality and has been approved by its participants.

This new enthusiasm gives hope on its continuity and possibly might take us to France or India in the near future.

In the journey over so many years, quite a few colleagues such as G. E. Andrews, N. Balakrishnan, W. Böhm, Ch. A. Charalambides, E. Csáki, I. M. Gessel, A. Godbole, A. W. Kemp, C. D. Kemp, C. Krattenthaler, A. Krinik, H. Niederhausen, and Kanwar Sen have decided to walk with me all along by my side. Some joined and left, and some others are still joining as time flows on. They have provided the strength for me to move on. They are more than professional colleagues; they are indeed true friends. I owe my deep gratitude to them, and without them, the story will remain incomplete. I also thank others who have decided to join with us supplying food and water to the walkers through their participation and professional and organizational help. The humble beginning, smallness of LP Conferences, and their structure have provided a close affinity among those who have been participating. Essentially, they have become members of what I call "Lattice Path" family. I wish a successful future journey and the well-being of the family.

Hamilton, ON, Canada
March 2018

Sri Gopal Mohanty

Preface

The 8th International Conference on Lattice Path Combinatorics and Applications took place from Monday, August 17, 2015, to Thursday, August 20, 2015, on the scenic and historic 1,400-acre campus of California State Polytechnic University, Pomona (Cal Poly Pomona), once the winter ranch of cereal magnate W. K. Kellogg, located about 30 miles east of downtown Los Angeles. During this four-day period, 42 talks were presented; see our schedule of speakers given below. Following the traditional organization of the seven previous lattice path combinatorics conferences (see *Lattice Path Conference: A Journey* by SRI GOPAL MOHANTY in our Foreword), these presentations were given sequentially. Most of our talks were 20 min in duration and took place on the first floor of the Cal Poly Pomona Library in room 1807 (see the pictures on the next page). Presentations were separated by 10 min to encourage questions and brief discussions, to facilitate the transition between speakers, and to provide participants short breaks.

The 8th International Conference on Lattice Path Combinatorics and Applications was dedicated to SHREERAM SHANKAR ABHYANKAR (1930–2013) and PHILIPPE FLAJOLET (1948–2011). Both of these outstanding mathematicians had a strong influence on the subject of lattice path combinatorics, and each had unfortunately passed away during the intervening years between the 7th and 8th lattice path combinatorics conferences, so our 8th International Conference on Lattice Path Combinatorics and Applications was a natural time to acknowledge their seminal contributions and honor their memory. We also took the opportunity during our conference to recognize and pay tribute to the many significant contributions of GEORGE ANDREWS and LAJOS TAKÁCS to lattice path combinatorics and its applications. We were pleased to be able to schedule several prominent researchers who had firsthand information or personal knowledge of the preceding mathematician's work and biographical details of their life. These speakers were well-positioned, skillful, and enthusiastic in presenting tributes and describing many of the major mathematical achievements of ABHYANKAR, FLAJOLET, ANDREWS, and TAKÁCS as well as providing biographical glimpses of the personal lives and personalities of these mathematicians.

Conference photograph, taken in Room 1807
Row 1: Anant Godbole, Tri Lai, Yan Zhuang, Ranjan Rohatgi, Larry Ericksen, Devadatta Kulkarni, Juan D. Gil, Malvina Vamvakari, Juan B. Gil, Samuel Houk, Benson Chen; row 2: Michael Wallner, Cyril Banderier, Krishnaswami Alladi, Ryan Kmet, Daniel Birmajer, M. I. A. Ageel, Gopalan Nair, Sudhir Ghorpade, Alan Krinik, Meesue Yoo; row 3: Rika Yatchak, Michael Weiner, David Nguyen, Erik Slivken, Jordan Tirrell, Christian Krattenthaler, Heinrich Niederhausen, Barbara Margolius, Dennis Eichhorn, Gregory Morrow

Room 1807, with scattered participants of the conference
Row 1: Alan Krinik, Christian Krattenthaler, Larry Ericksen, David Nguyen; row 2: Malvina Vamvakari, Sudhir Ghorpade, Devadatta Kulkarni, M. I. A. Ageel, N. N., N. N.; row 3: Ranjan Rohatgi, Tri Lai, Heinrich Niederhausen, Erik Slivken; row 4: Dennis Eichhorn, Rika Yatchak, Meesue Yoo, Krishnaswami Alladi, Daniel Birmajer, Michael Weiner; row 5: Gregory Morrow, Barbara Margolius, Cyril Banderier, Michael Wallner, Ryan Kmet, Gopalan Nair; row 6: Jordan Tirrell, Yan Zhuang; standing: Anant Godbole

Sri Gopal Mohanty, Montazer Haghighi, and Alan Krinik during the conference

This edited volume of 17 refereed articles accurately captures the commemorative and creative spirit of the 8th International Conference on Lattice Path Combinatorics and Applications. In their compelling article, *Professor Lajos Takács: A Tribute*, Sri Gopal Mohanty and Aliakbar Montazer Haghighi share their personal interactions, memories, and pictures of the Hungarian mathematician Lajos Takács as they describe Takács' many impressive contributions in probability theory, queueing theory, and combinatorics. Lajos Takács was invited to attend the 8th International Conference on Lattice Path Combinatorics and Applications, but he declined due to failing health. He passed away three and a half months later on December 4, 2015, at the age of 91 years. Sri Gopal Mohanty was able to obtain from Lajos Takács' wife, Dalma (who has since passed away on June 24, 2016), a previously unpublished research article by Lajos Takács entitled *The Distribution of the Local Time of Brownian Motion with Drift*. Professor Mohanty also obtained Dalma's permission to consider this article for publication. We are delighted to honor Lajos Takács by including this interesting article in our edited volume.

The next two articles are eloquent historical tributes by Krishnaswami Alladi entitled: *Reflections on Shreeram Abhyankar* and *My Association and Collaboration with George E. Andrews, Torchbearer of Ramanujan and Partitions*. Alladi's insider perspective should intrigue readers to understand the personal circumstance and mathematical accomplishments of these two giants in mathematics. Here is an excerpt from each article:

> Shreeram Abhyankar (22 July, 1930 – 2 Nov., 2012) was one of world's most eminent algebraic geometers. He ranked among the ten greatest mathematicians of India in the twentieth century. He belonged to the Chitpavan Brahmin community of Maharashtra and was proud of its illustrious lineage. Abhyankar's PhD thesis on the resolution of singularities problem is a classic and is among his most important contributions. I was fortunate to get to know him from my boyhood because he was a close friend of my father. Abhyankar and his wife Yvonne were our house guests in India in the sixties. Abhyankar was a fascinating, colorful, and engaging personality. He would grab your attention with his warmth, his open frankness, and his firm opinions on various matters—mathematical and non-mathematical.
>
> I could say so much more about Andrews' work on partitions, q-series, and Ramanujan, but here I chose to focus on an aspect of our joint work that shows that in manipulating q-hypergeometric series, he has no match in our generation. Even though he towers head and shoulders above the rest in the world of partitions, q-series, and Ramanujan, he is a perfect gentleman always willing to help. It is a pleasure and a privilege for me to be his friend and collaborator.

George Andrews was the only one of our honorees who physically attended our 8th International Conference on Lattice Path Combinatorics and Applications. It is hard to overstate the importance of his presence to the success of our conference and how much we all appreciated having George take part. It gave his close friend and collaborator Krishnaswami Alladi (University of Florida) the opportunity to enumerate, to explain, and to acknowledge George's many lifetime accomplishments during a Wednesday, August 19, 2015, evening dinner talk from 7–10 PM on the large outdoor patio beside Cal Poly Pomona Kellogg ranch house. Krishnaswami Alladi's talk that evening was inspiring and included a fascinating collection of personal photographs and anecdotes of George and other mathematical luminaries dating back several decades. Dr. Alladi summarized his dinner talk that evening as follows:

> George E. Andrews is the unquestioned leader in the theory of partitions and on the work of the Indian mathematical genius Srinivasa Ramanujan. My first contact with him was in 1981 in connection with his first visit to India when I put him in touch with my father who hosted him there. From then on, our friendship grew and was strengthened by his visits to India and Florida as our guest, and my frequent visits to Penn State. We collaborated on some of the most appealing problems in the theory of partitions, and I had the opportunity to observe this genius at work. I will also recall some wonderful incidents ranging from Andrews' visit to the Ramanujan Centennial in India in 1987, to his getting honorary doctorates at the University of Florida in 2002 and at SASTRA University in India in 2012 for Ramanujan's 125th Birth Anniversary.

Earlier that Wednesday afternoon George Andrews presented a warmly received talk entitled *Congruences for the Fishburn Numbers* (with colleague James Sellers listed as his co-author). However, for our edited volume, George submitted the paper *A Refinement of the Alladi–Schur Theorem*. George's conclusion from this article states:

> First, Alladi's addition to Schur's Theorem [1] given in Theorem 5.1 merits much closer study than it has received to date. Second, the conjectures of Kanade and Russell [6] suggest that the q-difference equation techniques, as initiated in [2], [3], need to be extended beyond partitions in which all parts are distinct. Part of the motivation for this paper was to show that such an extension is feasible.

Finally as a member of the Lattice Path Combinatorics and Applications Conference scientific committee, George Andrews suggested publishing articles from our 8th International Conference on Lattice Path Combinatorics and Applications with Springer-Verlag and through Krishnaswami Alladi. This edited volume owes a lot to the close Andrews–Alladi relationship. And George Andrews, once again, stepped up again by agreeing to become (along with Christian Krattenthaler and Alan Krinik) co-editor of this edited volume.

The next paper in our edited volume is entitled *Explicit Formulas for Enumeration of Lattice Paths: Basketball and the Kernel Method* and is unusual in having seven co-authors: Cyril Banderier, Christian Krattenthaler, Alan Krinik, Dmitry Kruchinin, Vladimir Kruchinin, David Nguyen, and Michael Wallner. This article grew through an exciting collaborative effort of those in attendance at our conference. The story begins with a presentation of a work in progress entitled *Counting Lattice Paths having Step Sizes of* $\{-2,-1,1,2\}$ *from j to k, where j, k are Natural Numbers and the Path Never Touches nor Goes Below the x-Axis* presented by Alan Krinik and David Nguyen (having co-authors Dmitry Kruchinin and Vladimir Kruchinin) at the 8th International Conference on Lattice Path Combinatorics and Applications. Some results were stated and explained but were not proved. By the end of the presentation, there was a lot of interest and discussion on how to complete this work. The presenters encouraged collaborative discussions. In the end, the conference participants Cyril Banderier, Christian Krattenthaler, and Michael Wallner were added to the original collection of four co-authors. After months of revisions, the original ideas evolved into a much improved article that was more general and employed the kernel method to supply the previously missing proofs.

The next article is *The Kernel Method for Lattice Paths below a Line of Rational Slope* by Cyril Banderier and Michael Wallner. Starting point for this work was a curious problem posed by Donald Knuth at the conference AofA'2014 in Paris. As with the previous paper, the kernel method is the approach that allows the authors to solve the problem, by analyzing enumerative and asymptotic properties of lattice paths below a boundary line of rational slope. We mention that Cyril Banderier, in addition to being a significant co-author on the last two long research articles, also gave an engrossing dinner presentation on Monday, August 17, 2018 7:00 pm–9:30 pm in the Kellogg West dining area, on his beloved doctoral advisor, Philippe Flajolet, entitled *The Analytic Combinatorics Point of View of Philippe Flajolet on Lattice Paths.*

The next research article in our edited volume is titled *Enumeration of Colored Dyck Paths Via Partial Bell Polynomials* by Daniel Birmajer, Juan B. Gil, Peter R. W. McNamara, and Michael D. Weiner. The authors consider a class of lattice paths with certain restrictions on their ascents and down steps and use them as building blocks to construct various families of Dyck paths. They let every building block P_j take on c_j colors and count all of the resulting colored Dyck paths of a given semilength. Their approach is to prove a recurrence relation of the convolution type, which yields a representation in terms of partial Bell polynomials that simplifies the handling of different colorings. This allows them to recover multiple known formulas for Dyck paths and related lattice paths in a unified manner. It is interesting to note that during our 8th International Conference on Lattice Path Combinatorics and Applications, we had two polished talks presented by two exceptional high school students, Samuel Houk and Juan D. Gil; see the program schedule below. Juan D. Gil talked on Dyck paths colored by Catalan numbers. Juan D. Gil is the son of Juan B. Gil who also delivered his own separate talk, A Family of Bell Transformations, during our conference. So we had a father and son each give excellent separate presentations during our 8th International Conference on Lattice Path Combinatorics and Applications. This may be a first time event in the history of lattice path combinatorics conferences.

Discrete distributions have always been a prominent topic discussed at lattice path conferences. The background is random walks and statistics on random walks, which give rise to discrete distributions. A more recent development is the study of q-analogues of discrete distributions, somewhat analogous to generalizing, say, binomial coefficients to q-binomial coefficients. This leads us to the so-called discrete q-distributions, which enjoyed many presentations during the past lattice path conferences. We are very pleased that, for the present volume, Charalambos Charalambides has written an extremely informative survey on discrete q-distributions: *A Review of the Basic Discrete q-Distributions*. Not immediately following it, the reader finds a research article on this topic: *Asymptotic Behaviour of Certain q-Poisson, q-Binomial and Negative q-Binomial Distributions* by Andreas Kyriakoussis and Malvina Vamvakari. In this article, the authors present an asymptotic analysis of some "classical" discrete q-distributions, in particular the Heine distribution and the Euler distribution, which are themselves limits of certain q-Binomial distributions.

Back to the order of articles as they appear in this volume, then next is *Families of Parking Functions Counted by the Schröder and Baxter Numbers* by Robert Cori, Enrica Duchi, Simone Rinaldi and Veronica Guerrini. Here, two new families of parking functions are introduced, one enumerated by the Schröder numbers, the other enumerated by the Baxter numbers. The link to lattice path combinatorics comes from the fact that both contain non-decreasing parking functions as special cases, which via a straightforward map are in bijection with the classical Dyck paths. Combinatorial properties of these parking functions are investigated, and bijections between these two families and classes of lattice paths are constructed. All this is as well linked to pattern-avoiding permutations, producing a rich panorama of various combinatorial objects.

Not very long ago, the so-called Stern sequence—a sequence (a_n) satisfying a system of linear recurrences implying that the value a_n will strongly depend on the binary expansion of n—was generalized to polynomial sequences in two different ways. The purpose of *Some Tilings, Colorings and Lattice Paths via Stern Polynomials* by KARL DILCHER and LARRY ERICKSEN is to tie these new polynomial sequences with lattice path enumeration. This is done via tilings of a strip of finite length.

From the outset, rook theory—going back to IRVING KAPLANSKY and JOHN RIORDAN in the 1940s—has no direct link to lattice paths. Yet, paths play an important role in *p-Rook Numbers and Cycle Counting in $C_p \wr S_n$* by JIM HAGLUND, JEFF REMMEL (who attended our conference and who has unfortunately passed away far too soon on September 29, 2017, at the age of 68 years) and MEESUE YOO, though these are paths in certain graphs and not lattice paths. In the article, the "ordinary" rook configuration counting is refined to cycle *and* q-counting—which had been done earlier in unpublished work of RICKARD EHRENBORG, JIM HAGLUND, and MARGARET READDY—and then to the context of wreath products $C_p \wr S_n$ (instead of just S_n).

With BARBARA MARGOLIUS' article *Asymptotic Estimates for Queueing Systems with Time-Varying Periodic Transition Rates*, we move to queuing theory. Obviously, there is a close connection to lattice paths since the evolution of a queue can be conveniently encoded in terms of a lattice path. The $M_t/M_t/1$-queue, the multi-server queue $(M_t/M_t/c_t)$, and queues with jumps of size one and two are considered in the above article. Estimates are derived which are asymptotic in the length of the queue. The results highlight the connections between the asymptotic periodic distribution of a stable queue with time-varying rates and the same type of queue with constant rates. The subsequent article within our edited volume is *A Combinatorial Analysis of the M/M $^{[m]}$/1 Queue*, written by GUVEN MERCANKOSK and GOPALAN NAIR. It is Markov chain techniques which play an important role here. They lead to a reformulation that allows for the application of combinatorial tools. In the case of the $M/M^{[m]}/1$ queue, an explicit expression in terms of hypergeometric series is derived.

A different kind of Markov process is in the focus of *Laws Relating Runs, Long Runs, and Steps in Gambler's Ruin, with Persistence in Two Strata* by GREGORY MORROW: the gambler's ruin process. A weighted average of runs, long runs, and steps in the path representation of the process is considered, and the limiting distribution as the amount in the "base capital" tends to infinity is computed.

The last (but certainly not least) article in the present volume is *Paired Patterns in Lattice Paths* by Ran Pan and—again—JEFF REMMEL. Paired patterns define a certain kind of "pattern containment" in lattice paths consisting of north and east steps. In the above article, paired patterns of length 4 are considered, and sets of these. The number of pattern matches is given natural combinatorial interpretations, which one finds intrinsically in the path structure. Furthermore, the corresponding generating functions are explicitly given.

Not surprisingly, this edited volume came together with some significant help and work from many people. Our appreciation begins with the fact that 33 authors associated with the 8th International Conference on Lattice Path Combinatorics and

Applications took the time and made the effort to compose and contribute the seventeen articles that comprise this edited volume. We gratefully acknowledge the important contributions of our referees for improving our articles. We thank our Springer Series Editor, Krishnaswami Alladi, for his vision and constant encouragement and our Executive Editor, Elizabeth Loew (Mathematics, Springer), for her reliable assistance, persistence, and patience and the Springer staff for their professional editing. Finally, it is a great pleasure to recognize and honor Professor Emeritus Sri Gopal Mohanty for being the founder and leader of the Lattice Path Combinatorics International Conferences and for his lifetime effort to organize, foster, and promote research in lattice path combinatorics.

University Park, USA — George E. Andrews
Vienna, Austria — Christian Krattenthaler
Pomona, USA — Alan Krinik
March 2018

Program Schedule

8th International Conference on Lattice Path Combinatorics and Applications

- **Monday, August 17, 2015**

  9:00 am–9:20 am Samuel Houk: *Dyck Paths Colored by Fibonacci Numbers*
  Juan D. Gil: *Dyck paths colored by Catalan numbers*

  9:30 am–9:50 am Thu Dinh: *The Repeated Sums of Integers*

  10:00 am–10:15 am Welcome from Dean Brian Jersky, College of Science, Cal Poly Pomona

  10:30 am–10:50 am Tri Lai: *Lozenge Tilings of a Hexagon with Three Holes*

  11:00 am–11:20 am Ranjan Rohatgi: *Lozenge Tilings of Halved Hexagons with Defects*

  Noon–1:30 pm Lunch at Kellogg West

  2:00 pm–2:20 pm Michael Weiner: *Counting Colored Paths Using Partial Bell Polynomials*

  2:30 pm–2:50 pm Juan B. Gil: *A Family of Bell Transformations*

  3:00 pm–3:20 pm Daniel Birmajer: *On Restricted Words and Colored Compositions*

  3:50 pm–4:10 pm Michael Schlosser and Meesue Yoo: *An Elliptic Analogue of the Rook Numbers*

  4:20 pm–5:10 pm Christian Krattenthaler: *Tutorial on Non-Intersecting Lattice Paths, Classical Group Characters and Multivariate (Basic) Hypergeometric Series*

  7:00 pm–9:30 pm Dinner and *The Analytic Combinatorics Point of View of Philippe Flajolet on Lattice Paths*, Cyril Banderier, Kellogg West

- **Tuesday, August 18, 2015**

  9:00 am–9:20 am YAN ZHUANG: *The Goulden–Jackson Cluster Method for Monoid Networks and an Application to Lattice Path Enumeration*

  9:30 am–9:50 am CHRISTOPHER HOFFMAN, DOUGLAS RIZZOLO AND ERIK SLIVKEN: *Fixed Points of Pattern-Avoiding Permutations*

  10:30 am–10:50 am KARL DILCHER AND LARRY ERICKSEN: *Lattice Paths and Tilings Using Stern Polynomials*

  11:00 am–11:20 am DEVADATTA KULKARNI: *Hilbert Polynomial of Ladder Determinantal Ideals: A Perspective Noon*–1:30 pm Lunch at Kellogg West

  2:00 pm–2:20 pm GERARDO RUBINO: *Analyzing Extreme Values and Loss Parameters of Queuing Models in a Finite Time Interval*

  2:30 pm–2:50 pm BARBARA MARGOLIUS: *Asymptotic Estimates for the Level Distribution of Stable Queues with Time-Varying Periodic Transition Rates*

  3:00 pm–3:20 pm ALAN KRINIK, DMITRY KRUCHININ,VLADIMIR KRUCHININ AND DAVID NGUYEN: *Counting Lattice Paths having Step Sizes of* $\{-2,-1,1,2\}$ *from j to k, where j, k are Natural Numbers and the Path Never Touches nor Goes Below the x-axis*

  3:50 pm–4:10 pm GOPALAN NS: *Generalised Ballot Theorem in a Combinatorial Analysis of* $M/M[m]/1$ *Queue*

  4:20 pm–5:10 pm ALIAKBAR HAGHIGHI AND SRI GOPAL MOHANTY: *Professor Lajos Takács: Life and Contribution to Combinatorics*

  7:00 pm–9:30 pm Dinner and *Mathemagics*, ARTHUR BENJAMIN, Kellogg West

- **Wednesday, August 19, 2015**

  8:30 am–8:50 am RAN PAN AND JEFFREY REMMEL: *Distributed Patterns in Paths*

  9:00 am–9:20 am ANTONIO BLANCA, SAMUEL CONNOLLY, ZACHARY GABOR AND ANANT GODBOLE: *Two Recent Applications of Lattice Path Theory to Other Areas of Combinatorics*

  9:30 am–9:50 am SUBIR GHOSH: *The World of Jagdish Narayan Srivastava*

  9:50 am–10:00 am SRI GOPAL MOHANTY: *Remembering Professor Joti Lal Jain*

  10:30 am–10:50 am ARTHUR BENJAMIN AND ELIZABETH REILAND: *Combinatorial Proofs of Fibonomial Identities*

  11:00 am–11:20 am SUDHIR GHORPADE: *Number of Points of Schubert Varieties over Finite Fields*

  Noon–1:30 pm Lunch at Kellogg West

2:00 pm–2:20 pm Krishnaswami Alladi: *Göllnitz's (Big) Partition Theorem and a New Companion*
2:30 pm–2:50 pm Cyril Banderier: *On a Problem by Don Knuth: Lattice paths of Slope* $2/5$

3:00 pm–3:50 pm Sudhir Ghorpade and Devadatta Kulkarni: *Shreeram Abhyankar and His Work on Enumerative Combinatorics*

4:10 pm–4:30 pm George Andrews and James Sellers: *Congruences for the Fishburn Numbers*

4:40 pm–5:00 pm Manuel Kauers and Rika Yatchak: *Walks in the Quarter-Plane with Multiple Steps*

7:00 pm–10:00 pm Dinner and talk *On George E. Andrews and Shreeram Abhyankar*, Krishnaswami Alladi, Concluding Remarks by Alan Krinik at the Kellogg House

- **Thursday, August 20, 2015**

8:30 am–8:50 am Andreas Kyriakoussis and Malvina Vamvakari: *Asymptotic Behaviour of the q-Poisson Distributions Heine and Euler by Pointwise Convergence*

9:00 am–9:20 am Felix Breuer, Dennis Eichhorn, and Brandt Kronholm: *The Combinatorics Governing the Periodicity of* $p(n, d)$ *modulo M*

9:30 am–9:50 am Jordan Tirrell: *Orthogonal Polynomials and Motzkin Paths with Peak and Flat Restrictions*

10:30 am–10:50 am Gregory J. Morrow: *Laws Relating Runs, Long Runs, and Steps in Gambler's Ruin, with Persistence in Two Strata*

11:00 am–11:20 am Michael Wallner: *A Half-Normal Limit Distribution Scheme*

Noon–1:30 pm Lunch at Kellogg West

2:00 pm–2:20 pm Alan Krinik and Gerardo Rubino: *The Concept of Pseudo-Dual of a System of Differential Equations*

2:30 pm–2:50 pm Tewodros Amdeberhan and Emily Leven: *A New Family of Lattice Paths Enumerating Cores*

3:00 pm–3:20 pm Anastasia Chavez and Nicole Yamzon: *The Dehn–Somerville Relations and the Catalan Matroid*

3:30 pm–3:50 pm Alan Krinik and Satori Schweitz: *The Restart Queue Model*

Contents

Contributors

Krishnaswami Alladi Department of Mathematics, University of Florida, Gainesville, FL, USA

George E. Andrews Department of Mathematics, The Pennsylvania State University, University Park, PA, USA

Cyril Banderier Laboratoire d'Informatique de Paris-Nord, CNRS/Université Paris 13, Villetaneuse, France

Daniel Birmajer Department of Mathematics, Nazareth College, Rochester, USA

Ch. A. Charalambides Department of Mathematics, University of Athens, Athens, Greece

Robert Cori Labri, Université Bordeaux, Bordeaux, France

Karl Dilcher Department of Mathematics and Statistics, Dalhousie University, Halifax, Nova Scotia, Canada

Enrica Duchi LIAFA, Université Paris 7, Paris, France

Larry Ericksen Millville, NJ, USA

Juan B. Gil Penn State Altoona, Altoona, USA

Veronica Guerrini Dipartimento di Ingegneria dell'Informazione e Scienze Matematiche, University of Siena, Siena, Italy

Aliakbar Montazer Haghighi Department of Mathematics, Prairie View A&M University, Prairie View, TX, USA

James Haglund Department of Mathematics, University of Pennsylvania, Philadelphia, PA, USA

Christian Krattenthaler Fakultät für Mathematik, Universität Wien, Vienna, Austria

Alan Krinik Department of Mathematics and Statistics, California State Polytechnic University, Pomona, Pomona, CA, USA

Dmitry Kruchinin Tomsk State University of Control Systems and Radio Electronics, Tomsk, Russia

Vladimir Kruchinin Tomsk State University of Control Systems and Radio Electronics, Tomsk, Russia

Andreas Kyriakoussis Department of Informatics and Telematics, Harokopion University, Tavros, Greece

Barbara Margolius College of Sciences and Health Professions, Cleveland State University, Cleveland, OH, USA

Peter R. W. McNamara Department of Mathematics, Bucknell University, Lewisburg, USA

Guven Mercankosk The University of Western Australia, Crawley, Perth, WA, Australia

Sri Gopal Mohanty Department of Mathematics and Statistics, McMaster University, Hamilton, Ontario, Canada

Gregory J. Morrow University of Colorado, Colorado Springs, CO, USA

Gopalan M. Nair The University of Western Australia, Crawley, Perth, WA, Australia

David Nguyen Department of Mathematics, University of California, Santa Barbara, CA, USA

Ran Pan Department of Mathematics, University of California, San Diego, La Jolla, CA, USA

Jeffrey B. Remmel Department of Mathematics, University of California, San Diego, La Jolla, CA, USA

Simone Rinaldi Dipartimento di Ingegneria dell'Informazione e Scienze Matematiche, University of Siena, Siena, Italy

Lajos Takács Case Western Reserve University, Cleveland, OH, USA

Malvina Vamvakari Department of Informatics and Telematics, Harokopion University, Tavros, Greece

Michael Wallner Institute of Discrete Mathematics and Geometry, TU Wien, Austria;
Laboratoire Bordelais de Recherche en Informatique, Université de Bordeaux, Bordeaux, France

Michael D. Weiner Penn State Altoona, Altoona, USA

Meesue Yoo Applied Algebra and Optimization Research Center, Sungkyunkwan University, Suwon, Republic of Korea

Professor Lajos Takács: A Tribute

Aliakbar Montazer Haghighi and Sri Gopal Mohanty

Abstract The paper consists of four parts. The first part is a description of the first author's recollection of Takács - their acquaintance, him as a teacher and a mentor. In the second part, he recounts Takács' life from his childhood to schools, university, employment and retirement, and his publications and achievements. Similar to the first part, the third part is also an account of the second author's interaction with Takács. Finally, in the fourth part, he presents a specific aspect of Takács' contribution to combinatorics.

Keywords Lajos Takács · Ballot problem

2010 Mathematics Subject Classification Primary 01A70 · Secondary 60-03 60C05

This paper is based on a presentation to honor Professor Lajos Takács on his 91st birthday, which has the title "Professor Lajos Takács: Life and Contribution to Combinatorics" and was made by Aliakbar Montazer Haghighi and Sri Gopal Mohanty at the 8th International Conference on Lattice Path Combinatorics and Applications, on August 18, 2015, at California Polytechnic State University, Pomona, California, United States of America. A different version of the presentation was published as the "Preface" in the Appl. Appl. Math. (AAM), vol. 10, Issue 2 (December 2015). Parts 1 and 2 are written by Aliakbar Montazer Haghighi, while Parts 3 and 4 are written by Sri Gopal Mohanty.

A. M. Haghighi (✉)
Department of Mathematics, Prairie View A&M University, Prairie View, TX, USA
e-mail: amhaghighi@pvamu.edu

S. G. Mohanty
Department of Mathematics and Statistics, McMaster University, Hamilton, Ontario, Canada
e-mail: mohanty@mcmaster.ca

G. E. Andrews et al. (eds.), *Lattice Path Combinatorics and Applications*, Developments in Mathematics 58, https://doi.org/10.1007/978-3-030-11102-1_1

Lajos Takács: Born in Maglód, Hungary, August 21, 1924
Died in Cleveland, Ohio, USA, December 4, 2015
Professor Emeritus of Mathematics & Statistics
Case Western Reserve University, Cleveland, Ohio, at the time of his death

1 In Memory of Lajos Takács

1.1 How Did I Come to Know Takács?

At San Francisco State University in northern California, where I was doing my graduate work during 1966–1968, I heard Lajos Takács' name in the class with Dr. Siegfried F. Neustadter (1924–2012), known as Fred. Professor Neustadter received his Ph.D. from Berkeley and spent several years at Harvard and MIT before joining SF State in 1958. He was a Mathematician consultant at Sylvania Electronics in Waltham, Massachusetts during which period Takács was working at Bell Lab. Neustadter knowing Takács, brought his 1962 celebrated book, Introduction to the Theory of Queues, to the attention of the class and taught the single-server queue, $M/M/1$, from that book.

1.2 Left the USA, Back to Iran

I left the USA in 1968 before completing my master's degree due to my mother's health problem. However, because of my political activities against the regime of late Shah of Iran while I was a student at SF State, Shah's Secret Service called "Savak" put three sanctions against me, one being that I could not leave the country for 5 years.

After some communication with the Department of Mathematics at SF State, I completed my master's degree by taking a written comprehensive exam at the US Embassy in Tehran under the supervision of late Dr. Mohsen Hashtroodi, a professor at University of Tehran.

During that 5-year period, besides writing and translating books, I guided one of my students to translate a part of Takács' book as part of her master thesis.

1.3 At Case Western Reserve University

As the duration for the sanction was coming to the end, I thought seriously of working with Takács. Finding out that Takács was at Case Western Reserve University (CWRU) in Cleveland, USA, I applied for Ph.D. program there in the Department of Operations Research (OR) and got accepted. Since there was no diplomatic relation between Iran and USA, I had to go to the Netherlands to get US visa. I along with my family came to USA in 1973 for my study at CWRU.

While I was registered in OR Department, I realized soon that Takács was in the Department of Mathematics and Statistics. (Of course, those days there was no Google to search for the location of Takács' Department.) However, after one semester in the OR Department, Professor Shelemyahu Zacks, the then Chair of the Department of Mathematics and Statistics (now Professor Emiratous at SUNY-Binghamton University), accepted me to transfer my study to his department. There, I enrolled in stochastic processes course, which was given by Takács.

Before taking my Graduate Record Exam (GRE), Professor Takács gave me the privilege of starting to work on my dissertation under his guidance. It seemed he trusted that I would pass GRE. And I did. I continued my dissertation and finished my work for Ph.D. degree by January 1976, which I received in June of that year.

1.4 Left the USA, Back to Iran Again!

Only a few days after my passing the dissertation's defense, I and my family returned home. Just after two years, in 1978 Iran's revolution occurred. During the first few years of revolution, I continued teaching and even held some highest ranks in university administration. In the meantime, the war between Iran and Iraq was becoming ugly and many unpleasant things were happening in universities. In the war, boys of fourteen and above were mandatory to join and my son was soon to become fourteen. Such circumstances made me to think of leaving the country again for a peaceful atmosphere to raise my family. So I decided to contact Takács for help and for financial support.

1.5 *Meeting Professor Mohanty in Canada*

Takács was not a fan of obtaining research financial support and grant funds, as he believed that the work completed should be awarded rather than a work being promised for future completion. So, he referred me to Professor Sri Gopal Mohanty at McMaster University and asked me to contact him. To him, I guess Mohanty was a "bank" with much grant funds.

After contacting Professor Mohanty, I was graciously offered three months teaching and research during summer 1984 at McMaster University.

At that time, two other professors, Professor J. P. Medhi from India and Dr. E. Csáki from Hungary were associated with Professor Mohanty. As a result of this acquaintance and friendship, research collaboration among us started.

(From left: Mohanty, Haghighi, Csáki, Medhi)

(From left: Shanti, Shahin, Ali, Mahyar and Mahroo in front)

We also established family friendship with Mohanty's family which is continuing till now. His wife, Shanti, is a kind and hospitable host. She makes good food and excellent chai (tea)! My entire family (wife – Shahin, myself – Ali Haghighi, son – Mahyar and daughter – Mahroo) have enjoyed the relationship.

1.6 Back to the USA and Started a Lifelong Living!

Visiting Mohanty was the first step for me and my family in preparation to leave Iran. We left the country and arrived at Los Angeles in February of 1985 to spend my sabbatical leave at California State University Fullerton. After 6 months, we moved to Columbia, South Carolina, stayed there until our both children received their first university degrees and left us. After 17 years in Columbia, my wife and I moved to Houston, Texas, where we currently live. After his education, our son, Micheal Haghighi, practices medicine and lives with his wife (Roshni Patel, also a physician) and two children (Maya and Kayvan) in Jacksonville, Florida. Our daughter received her MBA and is working and living with her daughter (Leila) in Charlotte, North Carolina.

1.7 Remembrances

Now, I will give you some particular remembrances from the time I was a student of Takács:

- Television weather forecasters (Meteorologists) in Cleveland were using the word "chance" in describing the weather, say the "chance" of rain or snow, as it was commonly used terminology for the purpose. However, during my stay, June 1973–January 1976, the use of "chance" changed to the word "probability." This change of word made Professor Takács very excited that the word "probability" finally found it place in media's vocabulary, at least among Television weather forecasters (meteorologists) in Cleveland.
- Takács was often criticized because he did not present real-life examples for his deep theoretical work to make the theory easier to understand. His response, as he used to tell me, was an analogy that his work was like new medicines available for doctors to recognize their appropriate use for patients. When I told this response to Professors Joe Gani and the late Marcel Neuts at a conference in Honor of Joe Gani in Athens, Greece, 1995, they simply laughed and enjoyed Takács' response.
- As Takács entered our classroom at Case, he would go to the blackboard (there was no whiteboard those days) and start writing from left to write and from top to bottom of the board, erase and continue writing all over the board, without looking at any note or at us, the students. It was up to us, the students, to figure out later what he was teaching. I later discovered that his lectures were based on published research papers and he usually offered no detail and explanation to us. Once in one of my queueing theory classes, Professor Otomar Hájek from Department of Mathematics and Statistics was sitting to audit the course. (Emeritus Professor Otomar Hájek was originally from Czech. He is a recipient of von Humboldt award at the TH Darmstadt, Fachbereich Mathematik and is known for his contributions to dynamical systems, game theory, and control theory.) As Takács was writing and writing and writing, suddenly he stopped. Professor Hájek asked Takács what

was it that he forgot. Takács turned his face to him, just looked at him for a few seconds, went back to the board and continued writing. Suddenly, he remembered what he had forgotten. That was the only time I noticed him forgetting something during teaching without note.

- As mentioned, I attended CWRU in June 1973. I was with my, then, 2-year old son, Mahyar (now called Micheal), and my wife, Shahin. We were staying in a graduate residence hall, right in the heart of the downtown where crimes and particularly gun shooting were high.
  After two years of frustration, my wife decided to return home, Tehran, with my son, after his 4th birthday, so that I could rush to finish and they could feel safe. After they left, I practically lived in my office, mostly with very brief naps at my desk and trying to finish my dissertation.
  During the Christmas Holidays of 1975, when I was preparing defense of my dissertation, I needed help. I called Professor Takács and ask him for help. He responded without hesitation, went to his office, and helped me out. This was one of many caring experiences I remember from this great teacher of mine at CWRU.
  As a graduate student of Lajos Takács at CWRU, who did his dissertation under his guidance and graduated in 1976, I am extremely thankful to him, among other things, for being a caring professor. With all his academic brilliance that could have made him arrogant (as some of us professors are!) he was humble, caring, and a down-to-earth person. His loss on December 4, 2015, is a loss of a great human being, a teacher, a scholar, and a world-leading scientist.

2 The Life of Lajos Takács

2.1 A Biographical Sketch of Lajos Takács

2.1.1 Takács' Childhood

Lajos was born on August 21, 1924, in Maglód (a little town 16 miles from Budapest), Hungary. His father had a general store with his mother helping him. Once in store, at age 4, with his mother, a neighbor walked in and told her mother that she sold her pig for 40 pengös (Hungarian currency during 1927–1946). Lajos, who could multiply at that age, asked her mother how many fillérs was in a pengö? After her mother responded 100, he thought a moment and told her that the neighbor has 4000 fillérs! So, he became well-known mathematician at age 4 in his small town, Maglód!

2.1.2 Takács in Elementary School

While in elementary school, Lajos took note of all of his learning and carried the notebook with him. He was interested in technology, arithmetic, electronic, and radio. A mathematical problem Lajos solved while in elementary school was the magic property of π in calculating the area of a circle and consequently the exact volume of a cylindrical barrel in his yard.

Takács calculated the chess game's grain sum reward while he was in elementary school as one of his mathematical problems at the young age. The problem posed is the following:

Suppose the chess game inventor asked that as a reward he be given one grain of wheat for the first square of the chessboard, two for the second, and double the previous one for the subsequent square. At that age he was not aware of how to calculate sum of geometric series, to find out the total number of grains to be awarded. So, he just added the numbers for all 64 squares and found the exact sum in kilograms and sacks. The question is how Takács calculated the number of sacks needs? So, I asked Takács the following question in an e-mail:

In elementary school, how did you know the weight of a wheat grain and a sack weight capacity to figure out how many sacks of grain the reward would be?

Here is Takács' response and his wish for this conference in his e-mail [9] on July 23, 2015:

To answer your question, I had no way to measure the weight of one grain. My mother had a general store, not a pharmacy! But she did have an old-fashioned scale with a collection of weights of various sizes. I put the 100 gram weight in one tray of the scale, and enough wheat grains in the other to tip the scale. Then I counted the number of grains in 100 grams, and divided 100 by the number of grains to find the weight of one grain of wheat. I already knew that the standard weight capacity of a sack of grain was 80 kilograms.

I checked the accuracy of my calculations by taking into consideration that $2^{10} = 1024$.

2.1.3 Takács in Secondary School

By the time Takács attended the secondary school, his mentioned notebook was full of everything he had learned about technology and arithmetic.

To attend his secondary school, Takács had to commute with train to Budapest. Riding train was an opportunity for him to borrow books from older students and study them. He studies many books on his own, including the Differential and Integral Calculus by Manó (Emanuel) Beke that he bought.

At the age of 14, Takács lost his father and had to help his mother at the store. Hence, he had to take care of errands of the business in Budapest. Thus, during the high school period, he was the manager and buyer for his mother's business.

At age 15, Takács approached his mathematics teacher, Dezső Vörös, asking a mathematics book to read and the teacher suggested Euler's Algebra. Euler's Algebra [11] is an extremely complicated and difficult German language book with old-fashion Gothic letters (relating to the Goths or their extinct East Germanic language). Nonetheless, under insistence of the teacher, Takács purchased the book and spent Christmas vacation of 1939 to study this book. As soon as he mastered the Euler's syntax, he was amazed by the contents of the book of this Swiss Mathematician such as solution of cubic and quadratic equations, Diophantine equation, and solution of Fermat's problem for $n = 3$ and $n = 4$.

Throughout high school, Takács spent his free times and vacations reading and studying many books with different topics, including Diophantine equations in Euler and notes of Vörös on Gusztáv Rados' lectures on number theory and some other books on number theory, including [6, 7]. He became, particularly, interested in Number Theory and out of his readings; he made some discoveries at the time. For instance, in a book Takács read that the number of branches on a tree increases annually according to the sequence 1, 2, 3, 5, 8, 13, ... Assuming this statement is true, while he had no knowledge of "Calculus of Finite Differences", he wanted to find a general formula for the number of branches of a tree in the nth year.

Takács solved the problem posed in a *roundabout* way by observing that the numbers in the sequence 1, 2, 3, 5, 8, ...are positive integers in y of the Diophantine equation $x^2 - 5y^2 = \pm 4$ that he could solve. However, he was not aware that solution of the problem already existed in the literature of Fibonacci numbers.

During high school years, Takács was familiar with *combinatorics* and solved many probability problems. However, he considered probability as a branch of combinatorics. Much later, he realized the importance of the probability theory and its role in describing the physical world. He had particular interest in radio technology. Hence, he bought the monthly journal *Rádió Tecknika* and constructed different kinds of electronic equipment. He learned logarithm, trigonometry, and complex numbers from a mathematical column of this journal.

Although during high school years, Takács was spending most of his times studying mathematics books, he also studied classical and modern physics. However, he did not ignore physical activities. He was proud of his accomplishment in high jump, long jump, running, swimming, and skating. He also was an entertainer for his friends by mathematical puzzles and tricks.

In 1943, when graduated from high school, Takács won the second prize in the 47-th *Loránd Eötvös* mathematical and physical society competition for high school graduates in Hungary. The award was established in 1894, given to two high school students annually. Including in the list of winners are great mathematicians such as the following:

Loránd Eötvös

First Name	Last Name	Year Won	Year Born	Year Died
Lipót	Fejér	1897	1880	1959
Theodor	Kármán	1898	1881	1963
Dénes	König	1902	1884	1944
Alfréd	Haar	1903	1885	1933
Marcel	Riesz	1904	1886	1969
Gábor	Szegő	1912	1895	1985
Tibor	Radó	1913	1895	1965
László	Rédei	1918	1900	1967
László	Kalmár	1922	1905	1976
Edward	Teller	1925	1908	2003
Lajos	Takács	1943	1924	2015

Problems and their solutions for the 47th competition may be found online at http://www.math-olympiad.com/47th-eotvos-competition-1943-problems-solutions.htm.

2.1.4 Takács in University

Takács studied at the Technical University of Budapest from 1943 to 1948. While studying at the Technical University of Budapest, he was also attending classes at the Pázmány University. At the end of 1943 school year, the Second World War reached the door step of the university, and as a result, academic activities were interrupted until the fall of 1945. Thus, Takács was spending his time on studying by himself.

Back to the university that resumed in 1945, Takács took courses in probability theory and mathematical statistics with Charles (Károly) Jordán (1871–1959); courses that made Takács' mind to take the area as his career. Jordán became Takács' mentor during the years he was a student. Takács not only learned from Jordán, but contributed to his class too.

Charles Jordán

An example of his contribution is when Jordán was discussing "matching." A match occurs when n cards marked individually with one of $1, 2, \ldots, n$ are drawn at random without replacement and at the ith draw the card marked with i appears, $i = 1, 2, \ldots, n$. In his class, Jordán presented the problem of finding the probability of at least one match to occur from Montmort's book [23] and stated its solution by Montmort who did not provide any proof for his result (see [23, pp. 54–64] and [24, pp. 130–143]). In 1946, Takács found the solution by considering the number of permutations in which no matches occur and expressing it in terms of permanent of a specific type of matrix. Jordán added this as a short note in already finished manuscript of his book on probability theory.

Incidentally, Takács presented the origin, history, and various versions of the problem of coincidences (matches, rencontres) in the theory of probability (see [33]) which was communicated by B.L. Van Der Waerden, Department of Mathematics & Statistics, Case Western Reserve University, Cleveland, Ohio, on September 14, 1979.

Later, Takács [30] wrote an article in Charles Jordán's memory, while he was at Columbia University in New York.

At the age of 21, Takács accepted an offer by Professor Zoltán Bay, a professor of atomic physics, as a student assistant to him. During the years 1945 to 1948, Takács participated in Bay's famous experiment of receiving microwave echoes from the moon.

Zoltán Lajos Bay (1900–1992) was a Hungarian physicist, and engineer who developed microwave technology, including tungsten lamps. Bay was the second person to observe radar echoes from the Moon.

Zoltán Lajos Bay

As a student assistant, Lajos was appointed as the consultant to Bay's research laboratory, the lab that consisted of mostly university researchers. Takács' job at the lab was to calculate the position of the moon at 15-minutes interval and participate in the nightly experiment. The success arrived in the evening of February 6, 1946, also a triumph of probability theory, Takács calls it in his Chance or Determinism, see [34]. The experiment showed that *the reflection of radar beams aimed at the moon, which was considered revolutionary in space research at that time.*

Takács received a doctorate degree (Ph.D.) in 1948 with his thesis title as *"Probability Theoretical Investigation of Brownian Motion"*, that was refereed by Charles Jordan. However, he continued his registration at the university to become a teacher.

Takács won the first prize in the Miklós Schweitzer Prize mathematical competition for university graduates in Hungary in 1949. The Miklós Schweitzer Competition (*Schweitzer Miklós Matematikai Emlékverseny*) is an annual Hungarian mathematics competition for university recent graduates, established in 1949. The *Schweitzer* Competition is uniquely high level among mathematics competitions. The problems on the competition can be classified roughly in the following categories: algebra, combinatorics, theory of functions, geometry, measure theory, number theory, operators, probability theory, sequences and series, topology, and set hteory. For sample questions, see:
http://www.artofproblemsolving.com/community/c3253_mikls_schweitzer.

Takács won the second *Géza Grünwald Prize* awarded by the János Bolyai Mathematical Society in 1952 for his paper *"Investigation of waiting time problems by reduction to Markov processes"*. See [28]. The János Bolyai Mathematical Society (Bolyai János Matematikai Társulat, BJMT), founded in 1947, is the Hungarian mathematical society, named after János Bolyai (1802–1860), a 19th-century Hungarian mathematician, a co-discoverer of non-Euclidean geometry. (Portrait by Ferenc Márkos -2012). The paper was presented by Alfréd Rényi in Budapest, who is a Hungarian mathematician who made contributions in combinatorics, graph theory, number theory and in probability theory. With the win of this paper, Takács posed himself to be considered among the frontiers in queueing theory.

János Bolyai

In 1957, Takács received the academic doctor's degree (D. Sc.) in Mathematics with his dissertation entitled *"Stochastic processes arising in the theory of particle counters"*, at the Department of Mathematics of the L. Eötvös University.

2.2 Takács' Employments and Further Achievements

Takács, who was one of the first to introduce semi-Markov processes in queueing theory in 1952, worked as a mathematician at the Tungsram Research Laboratory (1948–1955). While at the Tungsram Research Laboratory, Takács also accepted a staff position at the newly created Research Institute for Mathematics of the Hungarian Academy of Sciences (1950–1958), during which he published several papers on queueing theory, involving applications to telephone traffic, inventories, dams, and insurance risk. During this period, Takács developed the theory of point processes and introduced the process, which later introduced as *semi-Markov processes* by Paul Lévy. Takács also gave the generalization of Agner Krarup Erlang's telephone traffic congestion formula that later was discussed in his celebrated book [31], Introduction to the Theory of Queues.

Takács was an associate professor in the Department of Mathematics of the L. Eötvös University (1953–1958), formerly called Pázmány University. He took a lecturer appointment at Imperial College in London and London School of Economics (1958), where he lectured on the theory of stochastic processes and queueing theory

Between 1954 and 1958, Professor Takács published 55 research papers on various topics in stochastic processes and the foundations of modern queueing theory. His early research in queueing theory was summarized in one of his finest works, "Some Probability Questions in the Theory of Telephone Traffic," [29]. This paper just recently has been cited by [17]. Interestingly, Kim's paper is focused on available server management in the Internet-connected network environments, in which local backup servers are hooked up by local area network (LAN) and remote backup servers are hooked up by virtual private network (VPN) with high-speed optical network.

The year 1958, while at the Imperial College of London, was a turning point in Takács' life. He decided to leave Hungary forever and move to the USA. As his first job in the USA, in 1959, Takács received an offer from Columbia University in New York as assistant professor that he accepted and after a year, in 1960, he was promoted to the rank of associate professor. Takács stayed at Columbia University for the next 7 years teaching probability theory and stochastic processes. While at Columbia University, Takács had a consulting job at Bell Laboratories and at IBM.

In the early 1960s, Takács developed the time-dependent behavior of various queuing processes, specifically, the virtual waiting-time process, that now is referred to as the Takács Process. In summer of 1961, he had a visiting position at Boeing Research Laboratories in Seattle, Washington.

Initially, Takács' publications were in Hungarian, and later, they were translated into various languages by different authors.

Takács developed a large variety of multichannel queueing systems, applying his extraordinary fluency in combinatorial and continuous mathematics, including primarily results for embedded queueing processes. Their extensions to continuous-time-parameter queueing processes were later included in his celebrated monograph [31], Introduction to the Theory of Queues that appeared in 1962.

While at Columbia University, attending many conferences focusing on variety of topics in queueing theory, Takács found *a generalization of the classical ballot theorem* of Bertrand, which made it possible to solve many problems in queueing theory, in the theory of dams, and in order statistics. More on this topic will be discussed later in this paper. He also developed queues with feedback, balking, various orders of service, and priority queues. Multi-server with Feedback was, indeed, the title of my dissertation. Finally, while at Columbia University, Takács advised nine Ph.D. students with their dissertations, as follows:

1. Paul J. Burke
2. Ora Engelberg
3. Joseph Gastwirth
4. Peter Linhart
5. Clifford Marshall
6. Lloyd Rosenberg
7. Saul Shapiro
8. Lakshmi Venkataraman
9. Peter Welch

Takács spent part of his sabbatical, in 1966, at Stanford University working on his second bestseller book [32], *Combinatorial Methods in Theory of Stochastic Processes*. In that year, he accepted the appointment as Professor of Mathematics at Case Western Reserve University in Cleveland, Ohio, where he held this position until 1987, when he retired as a Professor Emeritus.

During his tenure at Case Western Reserve University, Takács wrote over 100 monographs and research papers. By this time, Takács' major research areas became sojourn time, fluctuation theory, and random trees. Though he was not in favor of writing grant proposal for research funding as he believed awards should be given after a valuable achievement has occurred rather than pay in advance for something that its result is not guaranteed, as a result of a research grant from the National

Science Foundation, he published a paper on random tree. In it, he proved several theorems on random tree including the height of a tree.

At Case, Takács guided additional 14 Ph.D. students, as follows:

1. Roberto Altschul, 1973
2. John Bushnell
3. Daniel Michael Cap, 1985
4. Jin Yuh Chang, 1976
5. Sara Debanne, 1977
6. Nancy (Mailyn) Geller
7. Aliakhbar Montazer-Haghighi, 1976
8. Andreas Papanicolau
9. Pauline Ramig
10. Josefina De Los Reyes
11. Douglas Rowland
12. Elizabeth Van Vought
13. Enio E. Velazco
14. Fabio Vincentini

2.3 Takács' Publications

2.3.1 Papers

In summary, with his own count, as of May 23, 2015, Takács has published 225 papers,[1] the last of which appeared in 1999, and many of which have had a huge impact on the contemporary theory of probability and stochastic processes. Topics he worked on may be summarized as:

1. Combinatorial Problems
2. Ballot Theorems
3. Random Walks
4. Random Graphs
5. Point Processes
6. Queueing Processes
7. Binomial Moments
8. Sojourn Time Problems
9. Branching Processes
10. Fluctuation Theory
11. Order Statistics

The list of Takács' papers up to 1994 has appeared in the *Journal of Applied Mathematics and Stochastic Analysis*, volume 7, a small sample of which is listed below. The list of all 226 follows this sample. The numbers are indicated in the brackets. For instance, the first 211 papers are the ones listed in the *Journal of Applied Mathematics and Stochastic Analysis*, and the rest are published as indicated across their corresponding numbers.

Sample from the First 211 Papers:

1954. Some investigations concerning recurrent stochastic processes of a certain kind, Magyar Tud. Akad. Alk. Mat.Int. Kozl. 3, 115–128.
1955. Investigations of waiting time problems by reduction to Markov processes, Acta Math. Acad. Sci. Hung. 6, 101–129.

[1] An impressive number, which becomes 226 with the posthumously published paper in this volume

1970. On the distribution of the supremum for stochastic processes, Annales de l'Institut Henri Poincaré (B) Probabilités et Statistiques 6, 237–247.
1977. An identity for ordered partial sums, J. Combin. Theory Ser. A 23, 364–365.
1981. On the "problème des ménages", Discrete Mathematics 36, 289–297.
1981. On a combinatorial theorem related to a theorem of G. Szegő, J. Combin. Theory Ser. A 30, 345–348.
1988. Queues, random graphs and branching processes, J. Appl. Math. Stochast. Anal. 1, 223–243.
1990. Counting forests, Discrete Mathematics 84(3): 323-326. 1990. On Cayley's formula for counting forests, J. Combin. Theory Ser. A 53, 321–323.
1990. A generalization of an inequality of Stepanov, J. Combin. Theory Ser. B 48, 289–293.
1990. On the number of distinct forests, SIAM J. Discrete Math. 3, 574–581.
1991. On a probability problem connected with railway traffic, J. Appl. Math. Stochast. Anal. 4, 1–27.
1991. Conditional limit theorems for branching processes, J. Appl. Math. Stochast. Anal. 4, 263–292.
1991. On the distribution of the number of vertices in layers of random trees, J. Appl. Math. Stochast. Anal. 4, 175–186.
1993. Limit distributions for queues and random rooted trees, J. Appl. Math. Stochast. Anal. 6, 189–216.
1994. On the total heights of random rooted binary trees, J. Combin. Theory Ser. B 61, 155–166.

The Entire List of Papers Thereafter

[1]–[211] (1994). J. Appl. Math. Stochast. Anal. 7, 229–237.
[212] (1995). On the local time of the Brownian motion. Ann. Appl. Probability 5, 741–756.
[213] (1995). Brownian local times. J. Appl. Math. Stochast. Anal. 8, 209–232.
[214] (1996). On a test for uniformity of a circular distribution. Math. Meth. Statist. 5, 77–78.
[215] (1996). On a three-sample test. In: Heyde, C.C., Prohorov, Y.V., Pyke, R., Rachev, S.T. (eds.) Athens Conference on Applied Probability and Time Series, vol. 1: Applied Probability; in honor of J.M. Gani, pp. 433–448. Springer–Verlag, New York.
[216] (1996). On a generalization of the arc-sine law. Ann. Appl. Probab. 6, 1035–1040.
[217] (1996). Sojourn times. J. Appl. Math. Stochast. Anal. 9, 415–426.
[218] (1996). In memoriam. Pál Erdős (March 26, 1913 – September 20, 1996). J. Appl. Math. Stochast. Anal. 9, 563–564.
[219] (1997). On the ballot problems. In: Balakrishnan, N. (ed.) Advances in Combinatorial Methods and Applications in Probability and Statistics, pp. 97–114. Birkhäuser, Boston
[220] (1997). Holdvisszhang 1946 február 6-án. Fizikai Szemle 47:1, 20–21.

[221] (1998). On the comparison of theoretical and empirical distribution functions. In: Barbara Szyszkowicz (ed.) Asymptotic Methods in Probability and Statistics; a volume in honor of Miklós Csörgö, pp. 213–231. Elsevier Science B.V., Amsterdam.
[222] (1998). Sojourn times for the Brownian motion. J. Appl. Math. Stochast. Anal. 11, 231–246.
[223] (1998). On cyclic permutations. Math. Sci. 23, 91–94.
[224] (1999). On the local time of the Brownian bridge. In: Shanthikumar, J.G., Sumita, U. (eds.) Applied Probability and Stochastic Processes, pp. 45–62. Kluwer Academic Publishers, Boston, M.A.
[225] (1999). The distribution of the sojourn time of the Brownian excursion. Meth. Comput. Appl. Probab. 1, 7–28.
[226] (2019). The distribution of the local time of Brownian motion with drift. This volume.

Takács has also published six books, three in Hungarian and three in English language. They are the following:

(1) Az Elektroncsö (The Vacuum Tube) with A. Dallos, Tankönyv Kiadó, Budapest, 1950.
(2) Valószinüségszámitás (Theory of Probability), With M. Ziermann, Tankönyv Kiadó, Budapest, 1955 (original publishing), 1967, 1972.
(3) Valószinüségszámitás (Theory of Probability), with P. Medgyessy (Part A: Probability Theory) and L. Takács (Part B: Stochastic Processes), Tankönyvkiadó, Budapest 1957 (original publishing), 1966, 1973.
(4) Stochastic Processes, Problems and Solutions, Methuen & CO LTD, 1960, Translated by P. Zádor (1962).
(5) Introduction to the Theory of Queues, Oxford University Press, 1962.
(6) Combinatorial Methods in the Theory of Stochastic Processes, John Wiley, 1967.

2.4 Awards

- 1993, Foreign Membership Magyar Tudomanyos Akademia, Matematikai es Fizikai Tudomanyok Osztalyanak Közlemenyei, Hungarian
- 1994, John von Neumann Theory Prize

  A Hungarian mathematician who made major contributions to a number of fields, including mathematics (foundations of mathematics, functional analysis, ergodic theory, geometry, topology, and numerical analysis), physics (quantum mechanics, hydrodynamics, and fluid dynamics), economics (game theory), computing [Von Neumann architecture, linear programming, self-replicating machines, stochastic computing], and statistics.

John von Neumann (1903–1957)

- 2002, Fellows Award. Institute for Operations Research and Management Sciences.

2.5 *Honors and Recognitions*

2.5.1 Hungary

Professor Lajos Takács is noted as "the most well-known, reputed and celebrated Hungarian" in the field of probability and stochastic processes.

2.5.2 Pioneer in the Field of Queueing Theory

Celebrating Takács' 70th birthday, as a Pioneer in the field of Queueing Theory and the author of Combinatorial Methods in the Theory of Stochastic Processes, many scientific institutions honored him in mid-1994, see [8]. They include the following:

- The Institute of Mathematical Statistics
- Operations Research Society of America
- The Institute of Management Sciences
- Hungarian Academy of Sciences
- A special volume [14], Studies in Applied Probability, 31A, edited by J. Galambos and J. Gani.

Additionally, some well-known authors in probability, statistics, and stochastic processes wrote about him to honor him and his work. These authors include Dshalalow, Syski, J. Galambos, and Joe Gani.

2.5.3 Takács' Contribution to Combinatorics

In Honor of Lajos Takács, a paper entitled "Professor Lajos Takács: Life and Contribution to Combinatorics" was presented by Aliakbar Montazer Haghighi and Sri Gopal Mohanty, at the 8th International Conference on Lattice Path Combinatorics and Applications, August 17–20, (2015), California Polytechnic State University, Pomona, CA.

2.6 *From Among Takács' Latest Correspondences*

2.6.1 Not Able to Travel

During the last months of his life, Takács carried his title as Professor Emeritus at Case Western Reserve University. At age 91, he was "confined at home," as he stated, with his wife Dalma. In an e-mail [9] to me just a few months before his death, 5/23/2015, Professor Lajos Takács wrote:

"As for me, I am doing reasonably well, but I am more or less confined to my home. Since travel is a serious problem for me, I am unfortunately unable to participate in meetings."

2.6.2 Waiting for a Right Publisher

A great part of his work was yet waiting for a right publisher who would have agreed to his terms and conditions. Here is what he described his unpublished work to me in his e-mail [9] dated May 2015 (original in the standard e-mail text format):

"An additional note: In July 1973 I completed my book **Theory of Random Fluctuations**, *which unfortunately is still in manuscript form. While I was working on the book, John Wiley was so interested that they offered me a contract before the book was finished. I did not like to sign a contract until my work was ready for the press. The finished manuscript turned out to be 1600 pages which Wiley considered too big for a book. They offered to publish it in lecture note form, which was unacceptable to me. I was also unwilling to shorten the MS, so it is still unpublished."*

2.7 *Takács' Family*

Takács' greatest achievement was on April 9, 1959, when he married Dalma Horváth in London with Ityszard Syski (1924–2007) as Takács' best man at his wedding.

Dalma is now 82. From their marriage, they have two daughters, Judith, an artist, and Susan, a legal assistant

Lajos, Dalma and their three grandsons: Eric, David and Mark

Dalma was a Professor of English Literature and chair of the English Department at Notre Dame College of Ohio and an author of historical family memoirs, plays and several fiction books as well.

Dalma, painting by Judy on Mother's day of 2013, Mother having a ball, while diagnosed with cancer!

Judy is a commercial artist and illustrator. She is a seven time Best of Show winner, her work has been exhibited in many places. She does Pastel Portraiture, Paintings and Drawings.

Judy Takács, Self-Portrait

And, finally, below are two photos of Professor Takács' 90th birthday.

3 Lajos Takács: As I Remember Him

Professor Lajos Takács was Professor A.M. Haghighi's supervisor and mentor too. I did not study under his guidance nor have the privilege to be a coworker. Nevertheless, he was my mentor from a distance.

After finishing the training in Statistics at Statistical Wing of Indian Council of Agricultural Research, Government of India in New Delhi, I was looking for a job for which I approached the Head of the Statistical Wing for his advice. He was very encouraging but asked me to wait for the right opportunity and to start, in the meantime, some research work with T.V. Narayana, a fresh Ph.D. from University of North Carolina, Chapel Hill, USA, who joined the Wing during that period. I was disappointed because I was only looking for a job. I was flabbergasted since to my embarrassment I did not know exactly how "research" was done.

Narayana gave me his thesis to read and asked me to create artificial samples by using random numbers from Fisher & Yates Tables. The work was nothing but to simulate samples. While preparing these samples, I observed patterns and became curious counting structures with certain patterns. Incidentally, unlike me a real research scholar pointed out Feller's Vol. 1 Probability Theory book [12] to me. The book was very fascinating, specially its Chap. 3 dealing with fluctuations in coin tossing and random walk. Subsequently my quest for learning more about counting led me to Riordan's book [27] on combinatorial analysis, which became an invaluable reference for me. I was delighted to learn about it since my previous exposure to combinatorics was limited to only solving discrete problems in probability theory.

TV as Narayana was called by his friends, left India soon but started what he termed as coin-tossing problems somewhat related to patterns that I was pointing out to him during our interaction in New Delhi. I became matured enough in understanding "research" so as to publish my first paper in 1955 with Narayana as the senior author (see [25]). In my belief, soon he developed a passion for combinatorics moving away from his earlier statistical work. This was evident from the fact that he guided me as his first Ph.D. student and my thesis title was “On some properties of compositions of an integer and their application to probability theory and statistics” in which I discussed the so-called ballot problem in [12] and used “lattice path” representation.

This was the beginning of the sixties when I coincidentally found a series of papers by Takács on the “urn problem” as a generalization of the ballot problem and its extensions and applications especially to queues. I made a humble contribution by providing a lattice path-based combinatorial proof for his urn problem (see [19]). Since then I have tried to follow his work in this particular direction. His application to queues drew my attention to learn about queues. It happened that I was a faculty member at University of Buffalo and that my friend and colleague Norm Severo was giving a seminar course on queues using Takács’ book [31]. I joined this course for some time and started corresponding with Takács.

It was serendipity that brought me to Takács closer but always from a distance. During my stay at Indian Institute of Technology (1966–1968), Professor J. L. Jain who was then a research scholar at University of Delhi and was working on Queueing Theory contacted me to study Takács’ work more particularly his use of combinatorics. This gave me another opportunity to go over his contribution while learning the subject of Theory of Queues. Eventually, our efforts paid. Takács’ generalization of the ballot problem was applied to queues involving batches (for instance, see [15, 20, 21]) about which not much were known in those days.

In the meantime, Takács’ celebrated book, *Combinatorial Methods in the Theory of Stochastic Processes* appeared in 1967. The book became a treasure and inspiration for me.

Takács indirectly created curiosity within me to learn more about Hungarian researchers in the field and in general the country itself. In the process, I found out the work of I. Vincze and E. Csáki directly befitting my interest and was able to apply lattice path combinatorics to nonparametric methods in Statistics. At some point starting from the seventies, I went on visiting Budapest several times either directly to the Mathematical Institute of Hungarian Academy of Sciences or otherwise, most often being a guest of I. Vincze. During my visits, I also came to know P. Revesz and G. Katona very well. Takács along with Csáki and Vincze have been amply cited in my book on lattice path counting (see [22]). Incidentally, I. Vince passed away in 1999, and the Fifth Conference on Lattice Path Combinatorics and Applications and the Special Issue coming out of the Conference were dedicated in his memory. The issue contains the article [5] *“István Vincze (1912–1999) and his contribution to lattice path combinatorics and statistics”* by E. Csáki.

Soon I realized that Hungary has been the land of combinatorics often through number theory where the exposure to the subject starts at a very early stage of education and excellence in it is highly recognized. At the same time, the study of probability theory is very much emphasized. In my assessment, almost all probabilists in Hungary have a tilt toward combinatorics by training which is reflected in their work. For instance, Alfréd Rényi, who happened to be an eminent Hungarian probabilist, decided to give a talk not on probability but on “Enumeration of search codes" while visiting McMaster University sometime at the beginning of seventies. This prompted me and Professor Chorneyko to write a paper on the subject (see [4]). Having been nurtured in that unique environment in Mathematics, Takács with his own brilliance could excel and became a famous mathematician.

In Takács' words, his book on combinatorial methods in stochastic processes consists of results as applications of a generalization of the classical ballot problem to a variety of situations arising in queues, dams, storage, insurance risk, order statistic, and others. The treatment in the book is essentially probabilistic rather than combinatorial and yet it uses the word "combinatorial" because of the nature of the ballot problem which can be solved by combinatorics, strictly speaking, by enumerative combinatorics. Enumerative combinatorics often use constructive methods to count structures that arise in the problem. It seems there is a tradition in Hungary to treat probability theory in the classical sense so as to crisscross over to combinatorics. (Note: There exists a school of combinatorics which solves problems in combinatorics by using the probabilistic method. It is a non-constructive method, primarily used in combinatorics and pioneered by Paul Erdős — most well-known Hungarian mathematician of recent time for proving the existence of a prescribed kind of mathematical object. See [10].

Whereas Takács' main focus was probability theory and he dealt with combinatorics whenever the situation arose but most comfortably by using probabilistic argument, my interest on the other hand has been primarily on enumeration and properties of certain combinatorial structures like lattice paths that have appeared in different branches including random walk, queues (such as discrete time queues and their transient behavior), and non-parametric methods in statistics. Nevertheless, there is an overlap in the application to queues by both of us.

However, Takács' contribution gave the incentive to include combinatorial methods applied to queues in Chapters 3, 6, and 9 of the book [16], which is unlike any other text book in the field. At the same time, the book has two chapters on computational methods suitable for applications.

Having been influenced by Takács' work, I wanted to meet him personally even though I might have talked to him on phone besides our exchange of correspondence. The opportunity came at the Conference on Mathematical Methods in Queueing Theory held at Western Michigan University in 1973, which we both attended. I was overly impressed by his simplicity, soft spoken, affectionate, and gentle manner. It became a strange coincidence that later when I met I. Vincze and E. Csáki and others in Budapest I had the same appreciation of the people that they were all soft spoken, kind, and caring.

Later I met Takács and his wife Dalma in a conference held in Toronto. Once I drove down to his place in Cleveland along with J.L. Jain. He participated in the First International Conference on Lattice Path Combinatorics held at McMaster University in 1984. Our final meeting was when he graciously participated in the Conference kindly organized by Professor N. Balakrishnan, my colleague and friend, at McMaster University in 1997, in connection with my retirement.

(From right: Galambos, Takács. Dalma Takács, Mohanty, Shanti Mohanty and their children at the Conference banquet)

I also felt privileged and honored when J. Galambos invited me to contribute to a special volume honoring Professor Takács and I did (see [2]).

I ask myself: What am I, a statistician or a combinatorialist? My basic training was in statistics. But by accident through T.V. Narayana, I started getting into enumerative combinatorics, but not quite. Again accidentally through L. Takács, I stumbled on Theory of Queues, but perhaps only on a segment of it. Were there doubts in my mind?

Yet Takács had a similar fate when he was asked by J. Gani and M. Neuts about the applications of his theories. (See Part 1, Sect. 1.7.)

I followed Takács' footsteps, but he was far ahead waiving his hand.

Oh, the distant mentor — did you say: Never mind, stay on course!

4 Contribution to Combinatorics

The combinatorial approach of Takács is based on a generalization of the classical ballot theorem which is stated as follows:

Theorem 1.1 (The Classical Ballot Theorem [35, Theorem 7.2.1]) *If in a ballot, candidate A scores a votes and candidate B scores b votes, with $a > b\mu$, where μ is a positive integer, then, the probability that throughout the number of votes registered for A is always greater than times the number of votes registered for B is given by*

$$P(a, b, \mu) = \frac{a - b\mu}{a + b}, \tag{1.1}$$

provided that all the possible voting records are equally probable.

Theorem 1.2 (Takács' generalization [35, Theorem 7.5.1]) *Let us suppose that a box contains n cards marked with nonnegative integers* $k_1, k_2, \ldots, k_n$ *such that* $k_1 + k_2 + \cdots + k_n = k < n$. *All the n cards are drawn without replacement from the box. Denote by* ν_r *the number obtained at the rth drawing,* $r = 1, 2, \ldots, n$. *Then,*

$$P\{\nu_1 + \nu_2 + \cdots + \nu_r < r,\ r = 1, 2, \ldots, n\} = \frac{n-k}{n}, \tag{1.2}$$

provided that all the possible results are equally probable.

Let there be a cards marked 0 and b cards marked $\nu + 1$. Suppose a card marked with 0 corresponds to a vote for A and a card marked with $\nu + 1$ corresponds to a vote for B. Then, it can be shown that the classical ballot theorem follows as a special case.

Note that letting x and y be the number of votes for A and B, respectively, at the rth count, then,

$$x + y = r, \quad a + b = n, \quad b(\nu + 1) = k. \tag{1.3}$$

Hence, the event in the theorem becomes

$$x(0) + y(\nu + 1) < x + y \quad \text{if and only if} \quad y\nu < x, \tag{1.4}$$

and the probability becomes $(n-k)/n = (a - \nu b)/(a + b)$ that checks the *Classical Ballot Theorem* in Theorem 1.1.

Drawing cards without replacement implies consideration of $n!$ permutations of n cards. The theorem is true if permutations are replaced by cyclic permutations and is stated as follows as a counting result.

Theorem 1.3 ([35, Theorem 7.5.3]) *Let us suppose that n cards are marked with non-negative integers* $k_1, k_2, \ldots, k_n$ *such that* $k_1 + k_2 + \cdots + k_n = k < n$. *Among the n cyclic permutations of the n cards, there are exactly* $n - k$ *in which the sum of the numbers on the first r cards is less than r for every* $r = 1, 2, \ldots, n$.

Its Continuous Version

Theorem 1.4 ([32, p. 1, Theorem 1]) *Let* $\phi(u)$, $0 \le u \le t$, *be a non-decreasing function for which* $\phi'(u) = 0$ *almost everywhere and* $\phi(0) = 0$. *Let* $\phi(t + u) = \phi(t) + \phi(u)$ *for* $0 \le u \le t$. *For* $0 \le u \le t$ *define*

$$\delta(u) = \begin{cases} 1, & \text{if } \phi(v) - \phi(u) \le v - u,\ u \le v \le u + 1, \\ 0, & \text{otherwise.} \end{cases} \tag{1.5}$$

Then,

$$\int_0^t \phi(u)\,du = t - \phi(t), \quad \text{for } \phi(t) \le t. \tag{1.6}$$

Theorem 1.3 can be formulated in the following more general way.

Theorem 1.5 ([35, Theorem 7.5.4]) *Let $\nu_1, \nu_2, \ldots, \nu_n$ be interchangeable or cyclically interchangeable discrete random variables which take on non-negative integers only. Write*

$$N_r = \nu_1 + \nu_2 + \cdots + \nu_r, \quad r = 1, 2, \ldots, n, \quad N_0 = 0. \tag{1.7}$$

Then, we have

$$P\{N_r < r,\ 1 \le r \le n,\ \text{and}\ N_n = n - i\} = \frac{i}{n} P\{N_n = n - i\}, \qquad 0 \le i \le n, \quad n = 1, 2, \ldots. \tag{1.8}$$

Theorem 1.6 ([35, Theorem 7.5.5]) *Let $\nu_1, \nu_2, \ldots, \nu_n$ be interchangeable or cyclically interchangeable discrete random variables which take on non-negative integers only. Write*

$$N_r = \nu_1 + \nu_2 + \cdots + \nu_r, \quad r = 1, 2, \ldots, n, \quad N_0 = 0. \tag{1.9}$$

We have

$$P\{N_r < r\ \text{for at least one}\ r = 1, 2, \ldots, n\} = \sum_{i=1}^{n} \frac{1}{i} P\{N_i = i - 1\}, \qquad n = 1, 2, \ldots. \tag{1.10}$$

Another Extension of the Ballot Theorem

While in the ballot problem we consider the number of votes for A is always greater than ν times the number of votes for B, we consider now the number to be greater exactly in j cases, $j = 1, 2, \ldots, a + b$, not always. This has been studied by Chao and Severo [3]. Denote by α_r and β_r the number of votes registered for A and B, respectively, among the first r votes counted. Let μ be a positive real number and define

$$P_j(a, b, \mu) = P\{\alpha_r > \beta_r \mu, \text{for } j \text{ subscripts } r = 1, 2, \ldots, a + b\}, \tag{1.11}$$

for $j = 0, 1, \ldots, a + b$. We can write this in the form

$$P_j(a, b, \mu) = \sum_{0<s<j} P\{\beta_j = s\}\, P_j(j - s, s, \mu) P_0(a + s - j, b - s, \mu), \tag{1.12}$$

where

$$P(\beta_j = s) = \frac{\binom{j}{s}\binom{a+b-j}{b-s}}{\binom{a+b}{b}} = \frac{\binom{a}{j-s}\binom{b}{s}}{\binom{a+b}{j}}, \tag{1.13}$$

whenever $0 \le s \le j$ and $j - a \le s \le b$ (see [35, Section 7.7]). This follows from the following auxiliary theorem.

Auxiliary Theorem ([35, Theorem 7.7.1]) *Let $\xi_1, \xi_2, \ldots, \xi_n$ be interchangeable real random variables. Define $\zeta_r = \xi_1 + \xi_2 + \cdots + \xi_r$ for $r = 1, 2, \ldots, n$ and $\zeta_0 = 0$. Denote by Δ_n the number of subscripts $r = 1, 2, \ldots, n$ for which $\zeta_r > 0$. Then,*

$$P\{\Delta_n = j\} = P\{\zeta_r < \zeta_j, 0 \le r < j, \text{ and } \zeta_r \le \zeta_j,\ j \le r \le n\}. \tag{1.14}$$

See [1, 13].

A recent generalization of Theorem 1.3 is due to Mercankosk, Nair and Soet [18], which the authors call Batch Ballot Theorem.

Theorem 1.7 (Batch Ballot Theorem [18, Theorem 2]) *Let $n_1, n_2, \ldots, n_k$ be nonnegative integers such that $n_1 + n_2 + \cdots + n_k = n < km$. For $0 \le d \le m-1$, let C_d denote the number of cyclic permutations of $(n_1, n_2, \ldots, n_k)$ for which the sum of the first s elements is less than $sm - d$ for $1 \le s \le k$. Then,*

$$C_0 + C_1 + \cdots + C_{m-1} = km - n. \tag{1.15}$$

Theorem 1.8 ([18, Theorem 3]) *Let $\nu_1, \nu_2, \ldots, \nu_k$ be cyclically interchangeable random variables taking on nonnegative integral values. Set $N_s = \nu_1 + \nu_2 + \cdots + \nu_s$, for $1 \le s \le k$, $N_0 = 0$. Then, we have*

$$\sum_{d=0}^{m-1} P\{N_s < sm - d, \text{ for } 1 \le s \le k \mid N_k = n\} = \begin{cases} \frac{km-n}{k}, & \text{if } 0 \le n \le km, \\ 0, & \text{otherwise.} \end{cases} \tag{1.16}$$

These results are useful in handling $M/G/1$-type queues as proposed by Neuts [26]:

$$P\{\zeta = 0 \mid \zeta_n = i\} = \sum_{j=0}^{n-i} \left(1 - \frac{j}{n}\right) P\{N_n = j\}, \quad i \ge 0. \tag{1.17}$$

Interesting Corollaries

As a corollary of Theorem 1.3, we obtain that the number of paths from the origin to $(n_0, n_1, \ldots, n_r)$ that do not touch the hyperplane

$$x_0 = \sum_{i=1}^{r} \mu_i x_i \tag{1.18}$$

is given by

$$\frac{\alpha}{\alpha + \sum_{i=1}^{r}(\mu_i + 1)n_i}\binom{\alpha + \sum_{i=1}^{r}(\mu_i + 1)n_i}{n_1, \ldots, n_r}, \tag{1.19}$$

where

$$\alpha = n_0 - \sum_{i=1}^{r} \mu_i n_i \tag{1.20}$$

and the μ_i's (≥ 0) are all different.

Another simple corollary of Theorem 1.5, which finds applications in batch queues, is the following:

$$P\left(\sum_{i=1}^{r} X_i < \left\lfloor \frac{r-1}{m} \right\rfloor + 1,\ r = 1, 2, \ldots, n \,\middle|\, X_r = k\right) = \begin{cases} 1 - \frac{mk}{n}, & \text{if } 0 \leq mk \leq n, \\ 0, & \text{otherwise.} \end{cases} \tag{1.21}$$

See [22].

Applications

In his celebrated 1967 book [32], Takács has discussed applications of these ballot related combinatorial results to processes arising in queues, dams, storage, and risk and to order statistics.

We give an example from a queueing process (see [32, p. 94/95]).

Denote by $\nu_1, \nu_2, \ldots, \nu_r$ the number of customers joining the queue during the 1-st, 2-nd, ..., rth, ...services, respectively, and write

$$N_r = \nu_1 + \nu_2 + \cdots + \nu_r, \quad r = 1, 2, \ldots, \quad N_0 = 0, \tag{1.22}$$

and ζ_n, $n = 1, 2, \ldots$, the queue size immediately after the nth departure; and ζ_0 is the initial queue size. In this case, we speak about a queueing process of type

$$Q = \{\zeta_0;\ N_r,\ r = 0, 1, 2, \ldots\}. \tag{1.23}$$

Theorem 1.9 ([32, p. 99, Theorem 1]) *If $\nu_1, \nu_2, \ldots, \nu_n$ are interchangeable random variables, then*

$$\begin{aligned} P\{\zeta_n \leq k \mid \zeta_0 = i\} &= P\{N_n \leq n + k - i\} \\ &- \sum_{j=1}^{n-i} \sum_{l=0}^{n-i-j} \left(1 - \frac{1}{n-j}\right) P\{N_j = j + k,\ N_n = j + k + l\}, \quad i \geq 1, \end{aligned} \tag{1.24}$$

$$P\{\zeta_n \leq k \mid \zeta_0 = 0\} = P\{\zeta_n \leq k \mid \zeta_0 = 1\}, \tag{1.25}$$

and, in particular,

$$P\{\zeta_n = 0 \mid \zeta_0 = i\} = \sum_{j=0}^{n-i} \left(1 - \frac{1}{n-j}\right) P\{N_n = j\}. \tag{1.26}$$

References

1. Andersen, E.S.: On the fluctuations of sums of random variables. II. Math. Scand. **2**, 195–223 (1954)
2. Böhm, W., Mohanty, S.G.: Transient analysis of $M/M/1$ queues in discrete time with general server vacations. Studies in Applied Probability. J. Appl. Probab. **31A**, 115–129 (1994)
3. Chao, C.C., Severo, N.C.: Distributions of ballot problem random variables. Adv. Appl. Probab. **23**, 586–597 (1991)
4. Chorneyko, I.Z., Mohanty, S.G.: On the enumeration of pseudo-search code. Studia Sci. Math. Hungar. **7**, 47–54 (1972)
5. Csáki, E.: István Vincze (1912–1999) and his contribution to lattice path combinatorics and statistics. J. Stat. Plan. Inference **135**, 3–17 (2005)
6. Dickson, L.E.: Introduction to the Theory of Numbers. The University of Chicago Press, Chicago (1929)
7. Dirichlet, P.G.L.: In: Dedekind, R. (ed.) Vorlesungen über Zahlentheorie. Friedrich Vieweg und Sohn, Braunschweig (1871)
8. Dshalalow, J.H., Syski, R.: Lajos Takács and his work. J. Appl. Math. Stoch. Anal. **7**, 215–237 (1994)
9. E-mail communications between L. Takács and Aliakbar Montazer Haghighi during May to July of 2015, all are archived on PVAMU's server under http://amhaghighi@pvamu.edu
10. Erdős, P., Spencer, J.: Probabilistic Methods in Combinatorics. Academic (1974)
11. Euler, L.: Vollständige Anleitung zur Algebra. St. Petersburg, Kaiserliche Academie der Wissenschaften (1770). http://www.sophiararebooks.com/pages/books/2710/leonhard-euler/vollstandige-anleitungzur-algebra/?soldItem=true
12. Feller, W.: An Introduction to Probability Theory and Its Applications, vol. 1, 2nd edn. Wiley, New York (1957)
13. Feller, W.: On combinatorial methods in fluctuation theory. In: Grenander, U. (ed.) Probability and Statistics: The Harald Cramér Volume. Almqvist and Wiksell, Stockholm. Wiley, New York (1959)
14. Galambos, J., Gani, J. (eds.): Studies in Applied Probability. Essays in honor of Lajos Takács. J. Appl. Probab. **31A** (1994)
15. Jain, J.L., Mohanty, S.G.: Busy period distributions for two heterogeneous queueing models involving batches. Inf. Syst. Oper. Res. **19**, 133–139 (1981)
16. Jain, J.L., Mohanty, S.G., Böhm, W.: A Course on Queueing Models. Chapman and Hall/CRC Press, London (2007)
17. Kim, S.-K.: Stochastic design of enhanced network management architecture and algorithmic implementations. Am. J. Oper. Res. **3**, 87–93 (2014)
18. Mercankosk, G., Nair, G.M., Soet, W.J.: Combinatorial techniques for $M/G/1$-type queues. J. Appl. Probab. **38**, 722–736 (2001)
19. Mohanty, S.G.: An urn problem related to the ballot problem. Am. Math Mon. **73**, 526–528 (1966)
20. Mohanty, S.G., Jain, J.L.: On two types of queueing process involving batches. Canad. Oper. Res. Soc. J. **8**, 38–43 (1970)
21. Mohanty, S.G.: On queues involving batches. J. Appl. Probab. **9**, 430–435 (1972)
22. Mohanty, S.G.: Lattice Path Counting and Applications. Academic, New York (1979)
23. Montmort, P.R.: Essay d'analyse sur les jeux de hazard. Paris, Quillau (1708)

24. Montmort, P.R.: Essay d'analyse sur les jeux de hazard, 2nd edn. Paris, Quillau (1713)
25. Narayana, T.V., Mohanty, S.G.: A note on difference equations and combinatorial identities arising out of coin-tossing problems. Indian Soc. Agric. Stat. **17**, 83–86 (1955)
26. Neuts, M.F.: Structured Stochastic Matrices of $M/G/1$ Type and Their Applications. Marcel Dekker, New York (1989)
27. Riordan, J.: An Introduction to Combinatorial Analysis. Wiley, New York (1958)
28. Takács, L.: Investigation of waiting time problems by reduction to Markov processes. Acta Math. Acad. Sci. Hungar. **6**, 101–129 (1955)
29. Takács, L.: Some probability questions in the theory of telephone traffic. Magyar Tudomanyos Akademia. Matematikai es Fizikai Osztaly Kozleményei, vol. 8, pp. 155–175 (1958)
30. Takács, L.: Charles Jordan, 1871–1959. Ann. Math. Stat. **32**, 1–11 (1961)
31. Takács, L.: Introduction to the Theory of Queues. Oxford University Press, New York (1962)
32. Takács, L.: Combinatorial Methods in the Theory of Stochastic Processes. Wiley, New York (1967)
33. Takács, L.: Problem of coincidences. Arch. Hist. Exact Sci. **21**, 229–244 (1979/1980)
34. Takács, L.: Chance or determinism? In: Gani, J. (ed.) The Craft of Probabilistic Modelling: A Collection of Personal Accounts, pp. 139–149. Springer-Verlag, New York (1986)
35. Takács, L.: On the ballot theorems. In: Balakrishnan, N. (ed.) Advances in Combinatorial Methods and Applications in Probability and Statistics, pp. 97–114. Birkhäuser, Boston (1997)
36. The Internet, Google Search, https://www.google.com/, and in particular, Wikipedia, the free encyclopedia

The Distribution of the Local Time of Brownian Motion with Drift

Lajos Takács

Abstract In this paper Brownian motion with drift is considered, and explicit formulas are given for the distribution function, the density function, and the moments of the local time of the process and of the local time of the absolute value of the process.

Keywords Random walks · Brownian motion · Local time · Moments

2010 Mathematics Subject Classification Primary: 60J65 · Secondary: 60J50 60J55

1 Introduction

Let $\{\xi(u), u \geq 0\}$ be a standard Brownian motion process. We have $\mathbb{P}\{\xi(u) \leq x\} = \Phi(x/\sqrt{u})$ for $u > 0$, where

$$\Phi(x) = \frac{1}{\sqrt{2\pi}} \int_{-\infty}^{x} e^{-u^2/2}\, du \tag{2.1}$$

is the normal distribution function. We shall consider the process $\{\xi(u) + mu, u \geq 0\}$, where m is a real number. Let us define

$$\tau(\alpha, m) = \lim_{\varepsilon \to 0} \frac{1}{\varepsilon} \text{measure}\{u : \alpha \leq \xi(u) + mu < \alpha + \varepsilon, 0 \leq u \leq 1\} \tag{2.2}$$

for any real α and m. The limit (2.2) exists with probability one, and $\tau(\alpha, m)$ is a nonnegative random variable which is called the local time at level α. The concept of local time was introduced by Lévy [4, 5]. See also [2, 10].

Lajos Takács is deceased on 4 December 2015.

L. Takács (✉)
Case Western Reserve University, Cleveland, OH, USA
e-mail: judytakacs@me.com

G. E. Andrews et al. (eds.), *Lattice Path Combinatorics and Applications*, Developments in Mathematics 58, https://doi.org/10.1007/978-3-030-11102-1_2

In this paper we determine the distribution of $\tau(\alpha, m)$, the local time of $\{\xi(u) + mu, 0 \leq u \leq 1\}$ at level α, and the distribution of $\tau(\alpha, m) + \tau(-\alpha, m)$, the local time of $\{|\xi(u) + mu|, 0 \leq u \leq 1\}$ at level $|\alpha| > 0$. Our approach is based on a random walk $\{\zeta_r, r \geq 0\}$, where $\zeta_r = \xi_1 + \xi_2 + \cdots + \xi_r$ for $r \geq 1$, $\zeta_0 = 0$, and $\{\xi_r, r \geq 1\}$ is a sequence of independent and identically distributed random variables for which

$$\mathbb{P}\{\xi_r = 1\} = p \quad \text{and} \quad \mathbb{P}\{\xi_r = -1\} = q, \tag{2.3}$$

where $p > 0$, $q > 0$, and $p + q = 1$.

We define

$$\tau_n(a) = \#\{r = 1, 2, \ldots, n \text{ for which } \zeta_{r-1} = a - 1 \text{ and } \zeta_r = a\} \tag{2.4}$$

for $a = 1, 2, \ldots$, and

$$\tau_n(-a) = \#\{r = 1, 2, \ldots, n \text{ for which } \zeta_{r-1} = -a + 1 \text{ and } \zeta_r = -a\} \tag{2.5}$$

for $a = 1, 2, \ldots$.

If we assume that in the random walk

$$p = p_n = \frac{1}{2} + \frac{m}{2\sqrt{n}} \quad \text{and} \quad q = q_n = \frac{1}{2} - \frac{m}{2\sqrt{n}} \tag{2.6}$$

for $n > m^2$, then the process $\{\zeta_{[nu]}/\sqrt{n}, 0 \leq u \leq 1\}$ converges weakly to the process $\{\xi(u) + mu, 0 \leq u \leq 1\}$ as $n \to \infty$. By using the same argument as Knight [3] used for the Brownian motion process, we can conclude that

$$\lim_{n \to \infty} \mathbb{P}\left\{\frac{2\tau_n([\alpha\sqrt{n}])}{\sqrt{n}} \leq x\right\} = \mathbb{P}\{\tau(\alpha, m) \leq x\} \tag{2.7}$$

for any α and $x > 0$, and also

$$\lim_{n \to \infty} \mathbb{P}\left\{\frac{2\tau_n([\alpha\sqrt{n}]) + 2\tau_n(-[\alpha\sqrt{n}])}{\sqrt{n}} \leq x\right\} = \mathbb{P}\{\tau(\alpha, m) + \tau(-\alpha, m) \leq x\} \tag{2.8}$$

for $\alpha > 0$ and $x > 0$.

We shall determine the distributions and the moments of $\tau_n(a)$ and $\tau_n(a) + \tau_n(-a)$, and by a suitable limiting process we shall find the distributions and the moments of $\tau(\alpha, m)$ and $\tau(\alpha, m) + \tau(-\alpha, m)$. In the particular case where $m = 0$, the distributions and the moments of $\tau(\alpha, m)$ and $\tau(\alpha, m) + \tau(-\alpha, m)$ have already been determined. See [8, 9]. However, the asymmetric case, $m \neq 0$, is much more complicated than the symmetric case, $m = 0$, and indeed it is surprising that we can obtain explicit formulas for the distributions and moments of the local time.

2 A Random Walk

Let us recall some results for the random walk $\{\zeta_r, r \geq 0\}$ which we need in this paper. See [7]. We have

$$\mathbb{P}\{\zeta_r = 2j - r\} = \binom{r}{j} p^j q^{r-j} \tag{2.9}$$

for $j = 0, 1, \ldots, r$. Let us define $\rho(k)$ as the first passage time through k, that is,

$$\rho(k) = \inf\{r : \zeta_r = k \text{ and } r \geq 0\} \tag{2.10}$$

for $k = 0, \pm 1, \pm 2, \ldots$. Clearly, $\rho(0) = 0$.

If $1 \leq k \leq n$, then

$$\mathbb{P}\{\rho(k) \leq n\} = \mathbb{P}\{\zeta_n \geq k\} + \left(\frac{p}{q}\right)^k \mathbb{P}\{\zeta_n < -k\}. \tag{2.11}$$

This can be proved simply by applying the reflection principle to the random walk $\{\zeta_r, r \geq 0\}$. It follows by symmetry that

$$\mathbb{P}\{\rho(-k) = j\} = \left(\frac{q}{p}\right)^k \mathbb{P}\{\rho(k) = j\} \tag{2.12}$$

for $j \geq 0$. By (2.11), we have

$$\mathbb{P}\{\rho(k) = k + 2j\} = \frac{k}{k + 2j}\binom{k + 2j}{j} p^{k+j} q^j \tag{2.13}$$

for $k \geq 1$ and $j \geq 0$. Furthermore, the identity

$$\sum_{j=0}^{n} \mathbb{P}\{\rho(k) = j\}\mathbb{P}\{\rho(\ell) = n - j\} = \mathbb{P}\{\rho(k + \ell) = n\} \tag{2.14}$$

is valid for any $k \geq 1$, $\ell \geq 1$ and $n \geq 1$. By (2.13), we have

$$\sum_{j=0}^{\infty} \mathbb{P}\{\rho(k) = k + 2j\} w^j = [G(w)]^k \tag{2.15}$$

for $k \geq 1$ and $|4pqw| \leq 1$, where $G(0) = p$ and

$$G(w) = \frac{1 - \sqrt{1 - 4pqw}}{2qw} \tag{2.16}$$

for $0 < |4pqw| \le 1$. We note also that

$$\sum_{j=0}^{\infty} \binom{2j}{j} p^j q^j w^j = R(w) \tag{2.17}$$

for $|4pqw| < 1$, where

$$R(w) = (1 - 4pqw)^{-1/2}. \tag{2.18}$$

Finally, we note that, if $p = p_n$ and $q = q_n$ are defined by (2.6), then, by the central limit theorem,

$$\lim_{n\to\infty} \mathbb{P}\left\{\frac{\zeta_n}{\sqrt{n}} \le x\right\} = \Phi(x - m) \tag{2.19}$$

for any x, where $\Phi(x)$ is defined by (2.1).

3 The Distribution of $\tau_n(a)$

If we know the distribution of $\tau_n(a)$, where $a = 1, 2, \dots$, then the distribution of $\tau_n(-a)$ can be obtained by interchanging the roles of p and q. Thus it is sufficient to consider the case where $a = 1, 2, \dots$ and $n = 1, 2, \dots$.

Theorem 2.1 *If* $a = 1, 2, \dots$ *and* $k = 1, 2, \dots$, *then*

$$\mathbb{P}\{\tau_n(a) \ge k\} = \left(\frac{q}{p}\right)^{k-1} \mathbb{P}\{\rho(a + 2k - 2) \le n\}, \tag{2.20}$$

where the right-hand side is determined by (2.11).

Proof Let $a \ge 1$ and denote by $\theta_1, \theta_1 + \theta_2, \dots, \theta_1 + \theta_2 + \dots + \theta_r, \dots$ the successive subscripts $r = 1, 2, \dots$ for which $\zeta_{r-1} = a - 1$ and $\zeta_r = a$. Then $\theta_1, \theta_2, \dots, \theta_r, \dots$ are independent random variables. We have $\mathbb{P}\{\theta_1 = j\} = \mathbb{P}\{\rho(a) = j\}$ for $j \ge a$ and

$$\mathbb{P}\{\theta_k = j\} = \frac{q}{p}\mathbb{P}\{\rho(2) = j\} \tag{2.21}$$

for $j = 2, 3, \dots$ and $k \ge 2$. Obviously, θ_k $(k \ge 2)$ has the same distribution as $\rho(1) + \rho(-1)$, where $\rho(1)$ and $\rho(-1)$ are independent, and the distribution of $\rho(-1)$ is given by (2.12). Now, by (2.14),

$$\mathbb{P}\{\tau_n(a) \ge k\} = \mathbb{P}\{\theta_1 + \dots + \theta_k \le n\} = \left(\frac{q}{p}\right)^{k-1} \mathbb{P}\{\rho(a + 2k - 2) \le n\} \tag{2.22}$$

for $k = 1, 2, \dots$.

By (2.20), we can calculate the rth binomial moment of $\tau_n(a)$ for $r = 1, 2, \ldots, n$. We have

$$\begin{aligned}\mathbb{E}\left\{\binom{\tau_n(a)}{r}\right\} &= \sum_{k=r}^{n}\binom{k}{r}\mathbb{P}\{\tau_n(a)=k\} = \sum_{k=r}^{n}\binom{k-1}{r-1}\mathbb{P}\{\tau_n(a)\geq k\}\\ &= \sum_{k=r}^{n}\binom{k-1}{r-1}\left(\frac{q}{p}\right)^{k-1}\mathbb{P}\{\rho(a+2k-2)\leq n\}\\ &= \sum_{k=r}^{n}\binom{k-1}{r-1}\left(\frac{q}{p}\right)^{k-1}\Bigg(\mathbb{P}\{\zeta_n\geq a+2k-2\}\\ &\qquad + \left(\frac{p}{q}\right)^{a+2k-2}\mathbb{P}\{\zeta_n< -a-2k+2\}\Bigg)\end{aligned} \tag{2.23}$$

for $a \geq 1$. The last equality follows from (2.11).

We can also express (2.23) in the following way.

Theorem 2.2 *We have*

$$\mathbb{E}\left\{\binom{\tau_n(a)}{r}\right\} = pq^{r-1}\sum_{0\leq i\leq (n-a-2r+2)/2}(-1)^i\binom{-r/2}{i}(4pq)^i \cdot\mathbb{P}\{\rho(a+r-2)\leq n-r-2i\} \tag{2.24}$$

for $a = 1, 2, \ldots, n$ *and* $r = 1, 2, \ldots, n$, *where* $\mathbb{P}\{\rho(a+r-2)\leq n-r-2i\}$ *is determined by* (2.11).

Proof Denote by A_i, $i = 1, 2, \ldots, n$, the event that $\zeta_{i-1} = a-1$ and $\zeta_i = a$, where $a = 1, 2, \ldots$. Then the r-th binomial moment of $\tau_n(a)$ can be expressed in the form

$$\mathbb{E}\left\{\binom{\tau_n(a)}{r}\right\} = \sum_{0\leq j_1<j_2<\cdots<j_r\leq (n-a)/2}\mathbb{P}\{A_{a+2j_1}A_{a+2j_2}\ldots A_{a+2j_r}\}. \tag{2.25}$$

It is easy to see that

$$\mathbb{P}\{A_{a+2j}\} = p\sum_{0\leq i\leq j}\mathbb{P}\{\rho(a-1)=a-1+2j\}\binom{2j-2i}{j-i}(pq)^{j-i} \tag{2.26}$$

for $j \geq 0$. Thus, by (2.15) and (2.17), we have

$$\sum_{j=0}^{\infty}\mathbb{P}\{A_{a+2j}\}w^j = p[G(w)]^{a-1}R(w) \tag{2.27}$$

for $|4pqw| < 1$, where $G(w)$ is given by (2.16) and $R(w)$ by (2.18). In a similar way, we obtain that

$$\mathbb{P}\{A_{a+2i+2j}|A_{a+2i}\} = p \sum_{0\leq s\leq j-1} \mathbb{P}\{\rho(-1) = 1+2s\}\binom{2j-2s-2}{j-s-1}(pq)^{j-s-1} \tag{2.28}$$

for $j \geq 1$. Hence

$$\sum_{j=1}^{\infty} \mathbb{P}\{A_{a+2i+2j}|A_{a+2i}\}w^{j-1} = qG(w)R(w) \tag{2.29}$$

for $|4pqw| < 1$.

Since the random walk $\{\zeta_r, r \geq 0\}$ possesses the Markov property, we can determine (2.25) by (2.27) and (2.29). If, in the generating function

$$pq^{r-1}[G(w)]^{a+r-2}[R(w)]^r \tag{2.30}$$

we extract the coefficient of w^i and sum these coefficients over all i with $0 \leq i \leq (n-a-2r+2)/2$, then we obtain (2.24).

4 The Distribution of $\tau_n(a) + \tau_n(-a)$

The distribution $\tau_n(a) + \tau_n(-a)$ is determined by its binomial moments. We have

$$\mathbb{P}\{\tau_n(a) + \tau_n(-a) = k\} = \sum_{r=k}^{n}(-1)^{r-k}\binom{r}{k}\mathbb{E}\left\{\binom{\tau_n(a)+\tau_n(-a)}{r}\right\} \tag{2.31}$$

for $k = 0, 1, 2, \ldots$.

Theorem 2.3 *If* $r = 1, 2, \ldots, n$ *and* $a = 1, 2, \ldots, n$, *we have*

$$\mathbb{E}\left\{\binom{\tau_n(a)+\tau_n(-a)}{r}\right\} = \left(1+\left(\frac{q}{p}\right)^a\right)\sum_{\ell=1}^{r}\binom{r-1}{\ell-1}\left(\frac{q}{p}\right)^{(a-1)(\ell-1)} \cdot\mathbb{E}\left\{\binom{\tau_n(2(a-1)(\ell-1)+a)}{r}\right\}, \tag{2.32}$$

where the right-hand side is determined by (2.24).

Proof For $a = 1, 2 \ldots$ we define A_i, $i = 1, 2, \ldots, n$, in the same way as in the proof of Theorem 2.2. In addition, we define B_i, $i = 1, 2, \ldots, n$, as the event that $\zeta_{i-1} = -a+1$ and $\zeta_i = -a$. Let $C_i = A_i \cup B_i$ for $i = 1, 2, \ldots, n$. Then

$$\mathbb{E}\left\{\binom{\tau_n(a)+\tau_n(-a)}{r}\right\} = \sum_{0\le i_1<i_2<\cdots<i_r\le(n-a)/2} \mathbb{P}\{C_{a+2i_1}C_{a+2i_2}\ldots C_{a+2i_r}\}. \quad (2.33)$$

Similarly to (2.26), we have

$$\mathbb{P}\{B_{a+2j}\} = q\sum_{0\le i\le j}\mathbb{P}\{\rho(-a+1)=a-1+2i\}\binom{2j-2i}{j-i}(pq)^{j-i} \quad (2.34)$$

for $j \ge 0$. Hence

$$\sum_{j=0}^{\infty}\mathbb{P}\{B_{a+2j}\}w^j = p\left(\frac{q}{p}\right)^a [G(w)]^{a-1}R(w) \quad (2.35)$$

for $|4pqw| < 1$, where $G(w)$ is given by (2.16) and $R(w)$ by (2.18). Obviously,

$$\mathbb{P}\{B_{a+2i+2j}|B_{a+2i}\} = \mathbb{P}\{A_{a+2i+2j}|A_{a+2i}\} \quad (2.36)$$

for $j \ge 1$ and

$$\mathbb{P}\{B_{3a+2i+2j}|A_{a+2i}\} = q\sum_{0\le s\le j}\mathbb{P}\{\rho(-2a+1)=2a-1+2s\}\binom{2j-2s}{j-s}(pq)^{j-s} \quad (2.37)$$

for $j \ge 0$. By (2.15) and (2.17), we have

$$\sum_{j=0}^{\infty}\mathbb{P}\{B_{3a+2i+2j}|A_{a+2i}\}w^j = p\left(\frac{q}{p}\right)^{2a} [G(w)]^{2a-1}R(w) \quad (2.38)$$

for $|4pqw| < 1$. In the same way, we obtain that

$$\mathbb{P}\{A_{3a+2i+2j}|B_{a+2i}\} = p\sum_{0\le s\le j}\mathbb{P}\{\rho(2a-1)=2a-1+2s\}\binom{2j-2s}{j-s}(pq)^{j-s} \quad (2.39)$$

for $j \ge 0$ and

$$\sum_{j=0}^{\infty}\mathbb{P}\{A_{3a+2i+2j}|B_{a+2i}\}w^j = p[G(w)]^{2a-1}R(w) \quad (2.40)$$

for $|4pqw| < 1$.

The probabilities (2.26), (2.28), (2.34), (2.36), (2.37), and (2.39) completely determine

$$\mathbb{P}\{C_{a+2i_1}C_{a+2i_2}\ldots C_{a+2i_r}\} \quad (2.41)$$

in (2.33). In (2.41) let us replace each C_i by $A_i \cup B_i$. If we perform the operations indicated, the event $C_{a+2i_1} C_{a+2i_2} \dots C_{a+2i_r}$ becomes the union of 2^r mutually exclusive events, and consequently (2.41) can be expressed as the sum of 2^r probabilities, each probability having the form

$$\mathbb{P}\{D_{a+2i_1} D_{a+2i_2} \dots D_{a+2i_r}\}, \tag{2.42}$$

where D_{a+2i_s} is either A_{a+2i_s} or B_{a+2i_s}. To find

$$\sum_{0 \le i_1 < i_2 < \cdots < i_r \le (n-a)/2} \mathbb{P}\{D_{a+2i_1} D_{a+2i_2} \dots D_{a+2i_r}\}, \tag{2.43}$$

we count the number of alternations in the sequence of r events in (2.42). We speak of an alternation if an event A_i is followed by an event B_j or an event B_i is followed by an event A_j. There are $\binom{r-1}{\ell-1}$ possible sequences in which the number of alternations is $\ell - 1$ and the first event is of the type A_i, and there are also $\binom{r-1}{\ell-1}$ possible sequences in which the number of alternations is $\ell - 1$ and the first event is of type B_j. The total number of possible sequences is

$$2 \sum_{\ell=1}^{r} \binom{r-1}{\ell-1} = 2^r, \tag{2.44}$$

as it should be.

Now we shall prove that, if in (2.42) the number of alternations in the sequence of events is $\ell - 1$, an even number, and if the first event in the sequence is of type A_i, then (2.43) is equal to

$$\left(\frac{q}{p}\right)^{(a-1)(\ell-1)} \mathbb{E}\left\{\binom{\tau_n(2(a-1)(\ell-1)+a)}{r}\right\}, \tag{2.45}$$

where the right-hand side is determined by (2.24). If in (2.42) the number of alternations in the sequence of events is $\ell - 1$, an odd number, and if the first event in the sequence is of type B_i, then (2.43) is again equal to (2.45).

If in (2.42) the number of alternations in the sequence of events is $\ell - 1$, an even number, and if the first event in the sequence is of type B_j, then (2.43) is equal to

$$\left(\frac{q}{p}\right)^{(a-1)(\ell-1)+a} \mathbb{E}\left\{\binom{\tau_n(2(a-1)(\ell-1)+a)}{r}\right\}, \tag{2.46}$$

where the right-hand side is determined by (2.24). If in (2.42) the number of alternations in the sequence is $\ell - 1$, an odd number, and if the first event in the sequence is of type A_i, then (2.43) is again equal to (2.46).

If we multiply both (2.45) and (2.46) by $\binom{r-1}{\ell-1}$, and if we form the sum of the products for all $\ell = 1, 2, \dots, r$, then we obtain (2.32). It remains to prove (2.45), and

Table 1 Various sequences of events

Case	ℓ	(A)	(B)	(AB)	(BA)	(AA) and (BB)
(i)	Odd	1	0	$(\ell-1)/2$	$(\ell-1)/2$	$r-l$
(ii)	Odd	0	1	$(\ell-1)/2$	$(\ell-1)/2$	$r-l$
(iii)	Even	1	0	$\ell/2$	$(\ell-1)/2$	$r-l$
(iv)	Even	0	1	$(\ell-2)/2$	$\ell/2$	$r-\ell$
G.F.		(2.27)	(2.35)	(2.38)	(2.40)	(2.29)

(2.46). It is convenient to use the generating functions (2.27), (2.29), (2.35), (2.38), and (2.40). In the sequence of events in (2.42), denote the numbers of occurrences of the pairs of type AB, BA, AA, BB by (A, B), (B, A), (A, A), (B, B), respectively. Let $(A) = 1$ if the first event is of type A, and $(A) = 0$ if it is of type B. Also let $(B) = 1$ if the first event is of type B, and $(B) = 0$ if it is of type A. For any $\ell = 1, 2, \ldots, r$, Table 1 contains all the possible cases.

First let us consider the case (i) when $\ell =$ odd and the first event in the sequence is of type A_i. To obtain (2.43), we form the product of r generating functions. In Table 1, the last line indicates the corresponding formulas for the appropriate generating functions, and the entries in line (i) indicate how many times we should take each generating function. If we perform all the r multiplications, the result is the generating function

$$(q/p)^{(a-1)(\ell-1)}pq^{r-1}[G(w)]^{2a\ell-2\ell-a+r}[R(w)]^r, \tag{2.47}$$

where $G(w)$ is given by (2.16) and $R(w)$ by (2.18). If we extract the coefficient of w^s in (2.47) and sum these coefficients for $0 \le s \le [n - 2\ell(a-1) + a - 2r]/2$, then by (2.24) and (2.30) we obtain (2.45). If instead of (i), we consider (iv), we obtain again (2.47) and this yields (2.45). If we consider line (ii), then following the same procedure as before, we obtain the following generating function

$$(q/p)^{(a-1)(\ell-1)+a}pq^{r-1}[G(w)]^{2a\ell-2\ell-a+r}[R(w)]^r. \tag{2.48}$$

If we extract the coefficient of w^s in (2.48) and sum these coefficients for $0 \le s \le [n - 2\ell(a-1) + a - 2r]/2$, then we obtain (2.46). If, instead of (ii), we consider (iii), we obtain again (2.48), and this yields (2.46). Formulas (2.45) and (2.46) imply (2.32). This completes the proof of Theorem 2.3.

5 The Distribution of $\tau(\alpha, m)$

Since $\tau(-\alpha, m)$ has the same distribution as $1 - \tau(\alpha, -m)$, it is sufficient to determine the distribution of $\tau(\alpha, m)$ for $\alpha > 0$ and $m \in (-\infty, \infty)$. Let us assume now that $p = p_n$ and $q = q_n$ are given by (2.6) for $n > m^2$. Then we have the following limit theorem.

Theorem 2.4 *If* $\alpha > 0$ *and* $x > 0$, *then*

$$\lim_{n\to\infty} \mathbb{P}\left\{\frac{2\tau_n([\alpha\sqrt{n}])}{\sqrt{n}} > x\right\} = 1 - F(x, \alpha, m), \tag{2.49}$$

where

$$F(x, \alpha, m) = 1 - e^{-mx}[1 - \Phi(x + \alpha - m)] - e^{m(x+2\alpha)}[1 - \Phi(x + \alpha + m)], \tag{2.50}$$

and $\Phi(x)$ *is given by* (2.1).

Proof Now

$$\lim_{n\to\infty} \left(\frac{q_n}{p_n}\right)^{\sqrt{n}} = e^{-2m}. \tag{2.51}$$

If, in (2.20), we put $a = [\alpha\sqrt{n}]$, where $\alpha > 0$ and $k = [x\sqrt{n}/2]$ with $x > 0$ and let $n \to \infty$, then by (2.11) and (2.19) we obtain (2.49).

By (2.7), this proves that

$$\mathbb{P}\{\tau(\alpha, m) \le x\} = F(x, \alpha, m) \tag{2.52}$$

for $x \ge 0$. Clearly, we have $\mathbb{P}\{\tau(\alpha, m) \le x\} = 0$ if $x < 0$.

The moments

$$\mathbb{E}\{[\tau(\alpha, m)]^r\} = \mu_r(\alpha, m) \tag{2.53}$$

exist for $r \ge 0$. We have $\mu_0(\alpha, m) = 1$ and

$$\mu_r(\alpha, m) = r\int_0^\infty [1 - F(x, \alpha, m)]x^{r-1}dx \tag{2.54}$$

for $r \ge 1$.

Theorem 2.5 *If* $\alpha > 0$, *then*

$$\lim_{n\to\infty} \mathbb{E}\left\{\left[\frac{2\tau_n([\alpha\sqrt{n}])}{\sqrt{n}}\right]^r\right\} = \mu_r(\alpha, m) \tag{2.55}$$

for $r \ge 0$, *where* $\mu_r(\alpha, m)$ *is given by* (2.54).

Proof If we assume that $p = p_n$ and $q = q_n$ are given by (2.6) for $n > m^2$, and if in (2.23) we put $a = [\alpha\sqrt{n}]$, where $\alpha > 0$, then, by letting $n \to \infty$, we obtain (2.55) by (2.19).

6 The Distribution of $\tau(\alpha, m) + \tau(-\alpha, m)$

Let us assume that $p = p_n$ and $q = q_n$ are given by (2.6) for $n > m^2$. Then we have the following limit theorem.

Theorem 2.6 *If $\alpha > 0$ and $r \geq 1$, then the limit*

$$\lim_{n\to\infty} \mathbb{E}\left\{\left[\frac{2\tau_n([\alpha\sqrt{n}]) + 2\tau_n(-[\alpha\sqrt{n}])}{\sqrt{n}}\right]^r\right\} = M_r(\alpha, m) \tag{2.56}$$

exists, and

$$M_r(\alpha, m) = (1 + e^{-2m\alpha}) \sum_{\ell=1}^{r} \binom{r-1}{\ell-1} e^{-2m\alpha(\ell-1)} \mu_r((2\ell-1)\alpha, m), \tag{2.57}$$

where $\mu_r(\alpha, m)$ is given by (2.54).

Proof If in (2.32) we put $a = [\alpha\sqrt{n}]$, where $\alpha > 0$, and if we let $n \to \infty$ by (2.55) we obtain (2.57).

Theorem 2.7 *If $\alpha > 0$, then there exists a distribution function $L_\alpha(x, m)$ of a non-negative random variable such that*

$$\lim_{n\to\infty} \mathbb{P}\left\{\frac{2\tau_n([\alpha\sqrt{n}]) + 2\tau_n(-[\alpha\sqrt{n}])}{\sqrt{n}} \leq x\right\} = L_\alpha(x, m) \tag{2.58}$$

in every continuity point of $L_\alpha(x, m)$. The distribution function $L_\alpha(x, m)$ is uniquely determined by its moments

$$\int_{-\infty}^{\infty} x^r L_\alpha(x, m) dx = M_r(\alpha, m) \tag{2.59}$$

for $r \geq 0$, where $M_0(\alpha, m) = 1$ and $M_r(\alpha, m)$ for $r \geq 1$ is given by (2.57).

Proof Since $\mu_r(\alpha, m)$ is a decreasing function of α if $\alpha > 0$, we have

$$M_r(\alpha, m) \leq (1 + e^{-2m\alpha})^r \mu_r(\alpha, m) \tag{2.60}$$

for $r \geq 1$. Consequently, the sequence of moments $\{M_r(\alpha, m)\}$ uniquely determines $L_\alpha(x, m)$, and $L_\alpha(x, m) = 0$ for $x < 0$. By the moment convergence theorem of Fréchet and Shohat [1], we can conclude that (2.59) implies (2.58).

By (2.8) and (2.58), we have

$$\mathbb{P}\{\tau(\alpha, m) + \tau(-\alpha, m) \le x\} = L_\alpha(x, m), \tag{2.61}$$

and, by (2.59),

$$\mathbb{E}\{[\tau(\alpha, m) + \tau(-\alpha, m)]^r\} = M_r(\alpha, m) \tag{2.62}$$

for $\alpha > 0$ and $r \ge 0$. Formula (2.57) is a surprisingly simple expression for the r-th moment of $\tau(\alpha, m) + \tau(-\alpha, m)$. If we know the r-th moment of $\tau(\alpha, m)$ for $\alpha > 0$, then by (2.57) the r-th moment of $\tau(\alpha, m) + \tau(-\alpha, m)$ can immediately be determined for $\alpha > 0$. Moreover, formula (2.57) makes it possible to determine $L_\alpha(x, m)$ explicitly.

Theorem 2.8 *If $x \ge 0$, $\alpha > 0$, and $m \in (-\infty, \infty)$, then we have*

$$L_\alpha(x, m) = 1 + (1 + e^{2m\alpha}) \sum_{\ell=1}^{\infty} \frac{(-1)^\ell e^{-2\ell m\alpha}}{(\ell-1)!} \left(\frac{d^{\ell-1} x^{\ell-1}[1 - F(x, (2\ell-1)\alpha, m)]}{d\, x^{\ell-1}} \right), \tag{2.63}$$

and, if $x > 0$, $\alpha > 0$, and $m \in (-\infty, \infty)$, then we have

$$\frac{dL_\alpha(x, m)}{dx} = (1 + e^{2m\alpha}) \sum_{\ell=1}^{\infty} \frac{(-1)^\ell e^{-2\ell m\alpha}}{(\ell-1)!} \left(\frac{d^{\ell} x^{\ell-1}[1 - F(x, (2\ell-1)\alpha, m)]}{d\, x^{\ell}} \right), \tag{2.64}$$

where $F(x, \alpha, m)$ is given by (2.50).

Proof For $\alpha > 0$ the Laplace–Stieltjes transform

$$\Psi_\alpha(s) = \int_{-\infty}^{\infty} e^{-sx} L_\alpha(x, m) dx \tag{2.65}$$

can be expressed as

$$\Psi_\alpha(s) = \sum_{r=0}^{\infty} (-1)^r M_r(\alpha, m) s^r / r!, \tag{2.66}$$

and the series is convergent on the whole complex plane. Here, $M_r(\alpha, m)$ is given by (2.57). If we substitute (2.57) into (2.66), express $\mu_r(\alpha, m)$ by (2.54), and interchange summations with respect to r and ℓ, we obtain that

$$\Psi_\alpha(s) = 1 + (1 + e^{2m\alpha}) \sum_{\ell=1}^{\infty} \frac{e^{-2\ell m\alpha}}{(\ell-1)!} \int_0^{\infty} \left(\frac{d^\ell e^{-sx}}{dx^\ell} \right) [1 - F(x, (2\ell-1)\alpha, m)] x^{\ell-1} dx. \tag{2.67}$$

By integrating by parts, we get

$$\Psi_\alpha(s) = 1 + (1 + e^{2m\alpha}) \sum_{\ell=1}^{\infty} (-1)^\ell e^{-2\ell m\alpha} [1 - F(0, (2\ell - 1)\alpha, m)]$$
$$+ (1 + e^{2m\alpha}) \sum_{\ell=1}^{\infty} \frac{(-1)^\ell e^{-2\ell m\alpha}}{(\ell - 1)!} \int_0^\infty e^{-sx} \left(\frac{d^\ell [1 - F(x, (2\ell - 1)\alpha, m)] x^{\ell-1}}{dx^\ell} \right) dx. \tag{2.68}$$

Hence we can conclude that (2.63) and (2.64) hold. By (2.68), we have

$$L_\alpha(0, m) = 1 + (1 + e^{2m\alpha}) \sum_{\ell=1}^{\infty} (-1)^\ell e^{-2\ell m\alpha} [1 - F(0, (2\ell - 1)\alpha, m)]. \tag{2.69}$$

Obviously,

$$L_\alpha(0, m) = \mathbb{P}\{ \sup_{0 \le u \le 1} |\xi(u) + mu| < \alpha \} \tag{2.70}$$

for $\alpha > 0$. We have

$$\mathbb{P}\{ \sup_{0 \le u \le 1} |\xi(u) + mu| < \alpha \} = \sum_j e^{-4j\alpha m} [\Phi((4j + 1)\alpha - m) - \Phi((4j - 1)\alpha - m)]$$
$$- \sum_j e^{(4j+2)\alpha m} [\Phi((4j + 3)\alpha + m)] - \Phi((4j + 1)\alpha + m), \tag{2.71}$$

where $\Phi(x)$ is defined by (2.1). See [6, p. 226]. Formulas (2.69) and (2.71) are in agreement.

In (2.63) and (2.64), the derivatives can be expressed explicitly by Hermite polynomials. If we write

$$\varphi(x) = \frac{1}{\sqrt{2\pi}} e^{-x^2/2} \tag{2.72}$$

for the density function of the normal distribution, then

$$\frac{d^n \varphi(x)}{dx^n} = (-1)^n \varphi(x) H_n(x) \tag{2.73}$$

for $n = 0, 1, 2, \ldots$, where

$$H_n(x) = n! \sum_{j=0}^{[n/2]} \frac{(-1)^j x^{n-2j}}{2^j j!(n - 2j)!} \tag{2.74}$$

is the n-th Hermite polynomial. We have

$$H_n(x) = xH_{n-1}(x) - (n-1)H_{n-2}(x) \tag{2.75}$$

for $n \geq 2$, where $H_0(x) = 1$ and $H_1(x) = x$.

References

1. Fréchet, M., Shohat, J.: A proof of the generalized second-limit theorem in the theory of probability. Trans. Am. Math. Soc. **33**, 533–543 (1931)
2. Itô, K., McKean Jr., H.P.: Diffusion Processes and Their Sample Paths. Springer, Berlin (1965)
3. Knight, F.B.: Random walk and a sojourn density process of Brownian motion. Trans. Am. Math. Soc. **109**, 56–86 (1965)
4. Lévy, P.: Sur certains processus stochastiques homogènes. Compos. Math. **7**, 283–339 (1939)
5. Lévy, P.: Processus Stochastiques et Mouvement Brownien, 2nd edn. Gauthier-Villars, Paris (1965)
6. Takács, L.: Combinatorial Methods in the Theory of Stochastic Processes. Wiley, New York (1967)
7. Takács, L.: Fluctuation problems for Bernoulli trials. SIAM Rev. **21**, 222–228 (1979)
8. Takács, L.: On the local time of the Brownian motion. Ann. Appl. Probab. **5**, 741–756 (1995)
9. Takács, L.: Brownian local times. J. Appl. Math. Stoch. Anal. **8**, 209–232 (1995)
10. Trotter, H.: A property of Brownian motion paths. Ill. J. Math. **2**, 425–433 (1958)

Reflections on Shreeram Abhyankar

Krishnaswami Alladi

Abstract These are some reflections on the great algebraic geometer Shreeram Abhyankar given at the Banquet at the International Conference on Lattice Path Combinatorics and Applications, California Polytechnic University, Pomona, on Thursday, August 20, 2015.

Keywords Shreeram Abhyankar

2010 Mathematics Subject Classification Primary: 01A70 · Secondary: 14–03

Shreeram Abhyankar (22 July, 1930–2 Nov., 2012) was one of the world's most eminent algebraic geometers. He ranked among the ten greatest mathematicians of India in the twentieth century. He belonged to the Chitpavan Brahmin community of Maharashtra and was proud of its illustrious lineage. After completing his undergraduate studies in India, he went to Harvard University and did his doctoral work there under the direction of Oscar Zariski, one of the most influential figures in algebraic geometry. Abhyankar's Ph.D. thesis on the resolution of singularities problem is a classic and is among his most important contributions. I was fortunate to get to know him from my boyhood because he was a close friend of my father. Abhyankar and his wife Yvonne were our house guests in India in the sixties. Over the years, we have had several meetings, first in Madras, India, where we hosted him, then in Purdue when my parents and I were his house guests, and finally at the University of Florida where I had the opportunity to host him during my term as Chair. Abhyankar was a fascinating, colorful, and engaging personality. He would grab your attention with his warmth, his open frankness, and his firm opinions on various matters—mathematical and non-mathematical. I have observed him in close quarters, and I will now share a few anecdotes to convey his unusual and engaging personality.

K. Alladi (✉)
Department of Mathematics, University of Florida, Gainesville, FL 32611, USA
e-mail: alladik@ufl.edu

G. E. Andrews et al. (eds.), *Lattice Path Combinatorics and Applications*,
Developments in Mathematics 58, https://doi.org/10.1007/978-3-030-11102-1_3

Shreeram Abhyankar, Alladi Ramakrishnan, Krishnaswami Alladi, at the Alladis' home in January 2001

My late father Professor Alladi Ramakrishnan had founded MATSCIENCE, The Institute of Mathematical Sciences, Madras, India, in 1962, and was its Director until his retirement in 1983 at the age of sixty. The Institute was a realization of a dream of my father and a direct consequence of a Theoretical Physics Seminar that he conducted at our family home in Madras. Abhyankar had a strong attachment to India and a great regard for Indian culture and scientific heritage. Thus, he admired my father's efforts in creating such an institute for higher learning and so he visited MATSCIENCE several times in the sixties, his first visit being in August 1963 when he was at Johns Hopkins University, before moving to Purdue. His wife Yvonne always accompanied him, and we admired the way she wore the sari—so naturally and elegantly like an Indian lady. During one such visit in January 1968, my father requested Abhyankar to give a public lecture on Ramanujan—undoubtedly the greatest mathematician India has ever produced. Out of his profound regard for Ramanujan, Abhyankar readily agreed, even though he was not an expert on the mathematics of Ramanujan. The venue for the lecture was the C.P. Ramaswami Iyer Foundation, close to our house.

Sir C.P. Ramaswami Iyer was a very eminent lawyer, an illustrious statesman, and an orator par excellence. He was a contemporary and close friend of my grandfather Sir Alladi Krishnaswami Iyer, who was one of the greatest lawyers of India. After Sir C. P. died in 1966, a foundation was created in his name as he desired, and this foundation held public lectures on the spacious lawns of Sir C.P.'s mansion called "The Grove". As a 12-year-old boy, I attended Abhyankar's brilliant lecture on Ramanujan at the CP Foundation. He held the audience of more than two hundred citizens of Madras in various walks of life in rapt attention as he described some of Ramanujan's most important contributions in his unique and powerful lecturing style. The text of Abhyankar's lecture on Ramanujan appeared in a volume published by the Plenum Press and edited by my father [1].

Abhyankar was proud of his Indian roots and legacy. He valued and emphasized the classical approach to mathematics and did not care for abstraction. He therefore did not agree with the views of the mathematicians of the Tata Institute of Fundamental Research (TIFR), especially with their adoption of the Grothendieck program in algebraic geometry. He started an institute in 1976 called *Bhaskaracharya Prathishthana* in his native town of Poona (now Pune) in the state of Maharashtra—an institute inspired by the legacy of the great mathematical Guru Bhaskara.

Abhyankar visited the University of Florida regularly from the 1990s to interact with the famous group theorist John Thompson who was Graduate Research Professor in our department. During one such visit, I invited him to address our undergraduate mathematics club $\pi\mu\varepsilon$. He readily agreed and gave a lovely lecture entitled "An introduction to algebraic geometry". In his talk, he stressed that the foundations of algebraic geometry are in classical Cartesian analytic geometry. He lamented that not enough time is spent nowadays in high schools or undergraduate classes to discuss analytic geometry in detail with proofs. He said that his father (also a mathematician) had three years of analytic geometry, but he had only two, and that the younger generation has one year or less on analytic geometry. This decrease in the amount of time spent on analytic geometry worried him. His article "Historical ramblings in algebraic geometry" that appeared in the American Mathematical Monthly in 1976 [2] stresses elementary reasoning in algebraic geometry. His fundamental thesis in this paper is: "The method of high school algebra is powerful. So let us not be overwhelmed by groups-rings-fields or functorial arrows of the other two algebras[1] and thereby lose sight of the power of the explicit algorithmic process given to us by Newton, Tschirnhausen, Kronecker, and Sylvester." He received the Chauvenet Prize of the MAA in 1978 for this paper.

After his talk to $\pi\mu\varepsilon$, there was a dinner in his honor at my house and there we had a discussion on elementary approaches to deep mathematical problems. I mentioned that in number theory, Paul Erdős was the ultimate champion of the elementary method. To my surprise—or should I say to my shock!—he immediately shot back and said that he did not consider Erdős to be a great mathematician. I asked him why, and he responded saying that Erdős had written some very simple papers. I then said that a mathematician should be judged by his very best work and total contributions and not by his least significant paper. We argued. I gave examples of some ingenious elementary proofs of Erdős following which he asked me how I knew so much about Erdős and his mathematics. I said that Erdős was like a mentor to me and that I

[1] Abhyankar classified algebra into three types: (i) high school algebra (polynomials, power series): Bhaskara (1114), Cardano (1530), Ferrari (1540), Newton (1680), Tschirnhausen (1683), Euler (1748), Sylvester (1840), Cayley (1870), Kronecker (1882), Mertens (1886), König (1903), Perron (1905), Hurwitz (1913), Macaulay (1916), Zariski (1941), Hironaka (1964); (ii) college algebra (rings, fields, and ideals): Dedekind (1882), Noether (1925), Krull (1930), Zariski (1941), Chevalley (1943), Cohen (1946), Nagata (1960); and (iii) university algebra (functors): Serre (1955), Cartan (1956), Eilenberg (1956), Grothendieck (1960), Mumford (1965).

had collaborated with him. He immediately exclaimed: "Oh! He is your Guru. Thus I apologize and completely withdraw everything I said because you have every right to defend your Guru, and I should not criticize your Guru in your presence. For example, I would defend my Guru Oscar Zariski against anyone."

I mentioned earlier that Abhyankar had a great regard for Indian culture. In the Hindu tradition, your Guru is like a God and so should be worshipped. Thus, he had unbounded love and respect for Oscar Zariski, his Ph.D. advisor at Harvard. Abhyankar withdrew his arguments not because I was correct about Erdős, but because of his deep respect for the Guru.

Even after my father's retirement as Director of MATSCIENCE, Abhyankar made it a point to visit our home every time he was in Madras, and call on my parents. After my father died, when I edited a volume in his memory, Abhyankar contributed a massive paper [3] to that volume dedicated both to his father and my father. This paper was originally intended for the Journal of Algebra, but he decided to submit to the volume in memory of my father, and for this I am most grateful.

Even though he was blunt and brutally frank, he was a man of deep feelings and great kindness. He was an eminent mathematician and a fine person.

Acknowledgements I am thankful to the organizers of the International Conference on Lattice Paths Combinatorics and Applications and in particular to Alan Krinik, for not only inviting me to speak in one of the technical sessions, but also to be the Banquet Speaker at this conference which honored the memory of Shreeram Abhyankar and Philippe Flajolet.

References

1. Abhyankar, S.S.: The contributions of Ramanujan to mathematics. In: Ramakrishnan, Alladi (ed.) Symposia on Theoretical Physics and Mathematics, vol. 10, pp. 1–6. Plenum, New York (1970)
2. Abhyankar, S.S.: Historical ramblings in algebraic geometry and related algebra. Am. Math. Mon. **83**, 409–448 (1976)
3. Abhyankar, S.S.: Inversion and invariance of characteristic terms: Part I. In: Alladi, K., Klauder, J.R., Rao, C.R. (eds.) The Legacy of Alladi Ramakrishnan in the Mathematical Sciences, pp. 93–168. Springer, New York (2010)

My Association and Collaboration with George Andrews

Torchbearer of Ramanujan and Partitions

Krishnaswami Alladi

Abstract These are personal reflections on my association, and a report of my collaboration, with George Andrews given at the International Conference on Lattice Path Combinatorics and Applications, California Polytechnic University, Pomona.

Keywords George Andrews · Integer partitions · Weighted words · q-series · Rogers–Ramanujan identities

2010 Mathematics Subject Classification Primary: 01A70, 0503, 05A15, 05A17, 11P84

1 Introduction

George Andrews is the undisputed leader on partitions and the work of Ramanujan combined. After Hardy and Ramanujan, he, more than anyone else in the modern era, is responsible for making the theory of partitions a central area of research. His book on partitions [14] published first in 1976 as Volume 2 of the Encyclopedia of Mathematics (John Wiley), is a bible in the field, and his NSF-CBMS Lectures [15] of 1984–85 highlight the fundamental connections between partitions and Ramanujan's work with many allied fields. We definitely owe to him our present understanding of many of the deep identities in Ramanujan's Lost Notebook. I had the good fortune to collaborate with him and also interact with him very closely both at Penn State University (his home turf) where I visited often, and at the University of Florida, where he has spent the Spring term every year since 2005. I also have had the pleasure of hosting him in India several times. Thus I have come to know him really well as a mathematician, colleague, and friend. Here I will first share with you (in

Section 2 on personal recollections is based on the speech given at the banquet, while Sect. 3 on collaboration is based on a talk in one of the technical sessions of the conference.

K. Alladi (✉)
Department of Mathematics, University of Florida, Gainesville, FL 32611, USA
e-mail: alladik@ufl.edu

G. E. Andrews et al. (eds.), *Lattice Path Combinatorics and Applications*, Developments in Mathematics 58, https://doi.org/10.1007/978-3-030-11102-1_4

Sect. 2) my observations of him as a man and mathematician. I will then describe (in Sect. 3) some aspects of our joint work that will highlight his vast knowledge and brilliance. In Sect. 2, I will describe events chronologically rather than thematically. In Sect. 3, I will discuss my joint work with him on the Capparelli and the Göllnitz theorems.

George Andrews delivering a lecture on the story of Ramanujan's Lost Notebook at the Alladi residence in Madras, India, during the Ramanujan Centennial in December 1987.

2 Personal Recollections

First Visit to India: Even though Andrews has been studying Ramanujan's work since the sixties and had been "introduced to India" through the writings of, and on, Ramanujan, his first visit to India was only in Fall 1981. That academic year, I was visiting the Institute for Advanced Study in Princeton, and he contacted me saying that he was planning a visit to India, and to Madras in particular, and would appreciate any suggestions I would have. My father, the late Professor Alladi Ramakrishnan, was Director of MATSCIENCE, the Institute of Mathematical Sciences that he had founded in 1962, and so I put him in touch with my father who hosted him in Madras, and helped arrange a meeting for Andrews with Mrs. Janaki Ammal Ramanujan. Upon return from India, Andrews called me from Penn State, told me that it was an immensely enjoyable and fruitful visit, and that he appreciated my father's help and hospitality. To reciprocate, Andrews invited me to a Colloquium at Penn State where he was Department Chair at that time. Andrews is always a gracious host, but in his capacity as Chair, he rolled out the red carpet for me! He hosted a party for me at his house during that visit and that is how our close friendship began.

Andrews signing the Visitors Book in the office of Alladi Ramakrishnan after his lecture at the Alladi residence in December 1987

I was working at that time in analytic number theory but I wanted to learn partitions and q-series, and that aspect of the work of Ramanujan. So after I returned to India from Princeton, I wrote to Andrews and asked him for his papers. Promptly, I received two large packages containing more than 100 of his reprints. So I started studying them along with his Encyclopedia and gave a series of lectures at MATSCIENCE in Madras, the notes of which I still use today. Even after this course of lectures, I was unsure whether to venture into partitions and q-series. The infinite series formulae were beautiful, but daunting. The decision to change my field of research to the theory of partitions and q-series came during the Ramanujan Centennial in Madras in December 1987.

The Ramanujan Centennial: The Ramanujan Centennial was an occasion when mathematicians from around the world gathered in India to pay homage to the Indian genius. Among the mathematical luminaries at the conference, there was a lot of attention on Andrews, Richard Askey, and Bruce Berndt — jokingly referred to in the USA as the "Gang of Three" in the world of Ramanujan. I prefer to refer to them as the "Great Trinity" of the Ramanujan world, like Brahma, Vishnu, and Shiva, the three premier Hindu gods! The Great Trinity along with Nobel Laureate Astrophysicist Subrahmanyam Chandrasekhar and Fields Medalist Atle Selberg, were the stars of the Ramanujan Centennial. But Andrews occupied a special place in this elite group, because the Lost Notebook that he unearthed at the Wren Library in Cambridge University, was released in published form [21] at a grand public function in Madras on December 22, 1987, Ramanujan's 100-th birthday, by India's Prime Minister Rajiv Gandhi, who handed one copy to Janaki Ammal and another to Andrews. That definitely was a high point in the academic life of Andrews. Andrews has written a marvelous Preface to that book published by Narosa, which at that time was part of Springer, India.

December 1987 was a politically tense time in Madras because the Chief Minister of Madras, M. G. Ramachandran — MGR as he was affectionately known — a former cine hero to the millions, was terminally ill. There were several conferences in India around Ramanujan's 100-th birthday, and Andrews was a speaker in every one of them. He therefore arrived in Madras about a week before the 100-th birthday of Ramanujan and spent the first night at my house before traveling by road to other conferences. I told him that he should be very careful traveling by road in such a tense time, but he held my hand and said: "Krishna, do not worry. I am on a pilgrimage here to pay homage to Ramanujan. I will not let anything perturb me." As it turned out, one day as he, Askey, and Berndt were traveling by car a couple of hundred miles south of Madras, the car was suddenly encircled by a crowd of excited political activists. The car was stopped. Askey and Berndt were very nervous. But Andrews, cool as a cucumber, rolled down the window, and threw a load of cash into the air! The crowd cheered and let the car through because the foreigners had supported their cause. Andrews acted like James Bond, with tremendous presence of mind! Anyway, everyone made it safely to Madras for the December 22 function presided by Prime Minister Rajiv Gandhi.

The talks that Andrews gave at various conferences, including the one that I organized at Anna University on December 21, one day before the 100-th birthday of Ramanujan, were all for expert audiences. Since Andrews is a charismatic speaker, I wanted him to give a lecture to a general audience. So my father and I arranged a talk by him at our home on December 23, under the auspices of the Alladi Foundation that my father started in 1983 in memory of my grandfather Sir Alladi Krishnaswami Iyer, one of the most eminent lawyers of India. We invited the Consul General of the USA to preside over the lecture which was attended by prominent citizens of Madras in various walks of life — lawyers, judges, aristocrats, businessmen, college teachers and students. Andrews charmed them all with his inimitable description of the story of the discovery of Ramanujan's Lost Notebook. But something sensational happened that night after Andrews' lecture: Following the talk, many of us assembled at the Taj Coromandel Hotel for a dinner in honor of the conference delegates hosted by Mr. N. Ram, Editor of *The Hindu*, India's National Newspaper, based in Madras. (Ram's connection with Andrews was that in 1976, shortly after the Lost Notebook was discovered, he published a full page interview with Andrews in The Hindu.) After dinner, while we chatting over cocktails and dessert, the news came in whispers that MGR had passed away, and so the city would come to a standstill by daybreak once the general public would hear this news. So under the cover of darkness, we were asked to quietly make our way back to our hotels. And yes, as predicted, there was a complete shutdown and the Ramanujan Centenary Conference did not take place on December 24; instead all talks were squeezed into the next two days. Fortunately, Andrews had spoken at the conference on December 23. The Goddess of Namakkal had made sure that the Ramanujan Centenary celebration on December 22, and the talks the next day by the Great Trinity, would not be affected by such a tragedy!

The Frontiers of Science Lecture in Florida: At the University of Florida in Gainesville, there was a public lecture series called *Frontiers of Science*. This was organized by the physics department, and students received 1 (hour) course credit for attending these lectures. Many world famous scientists spoke in this lecture series such as group theorist John Conway, and Johansson, the discoverer of the "Lucy" skeleton. So after my return from the Ramanujan Centennial, I suggested to the organizers to invite George Andrews. I never heard back from them and so I felt they were not interested. Quite surprisingly, three years later, in Fall 1990, they contacted me and expressed interest in Andrews delivering a Frontiers of Science Lecture. So Andrews gave such a talk in November 1990, and held the 1000 or more members of the audience in the University Auditorium in rapt attention as he described the story of the discovery of the Lost Notebook. That was his first visit to Florida, but in that visit, our collaboration began in a remarkable way. I will now relate this fascinating story that will reveal the genius of this man.

In early 1989, I got a phone call from Basil Gordon, one of my former teachers at UCLA where I did my Ph.D. work. Gordon said that he would be on a fully paid sabbatical in 1989–90, and that he would like to spend the Fall of 1989 in Florida. After the Ramanujan Centennial, I attempted some research on partitions and q-series, but the visit of Gordon provided me a real opportunity because Gordon was a dominant force in this domain; in the 1960s he had obtained a far-reaching generalization of the Rogers–Ramanujan identities to odd moduli. Gordon and I first obtained a significant generalization of Schur's famous 1926 partition theorem by a new technique which we called the *method of weighted words*. We then extended this method to obtain a generalization and refinement of a deep 1967 partition theorem of Göllnitz. We cast this generalization in the form of a remarkable three parameter q-hypergeometric *key identity* which we were unable to prove. When Andrews arrived in Florida for the Frontiers of Science Lecture, I went to the airport to receive him. I did not waste any time and showed him the identity right there. He said it was fascinating. During his three-day stay in Gainesville, he thought of nothing else. He focused solely on the identity. In the visitors office that he occupied in our department, I saw him working on the identity, every day, and every hour. On the last day, on the way to the airport, he handed me an eight-page proof of this key identity by

q-hypergeometric techniques that only he could wield with such power. That is how my first paper with him (jointly also with Gordon) came about.

Sabbatical at Penn State, 1992–93: I was having my first sabbatical in 1992–93 and Andrews invited me to Penn State for that entire year. So I went to State College, Pennsylvania with my family. It was the most productive year of my academic life — I completed work on five papers of which two were in collaboration with Andrews. He and his wife Joy were gracious hosts. They showed us around State College and we got together as families for picnics. Most importantly, Andrews gave a year long graduate course on the theory of partitions that I attended. Although I was doing research in the theory of partitions, I never had a course on partitions and q-hypergeometric series as a student and so it was a treat for me to learn from the master. Dennis Eichhorn and Andrew Sills were also taking this course as graduate students.

The sabbatical year at Penn State gave me time to also write up work I had done previously. It was there that I finished writing my first joint paper with Andrews on the Göllnitz theorem. The story of my second joint paper with Andrews written at Penn State on the Capparelli conjecture is also equally remarkable and demonstrates once again Andrews' power in the area of partitions and q-hypergeometric series, and so I will relate this now.

In the summer of 1992, the Rademacher Centenary Conference was held at Penn State. Andrews was a former student of Rademacher, and so he was the lead organizer of this conference. On the opening day of the conference, Jim Lepowsky gave a talk on how Lie algebras could be used to discover, and in some instances, prove, various Rogers–Ramanujan type partition identities. During the talk, he mentioned a pair of partition identities that his student Stefano Capparelli had discovered in the study of vertex operators of Lie algebras but was unable to prove. Even though Andrews was the main conference organizer, he went into hiding during the breaks to work on the Capparelli Conjecture. By the end of the conference, he had proved the conjecture; so on the last day, he changed the title of his talk and spoke about a proof of the Capparelli conjecture. This story bears similarity to the way in which he proved the three parameter identity for the Göllnitz theorem that Gordon and I had found but could not prove.

I was not present at the Rademacher Centenary Conference since I was in India at that time, just two months before reaching Penn State for my sabbatical. But Basil Gordon was at that conference and he told me this story. Actually, during Lepowsky's lecture, Gordon realized that our method of weighted words would apply to the Capparelli partition theorems and he expressed this view to me in a telephone call soon after I arrived at Penn State. So during my sabbatical, I worked out the details of this approach to obtain a two parameter refinement of the Capparelli theorems, and in that process got a combinatorial proof as well. This led to my second joint paper with Andrews, with Gordon also as a co-author.

The mathematicians associated with the Capparelli partition conjecture and its resolution: Seated — Jim Lepowsky (left) and Basil Gordon. Standing — Stefano Capparelli (left), George Andrews (middle), and Krishnaswami Alladi (right) — at the Alladi House in Gainesville, Florida, in fall 1994

Honorary Doctorate at UF in 2002: In view of his fundamental research and his contributions to the profession, Andrews is the recipient of numerous honors. He has received honorary doctorates from the University of Illinois and the University of Parma. In 2002, he was awarded an Honorary Doctorate by the University of Florida. I was Department Chair at that time, and it was then that we formalized the arrangement to have him as a Distinguished Visiting Professor, so that he would spend the entire Spring Term each year at the University of Florida.

Visit to SASTRA University, 2003: In 2003, the recently formed SASTRA University, purchased Ramanujan's home in Kumbakonam, renovated it, and decided to maintain it as a museum. This was a major event in the preservation of Ramanujan's legacy for posterity. To mark the occasion, SASTRA decided to have an International Conference at their newly constructed Srinivasa Ramanujan Centre in Kumbakonam to coincide with Ramanujan's birthday, December 22. I was invited to organize the technical session and given funds to bring a team of mathematicians to Kumbakonam. SASTRA was a new entry in the Ramanujan world, but this conference seemed to me interesting and promising. But how to make a success of this? So I called Andrews and told him that something exciting is happening in Ramanujan's hometown, and I would like him to give the opening lecture at this conference. He readily agreed. Once he accepted, I called other mathematicians and told them that Andrews will be there. So they too accepted the invitation to the First SASTRA Conference. That shows Andrews' drawing power! That conference was inaugurated by India's President Abdul Kalam who also declared open Ramanujan's home as a museum and national treasure.

Krishnaswami Alladi and Joy Andrews with George Andrews after he received an honorary doctorate from the University of Florida, in 2002

Ramanujan 125, Honorary Doctorate at SASTRA: Many things developed after that 2003 SASTRA conference — the conferences at SASTRA became an annual event that I help organize, and in 2005 the SASTRA Ramanujan Prize was launched. SASTRA invited me to be Chair of the Prize Committee. I felt that Andrews' input would be crucial for the success of the prize. So I invited him to be on the Prize Committee during the first year, and he readily agreed. I then informed others about the prize and that Andrews was on the Prize Committee, and they too agreed enthusiastically. The prize as you know has become one of the most prestigious in the world, and I am grateful to Andrews for agreeing to serve on the Prize Committee during the first year.

In view of the annual conferences and the prize, SASTRA had become a major force in the world of Ramanujan by the time Ramanujan's 125-th Anniversary was celebrated in December 2012. So I suggested to the Vice-Chancellor of SASTRA, that the three greatest figures in the world of Ramanujan — namely the Trinity — should be recognized by SASTRA with honorary doctorates in Ramanujan's hometown, Kumbakonam. The Vice-Chancellor liked this suggestion, and so Andrews, Askey and Berndt were awarded honorary doctorates in a colorful ceremony with traditional Indian music being played as the recipients walked in.

Birthday Conferences Every Five Years: Andrews has remained productive defying the passage of time. In view of his enormous influence, and his charm, conferences in his honor have been organized every five years starting from his 60-th birthday, and I have had the privilege of participating in every one of them — in

Maratea, Italy in 1998 for his 60-th, in Penn State in 2003 and 2008 for his 65-th and 70-th, and in Tianjin, China in 2013 for his 75-th. Even though this is not a milestone birthday, I am happy to have taken part in this conference on Lattice Paths where he was honored along with three other eminent mathematicians.

G.H. Hardy once said that he had the unique privilege of collaborating with Ramanujan and Littlewood in something like equal terms. Although I am no Hardy, I can say proudly that I am unique in having had a close collaboration with Paul Erdős and George Andrews, two of the most influential mathematicians of our time! I next describe my joint work with Andrews on the Göllnitz and Capparelli theorems.

3 Collaboration with Andrews

Before describing my joint work with Andrews, I need to briefly provide as background, my joint work with Gordon on Schur's theorem.

One of the first results in the theory of partitions that one encounters, is a lovely theorem of Euler, namely:

Theorem E *The number of partitions $p_d(n)$ of n into distinct parts, equals the number of partitions $p_o(n)$ of n into odd parts.*

Euler's proof of this was to consider the product generating functions of these two partition functions and show they are equal by using the trick

$$1 + x = \frac{1 - x^2}{1 - x}.$$

More precisely,

$$\sum_{n=0}^{\infty} p_d(n) q^n = \prod_{m=1}^{\infty} (1 + q^m) = \prod_{m=1}^{\infty} \frac{1 - q^{2m}}{1 - q^m} = \prod_{m=1}^{\infty} \frac{1}{1 - q^{2m-1}} = \sum_{n=0}^{\infty} p_o(n) q^n. \tag{4.1}$$

Let us think of partitions into distinct parts as those for which the gap between the parts is ≥ 1, and partitions into odd parts as those whose parts are $\equiv \pm 1 \pmod 4$. If Euler's theorem is viewed in this fashion, then the celebrated Rogers–Ramanujan partition theorem is the "next level" result with gap ≥ 1 replaced by gap ≥ 2 between parts, and the congruence mod 4 replaced by modulus 5. More precisely, the first Rogers–Ramanujan partition theorem is:

Theorem R1 *The number of partitions of an integer n into parts that differ by ≥ 2, equals the number of partitions of n into parts $\equiv \pm 1 \pmod 5$.*

In the second Rogers–Ramanujan partition theorem (R2) we consider partitions whose parts differ by ≥ 2 but do not have 1 as a part, and equate these with partitions into parts $\equiv \pm 2 \pmod 5$. The two Rogers–Ramanujan partition identities can be cast in an analytic form, namely

$$\sum_{n=0}^{\infty} \frac{q^{n^2}}{(q)_n} = \frac{1}{(q;q^5)_\infty (q^4;q^5)_\infty}, \tag{4.2}$$

and

$$\sum_{n=0}^{\infty} \frac{q^{n^2+n}}{(q)_n} = \frac{1}{(q^2;q^5)_\infty (q^3;q^5)_\infty}. \tag{4.3}$$

In (4.2) and (4.3) and in what follows, we have used the standard notation

$$(a;q)_n = (a)_n = \prod_{j=1}^{n} (1 - aq^{j-1}),$$

and

$$(a)_\infty = \lim_{n\to\infty} (a)_n, \quad \text{for} \quad |q| < 1.$$

When the base is q, then as on the left in (4.2) and (4.3), we do not mention it, but when the base is other than q, then we always mention it, as on the right in (4.2) and (4.3).

Although the Rogers–Ramanujan identities are the next level identities beyond Euler's theorem, they are much deeper. They also have a rich history that we will not get into here. We just mention that the analytic forms of the identities (4.2) and (4.3) were first discovered by Rogers and Ramanujan independently, and it was only later that MacMahon and Schur independently provided the partition version, namely Theorems R1 and R2. Neither Rogers nor Ramanujan mentioned the partition versions of (4.2) and (4.3). So in fairness, Theorems R1 and R2 should be called the MacMahon–Schur theorems.

In the theory of partitions and q-series, a Rogers–Ramanujan (R–R) type identity is a q-hypergeometric identity in the form of an infinite (possibly multiple) series equals an infinite product. The series is the generating function of partitions whose parts satisfy certain difference conditions, whereas the product is the generating function of partitions whose parts usually satisfy certain congruence conditions. Since the 1960s, Andrews has spearheaded the study of R–R type identities (see [14], for instance). R–R type identities arise as solutions of models in statistical mechanics as first observed by Rodney Baxter in his fundamental work. After noticing the role of R–R type identities in certain physical problems, Baxter and his group approached Andrews to provide insight into the structure of such identities. Andrews then collaborated with Baxter and Peter Forrester to determine all R–R type identities that arise as solutions of the Hard-Hexagon Model in statistical mechanics. For a discussion of a theory of R–R type identities, see Andrews [14, Chap. 9]. For a discussion of connections with problems in physics, see Andrews' CBMS Lectures [15].

The partition theorem which is the combinatorial interpretation of an R–R type identity, is called a Rogers–Ramanujan type partition identity. A q-hypergeometric R–R type identity is usually discovered first and then its combinatorial interpretation as a partition theorem is given. There are important instances of Rogers–Ramanujan type partition identities being discovered first and their q-hypergeometric versions given later. Perhaps the first such significant example is the 1926 partition theorem of Schur [22].

In emphasizing the partition version of (4.2) and (4.3), Schur discovered the "next level" partition theorem, namely:

Theorem S (Schur, 1926) *Let $T(n)$ denote the number of partitions of an integer n into parts* $\equiv \pm 1 \pmod 6$.

Let $S(n)$ denote the number of partitions of n into distinct parts $\equiv \pm 1 \pmod 3$.

Let $S_1(n)$ denote the number of partitions of n into parts that differ by ≥ 3, where the inequality is strict if a part is a multiple of 3. Then

$$T(n) = S(n) = S_1(n).$$

The equality $T(n) = S(n)$ is simple and follows easily by using Euler's trick on their product generating functions, namely

$$\sum_{n=0}^{\infty} T(n)q^n = \frac{1}{(q;q^6)_\infty(q^5;q^6)_\infty} = (-q;q^3)_\infty(-q^2;q^3)_\infty = \sum_{n=0}^{\infty} S(n)q^n. \tag{4.4}$$

Thus it is the equality $S(n) = S_1(n)$ which is the real challenge. In 1966, Andrews [10] gave a a new q-theoretic proof of $S(n) = S_1(n)$. This enabled him to discover two infinite families of identities [11, 12] modulo $2^k - 1$ emanating from Schur's theorem.

In 1989, in collaboration with Gordon, I obtained a generalization and two parameter refinement of the equality $S(n) = S_1(n)$ (see [6]). The main idea in [6] was to establish the *key identity*

$$\sum_{i,j} a^i b^j \sum_m \frac{q^{T_{i+j-m}+T_m}}{(q)_{i-m}(q)_{j-m}(q)_m} = (-aq)_\infty(-bq)_\infty, \tag{4.5}$$

and to view a two parameter refinement of the equality $S(n) = S_1(n)$ as emerging from (4.5) under the transformations

$$(\textit{dilation}) \quad q \mapsto q^3, \quad \text{and} \quad (\textit{translations}) \quad a \mapsto aq^{-2}, b \mapsto bq^{-1}. \tag{4.6}$$

In (4.5) and below, $T_m = m(m+1)/2$ is the m-th triangular number.

The interpretation of the product in (4.5) as the generating function of bi-partitions into distinct parts in two colors is clear. In [6] it was shown that the series in (4.5) is the generating function of partitions (= words with weights attached) into distinct parts occurring in three colors - two primary colors a and b, and one secondary color ab, and satisfying certain gap conditions. We describe this now.

We assume that the integer 1 occurs in two primary colors a and b, and that each integer $n \geq 2$ occurs in the two primary colors as well as in the secondary color ab. By a_n, b_n, and ab_n, we denote the integer n in colors a, b, and ab respectively. In order to discuss partitions, we need to impose an order on the colors, and the order that Gordon and I chose is

$$a_1 < b_1 < ab_2 < a_2 < b_2 < ab_3 < a_3 < b_3 < \cdots . \tag{4.7}$$

Thus for a given integer n, the order of the colors is

$$ab < a < b. \tag{4.8}$$

The transformations in (4.6) correspond to the replacements

$$a_n \mapsto 3n-2, \quad b_n \mapsto 3n-1, \quad \text{and} \quad ab_n \mapsto 3n-3, \tag{4.9}$$

Under (4.9), the ordering of the colored integers in (4.7) becomes

$$1 < 2 < 3 < 4 \cdots ,$$

the standard ordering among the positive integers. This is one of the reasons Gordon and I chose the ordering in (4.7).

Using the colored integers, Gordon and I gave the following partition interpretation for the series in (4.5). We defined *Type 1* partitions as those of the form $x_1 + x_2 + \cdots$, where the x_i are symbols from the sequence in (4.7) with the condition that the *gap* between x_i and x_{i+1}, namely the difference between the subscripts of the colored integers they represent, is ≥ 1, with strict inequality if

$$x_i \quad \text{has a lower order color compared to} \quad x_{i+1}, \tag{4.10a}$$

or

$$x_i, \quad x_{i+1} \quad \text{are both of secondary color.} \tag{4.10b}$$

In (4.10a), the order of the colors is as in (4.8).

Using (4.9) it can be shown that that the gap conditions of Type 1 partitions in (4.10a) and (4.10b) translate to the difference conditions of $S_1(n)$ in Schur's theorem. Two proofs of (4.5) were given in [6] — one combinatorial, and another using the q-Chu–Vandermonde Summation. Thus the R–R type identity for Schur's theorem came half a century later.

Gordon then suggested that we should apply the method of weighted words to generalize and refine the deep 1967 theorem of Göllnitz [18] which is:

Theorem G *Let $B(n)$ denote the number of partitions of n into parts $\equiv 2$, 5, or* 11 (mod 12).

Let $C(n)$ denote the number of partitions of n into distinct parts $\equiv 2$, 4, *or* 5 (mod 6).

Let $D(n)$ denote the number of partitions of n into parts that differ by ≥ 6, *where the inequality is strict if a part is* $\equiv 0$, 1, *or* 3 (mod 6), *and with* 1 *and* 3 *not occurring as parts. Then*

$$B(n) = C(n) = D(n).$$

The equality $B(n) = C(n)$ is easy because

$$\sum_{n=0}^{\infty} B(n)q^n = \prod_{m=1}^{\infty} \frac{1}{(1-q^{12m-10})(1-q^{12m-7})(1-q^{12m-1})}$$
$$= \prod_{m=1}^{\infty}(1+q^{6m-4})(1+q^{6m-2})(1+q^{6m-1}) = \sum_{n=0}^{\infty} C(n)q^n. \quad (4.11)$$

This is one reason that we focus on the deeper equality $C(n) = D(n)$, the second reason being that it is this equality which can be refined.

Göllnitz' proof of Theorem G is very intricate and difficult but he succeeded in proving Theorem G in the refined form

$$C(n;k) = D(n;k), \quad (4.12)$$

where $C(n;k)$ and $D(n;k)$ denote the number of partitions of the type counted by $C(n)$ and $D(n)$ respectively, with the extra condition that the number of parts is k, and with the convention that parts $\equiv 0$, 1, or 3 (mod 6) are counted twice. Andrews [13] subsequently provided a simpler proof. I think besides Göllnitz, Andrews is the only other person to have gone through the difficult details of Göllnitz' proof of Theorem G. In Chap. 10 of his famous CBMS Lectures [15], Andrews asks for a proof that will provide insights into the structure of the Göllnitz theorem.

In view of (4.12) and our work on Schur's theorem, Gordon suggested that we should look at Göllnitz' theorem in the context of the method of weighted words. To this end, Gordon and I first considered the product

$$(-aq)_\infty(-bq)_\infty(-cq)_\infty \quad (4.13)$$

and viewed the generating function of $C(n)$ as emerging out of (4.13) under the substitutions

$$(\textit{dilation}) \quad q \mapsto q^6, \quad \text{and } (\textit{translations}) \quad a \mapsto aq^{-4}, b \mapsto bq^{-2}, c \mapsto cq^{-1}. \quad (4.14)$$

The problem then was to find a series that would sum to this product, with the series representing the generating function of partitions into colored integers with gap conditions that would correspond to those governing $D(n)$. What Gordon and I did was to consider the integer 1 to occur in three primary colors a, b, and c, and integers $n \geq 2$ to occur in these three primary colors as well as in three secondary colors ab, ac, and bc. As before, the symbols $a_n, b_n, \cdots, bc_n$ represent n in colors $a, b, \cdots, bc$ respectively. Here too we need an ordering on the colored integers, and the one we chose is

$$a_1 < b_1 < c_1 < ab_2 < ac_2 < a_2 < bc_2 < b_2 < c_2 < ab_3 < \ldots . \tag{4.15}$$

The effect of the substitutions (4.14) is to convert the symbols to

$$\begin{cases} a_m \mapsto 6m-4,\ b_m \mapsto 6m-2,\ c_n \mapsto 6m-1, & \text{for } m \geq 1, \\ ab_m \mapsto 6m-6,\ ac_m \mapsto 6m-5,\ bc_n \mapsto 6m-3, & \text{for } m \geq 2. \end{cases} \tag{4.16}$$

so that the ordering (4.15) becomes

$$2 < 4 < 5 < 6 < 7 < 8 < 9 < 10 < 11 < 12 < \cdots , \tag{4.17}$$

This is one reason for the choice of the ordering of symbols in (4.15), because they convert to the natural ordering of the integers in (4.17) under the transformations (4.16). Notice that 1, and 3 are missing in (4.17), and this explains the condition that 1 and 3 do not occur as parts in the partitions enumerated by $D(n)$ in Theorem G.

To view Theorem G in this context, we think of the primary colors a, b, c as corresponding to the residue classes 2, 4 and 5 (mod 6) and so the secondary colors ab, ac, bc correspond to the residue classes $2+4 \equiv 6$, $2+5 \equiv 7$ and $4+5 \equiv 9 \pmod 6$. Note that integers of secondary color occur only when $n \geq 2$ and so ab_1, ac_1 and bc_1 are missing in (4.15). This is why integers $ac_1 = 1$ and $bc_1 = 3$ do not appear in (4.17). This explains the absence of 1 and 3 among the parts enumerated by $D(n)$ in Theorem G. Note that ab_1 corresponds to the integer 0, which is not counted as a part in ordinary partitions anyway.

In (4.15) for a given subscript, the ordering of the colors is

$$ab < ac < a < bc < b < c. \tag{4.18}$$

We use (4.18) to say for instance that ab is of *lower order* compared to a, or equivalently that a is of *higher order* than ab. With this concept of the order of colors, we can define *Type 1* partitions to be of the form $x_1 + x_2 + \ldots$, where the x_i are symbols from (4.15) with the condition that the gap between x_i and x_{i+1} is ≥ 1 with strict inequality if

$$x_i \quad \text{is of lower order (color) compared to} \quad x_{i+1}, \tag{4.19a}$$

or

$$\text{if} \quad x_i \quad \text{and} \quad x_{i+1} \quad \text{are of the same secondary color.} \tag{4.19b}$$

Under the transformations given by (4.16), the gap conditions of Type 1 partitions become the difference conditions governing $D(n)$. Gordon and I then showed that the generating function of Type 1 partitions is

$$\sum_{i,j,k} a^i b^j c^k \sum_{\substack{s=\alpha+\beta+\gamma+\delta+\varepsilon+\phi \\ i=\alpha+\delta+\varepsilon,\ j=\beta+\delta+\phi,\ k=\gamma+\varepsilon+\phi}} \frac{q^{T_s+T_\delta+T_\varepsilon+T_{\phi-1}}(1-q^\alpha(1-q^\phi))}{(q)_\alpha(q)_\beta(q)_\gamma(q)_\delta(q)_\varepsilon(q)_\phi} \tag{4.20}$$

Thus our three three parameter *key identity* for the generalization and refinement of Göllnitz' theorem is

$$\sum_{i,j,k} a^i b^j c^k \sum_{\substack{s=\alpha+\beta+\gamma+\delta+\varepsilon+\phi \\ i=\alpha+\delta+\varepsilon,\ j=\beta+\delta+\phi,\ k=\gamma+\varepsilon+\phi}} \frac{q^{T_s+T_\delta+T_\varepsilon+T_{\phi-1}}(1-q^\alpha(1-q^\phi))}{(q)_\alpha(q)_\beta(q)_\gamma(q)_\delta(q)_\varepsilon(q)_\phi}$$
$$= \sum_{i,j,k} \frac{a^i b^j c^k q^{T_i+T_j+T_k}}{(q)_i(q)_j(q)_k} = (-aq)_\infty(-bq)_\infty(-cq)_\infty\,, \tag{4.21}$$

The partition interpretation of (4.21) that Gordon and I had was:

Theorem 1 *Let $C(n; i, j, k)$ denote the number of vector partitions $(\pi_1; \pi_2; \pi_3)$ of n such that π_1 has i distinct parts all in color a, π_2 has j distinct parts all in color b, and π_3 has k distinct parts all in color c.*

Let $D(n; \alpha, \beta, \gamma, \delta, \varepsilon, \phi)$ denote the number of Type 1 partitions of n having α a-parts, β b-parts, ..., and ϕ bc-parts.

Then

$$C(n; i, j, k) = \sum_{\substack{i=\alpha+\delta+\varepsilon \\ j=\beta+\delta+\phi \\ k=\gamma+\varepsilon+\phi}} D(n; \alpha, \beta, \gamma, \delta, \varepsilon, \phi).$$

It is to be noted that in Theorem 1,

$$i + j + k = \alpha + \beta + \gamma + 2(\delta + \varepsilon + \phi)$$

and so the parts in secondary color are counted twice. This corresponds to the condition that parts $\equiv 0, 1, 3 \pmod 6$ are counted twice in (4.12).

The proof in [8] that the expression in (4.20) is the generating function of minimal partitions is quite involved and goes by induction on $s = \alpha + \beta + \gamma + \delta + \varepsilon + \phi$, the number of parts of the Type-1 partitions, and also appeals to *minimal partitions* whose generating functions are given by multinomial coefficients (see [8] for details). Thus everything fitted perfectly, but Gordon and I had a problem: *we could not prove the key identity* (4.21). This is where Andrews entered into the picture. The story of how he proved (4.21) is described in Part 1. His ingenious proof of the remarkable key identity (4.21) relied on the Watson's q-analogue of Whipple's transformation and the ${}_6\psi_6$ summation of Bailey. For the proof of (4.21), we refer the reader to [8].

Let me just say, that there is no one in the world who can match Andrews' power in proving multi-variable q-hypergeometric identities!

One of the great advantages of the method of weighted words is that it provides a key identity for a partition theorem at the base level, and from this one can extract several partition theorems by suitable dilations and translations. I investigated in detail a variety of partition theorems that emerge from (4.21) (see [1, 2]), but will report here only two major developments that involved Andrews.

As noted earlier, Göllnitz' theorem pertains to the dilation $q \mapsto q^6$ in (4.21), and so I wanted to investigate the effect under the transformations

$$(dilation) \quad q \mapsto q^3, \tag{4.22a}$$

and

$$(translations) \quad a \mapsto aq^{-2}, \quad b \mapsto bq^{-1}, \quad c \mapsto c. \tag{4.22b}$$

In this case the product in (4.21) becomes

$$\prod_{m=1}^{\infty}(1+aq^{3m-2})(1+bq^{3m-1})(1+q^{3m}),$$

which is the three parameter generating function of partitions into distinct parts, and therefore is very interesting. The dilation $q \mapsto q^6$ converts the six colors $a, b, \cdots, bc$ into the six different residue classes mod 6, and under the dilation in (4.22a), one gets partitions into parts that differ by ≥ 3 but these partitions have to be counted with a weight because each positive integer ≥ 3 occurs in two colors - one primary and one secondary. Two major consequences of this weighted partition identity were (i) a new proof of Jacobi's triple product identity for theta functions, and (ii) a combinatorial proof of a variant of Göllnitz' theorem which is equivalent to it. In the course of identifying this variant, I found a new *cubic key identity* that represents it, namely

$$\sum_{i,j,k} \frac{a^i b^j c^k (-c)_i (-c)_j (-\frac{ab}{c}q)_k (-cq)_{i+j} q^{T_{i+j+k}}}{(q)_i (q)_j (q)_k (-c)_{i+j}} = (-aq)_\infty (-bq)_\infty (-cq)_\infty. \tag{4.23}$$

As in the case of (4.21), I approached Andrews for a proof of (4.23), and he supplied it in a matter of a few days utilizing Jackson's q-analogue of Dougall's summation. This led to our second joint paper [3]. While (4.23) is quite deep, it is simpler in structure compared to (4.21).

Next I investigated the combinatorial consequences of (4.21) under the

$$(dilation) \quad q \mapsto q^4, \tag{4.24a}$$

but here there are four possible translations depending on which residue class modulo 4 one chooses to omit for the primary color. For example, the translations

$$a \mapsto aq^{-3}, \quad b \mapsto bq^{-1}, \quad c \mapsto cq^{-3}, \tag{4.24b}$$

omits the residue class 0 (mod 4) for the primary colors, and there are three other important dilations. Some very interesting weighted partition identities emerge (see [2]), but I focused on the translations in (4.24b) owing to the symmetry. This led me to the following *quartic key identity*:

$$\sum_{i,j,k,\ell} \frac{a^{i+\ell} b^j c^{k+\ell} q^{T_{i+j+k+\ell}+T_\ell} \left(-\frac{bc}{a}\right)_i \left(-\frac{abq}{c}\right)_k}{(q)_i (q)_j (q)_k (q)_\ell} \frac{(1+\frac{bc}{a}q^{2i-1})}{(1+\frac{bc}{a}q^{i-1})} = (-aq)_\infty (-bq)_\infty (-cq)_\infty, \tag{4.25}$$

Once again, I approached Andrews for a proof of (4.25), and he supplied it using Jackson's q-analogue of Dougall's summation. This led to my third paper with Andrews [4].

When Göllnitz proved his theorem in 1967, it was viewed as a next level result beyond Schur's theorem because the two residue classes 1, 2 (mod 3) for $S(n)$ in Schur's Theorem are replaced by three residue classes 2, 4, 5 (mod 6) for $C(n)$ in Göllnitz' theorem. Apart from this, it is not clear why Göllnitz' theorem can be considered as an extension of Schur's. But then, by our method of weighted words, one sees exactly how our generalized Göllnitz Theorem 1 is an extension of Schur's to the next level, because the key identity (4.5) for Schur's theorem is simply the special case $c = 0$ in the key identity (4.21) for Göllnitz' theorem.

So if Göllnitz' theorem is the "next level" result beyond Schur's theorem, why is it so much more difficult to prove? One reason for this is because in Göllnitz' theorem, when expanding the product in (4.21), we consider only the primary and secondary colors in the series and omit the ternary color abc. Actually, as early as 1968 and 69, Andrews [11, 12], had obtained two infinite hierarchies of partition theorems to moduli $2^k - 1$ when $k \geq 2$, where he starts with k residue classes (mod $2^k - 1$) and considers the complete set of residue classes (mod $2^k - 1$) for the difference conditions. We now describe his results.

For a given integer $r \geq 2$, let $a_1, a_2, \ldots, a_r$ be r distinct positive integers such that

$$\sum_{i=1}^{k-1} a_i < a_k, \quad 1 \leq k \leq r. \tag{4.26}$$

Condition (4.26) ensures that the $2^r - 1$ sums $\sum \varepsilon_i a_i$, where $\varepsilon_i = 0$ or 1, not all $\varepsilon_i = 0$, are all distinct. Let these sums in increasing order be denoted by $\alpha_1, \alpha_2, \ldots, \alpha_{2^r-1}$.

Next let $N \geq \sum_{i=1}^r a_i \geq 2^r - 1$ be a modulus, and A_N denote the set of all positive integers congruent to some a_i (mod N). Similarly, let A'_N denote the set of all positive integers congruent to some α_i (mod N) Also let $\beta_N(m)$ denote the least positive residue of m (mod N). Finally, if $m = \alpha_j$ for some j, let $\phi(m)$ denote the number of terms appearing in the defining sum of m and $\psi(m)$ the smallest a_i appearing in this sum. Then the first general theorem of Andrews [11] is:

Theorem A1 *Let $C^*(A_N; n)$ denote the number of partitions of n into distinct parts taken from A_N.*

Let $D^*(A'_N; n)$ denote the number of partitions of n into parts $b_1, b_2, \ldots, b_\nu$ from A'_N such that

$$b_i - b_{i+1} \geq N\phi(\beta_N(b_{i+1})) + \psi(\beta_N(b_{i+1})) - \beta_N(b_{i+1}). \tag{4.27}$$

Then

$$C^*(A_N; n) = D^*(A'_N; n).$$

To describe the second general theorem of Andrews (1969), let a_i, α_i and N be as above. Now let $-A_N$ denote the set of all positive integers congruent to some $-a_i$ (mod N), and $-A'_N$ the set of all positive integers congruent to some $-\alpha_i$ (mod N). The quantities $\beta_N(m), \phi(m), \psi(m)$ are also as above. We then have (Andrews [12]):

Theorem A2 *Let $C(-A_N; n)$ denote the number of partitions of n into distinct parts taken from $-A_N$.*

Let $D(-A'_N; n)$ denote the number of partitions of n into parts $b_1, b_2, \ldots, b_\nu$, taken from $-A'_N$ such that

$$b_i - b_{i+1} \geq N\phi(\beta_N(-b_i)) + \psi(\beta_N(-b_i)) - \beta_N(-b_i) \tag{4.28}$$

and also

$$b_\nu \geq N(\phi(\beta_N(-b_s) - 1)).$$

Then

$$C(-A_N; n) = D(-A'_N; n).$$

When $r = 2, a_1 = 1, a_2 = 2, N = 3 = 2^r - 1$, Theorems A1 and A2 both become Theorem S. Thus the two hierarchies emanate from Theorem S, and it is only when $r = 2$ that the hierarchies coincide. Thus Theorem S is its own dual. Conditions (4.27) and (4.28) can be understood better by classifying b_{i+1} (in Theorem A1) and b_i (in Theorem A2) in terms of their residue classes (mod N). In particular, with $r = 3$, $a_1 = 1$, a_2, $a_3 = 4$ and $N = 7 = 2^3 - 1$, Theorems A1 and A2 yield the following corollaries.

Corollary 1 *Let $C^*(n)$ denote the number of partitions of n into distinct parts* $\equiv 1$, 2 *or* 4 (mod 7).

Let $D^*(n)$ denote the number of partitions of n in the form $b_1 + b_2 + \cdots_\nu$ such that $b_i - b_{i+1} \geq 7, 7, 12, 7, 10, 10$ or 15 if $b_{i+1} \equiv 1, 2, 3, 4, 5, 6$ or 7 (mod 7). *Then*

$$C^*(n) = D^*(n).$$

Corollary 2 *Let $C(n)$ denote the number of partitions of n into distinct parts* $\equiv 3$, 5 *or* 6 (mod 7).

Let $D(n)$ denote the number of partitions of n in the form $b_1 + b_2 + \cdots + b_\nu$ such that $b_i - b_{i+1} \geq 10, 10, 7, 12, 7, 7$ or 15 if $b_i \equiv 8, 9, 3, 11, 5, 6$ or 14 (mod 7) *and $b_\nu \neq 1, 2, 4$ or 7. Then*

$$C(n) = D(n).$$

Andrews' proofs of Theorems A1 and A2 are extensions of his proof [11] of Theorem S and not as difficult as the proof of Göllnitz' theorem. During the 1998 conference in Maratea, Italy, for Andrews' 60-th birthday organized by Dominique Foata, I gave a talk outlining a method of weighted words approach generalization of Theorems A1 and A2. Dominique Foata then asked whether there is a hypergeometric key identity that corresponds to this generalization. Even though the proofs of Theorems A1 and A2 are simpler compared to the the proof of Theorem G, no hypergeometric key identity has yet been found to represent the Andrews hierarchies when $k \geq 3$.

In view of the fact that with a complete set of alphabets one gets an infinite hierarchy of theorems, Andrews raised as a problem in his CBMS Lectures, whether there exists a partition theorem beyond Göllnitz' theorem in the same manner as Göllnitz' theorem goes beyond Schur. In the language of the method of weighted words, this is the same as asking whether there exists a partition theorem starting with four primary colors a, b, c, d and using only a proper subset of the complete alphabet of 15 colors, that will yield Göllnitz' theorem when we set the parameter $d = 0$. The answer to this difficult problem was found by Alladi–Andrews–Berkovich in 2000, by noticing that ALL ternary colors have to be dropped but the quaternary color $abcd$ needs to be retained. This led to a remarkable identity in four parameters a, b, c, d that went beyond (4.21). Our paper [7] describes the ideas behind the construction of this four parameter identity and provides the proof as well. I just mention here a striking (mod 15) identity that emerges from this four parameter q-hypergeometric identity:

Theorem 1* *Let $P(n)$ denote the number of partitions of n into distinct parts* $\equiv -2^3, -2^2, -2^1, -2^0$ (mod 15).

Let $G(n)$ denote the number of partitions of n into parts $\not\equiv 2^0, 2^1, 2^2, 2^3$ (mod 15), *such that the difference between the parts is* ≥ 15, *with equality only if a part is* $\equiv -2^3, -2^2, -2^1, -2^0$ (mod 15), *parts which are* $\equiv \pm 2^0, \pm 2^1, \pm 2^2, \pm 2^3$ (mod 15) *are* > 15, *the difference between the multiples of* 15 *is* ≥ 60, *and the smallest multiple of* 15 *is*

$$\begin{cases} \geq 30 + 30\tau, & \text{if 7 is a part, and} \\ \geq 45 + 30\tau, & \text{otherwise,} \end{cases}$$

where τ is number of non–multiples of 15 *in the partition. Then*

$$G(n) = P(n).$$

One aspect of Göllnitz' Theorem G that escaped attention was whether it had a dual in the sense that Theorems A1 and A2 can be considered as duals. More precisely, the residue classes of Corollary 1 that constitute the primary colors are $1, 2, 4 \pmod 7$, whereas the residue classes that constitute the primary colors in Corollary 2 are $-1, -2, -4 \pmod 7$. Now one can view $2, 4, 5 \pmod 6$ as $-1, -2, -4 \pmod 6$. So the question is whether there is a dual result to Theorem G starting with $1, 2, 4 \pmod 6$. Andrews found such a theorem, namely:

Theorem A *Let $B^*(n)$ denote the number of partitions of n into parts* $\equiv 1$, 7, *or* 10 (mod 12).

Let $C^*(n)$ denote the number of partitions of n into distinct parts $\equiv 1$, 2, *or* 4 (mod 6).

Let $D^*(n)$ denote the number of partitions of n into parts that differ by at least 6, *where the inequality is strict if the larger part is* $\equiv 0$, 3, *or* 5 (mod 6), *with the exception that* $6 + 1$ *may appear in the partition. Then*

$$B^*(n) = C^*(n) = D^*(n).$$

Andrews provided a proof of Theorem A very similar to his proof of Theorem G in [13]. My role then was to construct a key identity that represented this dual, which I did. This key identity for the dual, although different from (4.21), is equivalent to it. This led to our most recent joint paper [5].

I conclude by describing my joint paper with Andrews on the Capparelli partition theorems.

In fundamental work [19, 20], Lepowsky and Wilson gave a Lie theoretic proof of the Rogers–Ramanujan identities and in that process showed how R–R type identities arise in the study of vertex operators in Lie algebras. Using vertex operator theory, Stefano Capparelli, a Ph.D. student of Lepowsky in 1992, "discovered" two new partition results which he could not prove and so he stated them as conjectures:

Conjecture C1 Let $C^*(n)$ denote the number of partitions of n into parts $\equiv \pm 2$, ± 3 *(mod* 12*)*.

Let $D(n)$ denote the number of partitions of n into parts > 1 with minimal difference 2, where the difference is ≥ 4 unless consecutive parts are both multiples of 3 or add up to a multiple of 6. Then

$$C^*(n) = D(n).$$

He had a second partition result, **Conjecture C2**, which we do not state here because the conditions are more complicated; also that is not essential to what we will describe here.

As mentioned in Part I, Lepowsky stated Conjecture C1 on the opening day of the Rademacher Centenary Conference at Penn State, and by the time that conference ended, Andrews had a proof using q-recurrences (see [16]).

The first thing I did on seeing Conjecture C1 was to replace $C^*(n)$ by $C(n)$, the number of partitions of n into distinct parts $\equiv 2, 3, 4$ or $6 \pmod 6$, and to note that

$$C(n) = C^*(n) \tag{4.29}$$

This is because by Euler's trick

$$\sum_{n=0}^{\infty} C^*(n)q^n = \frac{1}{(q^2;q^{12})_\infty(q^3;q^{12})_\infty(q^9;q^{12})_\infty(q^{10};q^{12})_\infty}$$
$$= (-q^2;q^6)_\infty(-q^4;q^6)_\infty(-q^3;q^3)_\infty = \sum_{n=0}^{\infty} C(n)q^n. \tag{4.30}$$

One reason for replacing $C^*(n)$ by $C(n)$ is that the equality in (4.29) can be refined. Another reason is that Conjecture C2 can be more elegantly stated by replacing $C(n)$ by the function $C'(n)$ which enumerates the number of partitions into distinct parts $\equiv 1, 3, 5$, or $6 \pmod 6$.

The refinement of the Capparelli Conjecture C1 that Andrews, Gordon and I [9] proved was:

Theorem 2 *Let $C(n; i, j, k)$ denote the number of partitions counted by $C(n)$ with the additional restriction that there are precisely i parts $\equiv 4 \pmod 6$, j parts $\equiv 2 \pmod 6$, and of those $\equiv 0 \pmod 3$, exactly k are $> 3(i + j)$.*

Let $D(n; i, j, k)$ denote the number of partitions counted by $D(n)$ with the additional restriction that there are precisely i parts $\equiv 1 \pmod 3$, j parts $\equiv 2 \pmod 3$, and k parts $\equiv 0 \pmod 3$. Then

$$C(n; i, j, k) = D(n; i, j, k).$$

To establish Theorem 2, we put it in the context of the method of weighted words. More precisely, let the integer 1 occur in two colors a and c and let integers ≥ 2 occur in three colors a, b and c. As before, the symbols a_j, b_j and c_j represent the integer j in colors a, b and c respectively. To discuss partitions the ordering of the symbols we used is

$$a_1 < b_2 < c_1 < a_2 < b_3 < c_2 < a_3 < b_4 < c_3 < \cdots . \tag{4.31}$$

The Capparelli problem corresponds to the transformations

$$a_j \mapsto 3j - 2, \qquad b_j \mapsto 3j - 4, \qquad c_j \mapsto 3j, \tag{4.32}$$

in which case the inequalities in (4.31) become

$$1 < 2 < 3 < 4 < 5 < \cdots ,$$

the natural ordering among the positive integers. With this we were able to generalize and refine Theorem 2 as follows:

Theorem 3 *Let $K(n; i, j, k)$ denote the number of vector partitions of n in the form (π_1, π_2, π_3) such that π_1 has distinct even a-parts, π_2 has distinct even b-parts, and π_3 has distinct c-parts such that $\nu(\pi_1) = i$, $\nu(\pi_2) = j$, and the number of parts of π_3 which are $> i + j$ is k.*

Let $G(n; i, j, k)$ denote the number of partitions (words) of n into symbols a_j, b_j, c_j each $> a_1$, such that the gap between consecutive symbols is given by the matrix below:

	a	b	c
a	2	2	1
b	0	2	0
c	2	3	1

Then

$$K(n; i, j, k) = G(n; i, j, k).$$

Note. The matrix above is to read row-wise. Thus if a_j is a part of the partition, and the next larger part has color b, then its weight (= subscript) must be $> j + 2$.

In [9] we gave a combinatorial proof of Theorem 2 by using some ideas of Bressoud, and another proof by first showing that it is equivalent to the following *key identity*

$$\begin{aligned}\sum_{i,j,k,n} K(n; i, j, k)a^i b^j c^k q^n &= \sum_{i,j} \frac{a^i b^j q^{2T_i+2T_j}(-q)_{i+j}(-cq^{i+j+1})_\infty}{(q^2; q^2)_i (q^2; q^2)_j} \\ &= \sum_{i,j,k,n} G(n; i, j, k)a^i b^j c^k q^n \\ &= \sum_{i,j,k} \frac{a^i b^j c^k q^{2T_i+2T_j+T_k+(i+j)k}}{(q)_{i+j+k}} \begin{bmatrix} i+j+k \\ i+j, k \end{bmatrix}_q \begin{bmatrix} i+j \\ i, j \end{bmatrix}_{q^2},\end{aligned} \tag{4.33}$$

and then proving this identity.

The main difficulty in (4.33) was to show that the series on the right is the generating function of partitions with gap conditions given by the entries in the above table. This required the study of minimal partitions having a part in a specified color as the smallest part. Once the generating function of the $G(n; i, j, k)$ was shown to be the series on the right in (4.29), it was not difficult to establish the equality of this with the series on the left. If we take $c = 1$, then the generating function on the left in (4.33) becomes a product, because

$$(-q)_\infty \sum_{i,j,k} \frac{a^i b^j q^{2T_i+2T_j}}{(q^2; q^2)_i (q^2; q^2)_j} = (-q)_\infty (-aq^2; q^2)_\infty (-bq^2; q^2)_\infty. \tag{4.34}$$

In (4.30) if we replace $q \mapsto q^3$, $a \mapsto q^{-2}$, $b \mapsto q^{-4}$, we get

$$\prod_{j=0}^{\infty}(1+q^{6j-2})(1+q^{6j-4})(1+q^{3j}) = \sum_{n=0}^{\infty} C(n)q^n,$$

and so Capparelli's conjecture follows.

I could say so much more about Andrews' work on partitions, q-series and Ramanujan, but here I chose to focus on an aspect of our joint work that shows that in manipulating q-hypergeometric series, he has no match in our generation. Even though he towers head and shoulders above the rest in the world of partitions, q-series and Ramanujan, he is a perfect gentleman always willing to help. It is a pleasure and a privilege for me to be his friend and collaborator.

Acknowledgements I thank the organizers of the Conference on Lattice Paths Combinatorics, especially Alan Krinik, not only to have invited me to speak in a technical session, but also to have given me the role as Banquet Speaker. I am also grateful to Christian Krattenthaler for a very careful reading of the manuscript and for help with the Latex file of this paper.

References

1. Alladi, K.: A combinatorial correspondence related to Göllnitz' (Big) partition theorem. Trans. Am. Math. Soc. **349**, 2721–2735 (1997)
2. Alladi, K.: On a partition theorem of Göllnitz and quartic transformations (with an appendix by Gordon, B.). J. Number Theory **69**, 153–180 (1998)
3. Alladi, K., Andrews, G.E.: A new key identity for Göllnitz' (Big) partition theorem. In: Proceedings of the 10th Anniversary Conference of Ramanujan Mathematical Society. Contemporary Mathematics, vol. 210, pp. 229–241 (1997)
4. Alladi, K., Andrews, G.E.: A quartic key identity for a partition theorem of Göllnitz. J. Number Theory **75**, 220–236 (1999)
5. Alladi, K., Andrews, G.E.: The dual of Göllnitz' (Big) partition theorem. Ramanujan J. **36**, 171–202 (2015)
6. Alladi, K., Gordon, B.: Generalizations of Schur's partition theorem. Manuscripta Math. **79**, 113–126 (1993)
7. Alladi, K., Andrews, G.E., Berkovich, A.: A new four parameter q-series identity and its partition implications. Invent. Math. **153**, 231–260 (2003)
8. Alladi, K., Andrews, G.E., Gordon, B.: Generalizations and refinements of a partition theorem of Göllnitz. J. Reine Angew. Math. **460**, 165–188 (1995)
9. Alladi, K., Andrews, G.E., Gordon, B.: Refinements and generalizations of Capparelli's conjecture on partitions. J. Algebr. **174**, 1591–1608 (1995)
10. Andrews, G.E.: On Schur's second partition theorem. Glas. Math. J. **19**, 127–132 (1967)
11. Andrews, G.E.: A new generalization of Schur's second partition theorem. Acta Arith. **14**, 429–434 (1968)
12. Andrews, G.E.: A general partition theorem with difference conditions. Am. J. Math. **91**, 18–24 (1969)
13. Andrews, G.E.: On a partition theorem of Göllnitz and related formulae. J. Reine Angew. Math. **236**, 37–42 (1969)
14. Andrews, G.E.: The Theory of Partitions. Encyclopedia of Mathemathics, vol. 2. Addison–Wesley, Reading (1976)
15. Andrews, G.E.: q-Series: Their Development and Application in Analysis, Number Theory, Combinatorics, Physics and Computer Algebra. NSF-CBMS Lectures, vol. 66. American Mathematical Society, Providence (1985)

16. Andrews, G.E.: Schur's theorem, Capparelli's conjecture and the q-trinomial coefficients. In: Proceedings of the Rademacher Centenary Conference 1992. Contemporary Mathematics, vol. 166, pp. 141–154 (1994)
17. Capparelli, S.: On some representations of twisted affine Lie algebras and combinatorial identities. J. Algebr. **154**, 335–355 (1993)
18. Göllnitz, H.: Partitionen mit Differenzenbedingungen. J. Reine Angew. Math. **225**, 154–190 (1967)
19. Lepowsky, J., Wilson, R.L.: A Lie-theoretic interpretation and proof of the Rogers–Ramanujan identities. Adv. Math. **145**, 21–72 (1982)
20. Lepowsky, J., Wilson, R.L.: The structure of standard modules I, universal algebras and the Rogers–Ramanujan identities. Invent. Math. **77**, 199–290 (1984)
21. Ramanujan, S.: The Lost Notebook and Unpublished Papers. Springer, New Delhi (1987)
22. Schur, I.: Zur Additiven Zahlentheorie. In: Gesammelte Abhandlungen, vol. 2, pp. 43–70. Springer, Berlin (1937)

A Refinement of the Alladi–Schur Theorem

George E. Andrews

Abstract K. Alladi first observed a variant of I. Schur's 1926 partition theorem. Namely, the number of partitions of n in which all parts are odd and none appears more than twice equals the number of partitions of n in which all parts differ by at least 3 and more than 3 if one of the parts is a multiple of 3. In this paper, we refine this result to one that counts the number of parts in the relevant partitions.

Keywords Partition identities · Schur's theorem

2010 Mathematics Subject Classification Primary: 05A17 · Secondary: 05A15 05A30 11P84

1 Introduction

In 1926, I. Schur [7] proved the following result:

Theorem 5.1 *Let $A(n)$ denote the number of partitions of n into parts congruent to* $\pm 1 \pmod 6$. *Let $B(n)$ denote the number of partitions of n into distinct nonmultiples of 3. Let $D(n)$ denote the number of partitions of n of the form $b_1 + b_2 + \cdots + b_s$ where $b_i - b_{i+1} \geq 3$ with strict inequality if $3|b_i$. Then*

$$A(n) = B(n) = D(n).$$

K. Alladi [1] has pointed out (cf. [4, p. 46, Eq. (1.3)]) that if we define $C(n)$ to be the number of partitions of n into odd parts with none appearing more than twice, then also

$$C(n) = D(n).$$

G. E. Andrews (✉)
Department of Mathematics, The Pennsylvania State University, 215 McAllister Building, University Park, PA 16802, USA
e-mail: gea1@psu.edu

G. E. Andrews et al. (eds.), *Lattice Path Combinatorics and Applications*, Developments in Mathematics 58, https://doi.org/10.1007/978-3-030-11102-1_5

George Andrews and Krishnaswami Alladi at the banquet of the 8th International Conference on Lattice Path Combinatorics and Applications on the campus of California State Polytechnic University, Pomona.

This follows immediately from the fact that

$$\begin{aligned}\sum_{n=0}^{\infty} C(n)q^n &= \prod_{n=1}^{\infty} \left(1 + q^{2n-1} + q^{4n-2}\right)\\ &= \prod_{n=1}^{\infty} \frac{(1-q^{6n-3})}{(1-q^{2n-1})}\\ &= \prod_{n=1}^{\infty} \frac{(1-q^{6n-3})}{(1-q^{6n-5})(1-q^{6n-3})(1-q^{6n-1})}\\ &= \prod_{n=1}^{\infty} \frac{1}{(1-q^{6n-5})(1-q^{6n-1})}\\ &= \sum_{n=0}^{\infty} A(n)q^n = \sum_{n=0}^{\infty} D(n)q^n.\end{aligned}$$

Rather surprisingly the following refinement has been overlooked:

Theorem 5.2 *Let $C(m,n)$ denote the number of partitions of n into m parts, all odd and none appearing more than twice. Let $D(m,n)$ denote the number of partitions of n into parts of the type enumerated by $D(n)$ with the added condition that the total* *number of parts plus the number of even parts is m (i.e., m is the weighted count of parts where each even part is counted twice). Then $C(m, n) = D(m, n)$.*

For example $C(4, 16) = 6$ with the relevant partitions being $11 + 3 + 1 + 1$, $9 + 5 + 1 + 1$, $9 + 3 + 3 + 1$, $7 + 7 + 1 + 1$, $7 + 5 + 3 + 1$, $5 + 5 + 3 + 3$ while $D(4, 16) = 6$ with the relevant partitions being $14 + 2$, $12 + 4$, $11 + 4 + 1$, $10 + 6$, $10 + 5 + 1$, $9 + 5 + 2$.

The above theorem is analogous to W. Gleissberg's comparable refinement of the assertion that $B(n) = D(n)$ [5], and the proof is analogous to the proof of Gleissberg's theorem given in [2].

2 Proof of Theorem 5.2

We define $d_N(x, q) = d_N(x)$ to be the generating function for partitions of the type enumerated by $D(m, n)$ with the added condition that all parts be $\leq N$.

We also define

$$\varepsilon(n) = \begin{cases} 1, & \text{if } n \text{ is odd,} \\ 2, & \text{if } n \text{ is even.} \end{cases}$$

Then for $n \geq 0$

$$d_{3n}(x) = d_{3n-1}(x) + x^{\varepsilon(3n)} q^{3n} d_{3n-4}(x), \tag{5.1}$$

$$d_{3n+1}(x) = d_{3n}(x) + x^{\varepsilon(3n+1)} q^{3n+1} d_{3n-2}(x), \tag{5.2}$$

$$d_{3n+2}(x) = d_{3n+1}(x) + x^{\varepsilon(3n+2)} q^{3n+2} d_{3n-1}(x), \tag{5.3}$$

with the initial condition $d_{-1}(x) = d_{-2}(x) = 1$, $d_{-4}(x) = 0$.

In preparation for the essential functional equations needed to prove Theorem 5.2, we note that by (5.3)

$$d_{3n+1}(x) = d_{3n+2}(x) - x^{\varepsilon(3n+2)} q^{3n+2} d_{3n-1}(x). \tag{5.4}$$

Thus substituting (5.1) and (5.4) (as well as (5.4) with n replaced by $n - 1$) into (5.2), we find

$$\begin{aligned} d_{3n+2}(x) = {} & \left(1 + x^{\varepsilon(3n+1)} q^{3n+1} + x^{\varepsilon(3n+2)} q^{3n+2}\right) d_{3n-1}(x) \\ & + \left(x^{\varepsilon(3n)} q^{3n} - x^{\varepsilon(3n+1)+\varepsilon(3n-1)} q^{6n}\right) d_{3n-4}(x). \end{aligned} \tag{5.5}$$

Consequently

$$d_{6n+2}(x) = \left(1 + xq^{6n+1} + x^2 q^{6n+2}\right) d_{6n-1}(x) + \left(x^2 q^{6n} - x^2 q^{12n}\right) d_{6n-4}(x), \tag{5.6}$$

and

$$d_{6n-1}(x) = \left(1 + x^2q^{6n-2} + xq^{6n-1}\right) d_{6n-4}(x) + \left(xq^{6n-3} - x^4q^{12n-6}\right) d_{6n-7}(x). \tag{5.7}$$

Lemma 5.1 *For $n \geq 1$,*

$$d_{6n+2}(x) = \left(1 + xq + x^2q^2\right) d_{6n-1}(xq^2), \tag{5.8}$$

$$d_{6n-1}(x) = \left(1 + xq + x^2q^2\right) \left\{d_{6n-4}(xq^2) + xq^{6n-1}(1 - qx)d_{6n-7}(xq^2)\right\}, \tag{5.9}$$

where $d_{-1}(x)$ is defined to be 1.

proof We define

$$F(n) = d_{6n+2}(x) - \left(1 + xq + x^2q^2\right) d_{6n-1}(xq^2), \tag{5.10}$$

$$G(n) = d_{6n-1}(x) - \left(1 + xq + x^2q^2\right) \left\{d_{6n-4}(xq^2) + xq^{6n-1}(1 - qx)d_{6n-7}(xq^2)\right\}. \tag{5.11}$$

To prove (5.8) and (5.9) we need only show that $F(n) = G(n) = 0$ for each $n \geq 1$. In light of the fact that

$$d_2(x) = 1 + xq + x^2q^2,$$

$$\begin{aligned} d_5(x) &= 1 + xq + x^2q^2 + xq^3 + x^2q^4 + xq^5 + x^3q^5 + x^2q^6 + x^3q^7 \\ &= (1 + xq + x^2q^2)\{d_2(xq^2) + xq^5(1 - xq)\}, \end{aligned}$$

$$\begin{aligned} d_8(x) &= (1 + xq + x^2q^2)\big(1 + xq^3 + xq^5 + x^2q^6 + xq^7 \\ &\qquad + x^2q^8 + x^2q^{10} + x^3q^{11} + x^3q^{13}\big) \\ &= \left(1 + xq + x^2q^2\right) d_5(xq^2), \end{aligned}$$

we see that

$$F(1) = G(1) = 0.$$

For simplicity in the remainder of the proof, we define

$$\lambda(x) = 1 + xq + x^2q^2.$$

We now replace x by xq^2 in (5.7) then multiply both sides of the resulting identity by $\lambda(x)$ and subtract from (5.6). The resulting identity simplifies to the following:

$$F(n) = \left(1 + xq^{6n+1} + x^2q^{6n+2}\right) G(n) + x^2q^{6n} \left(1 - q^{6n}\right) F(n-1). \tag{5.12}$$

We also require a second recurrence now for $G(n)$. We begin with (5.11):

$$\begin{aligned}
G(n) &= d_{6n-1}(x) - \lambda(x)d_{6n-4}(xq^2) - \lambda(x)q^{6n-1}x\,(1-xq)\,d_{6n-7}(xq^2)\\
&= \left(1+x^2q^{6n-2}+xq^{6n-1}\right)d_{6n-4}(x) + \left(xq^{6n-3}-x^4q^{12n-6}\right)d_{6n-7}(x)\\
&\qquad - \left(1+xq^{6n-3}+xq^{6n-1}\right)\lambda(x)d_{6n-7}(xq^2)\\
&\qquad - \left(x^2q^{6n-2}-x^2q^{12n-8}\right)\lambda(x)d_{6n-10}(xq^2)\\
&= \left(1+xq^{6n-1}+x^2q^{6n-2}\right)F(n-1) + \left(-xq^{6n-3}+x^2q^{6n-2}\right)\lambda(x)d_{6n-7}(xq^2)\\
&\quad + \left(xq^{6n-3}-x^4q^{12n-6}\right)d_{6n-7}(x) - \left(x^2q^{6n-2}-x^2q^{12n-8}\right)\lambda(x)d_{6n-10}(xq^2)\\
&= \left(1+xq^{6n-1}+x^2q^{6n-2}\right)F(n-1)\\
&\quad + \left(-xq^{6n-3}+x^2q^{6n-2}\right)\lambda(x)\{\left(1+x^2q^{6n-4}+xq^{6n-5}\right)d_{6n-10}(xq^2)\\
&\qquad\qquad + \left(xq^{6n-7}-x^4q^{12n-10}\right)d_{6n-13}(xq^2)\}\\
&\qquad + \left(xq^{6n-3}-x^4q^{12n-6}\right)d_{6n-7}(x) - \left(x^2q^{6n-2}-x^2q^{12n-8}\right)\lambda(x)d_{6n-10}(xq^2)\\
&\qquad\qquad \text{(by (5.7))}\\
&= \left(1+xq^{6n-1}+x^2q^{6n-2}\right)F(n-1)\\
&\quad + \lambda(x)d_{6n-10}(xq^2)\left(-xq^{6n-3}+x^4q^{12n-6}\right) + \left(xq^{6n-3}-x^4q^{12n-6}\right)d_{6n-7}(x)\\
&\quad + \left(-xq^{6n-3}+x^2q^{6n-2}\right)\left(xq^{6n-7}+x^4q^{12n-10}\right)\lambda(x)d_{6n-13}(xq^2)\\
&= \left(1+xq^{6n-1}+x^2q^{6n-2}\right)F(n-1)\\
&\quad + \left(xq^{6n-3}-x^4q^{12n-6}\right)\left(d_{6n-7}(x)-\lambda(x)d_{6n-10}(xq^2)\right)\\
&\quad + q^{-4}\lambda(x)\left(xq^{6n-3}-x^4q^{12n-6}\right)\left(-xq^{6n-3}(1-xq)d_{6n-13}(xq^2)\right)\\
&= \left(1+xq^{6n-1}+x^2q^{6n-2}\right)F(n-1)\\
&\quad + \left(xq^{6n-3}-x^4q^{12n-6}\right)\{d_{6n-7}(x)-\lambda(x)d_{6n-10}(xq^2)\\
&\qquad\qquad - xq^{6n-7}\lambda(x)(1-xq)d_{6n-13}(xq^2)\}.
\end{aligned}$$

Finally we see that the expression inside curly brackets is actually $G(n-1)$. Consequently,

$$G(n) = \left(1+xq^{6n-1}+x^2q^{6n-2}\right)F(n-1) + \left(xq^{6n-3}-x^4q^{12n-6}\right)G(n-1). \tag{5.13}$$

Finally the initial conditions $F(1) = G(1) = 0$ together with the recurrences (5.12) and (5.13) imply by mathematical induction that $F(n) = G(n) = 0$ for all $n \geq 1$, and this fact, as observed earlier, proves the lemma.

Lemma 5.2 *We have*

$$\lim_{n\to\infty} d_n(x) = \prod_{n=1}^{\infty} \left(1 + xq^{2n-1} + x^2 q^{4n-2}\right). \tag{5.14}$$

proof We note that $\lim_{n\to\infty} d_n(x)$ is the generating function for all the partitions defined in the first paragraph of Sect. 2. Consequently it is dominated by the generating function

$$\prod_{n=1}^{\infty} \frac{1}{(1 - xq^{2n-1})(1 - x^2 q^{2n})},$$

which is convergent for $|q| < 1$, $|x| < \frac{1}{|q|}$.

Hence, defining

$$A_n(x,q) = \lim_{n\to\infty} d_n(x),$$

then $A(x,q)$ is absolutely convergent provided $|q| < 1$ and $|x| < \frac{1}{|q|}$.

Consequently, we have

$$\begin{aligned} A(x,q) &= \lim_{n\to\infty} d_n(x) \\ &= \lim_{n\to\infty} d_{6n+2}(x) \\ &= \lim_{n\to\infty} \left(1 + xq + x^2 q^2\right) d_{6n-1}(xq^2) \qquad \text{(by Lemma 5.1)} \\ &= \left(1 + xq + x^2 q^2\right) A(xq^2, q). \end{aligned} \tag{5.15}$$

Iterating (5.15) we see that

$$\begin{aligned} A(x,q) &= A(0,q) \prod_{n=1}^{\infty} \left(1 + xq^{2n-1} + x^2 q^{4n-2}\right) \\ &= \prod_{n=1}^{\infty} \left(1 + xq^{2n-1} + x^2 q^{4n-2}\right), \end{aligned}$$

which is the desired result.

It is now an easy matter to deduce Theorem 5.2 from Lemma 5.1:

$$\begin{aligned} \sum_{n,m\geq 0} C(m,n) x^m q^n &= \prod_{n=1}^{\infty} \left(1 + xq^{2n-1} + x^2 q^{4n-2}\right) \\ &= A(x,q) \\ &= \lim_{n\to\infty} d_n(x) \\ &= \sum_{n,m\geq 0} D(m,n) x^m q^n, \end{aligned} \tag{5.16}$$

and comparing coefficients in the extremes of (5.16) we establish the assertion in Theorem 5.2.

3 Conclusion

There are a couple of relevant observations. First, Alladi's addition to Schur's Theorem [1] given in Theorem 5.1 merits much closer study than it has received to date. Indeed, it would appear that it has been referred to in print subsequently only in [4].

Second, the conjectures of Kanade and Russell [6] suggest that the q-difference equation techniques, as initiated in [2, 3] need to be extended beyond partitions in which all parts are distinct. Part of the motivation for this paper was to show that such an extension is feasible.

References

1. Alladi, K.: Personal Communication
2. Andrews, G.E.: On a theorem of Schur and Gleissberg. Arch. Math. **22**, 165–167 (1971)
3. Andrews, G.E.: q-Series: Their Development and Application in Analysis, Number Theory, Combinatorics, Physics and Computer Algebra. C.B.M.S. Regional Conference Series, vol. 66. American Mathematical Society, Providence (1986)
4. Andrews, G.E.: Schur's theorem, partitions with odd parts and the Al-Salam-Carlitz polynomials. Contemp. Math. **254**, 45–53 (2000)
5. Gleissberg, W.: Über einen Satz von Herrn I. Schur. Math. Zeit. **28**, 372–382 (1928)
6. Kanade, S., Russell, M.: Identity Finder and some new identities of Roger-Ramanujan type. Exp. Math. **24**, 419–423 (2015)
7. Schur, I.: Zur additiven Zahlentheorie. S.-B. Preuss. Akad. Wiss. Phys.-Math. Kl., pp. 488–495 (1926)

Explicit Formulas for Enumeration of Lattice Paths: Basketball and the Kernel Method

Cyril Banderier, Christian Krattenthaler, Alan Krinik, Dmitry Kruchinin, Vladimir Kruchinin, David Nguyen and Michael Wallner

Abstract This article deals with the enumeration of directed lattice walks on the integers with any finite set of steps, starting at a given altitude j and ending at a given altitude k, with additional constraints, for example, to never attain altitude 0 in-between. We first discuss the case of walks on the integers with steps $-h, \ldots, -1, +1, \ldots, +h$. The case $h = 1$ is equivalent to the classical Dyck paths, for which many ways of getting explicit formulas involving Catalan-like numbers are known. The case $h = 2$ corresponds to "basketball" walks, which we treat in full detail. Then, we move on to the more general case of walks with any finite set of steps, also allowing some weights/probabilities associated with each step. We show how a method of wide applicability, the so-called kernel method, leads to explicit formulas for the number of walks of length n, for any h, in terms of nested sums of binomials. We finally relate some special cases to other combinatorial problems, or to problems arising in queuing theory.

Keywords Lattice paths · Dyck paths · Motzkin paths · Kernel method · Analytic combinatorics · Computer algebra · Generating function · Singularity analysis · Lagrange inversion · Context-free grammars · $\mathbb{N}$-algebraic function

---

C. Banderier
Laboratoire d'Informatique de Paris-Nord, CNRS/Université, Paris 13, Villetaneuse, France

C. Krattenthaler
Fakultät für Mathematik, Universität Wien, Vienna, Austria

A. Krinik
Department of Mathematics and Statistics, California State Polytechnic University, Pomona, Pomona, CA, USA

D. Kruchinin · V. Kruchinin
Tomsk State University of Control Systems and Radio Electronics, Tomsk, Russia

D. Nguyen
Department of Mathematics, University of California, Santa Barbara, CA, USA

M. Wallner (✉)
Institute of Discrete Mathematics and Geometry, TU Wien, Austria
e-mail: Michael.Wallner@tuwien.ac.at

M. Wallner
Laboratoire Bordelais de Recherche en Informatique, Université de Bordeaux, Bordeaux, France

G. E. Andrews et al. (eds.), *Lattice Path Combinatorics and Applications*,
Developments in Mathematics 58, https://doi.org/10.1007/978-3-030-11102-1_6

2010 Mathematics Subject Classification Primary: 05A15 · Secondary: 05A10 · 05A16 · 05A19

Dedicated to Sri Gopal Mohanty,
a pioneer in the field of lattice paths combinatorics,
on the occasion of his 84th birthday.

1 Introduction

While analyzing permutations sortable by a stack, Knuth [34, Ex. 1–4 in Sect. 2.2.1] showed they were counted by Catalan numbers and were therefore in bijection with Dyck paths (lattice paths with steps $(1, 1)$ and $(1, -1)$ in the plane integer lattice, from the origin to some point on the x-axis, and never running below the x-axis in-between). He used a method to derive the corresponding generating function (see [34, p. 536ff]) which Flajolet coined *"kernel method."* That name stuck among combinatorialists, although the method already existed in the folklore of statistics and statistical physics — without a name. The method was later generalized to enumeration and asymptotic analysis of directed lattice paths with any set of steps, and many other combinatorial structures enumerated by bivariate or trivariate functional equations (see, e.g., [6, 8, 19, 20, 26, 27]). We refer to the introduction of [11] for a more detailed history of the kernel method.

The emphasis in [8] is on asymptotic analysis, for which the derived (exact) enumeration results serve as a starting point. The latter are in a sense *implicit*, since they involve solutions to certain algebraic equations. They are nevertheless perfect for carrying out singularity analysis, which in the end leads to very precise asymptotic results.

In general, it is not possible to simplify the exact enumeration results from [8]. However, for models involving special choices of step sets, this is possible. These potential simplifications are the main focus of our paper.

Such models appear frequently in queuing theory. Indeed, birth and death processes and queues, like the one shown in Fig. 1, are naturally encoded by lattice

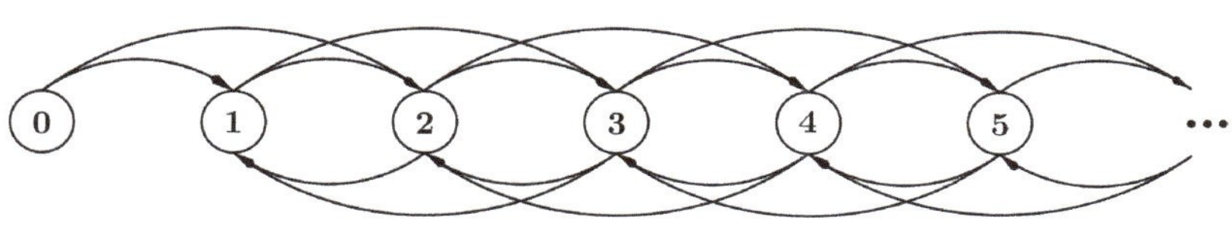

Fig. 1 A queue corresponding to the basketball walk model

paths (see [16, 29, 30, 35, 36, 41]). In this article, we solve a problem raised during the 2015 International Conference on "Lattice Path Combinatorics and Its Applications": to find closed-form formulas for the number of walks of length n from 0 to k for a full family of models similar to Fig. 1. As it turns out, the essential tool to achieve this goal is indeed the kernel method.

Our paper is organized as follows. We begin with some preliminaries in Sect. 2. In particular, we introduce the directed lattice paths that we are going to discuss here, we provide a first glimpse of the kernel method, and we briefly review the Lagrange–Bürmann inversion formula for the computation of the coefficients of implicitly defined power series. Section 3 is devoted to (old-time) "basketball walks," which, by definition, are directed lattice walks with steps from the set $\{(1, -2), (1, -1), (1, 1), (1, 2)\}$ which always stay above the x-axis. (They may be seen as the evolution of — pre-1984 — basketball games; see the beginning of that section for a more detailed explanation of the terminology.) We provide exact formulas (often several, not obviously equivalent ones) for generating functions and for the numbers of walks under various constraints. At the end of Sect. 3, we also briefly address the asymptotic analysis of the number of these walks. Section 4 then considers the more general problem of enumerating directed walks where the allowed steps are of the form $(1, i)$ with $-h \leq i \leq h$ (including $i = 0$ or not). Again, we provide exact formulas for generating functions — in terms of roots of the so-called kernel equation — and for numbers of walks — in terms of nested sums of binomials. All these results are obtained by appropriate combinations of the kernel method with variants of the Lagrange–Bürmann inversion formula. In the concluding Sect. 5, we relate basketball walks with other combinatorial objects, namely

- with certain trees coming from option pricing,
- with increasing unary-binary trees which avoid a certain pattern which arose in work of Riehl [39],
- and with certain Boolean bracketings which appeared in work of Bender and Williamson [13].

2 The General Setup and Some Preliminaries

In this section, we describe the general setup that we consider in this article. We use (subclasses of) so-called *Łukasiewicz paths* as main example(s) which serve to illustrate this setup. We recall here as well the main tools that we shall use in this article: the *kernel method* and the *Lagrange–Bürmann inversion formula.*

We start with the definition of the lattice paths under consideration.

Definition 6.1 A *step set* $\mathscr{S} \subset \mathbb{Z}^2$ is a finite set of vectors

$$\{(x_1, y_1), (x_2, y_2), \ldots, (x_m, y_m)\}.$$

An n-step *lattice path* or *walk* is a sequence of vectors $v = (v_1, v_2, \ldots, v_n)$, such that v_j is in $\mathcal{S}$. Geometrically, it may be interpreted as a sequence of points $\omega = (\omega_0, \omega_1, \ldots, \omega_n)$, where $\omega_i \in \mathbb{Z}^2$, $\omega_0 = (0, 0)$ (or another starting point), and $\omega_i - \omega_{i-1} = v_i$ for $i = 1, 2, \ldots, n$. The elements of $\mathcal{S}$ are called *steps*. The *length* $|\omega|$ of a lattice path is its number n of steps.

The lattice paths can have different additional constraints shown in Table 1.

We restrict our attention to *directed paths*, which are defined by the fact that, for each step $(x, y) \in \mathcal{S}$, one has $x \geq 0$. Moreover, we will focus only on the subclass of *simple paths*, where every element in the step set $\mathcal{S}$ is of the form $(1, b)$. In other words, these paths constantly move one step to the right. Thus, they are essentially one-dimensional objects and can be seen as walks on the integers. We introduce the abbreviation $\mathcal{S} = \{b_1, b_2, \ldots, b_n\}$ in this case. A *Łukasiewicz path* is a simple path where its associated step set $\mathcal{S}$ is a subset of $\{-1, 0, 1, \ldots\}$ and $-1 \in \mathcal{S}$.

Example 6.1 (*Dyck paths*) A Dyck path is a path constructed from the step set $\mathcal{S} = \{-1, +1\}$, which starts at the origin, never passes below the x-axis, and ends on the x-axis. In other words, Dyck paths are excursions with step set $\mathcal{S} = \{-1, +1\}$.

The next definition allows to merge the probabilistic point of view (*random walks*) and the combinatorial point of view (*lattice paths*).

Table 1 The four types of walks: unconstrained walks, bridges, meanders, and excursions

	ending anywhere	ending at 0
unconstrained (on $\mathbb{Z}$)	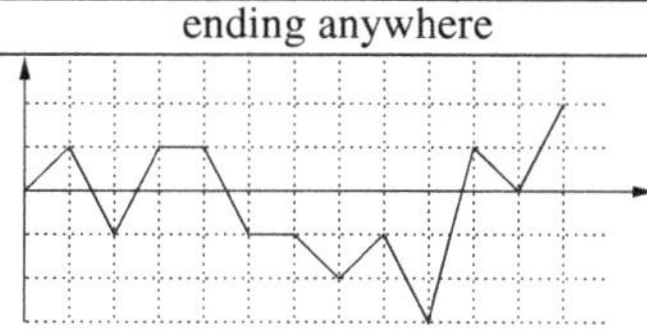 walk/path ($\mathcal{W}$)	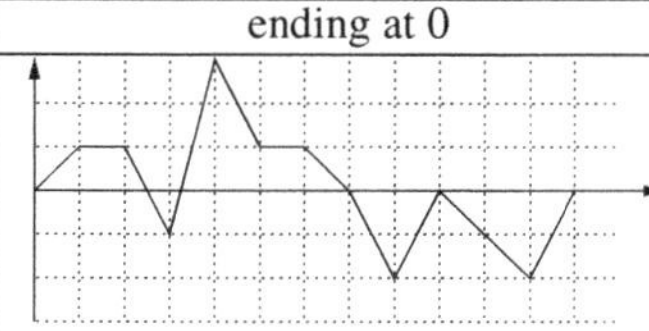 bridge ($\mathcal{B}$)
constrained (on $\mathbb{N}$)	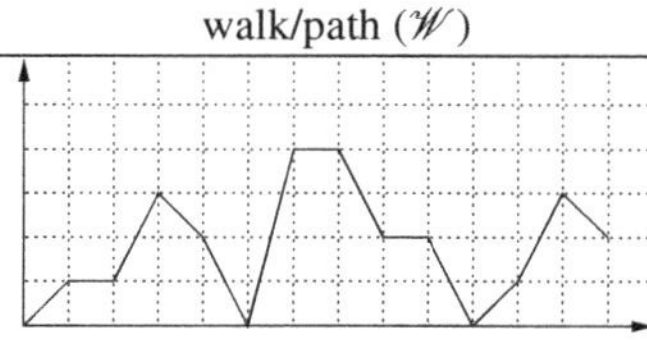 meander ($\mathcal{M}$)	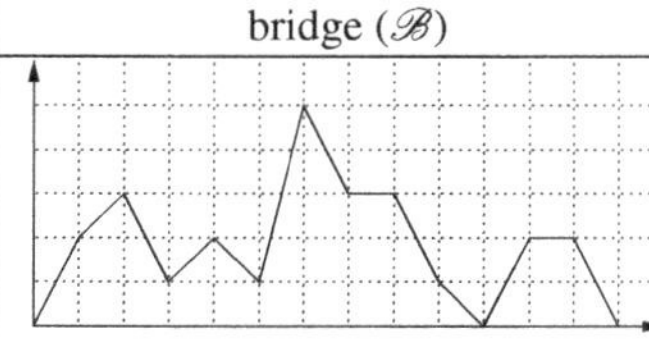 excursion ($\mathcal{E}$)

Definition 6.2 For a given step set $\mathcal{S} = \{s_1, s_2, \ldots, s_m\}$, we define the corresponding *system of weights* as $\{p_1, p_2, \ldots, p_m\}$, where $p_j > 0$ is the weight associated with step s_j for $j = 1, 2, \ldots, m$. The *weight of a path* is defined as the product of the weights of its individual steps.

Next, we introduce the algebraic structures associated with the previous definitions. The *step polynomial* of a given step set $\mathcal{S}$ is defined as the Laurent polynomial[1]

$$P(u) := \sum_{j=1}^{m} p_j u^{s_j}.$$

Let

$$c = -\min_j s_j \quad \text{and} \quad d = \max_j s_j \tag{6.1}$$

be the two extreme step sizes and assume throughout that $c, d > 0$. Note that for Łukasiewicz paths we have $c = 1$.

We start with the easy case of unconstrained paths. We define their bivariate generating function as

$$W(z, u) = \sum_{n=0}^{\infty} \sum_{k=-\infty}^{\infty} W_{n,k} z^n u^k,$$

where $W_{n,k}$ is the number of unconstrained paths ending after n steps at altitude k.

It is well known and straightforward to derive that

$$W(z, u) = \frac{1}{1 - zP(u)}. \tag{6.2}$$

We continue with the generating function of meanders:

$$F(z, u) := \sum_{n=0}^{\infty} \sum_{k=0}^{\infty} F_{n,k} z^n u^k,$$

where $F_{n,k}$ is the number of paths ending after n steps at altitude k, and constrained to be always at altitude ≥ 0 in-between. Note that we are mainly interested in solving the counting problem, i.e., determining the numbers $F_{n,k}$ for specific families of paths (see Table 1). The generating function encodes all information we are interested in.

---

[1] By a *Laurent polynomial* in u we mean a polynomial in u and u^{-1}.

We decompose $F(z, u)$ in two ways, namely

$$F(z, u) = \sum_{k \geq 0} F_k(z) u^k = \sum_{n \geq 0} f_n(u) z^n.$$

Here, the generating functions $F_k(z)$ enumerate paths ending at altitude k, i.e., $F_k(z) = \sum_{n \geq 0} F_{n,k} z^n$. In particular, the generating function for excursions is equal to $F_0(z)$. On the other hand, the polynomials $f_n(u)$ enumerate paths of length n. The power of u encodes their final altitude. We will use this decomposition for a step-by-step approach, similar to the one in the case of unconstrained paths.

For the sake of illustration, we show below how the kernel method can be used to find a closed form for the generating function of a given class of Łukasiewicz paths.

Theorem 6.1 *Let $\mathcal{S}$ be the step set of a class of Łukasiewicz paths, and let $P(u)$ be the associated step polynomial. Then, the bivariate generating function of meanders (where z marks length, and u marks final altitude) and excursions are*

$$F(z, u) = \frac{1 - zF_0(z)/u}{1 - zP(u)} \qquad \text{and} \qquad F_0(z) = \frac{u_1(z)}{z}, \tag{6.3}$$

respectively, where $u_1(z)$ is the unique small solution of the implicit equation

$$1 - zP(u) = 0,$$

that is, the unique solution satisfying $\lim_{z \to 0} u_1(z) = 0$.

Proof A meander of length n is either empty, or it is constructed from a meander of length $n - 1$ by appending a possible step from $\mathcal{S}$. However, a meander is not allowed to pass below the x-axis; thus, at altitude 0 it is not allowed to use the step -1. This translates into the relations

$$f_0(u) = 1, \qquad f_{n+1}(u) = \{u^{\geq 0}\} \left(P(u) f_n(u)\right),$$

where $\{u^{\geq 0}\}$ is the linear operator extracting all terms in the power series representation containing non-negative powers of u. Multiplying both sides of the above equation by z^{n+1} and subsequently summing over all $n \geq 0$, we obtain the functional equation

$$F(z, u) = 1 + zP(u)F(z, u) - \frac{z}{u} F_0(z).$$

Equivalently,

$$(1 - zP(u))F(z, u) = 1 - \frac{z}{u} F_0(z). \tag{6.4}$$

We write $K(z,u) := 1 - zP(u)$ and call this factor $K(z,u)$ the *kernel*. The above functional equation looks like an underdetermined equation as there are two unknown functions, namely $F(z,u)$ and $F_0(z)$. However, the special structure on the left-hand side will resolve this problem and leads us to the *kernel method*.

Using the theory of Newton polygons and Puiseux expansions (cf. [24, Appendix of Sect. 3]), we know that the *kernel equation*

$$1 - zP(u) = 0,$$

has $d+1$ distinct solutions in u (recall that $c = 1$, see Eq. (6.1)). One of them, say $u_1(z)$, maps 0 to 0. We call this solution the "small branch" of the kernel equation. It is in modulus smaller than the other d branches. These in turn grow to infinity in modulus while z approaches 0. Consequently, we call the latter the "large branches" and denote them by $v_1(z), v_2(z), \dots, v_d(z)$. Inserting the small branch into (6.4) (this is legitimate as we stay in the integral domain of Puiseux power series: substitution of the small branch always leads to series having a finite number of terms with negative exponents, even for intermediate computations), we get $F_0(z) = u_1(z)/z$. This proves our second claim. Using this result, we can solve (6.4) for $F(z,u)$ to get the first claim.

The formula (6.3) in the previous theorem implies that the number m_n of meanders of length n is directly related to the number e_n of excursions of length n via

$$m_n = P(1)^n - \sum_{k=0}^{n-1} P(1)^k e_{n-k-1}.$$

In the sequel, we therefore focus on giving explicit expressions for e_n.

A key tool for finding a formula for the coefficients of power series satisfying implicit equations is the Lagrange inversion formula [37], independently discovered in a slightly extended form by Bürmann [22] (see also [38]). In the statement of the theorem and also later, we use the *coefficient extractor* $[z^n]F(z) := f_n$ for a power series $F(z) = \sum f_n z^n$.

Theorem 6.2 (Lagrange–Bürmann inversion formula) *Let $F(z)$ be a formal power series which satisfies $F(z) = z\phi(F(z))$, where $\phi(z)$ is a power series with $\phi(0) \neq 0$. Then, for any Laurent*[2] *series $H(z)$ and for all non-zero integers n, we have*

$$[z^n]H(F(z)) = \frac{1}{n}[z^{n-1}]H'(z)\phi^n(z).$$

Proof See [28, Chap. A.6] or [49, Theorem 5.4.2].

[2] Here, by Laurent series we mean a series of the form $H(z) = \sum_{n \geq a} H_n z^n$ for some (possibly negative) integer a.

Table 2 Closed-form formulas for some famous families of lattice paths

Name and the associated step polynomial $P(u)$	Number e_n of excursions of length n
Dyck paths $P(u) = \frac{1}{u} + u$	$e_{2n} = \frac{1}{n+1}\binom{2n}{n}$
Motzkin paths $P(u) = \frac{1}{u} + 1 + u$	$e_n = \frac{1}{n+1}\sum_{k=0}^{\lceil \frac{n+1}{2} \rceil} \binom{n+1}{k}\binom{n+1-k}{k-1}$
Weighted Motzkin paths $P(u) = \frac{p_{-1}}{u} + p_0 + p_1 u$	$e_n = \frac{1}{n+1}\sum_{k=0}^{\lceil \frac{n+1}{2} \rceil} \binom{n+1}{k}\binom{n+1-k}{k-1}(p_1 p_{-1})^{k-1} p_0^{n+2-2k}$
Bicolored Motzkin paths $P(u) = \frac{1}{u} + 2 + u$	$e_{n+1} = \frac{1}{n+1}\binom{2n}{n}$
Łukasiewicz paths $P(u) = \frac{1}{u} + 1 + u + u^2 + \cdots$	$e_n = \frac{1}{n+1}\binom{2n}{n}$
d-ary trees $P(u) = \frac{1}{u} + u^{d-1}$	$e_{dn+1} = \frac{1}{(d-1)n+1}\binom{dn}{n}$
$\{1, 2, \dots, d\}$-ary trees $P(u) = \frac{1}{u} + 1 + \cdots + u^{d-1}$	$e_n = \frac{1}{n}\sum_{j=0}^{\lfloor \frac{n-1}{d+1} \rfloor} (-1)^j \binom{n}{j}\binom{2n-2-j(d+1)}{n-1}$
$\{d, d+1\}$-ary trees $P(u) = \frac{1}{u} + u^{d-1} + u^d$	$e_n = \frac{1}{n}\sum_{k=0}^{\lfloor \frac{n-1}{d} \rfloor} \binom{n}{k}\binom{k}{n-1-dk}$

Table 2 presents several applications of this Lagrange inversion formula to lattice path enumeration. It leads to the Catalan numbers for Dyck paths, and to the Motzkin numbers for the Motzkin paths, i.e., excursions associated with the step set $\mathscr{S} = \{-1, 0, +1\}$. They are two of the most ubiquitous number sequences in combinatorics, see [49, Ex. 6.19, 6.25, and 6.38] for more information. Table 2 also contains an example of weighted paths (namely weighted Motzkin paths and the special case of bicolored Motzkin paths), as well as an example with an infinite set of steps (namely the Łukasiewicz paths with all possible steps allowed).

All of the examples in Table 2 are intimately related to families of trees (as suggested by some of the namings in the table). In order to explain this, we recall that an *ordered tree* is a rooted tree for which an ordering of the children is specified for each vertex, and for which its arity (i.e., the outdegree, the number of children of each node) is restricted to be in a subset $\mathscr{A}$ of $\mathbb{N}$.[3] If $\mathscr{A} = \{0, 2\}$, this leads to the classical binary trees counted by the Catalan numbers; if $\mathscr{A} = \{0, 1, 2\}$, this leads to the unary-binary trees counted by Motzkin numbers, and if $\mathscr{A} = \mathbb{N}$, this gives the ordered trees (also called planted plane trees), which are also counted by Catalan numbers. Any ordered tree can be traversed starting from the root in *prefix order*: One starts from the root and proceeds depth-first and left-to-right. The listing of the outdegrees of nodes in prefix order is called the *preorder degree sequence*.

[3]In this article, by convention $0 \in \mathbb{N}$.

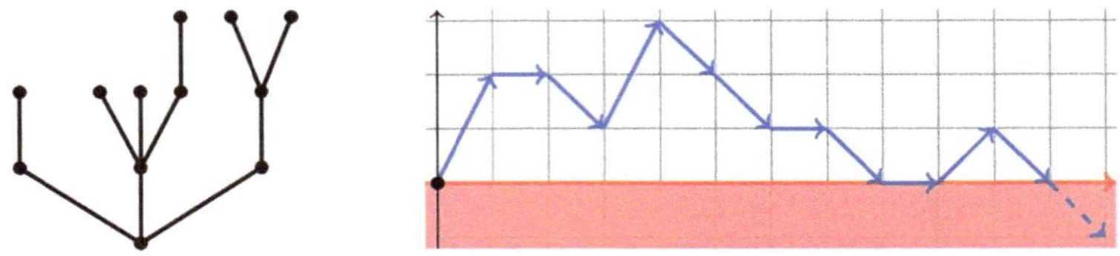

Fig. 2 The bijection between trees and Łukasiewicz paths. The preorder degree sequence $(3, 1, 0, 3, 0, 0, 1, 0, 1, 2, 0, 0)$ uniquely characterizes the tree and gives the corresponding Łukasiewicz path with step sequence $(2, 0, -1, 2, -1, -1, -1, 0, -1, 0, 1, -1, -1)$. Dropping the last -1 step yields an excursion

This characterizes a tree unambiguously, see Fig. 2, and it is best summarized by the following folklore proposition.

Proposition 6.1 (Łukasiewicz correspondence) *Ordered trees are in bijection with Łukasiewicz excursions.*

Proof Given an ordered tree with n nodes, the preorder sequence can be interpreted as a lattice path. Let $(\sigma_j)_{j=1}^n$ be a preorder degree sequence. With each σ_j we associate a step $(1, \sigma_j - 1) \in \mathbb{N} \times \mathbb{Z}$. Note that, as the minimal degree is 0, our smallest step is -1. Starting at the origin, we concatenate these steps for $j = 1, 2, \dots, n-1$, ignoring the last step. In this way, we obtain a Łukasiewicz excursion of length $n-1$.

As one can see, the combinatorics of the Łukasiewicz paths is well understood (see e.g. [28, 48]), and the true challenge is to analyze lattice paths with other negative steps than just -1. The smallest non-Łukasiewicz cases are the Duchon lattice paths (steps $\mathcal{S} = \{-2, +3\}$), and the Knuth lattice paths (steps $\mathcal{S} = \{-2, +5\}$). Their enumerative and asymptotic properties are the subject of another article in this volume [11]. For these two families of lattice paths, the asymptotics are tricky, because the generating functions involve several dominant singularities. In the next sections, we concentrate on closed formulas which appear for many other non-Łukasiewicz cases.

3 (Old-time) Basketball Walks: Steps $\mathcal{S} = \{-2, -1, +1, +2\}$

We now turn our attention to a class of lattice paths (lattice walks) with rich combinatorial properties: *the basketball walks*. They are constructed from the step set $\mathcal{S} = \{-2, -1, +1, +2\}$. This terminology was introduced by Arvind Ayyer and Doron Zeilberger [5], and these walks were later also considered by Mireille Bousquet-Mélou [18]. They can be seen as the evolution of the score during a(n old-time) basketball game (see Fig. 3).

Fig. 3 Since its creation in 1892 by James Naismith (November 6, 1861–November 28, 1939), the rules of basketball evolved. For example, since 1896, field goals and free throws were counted as two and one points, respectively. The international rules were changed in 1984 so that a "far" field goal was now rewarded by 3 points, while "ordinary" field goals remained at 2 points, a free throw still being worth one point

Ayyer, Zeilberger, and Bousquet-Mélou found interesting results on the shape of the algebraic equations satisfied by the excursion generating function, and similar properties when the height of the excursion is bounded. In this article, we analyze a generalization in which the starting point and the end point of the walks do not necessarily have altitude 0. Since, in that case, we lose a natural factorization happening for excursions, we are led to variations of certain parts in the kernel method. In addition, we are interested in closed-form expressions for the number of walks of length n. This is complementary to the results in [8] and in [11]. Moreover, contrary to the previous section, these walks are not Łukasiewicz paths any more. This makes them harder to analyze (the easy bijection with trees is lost, for example). Despite all that, the kernel method will strike again, thus illustrating our main motto:

"The kernel method is the method of choice for problems on directed lattice paths!"

3.1 Generating Functions for Positive (Old-time) Basketball Walks: The Kernel Method

We define *positive walks* as walks staying strictly above the x-axis, possibly touching it at the first or last step. Returning to the basketball interpretation, these correspond to the evolution of basketball scores where one team (the stronger team, the richer team?) is always ahead of the other team.

Let $G_{j,n,k}$ be the number of such walks running from $(0, j)$ to (n, k), and define by $G_j(z, u)$ the generating function of positive walks starting at $(0, j)$. We write

$$G_j(z, u) := \sum_{n,k\geq 0} G_{j,n,k} z^n u^k = \sum_{n=0}^{\infty} g_{j,n}(u) z^n = \sum_{k=0}^{\infty} G_{j,k}(z) u^k.$$

Similar to Sect. 2, we shall need the polynomial $g_{j,n}(u)$, the generating function for all walks with n steps, and the series $G_{j,k}(z)$, the generating function for all walks ending at altitude k. The bivariate generating function $G_j(z,u)$ is analytic for $|z| < 1/P(1)$ and $|u| \le 1$.

A walk is either the single initial point at altitude j, or a walk followed by a step not reaching altitude 0 or below. This leads to the functional equation

$$(1 - zP(u))G_j(z,u) = u^j - z\big(G_{j,1}(z) + G_{j,2}(z) + G_{j,1}(z)/u\big), \qquad j > 0, \tag{6.5}$$

where the *step polynomial* $P(u)$ is given by

$$P(u) := u^{-2} + u^{-1} + u + u^2.$$

Again, we call the factor $1 - zP(u)$ on the left-hand side of (6.5) the *kernel* of the equation and denote it by $K(z,u)$.

We refer to (6.5) as the *fundamental functional equation* for $G_j(z,u)$. The equation has a small problem though: This is *one* equation with *three* unknowns, namely $G_j(z,u)$, $G_{j,1}(z)$, and $G_{j,2}(z)$! The idea of the so-called *kernel method* is to equate the kernel $K(z,u)$ to 0, thus binding u and z in such a way that the left-hand side of (6.5) vanishes. This produces two extra equations.

To equate $K(z,u)$ to zero means to put

$$1 - zP(u) = 0 \quad \text{or equivalently} \quad u^2 - zu^2P(u) = 0. \tag{6.6}$$

We call this equation the *kernel equation*. As an equation of degree 4 in u, it has four roots. We call the two small roots (i.e., the roots which tend to 0 when z approaches 0) $u_1(z)$ and $u_2(z)$.

Then, on the complex plane slit along the negative real axis, we can identify the small roots $u_1(z)$ and $u_2(z)$ as

$$\begin{aligned}
u_1(z) &= -\frac{1}{4}\left(\frac{z - \sqrt{4z + 9z^2}}{z} + \sqrt{\frac{4 - 6z - 2\sqrt{4z + 9z^2}}{z}}\right)\\
&= \sqrt{z} + \frac{1}{2}z + \frac{1}{8}z^{3/2} + \frac{1}{2}z^2 + \frac{159}{128}z^{5/2} + O(z^3),\\
u_2(z) &= -\frac{1}{4}\left(\frac{z + \sqrt{4z + 9z^2}}{z} - \sqrt{\frac{4 - 6z + 2\sqrt{4z + 9z^2}}{z}}\right)\\
&= -\sqrt{z} + \frac{1}{2}z - \frac{1}{8}z^{3/2} + \frac{1}{2}z^2 - \frac{159}{128}z^{5/2} + O(z^3).
\end{aligned}$$

Moreover, their Puiseux expansions are related via the following proposition.

Proposition 6.2 (Conjugation principle for two small roots) *The small roots $u_1(z)$ and $u_2(z)$ of $1 - zP(u) = 0$ satisfy*

$$u_1(z) = \sum_{n\ge1} a_n z^{n/2} \text{ and } u_2(z) = \sum_{n\ge1} (-1)^n a_n z^{n/2}.$$

Proof The kernel equation yields

$$u = X(1 + u + u^3 + u^4)^{1/2},$$

with $X = z^{1/2}$ or $X = -z^{1/2}$. Since the above equation possesses a unique formal power series solution $u(X)$, the claim follows.

By substituting the small roots $u_1(z)$ and $u_2(z)$ of the kernel equation (6.6) into the fundamental functional equation (6.5), we see that the left-hand side vanishes. Subsequently, we solve for $G_{j,1}(z)$ and $G_{j,2}(z)$ and get[4]

$$G_{j,1}(z) = -\frac{u_1 u_2 (u_1^j - u_2^j)}{z(u_1 - u_2)}, \qquad j > 0, \tag{6.7}$$

$$G_{j,2}(z) = \frac{u_1 u_2 (u_1^j - u_2^j) + u_1^{j+1} - u_2^{j+1}}{z(u_1 - u_2)}, \qquad j > 0. \tag{6.8}$$

Substitution in the fundamental functional equation (6.5) then yields

$$G_j(z,u) = \frac{u^j - z(G_{j,1}(z) + G_{j,2}(z) + G_{j,1}(z)/u)}{1 - zP(u)}, \qquad j > 0. \tag{6.9}$$

By means of the kernel method, we have thus derived an explicit expression for the bivariate generating function $G_j(z,u)$ for walks starting at altitude $j > 0$.

In the following proposition, we summarize our findings so far. In addition, we express the generating function for walks from altitude j to altitude k (with $j, k > 0$) explicitly in terms of the small roots $u_1(z)$ and $u_2(z)$, and we also cover the special case $j = 0$, which offers some nice simplifications.

Proposition 6.3 *As before, let $G_{j,k}(z)$ be the generating function for positive basketball walks with steps $-2, -1, +1, +2$ starting at altitude j and ending at altitude k. Furthermore, let $u_1(z)$ and $u_2(z)$ be the small roots of the kernel equation $1 - zP(u) = 0$, with $P(u) = u^{-2} + u^{-1} + u + u^2$. Then, for $j, k > 0$, we have*

$$G_{0,k}(z) = \frac{u_1^{k+1}(z) - u_2^{k+1}(z)}{u_1(z) - u_2(z)}, \tag{6.10}$$

$$G_{j,k}(z) = -\frac{u_1(z)u_2(z)}{z} \sum_{i=0}^{j} \frac{u_1^{j-i+1}(z) - u_2^{j-i+1}(z)}{u_1(z) - u_2(z)} \frac{u_1^{k-i+1}(z) - u_2^{k-i+1}(z)}{u_1(z) - u_2(z)}, \tag{6.11}$$

Proof We start with the proof of (6.10). The first step of a walk can only be a step of size $+1$ or $+2$. Thus, removing this first step and shifting the origin, we have

$$G_{0,k}(z) = z\left(G_{1,k}(z) + G_{2,k}(z)\right),$$

---

[4]In this article, whenever we thought it could ease the reading, without harming the understanding, we write u_1 for $u_1(z)$, or F for $F(z)$, etc.

where $G_{1,k}(z)$ and $G_{2,k}(z)$ are the generating functions for positive walks running from altitude 1 to altitude k, respectively, from altitude 2 to altitude k. This decomposition is illustrated in Fig. 4.

By "time reversal" (due to the symmetry of our step set, i.e., $P(u) = P(u^{-1})$), we also have

$$G_{1,k}(z) = G_{k,1}(z), \quad \text{and} \quad G_{2,k}(z) = G_{k,2}(z),$$

where $G_{k,1}(z)$ and $G_{k,2}(z)$ are known from Eqs. (6.7) and (6.8). Now notice that

$$\begin{aligned} G_{k,2}(z) &= \frac{u_1u_2(u_1^k - u_2^k) + u_1^{k+1} - u_2^{k+1}}{z(u_1 - u_2)} = \frac{u_1u_2(u_1^k - u_2^k)}{z(u_1 - u_2)} + \frac{u_1^{k+1} - u_2^{k+1}}{z(u_1 - u_2)} \\ &= \frac{u_1^{k+1} - u_2^{k+1}}{z(u_1 - u_2)} - G_{k,1}(z). \end{aligned}$$

This leads directly to (6.10).

For computing $G_{j,k}(z)$ with $j, k > 0$, we use a first passage decomposition with respect to minimal altitude of the walk. Combining (6.10) with time reversal, we see that $h_m(z) := \frac{u_1^{m+1} - u_2^{m+1}}{u_1 - u_2}$ is the generating function for basketball walks starting at altitude m, staying always above the x-axis, but ending on the x-axis. Furthermore, by (6.7) with $j = 1$, the series $E(z) = -\frac{u_1u_2}{z}$ is the generating function for excursions (allowed to touch the x-axis). Then, the walks from altitude j to altitude k can be decomposed into three sets, as illustrated by Fig. 5:

1. The walk starts at altitude j and continues until it hits for the first time altitude i (the lowest altitude of the walk, so $1 \le i \le j$). This part is counted by $h_{j-i}(z)$.
2. The second part is the one from that point to the last time reaching altitude i. In other words, this part is an excursion on level i counted by $E(z)$.
3. The last part runs from altitude i to altitude j without ever returning to altitude i. By time reversal, one sees that this is counted by $h_{k-i}(z)$.

Summing over all possible i's, we get (6.11).

There is an alternative expression for the generating function $G_{j,k}(z)$, which we present in the next proposition.

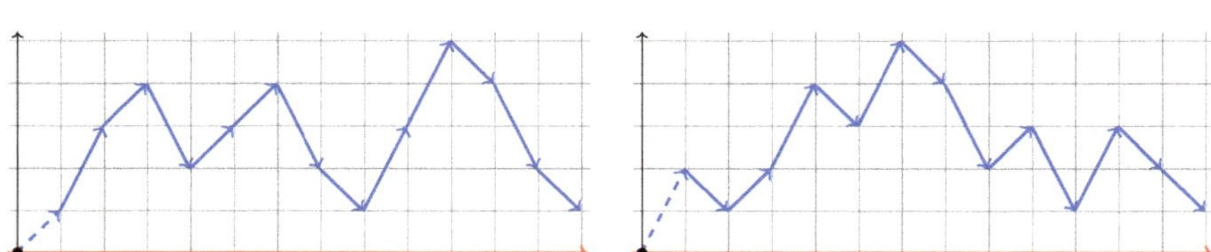

Fig. 4 Two different instances of walks counted by $G_{0,1}(z)$ showing the two possible first steps $+1$ and $+2$

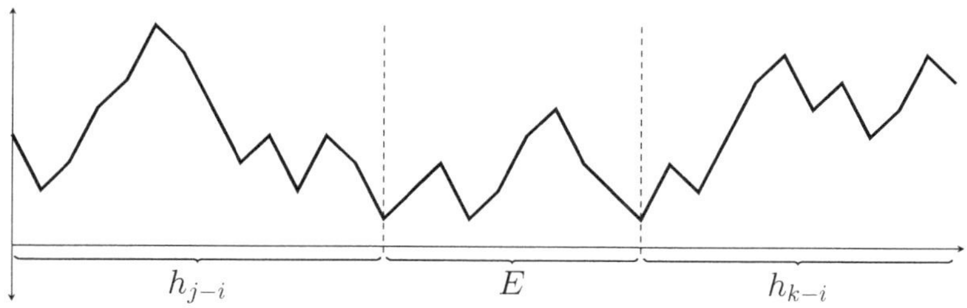

Fig. 5 The decomposition for $G_{j,k}$

Proposition 6.4 (Formula for walks from altitude j *to altitude* k) *Let* $u_1(z)$ *and* $u_2(z)$ *be the small roots of the kernel equation* $1 - zP(u) = 0$, *with* $P(u) = u^{-2} + u^{-1} + u + u^2$, *and let* $G_{j,k}(z)$ *be the generating function for positive basketball walks starting at altitude* j *and ending at altitude* k. *Then*

$$G_{j,k}(z) = W_{j-k} + h_j(u_1, u_2)W_{-k} + u_1u_2h_{j-1}(u_1, u_2)W_{-k+1}, \tag{6.12}$$

where

$$W_i(z) = z\left(\frac{u_1'}{u_1^{i+1}} + \frac{u_2'}{u_2^{i+1}}\right)$$

is the generating function of unconstrained walks starting at the origin and ending at altitude i, *and*

$$h_i(x_1, x_2) = \frac{x_1^{i+1} - x_2^{i+1}}{x_1 - x_2}$$

is the complete homogeneous symmetric polynomial of degree i *in* x_1 *and* x_2.

Proof Since $G_{j,k}(z) = G_{k,j}(z)$, without loss of generality we may assume that $j \leq k$. We start with (6.9). Extraction of the coefficient of u^k on the left-hand side gives $G_{j,k}(z)$. As coefficient extraction is linear, we need to find expressions for

$$[u^i]\frac{1}{1 - zP(u)}.$$

By (6.2), these are the generating functions $W_i(z)$ for unconstrained walks starting at the origin and ending at altitude i. For basketball walks, we have $P(u) = P(u^{-1})$, hence $W_i(z) = W_{-i}(z)$. Using a straightforward contour integral argument, using Cauchy's integral formula and the residue theorem, we have

$$W_i(z) = [u^i]\frac{1}{1 - zP(u)} = \frac{1}{2\pi\sqrt{-1}}\int_{\mathscr{C}} \frac{du}{u^{i+1}(1 - zP(u))} = z\left(\frac{u_1'(z)}{u_1^{i+1}(z)} + \frac{u_2'(z)}{u_2^{i+1}(z)}\right).$$

Thus, we obtain the claimed expression for $W_i(z)$ in terms of the small branches. Finally, the remaining factors in (6.12) are obtained by simplifications in (6.9).

Thus, by (6.10), walks starting at the origin are given by complete homogeneous symmetric polynomials in the small branches. In particular, we have

$$\begin{aligned} G_{0,1}(z) &= u_1(z) + u_2(z), \\ G_{0,2}(z) &= u_1^2(z) + u_1(z)u_2(z) + u_2^2(z). \end{aligned} \tag{6.13}$$

We now derive an explicit expression for $G_{0,1}(z)$ and $G_{0,2}(z)$. Note that, as (6.13) is not defined on the negative real axis, we apply analytic continuation in order to derive an expression which is defined for every $|z| < \frac{1}{4}$, which is the radius of convergence of $G_{0,1}(z)$. The function $G_{0,1}(z)$ is an algebraic function since it is the sum of two algebraic functions (namely, $u_1(z)$ and $u_2(z)$). Using a computer algebra package, it is easy to derive an algebraic equation for $G_{0,1}(z)$. For example, the following *Maple* commands (see [46] for more on these aspects) give the desired equation:

```
> AllRoots:=allvalues(solve(1-z*P(u),u)):
> u1:=AllRoots[2]: u2:=AllRoots[3]:
> algeq:=algfuntoalgeq(u1+u2,u(z));
```

$$zu^4 + 2zu^3 + (3z-1)u^2 + (2z-1)u + z. \tag{6.14}$$

In particular, $G_{0,1}(z)$ is uniquely determined by the previous equation and the fact that its expansion at $z = 0$ is a power series with non-negative coefficients. Solving this equation, we arrive at an analytic expression for $G_{0,1}(z)$ for $|z| < 1/4$:

$$\begin{aligned} G_{0,1}(z) &= -\frac{1}{2} + \frac{1}{2}\sqrt{\frac{2-3z-2\sqrt{1-4z}}{z}} \\ &= z + z^2 + 3z^3 + 7z^4 + 22z^5 + 65z^6 + 213z^7 + \cdots. \end{aligned} \tag{6.15}$$

Using a computer algebra package again, we find that $G_{0,2}(z)$ satisfies

$$z^3u^4 - 3z^2u^3 - (z^2 - 3z)u^2 + (z-1)u + z = 0. \tag{6.16}$$

Among its four branches, only one is a power series at $z = 0$ with non-negative coefficients, namely

$$\begin{aligned} G_{0,2}(z) &= \frac{3 - \sqrt{1-4z} - \sqrt{2 + 12z + 2\sqrt{1-4z}}}{4z} \\ &= z + z^2 + 4z^3 + 9z^4 + 31z^5 + 91z^6 + 309z^7 + \cdots. \end{aligned} \tag{6.17}$$

In order to undertake a small digression on complexity of computation, these explicit forms are not the fastest way to access the coefficients. A better way is to take advantage of the theory of holonomic functions (as, e.g., implemented in the *gfun Maple* package, see [46]). To begin with, the kernel method gave us an algebraic equation. Applying the derivative to both sides of this equation and using the obtained new relations, we are led to a linear differential equation satisfied by the function $G(z)$ (where we write $G(z)$ instead of $G_{0,1}(z)$ for short):

```
>   diffeq:=algeqtodiffeq(subs(u=G,algeq),
>   G(z),G(0)=0):
```

$$\begin{cases} G(0) = 0, \\ 6z + 6 + 12\,(z+1)\,G(z) + 2\,\left(162\,z^3 + 66\,z^2 + z - 3\right)\frac{d}{dz}G(z) \\ +z\,(9\,z+4)\,(4\,z-1)\,(6\,z+1)\,\frac{d^2}{dz^2}G(z) = 0 \end{cases}$$

Then, extraction of $[z^n]$ on both sides of the differential equation yields a linear recurrence satisfied by the coefficients $g(n)$ of G, namely

```
>   rec:=diffeqtorec(diffeq,G(z),g(n)):
```

$$\begin{cases} g(0) = 0,\, g(1) = 1,\, g(2) = 1\,, \\ 0 = 108\,n\,(2\,n+1)\,g(n) + 6\,\left(13\,n^2 + 35\,n + 24\right)g(n+1) \\ -\left(17\,n^2 + 49\,n + 18\right)g(n+2) - 2\,(2\,n+7)\,(n+3)\,g(n+3)\,. \end{cases}$$

From this recurrence, a binary splitting approach introduced by the Chudnovskys gives a procedure which surprisingly computes $g(n)$ in only $O(\sqrt{n})$ operations (and $O(n \ln n \ln(\ln n))$ bit complexity):

```
>   g:=rectoproc(rec,g(n)):
>   g(10^5):  #a 6014-digits number computed
>   #           in only 2 seconds!
```

The same approach applies to all our directed lattice path models. This approach is much faster than the naive approach by means of dynamic programming (which would compute the bivariate generating function and would then extract the desired $G(z)$ from it: This would cost $O(n^2)$ in time and $O(n^3)$ in memory).

We just saw how to efficiently compute $g(n)$, for any given value of n, but is there a closed-form formula holding for all n at once? We now further investigate this question.

3.2 How to Get a Closed Form for Coefficients: Lagrange–Bürmann Inversion

In Sect. 4, we present a closed form for the numbers of lattice walks with step polynomial $P(u) = u^{-h} + u^{-h+1} + \cdots + u^{h-1} + u^h$, for any h. In the case $h = 2$ that we are dealing with in the current section, a nice miracle occurs: A more ad hoc approach allows one to derive simpler expressions.

3.2.1 Closed Form for Coefficients of $G_{0,1}(z)$

The generating function $G_{0,1}(z)$ of walks starting at the origin, ending at altitude 1, and never touching the x-axis, satisfies the algebraic equation (6.14). We rewrite it in the form

$$G_{0,1}(z) + G_{0,1}^2(z) = z(1 + G_{0,1}(z) + G_{0,1}^2(z))^2.$$

Here, substitution of $G_{0,1}(z) + G_{0,1}^2(z)$ by $C(z) - 1$ gives the striking equation

$$1 + G_{0,1}(z) + G_{0,1}^2(z) = C(z), \tag{6.18}$$

where $C(z) = 1 + zC(z)^2$ is the generating function for Catalan numbers. A recursive bijection for this identity was found by Axel Bacher and (independently) by Jérémie Bettinelli and Éric Fusy (personal communication, see also [14]). It remains a challenge to find a more direct simple bijection. This identity is the key to get nice closed-form expressions for the coefficients, via the following variant of Lagrange inversion.

Lemma 6.1 (Lagrange–Bürmann inversion variant) *Let $F(z)$ and $H(z)$ be two formal power series satisfying the equations*

$$F(z) = z\phi(F(z)), \qquad H(z) = z\psi(H(z)),$$

where $\phi(z)$ and $\psi(z)$ are formal power series such that $\phi(0) \neq 0$ and $\psi(0) \neq 0$. Then,

$$[z^n]H(F(z)) = \frac{1}{n}\sum_{k=1}^{n} \left([z^{k-1}]\psi^k(z)\right)\left([z^{n-k}]\phi^n(z)\right). \tag{6.19}$$

Proof By the Lagrange–Bürmann inversion (Theorem 6.2), we have

$$[z^n]H(F(z)) = \frac{1}{n}[z^{n-1}]H'(z)\phi^n(z).$$

Now we apply the Cauchy product formula $[z^m]A(z)B(z) = \sum_{k=0}^{m} a_k b_{m-k}$ with $m = n-1$, $A(z) = H'(z)$, and $B(z) = \phi^n(z)$. This leads to

$$[z^n]H(F(z)) = \frac{1}{n}\sum_{k=0}^{n-1}\left([z^k]H'(z)\right)\left([z^{n-1-k}]\phi^n(z)\right)$$
$$= \frac{1}{n}\sum_{k=1}^{n}\left([z^{k-1}]H'(z)\right)\left([z^{n-k}]\phi^n(z)\right).$$

This gives Formula (6.19), after observing $[z^{k-1}]H'(z) = k[z^k]H(z) = [z^{k-1}]\psi^k(z)$, where we used Lagrange–Bürmann inversion again.

Proposition 6.5 *The number of basketball walks of length n from the origin to altitude 1 with steps in $\mathcal{S} = \{-2, -1, +1, +2\}$ and never returning to the x-axis equals*

$$\frac{1}{n}\sum_{k=1}^{n}(-1)^{k+1}\binom{2k-2}{k-1}\binom{2n}{n-k} = \frac{1}{n}\sum_{i=0}^{n}\binom{n}{i}\binom{n}{2n+1-3i}.$$

Proof Equation (6.18) implies that $G_{0,1}(z) = H(C(z)-1)$, where $H(z)$ is the functional inverse of the polynomial $x^2 + x$. Thus $H(z) = z\psi(H(z))$, with $\psi(z) = \frac{1}{1+z}$. Furthermore, it is well known that $C_0(z) := C(z) - 1$ satisfies $C_0(z) = z\phi(C_0(z))$ with $\phi(z) = (1+z)^2$. Hence, Eq. (6.19) yields

$$[z^n]G_{0,1}(z) = \frac{1}{n}\sum_{k=1}^{n}\left([z^{k-1}]\frac{1}{(1+z)^k}\right)\left([z^{n-k}](1+z)^{2n}\right)$$
$$= \frac{1}{n}\sum_{k=1}^{n}(-1)^{k+1}\binom{2k-2}{k-1}\binom{2n}{n-k}.$$

The alternative expression without the $(-1)^{k+1}$ factors comes from Formula (6.13), to which we apply the Lagrange–Bürmann inversion formula for u_1, remembering that u_1 satisfies $u^2 = zu^2P(u)$, and that the conjugation property of the small roots from Proposition 6.2 holds:

$$[z^n]G_{0,1}(z) = [z^n](u_1(z) + u_2(z)) = 2[z^n]u_1(z) = \frac{1}{n}\sum_{k=0}^{n}\binom{n}{k}\binom{n}{2n+1-3k}.$$

The last closed-form expression can also be explained via the so-called cycle lemma (cf. [49, Ex. 5.3.8]). Namely, by (6.2) combined with the factorization $u^{-2} + u^{-1} + u + u^2 = u^{-2}(1+u^3)(1+u)$, the number of unrestricted walks from 0 to 1 in n steps is given by

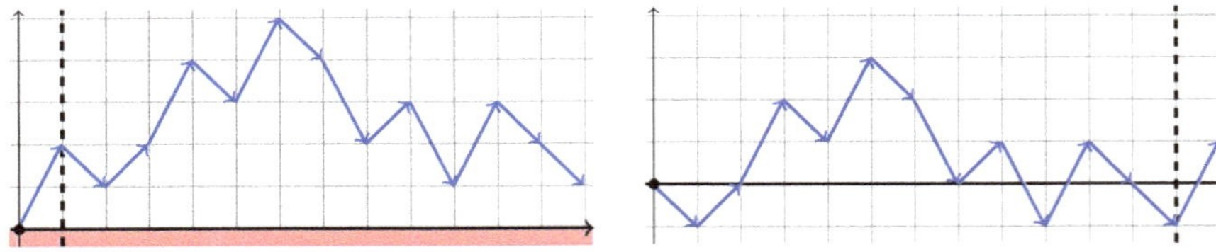

Fig. 6 Transforming a walk counted by $G_{0,1}(z)$ into a walk counted by $W_{0,1}(z)$

$$[u^1z^n]W(z,u) = [u^1]P(u)^n = [u^1]\left(\frac{(1+u^3)(1+u)}{u^2}\right)^n = \sum_{i=0}^{n}\binom{n}{i}\binom{n}{2n+1-3i}.$$

From the formulas, we see that $[z^n]G_{0,1}(z) = \frac{1}{n}[z^n]W_{0,1}(z)$. There exists indeed a 1-to-n correspondence between walks counted by $G_{0,1}(z)$ and those counted by $W_{0,1}(z)$. For each walk ω counted by $G_{0,1}(z)$, decompose ω into $\omega = \omega_\ell B\omega_r$ where B is any point in the walk. A new walk ω' counted by $W_{0,1}(z)$ is constructed by putting B at the origin and adjoining ω_ℓ at the end of ω_r, i.e., $\omega' = B\omega_r\omega_\ell$, see Fig. 6. If ω is of length n, then there are n choices for B. All these walks are different because there are no walks from altitude 0 to altitude 1 which are the concatenation of several copies of one and the same walk. (This is not true for walks from altitude 0 to altitude 2. For example, the walk (0, 2, 1, 3, 2) is the concatenation of two copies of the walk (0, 2, 1).)

Conversely, given a walk τ of length n counted by $W_{0,1}(z)$, we decompose τ into $\tau = \tau_\ell B\tau_r$, where B is the right-most minimum of τ. Then, $\tau' = B\tau_r\tau_\ell$ is a walk of length n counted by $G_{0,1}(z)$.

3.2.2 Closed Form for the Coefficients of $G_{0,2}(z)$

Recall that, by means of the kernel method, we derived a closed-form expression for the generating function $G_{0,2}(z)$ in (6.17).

Proposition 6.6 *The number of basketball walks of length n from the origin to altitude 2 with steps in* $\mathscr{S} = \{-2, -1, +1, +2\}$ *and never returning to the x-axis equals*

$$\frac{1}{2n+1}\sum_{k=0}^{n+1}(-1)^{n+k+1}\binom{2n+1}{n+k}\binom{n+2k-1}{k}.$$

Proof We define the series $F(z)$ by

$$-\frac{1}{F(z)} = G_{0,2}(z) - \frac{1}{z}. \tag{6.20}$$

It is straightforward to see from this equation that $F(z) = z + z^3 + \cdots$. The equation (6.16) translates into the equation

$$(F^3(z) - zF(z))(1 + F(z)) + z^2 = 0$$

for $F(z)$. We may rewrite this equation in the form

$$\left(F^2 - \frac{z}{2}\right)^2 = \frac{z^2}{4} \cdot \frac{1 - 3F(z)}{1 + F(z)}.$$

Next, we take the square root on both sides. In order to decide the sign, we have to observe that $F^2(z) = z^2 + \cdots$, hence

$$F^2(z) - \frac{z}{2} = -\frac{z}{2}\sqrt{\frac{1 - 3F(z)}{1 + F(z)}},$$

or, equivalently, $F(z)$ satisfies $F^2(z) = zB(F(z))$, where

$$B(z) = \frac{1}{2}\left(1 - \sqrt{\frac{1 - 3z}{1 + z}}\right).$$

It is straightforward to verify that $B(z)$ satisfies the equation $B(z) = zA(B(z))$ with $A(z) = \frac{1}{1-z} - z$, and it is the only power series solution of this equation. Hence, for $n \geq 1$, by (6.20), Lagrange–Bürmann inversion (Theorem 6.2) with $H(z) = z^{-1}$, we have

$$[z^n]G_{0,2}(z) = -[z^n]\frac{1}{F(z)} = \frac{1}{n}[z^{n-1}]z^{-2}\left(\frac{B(z)}{z}\right)^n = \frac{1}{n}[z^{2n+1}]B^n(z).$$

Now, we apply Lagrange–Bürmann inversion again, this time with $F(z)$ replaced by $B(z)$, n replaced by $2n + 1$, and $H(z) = z^n$. This yields

$$\begin{aligned} [z^n]G_{0,2}(z) &= \frac{1}{n(2n+1)}[z^{2n}]nz^{n-1}A^{2n+1}(z) \\ &= \frac{1}{2n+1}[z^{n+1}]\left(\frac{1}{1-z} - z\right)^{2n+1}. \end{aligned}$$

By applying the binomial theorem, we then obtain

$$[z^n]G_{0,2}(z) = \frac{1}{2n+1}[z^{n+1}]\sum_{k=0}^{2n+1}(-1)^{k+1}\binom{2n+1}{k}z^{2n+1-k}\left(\frac{1}{1-z}\right)^k.$$

Since

$$\left(\frac{1}{1-z}\right)^k = \sum_{\ell\geq 0} \binom{k+\ell-1}{\ell} z^\ell,$$

we get

$$\begin{aligned}
[z^n]G_{0,2}(z) &= \frac{1}{2n+1}[z^{n+1}]\sum_{\ell\geq 0}\sum_{k=0}^{2n+1}(-1)^{k+1}\binom{2n+1}{k}\binom{k+\ell-1}{\ell}z^{2n+1-k+\ell}\\
&= \frac{1}{2n+1}\sum_{k=n}^{2n+1}(-1)^{k+1}\binom{2n+1}{k}\binom{2k-n-1}{k-n}\\
&= \frac{1}{2n+1}\sum_{k=0}^{n+1}(-1)^{n+k+1}\binom{2n+1}{n+k}\binom{n+2k-1}{k},
\end{aligned}$$

as desired.

The idea of the above proof was to "build up" a chain of dependencies between the actual series of interest, $G_{0,2}(z)$, and several auxiliary series, namely the series $F(z)$, $B(z)$, and $A(z)$, so that repeated application of Lagrange–Bürmann inversion could be applied to provide an explicit expression for the coefficients of the series of interest. This raises the question whether this example is just a coincidence, or whether there exists a general method to transform a power series into a Laurent series with the same positive part, and a "nice" algebraic expression, allowing multiple Lagrange–Bürmann inversions to get "nice" closed forms for the coefficients. We have no answer to this question and therefore leave this to future research.

3.2.3 Closed Form for the Coefficients of Basketball Excursions

Here, we enumerate basketball *excursions*, that is, basketball walks which start at the origin, return to altitude 0, and in-between do not pass below the x-axis. A main difference to the previously considered *positive* basketball walks is that the excursions are allowed to touch the x-axis anywhere.

Proposition 6.7 (Enumeration of basketball excursions) *The number of basketball walks with steps in $\mathcal{S} = \{-2, -1, +1, +2\}$ of length n from the origin to altitude* 0 *never passing below the x-axis is*

$$e_n := \frac{1}{n+1}\sum_{k=0}^{n}(-1)^{n+k}\binom{2n+2}{n-k}\binom{n+2k+1}{k} = \frac{1}{n+1}\sum_{i=0}^{\lfloor n/2\rfloor}\binom{2n+2}{i}\binom{n-i-1}{n-2i}. \tag{6.21}$$

Remark 6.1 The first few values of the sequence defined by (6.21) are

$$1, 0, 2, 2, 11, 24, 93, 272, 971, 3194, 11293, 39148, 139687, 497756, \ldots$$

Proof (*of Proposition* 6.7) By the kernel method, we know that the generating function for excursions, $E(z)$ say, is given by $E(z) = -\frac{u_1 u_2}{z}$, and that it satisfies the algebraic equation

$$z^4 E^4 - (2z^3 + z^2)E^3 + (3z^2 + 2z)E^2 - (2z + 1)E + 1 = 0\,.$$

Among the branches of this algebraic equation, only one has a power series expansion. The equation may be rewritten in the form

$$zE(z) = z\left(\frac{1}{(1 - zE(z))^2} - \frac{2zE(z)}{1 - zE(z)} + z^2E^2(z)\right) = z\left(\frac{1}{1 - zE(z)} - zE(z)\right)^2.$$

This shows that we may apply Lagrange–Bürmann inversion (Theorem 6.2) with $\phi(z) = (\frac{1}{1-z} - z)^2$. So we have

$$\begin{aligned}[z^n]E(z) &= \frac{1}{n+1}[z^n]\phi^{n+1}(z) = \frac{1}{n+1}[z^n]\left(\frac{1}{1-z} - z\right)^{2n+2} \\ &= \frac{1}{n+1}\sum_{k=0}^{n}(-1)^{n+k}\binom{2n+2}{n-k}\binom{n+2k+1}{k}.\end{aligned}$$

It is possible to get an expression involving only positive summands by making use of the rewriting $\phi(z) = (1 + \frac{z^2}{1-z})^2$. This leads to (6.21).

The trick used in this proof can in fact be translated into an algorithm of wider use:

> `The "Lagrangean scheme" algorithm`
>
> `input:` an algebraic power series (given in terms of its algebraic equation $P(z, F) = 0$, plus the first terms of the expansion of F, so that we can uniquely identify the correct branch of the equation)
>
> `output:` a "Lagrangean equation" satisfied by F (i.e., $H(z^a F) = z\phi(z^a F)$, where $z^a F$ has valuation[a] 1.)
>
> `way to process:` if we assume that $H = H_1/H_2$ and $\phi = \phi_1/\phi_2$ are rational functions, then we identify them via an indeterminate coefficient approach, by substituting the *polynomials* H_1, H_2, ϕ_1, ϕ_2 in the equation $P(z, F) = 0$.
>
> [a]The valuation of a power series $\sum_{n\geq 0} f_n z^n$ is the least n such that $f_n \neq 0$.

This algorithm therefore provides a way to get multiple-binomial-sum representations. See [17, 25, 50] for other approaches not relying on the algebraic nature of F, but designed for the class of functions which can be written as diagonals of rational functions (these two classes coincide in the bivariate case). For example, Formula (6.21) for e_n has the following alternative representation:

$$(n+1)e_n = [t^n]\,\text{diagonal}\left(\frac{(1+u)^6ut^2}{1-(u(u+1)^2t+u(1+u)^4t^2)} + (u+1)^2\right).$$

The rational function on the right-hand side has the striking feature that its bivariate series expansion has only non-negative coefficients. In fact, it is even a bivariate $\mathbb{N}$-rational function (i.e., a function obtained as iteration of addition, multiplication, and quasi-inverse,[5] starting from polynomials in u and t with positive integer coefficients). Given a multivariate rational function, it is a hard task to write it as an $\mathbb{N}$-rational expression (an algorithm is known in the univariate case), so some human computations were needed here to get the above expression.

In fact (and we believe that it was not observed before), these multivariate rational functions appearing in the computation of diagonals related to nested sums of binomials are always $\mathbb{N}$-rational: This follows from the closure properties of $\mathbb{N}$-rational functions. It is an open question to give a combinatorial interpretation (in terms of the initial structure counted by the diagonal) of the other diagonals of this rational function. It is also not easy to extrapolate from this rational function a general pattern which could appear for more general sets of steps: We shall see in Sect. 4 which type of formulas generalizes the rich combinatorics that we had for $P(u) = u^{-2} + u^{-1} + u + u^2$.

3.3 How to Derive the Corresponding Asymptotics: Singularity Analysis

We close this section by briefly addressing how to find the asymptotics of numbers of basketball walks. Indeed, standard techniques from singularity analysis suffice to get the asymptotic growth of the coefficients of z^n in the generating functions that we consider here for $n \to \infty$. The interested reader is referred to [28] for more details on this subject (see Fig. VI.7 therein for an illustration of singularity analysis).

Theorem 6.3 *Let $G_{0,1}(z)$ and $G_{0,2}(z)$ be the generating functions for positive basketball walks with steps $-2, -1, +1, +2$ starting at the origin and ending at altitude* 1*, respectively, at* 2*. Then, as $n \to \infty$, the coefficients are asymptotically equal to*

[5]The *quasi-inverse* of a power series $f(z)$ of positive valuation is $1/(1 - f(z))$.

$$[z^n]G_{0,1}(z) = \frac{1}{\sqrt{5\pi}}\frac{4^n}{n^{3/2}}\left(1 - \frac{81}{200}\frac{1}{n} + O\left(\frac{1}{n^2}\right)\right),$$

$$[z^n]G_{0,2}(z) = \frac{5+\sqrt{5}}{10\sqrt{\pi}}\frac{4^n}{n^{3/2}}\left(1 - \frac{201+24\sqrt{5}}{200}\frac{1}{n} + O\left(\frac{1}{n^2}\right)\right).$$

Proof The asymptotic growth of the coefficients is governed by the location of the dominant singularity (the singularity closest to the origin). The dominant singularity of (6.15) and (6.17) is given by $1/4$, since the square root becomes singular at this point.

Next, we compute the singular expansion for $z \to 1/4$, which is a Puiseux series:

$$G_{0,1}(z) = -\frac{1-\sqrt{5}}{2} - \frac{2}{\sqrt{5}}\sqrt{1-4z} + O(1-4z),$$

$$G_{0,2}(z) = \left(3-\sqrt{5}\right) - \frac{5+\sqrt{5}}{5}\sqrt{1-4z} + O(1-4z).$$

Finally, we apply the standard function scale from [28, Theorem VI.1] and the transfer for the error term [28, Theorem VI.3] to get the asymptotics.

More generally, asymptotics for the number of walks from altitude i to altitude j in n steps can be obtained via singularity analysis of the small roots, similar to what was done in [8]. Note that it is easy to derive as many terms as needed in the asymptotic expansion of the coefficients by including more terms in the Puiseux expansion. We also want to point out that this process was implemented in *SageMath* (see [31]) or in *Maple* by Bruno Salvy (as a part of the `algolib` package). There, the `equivalent` command directly gives the above result:

```
>  equivalent(G01,z,n,3);
```

$$\frac{1}{5}\frac{\sqrt{5}\,4^n}{\sqrt{\pi}\,n^{3/2}} - \frac{81}{1000}\frac{\sqrt{5}\,4^n}{\sqrt{\pi}\,n^{5/2}} + O\left(\frac{4^n}{n^{7/2}}\right).$$

4 General Case: Lattice Walks with Arbitrary Steps

We first prove a theorem which holds for any symmetric set of steps, i.e., when the step polynomial satisfies $P(u) = P(1/u)$.

Theorem 6.4 (Positive walk enumeration) *Consider walks with a symmetric step polynomial $P(u)$. Let $G_{0,k}(z)$ be the generating function for positive walks, i.e., walks starting at the origin, ending at altitude k, and always staying strictly above the x-axis in-between, and let $M_{>0}(z)$ be the generating function of positive meanders, i.e., positive walks ending at any altitude > 0. Then*

$$M_{>0}(z) = \sum_{k>0} G_{0,k}(z) = \prod_{i=1}^{h} \frac{1}{1-u_i(z)},$$
$$G_{0,k}(z) = h_k\left(u_1(z), u_2(z), \ldots, u_h(z)\right),$$

where $u_1(z), u_2(z), \ldots, u_h(z)$ *are the small roots of the kernel equation* $1 - zP(u) = 0$, *and*

$$h_k(x_1, x_2, \ldots, x_h) = \sum_{\substack{i_1,\ldots,i_h \geq 0 \\ i_1+\cdots+i_h=k}} x_1^{i_1} x_2^{i_2} \cdots x_h^{i_h}$$

is the complete homogeneous symmetric polynomial of degree k *in the variables* $x_1, x_2, \ldots, x_h$.

Proof The formula for positive meanders follows from the expression for meanders (which are allowed to touch the x-axis!) in [8, Corollary 1],

$$M_{\geq 0}(z) = -\frac{1}{z} \prod_{i=1}^{h} \frac{1}{1-v_i(z)},$$

where $v_1(z), v_2(z), \ldots, v_h(z)$ are the large roots of $1 - zP(u) = 0$, i.e., those roots $v(z)$ for which $\lim_{z\to 0} |v(z)| = \infty$. Every meander starts with an initial excursion, and later never returns to the x-axis any more. This simple fact implies the generating function equation $M_{\geq 0}(z) = E(z)M_{>0}(z)$. Hence, we need to divide the above expression for $M_{\geq 0}(z)$ by the generating function for excursions — which, by [8, Theorem 2], is given by

$$E(z) = \frac{(-1)^{h-1}}{z} \prod_{i=1}^{h} u_i(z).$$

Finally, due to $P(u) = P(u^{-1})$, we have $u_i(z) = 1/v_i(z)$, which gives the final expression for $M_{>0}$, while the formula for $G_{0,k}(z)$ is proven in [10].

This proof shows, in particular, that generating functions for strictly positive walks, respectively, for weakly positive walks, are intimately related and are therefore given by similar expressions. (The price of positivity is a division by $E(z)$, which encodes the excursion prefactor.) The proof also extends to non-symmetric steps, but then the formulas involve one more factor. It is possible to deal with them exactly in the way we proceed for symmetric steps, but this leads to slightly less nice formulas.

In the sequel, we focus on positive walks with symmetric steps. We show in which way we can use the obtained expressions for the generating functions in order to get nice closed-form expressions for their coefficients.

4.1 Counting Walks with Steps in $\mathscr{S} = \{0, \pm 1, \ldots, \pm h\}$

In Sect. 3 on basketball walks, we had a taste of what the kernel method could do for us when combined with Lagrange–Bürmann inversion. This was, however, only for the case $\mathscr{S} = \{\pm 1, \pm 2\}$. In this section, we illustrate again the power of the kernel method, when applied to more general step sets $\mathscr{S}$. We first start with a generalization of Sect. 2 to $\mathscr{S} = \{0, \pm 1, \ldots, \pm h\}$. In order to have a convenient notation, we introduce *m-nomial coefficients* by defining

$$\binom{n}{k}_m := [u^k](1 + u + \cdots + u^{m-1})^n,$$

where k is between 0 and $(m-1)n$.

Proposition 6.8 *The m-nomial coefficient equals*

$$\binom{n}{k}_m = \sum_{i=1}^{n} (-1)^i \binom{n}{i} \binom{n+k-mi-1}{n-1}. \tag{6.22}$$

Proof Coefficient extraction in the defining expression for $\binom{n}{k}_m$ yields

$$\begin{aligned}\binom{n}{k}_m &= [u^k](1 + u + \cdots + u^{m-1})^n = [u^k](1-u^m)^n \frac{1}{(1-u)^n}\\ &= [u^k]\left(\sum_{i=0}^{n} \binom{n}{i}(-1)^i u^{mi}\right)\left(\sum_{j\geq 0} \binom{n+j-1}{n-1} u^j\right)\\ &= \sum_{i=0}^{\lfloor (n+k-1)/m \rfloor} (-1)^i \binom{n}{i}\binom{n+k-mi-1}{n-1}.\end{aligned}$$

The upper bound in the sum can be taken more naturally to be $i = n$, using the convention that binomials $\binom{n}{k}$ are 0 for $n < 0$ or $k > n$ (the reader should be warned that this is not the convention of *Maple* or *Mathematica*). This gives Formula (6.22).

Historical remark. These m-nomial coefficients appear in more than fifty articles (many of them focusing on trinomial coefficients) dealing with their rich combinatorial aspects (see, e.g., [2, 4, 12, 15]). We use the notation $\binom{n}{k}_m$ promoted by George Andrews [3]. It should be noted that they were previously called *polynomial coefficients* by Louis Comtet [23, p. 78], who is mentioning early work of Désiré André (with a typo in the date) and Paul Montel [1, 43], and who was himself using another notation for these numbers, namely $\binom{n,m}{k}$.

These coefficients have a direct combinatorial interpretation in terms of lattice walk enumeration.

Theorem 6.5 (Unconstrained walk enumeration) *The number of unconstrained*[6] *walks running from the origin to altitude k in n steps taken from* $\{0, \pm 1, \pm 2, \dots, \pm h\}$ *equals* $\binom{n}{k+hn}_{2h+1}$.

Proof By (6.2), the generating function for unconstrained walks is

$$W(z, u) = \frac{1}{1 - zP(u)} = \sum_{n=0}^{\infty} P^n(u) z^n .$$

Then, a simple factorization shows that

$$[u^k] P^n(u) = [u^k] \left(\sum_{i=-h}^{h} u^i \right)^n = [u^k] u^{-hn} \left(\sum_{i=0}^{2h} u^i \right)^n = \binom{n}{k+hn}_{2h+1} .$$

Now, we will see how to link these coefficients with *constrained* lattice walks. To this end, we first state the general version of the conjugation principle that we encountered in Proposition 6.2.

Proposition 6.9 (Conjugation principle for small roots) *Let*

$$P(u) = \sum_{i=-c}^{d} p_i u^i$$

be the step polynomial, and let $\omega = e^{2\pi i/c}$ *be a cth root of unity. The small roots* $u_i(z)$, $i = 1, 2, \dots, c$, *of* $1 - zP(u) = 0$ *satisfy*

$$u_i(z) = \sum_{n \ge 1} \omega^{n(i-1)} a_n z^{n/c}$$

for certain "universal" coefficients a_n, $n = 1, 2, \dots$.

Proof The kernel equation yields

$$u = X \left(p_{-c} + p_{-c+1} u + p_{-c+2} u^2 + \cdots + p_{d-1} u^{c+d-1} + p_d u^{c+d} \right)^{1/c} ,$$

with $X = \omega^j z^{1/c}$ for $j = 0, 1, \dots, c-1$. Since the above equation possesses a unique formal power series solution $u(X)$, the claim follows.

Next, we apply Lagrange–Bürmann inversion to the small roots given by the kernel method and combine it with the conjugation principle.

[6] Unconstrained means that the walks are allowed to have both positive and negative altitudes.

Proposition 6.10 (Explicit expansion of the roots u_i) *For lattice walks with step polynomial given by* $P(u) = u^{-h} + u^{-h+1} + \cdots + u^{h-1} + u^h$, *let* $U(z)$ *be the root of* $1 - z^h P(U) = 0$ *whose Taylor expansion at* 0 *starts* $U(z) = z + \cdots$. *The series* $U(z)$ *is a power series, not a genuine Puiseux series. Then all small and large roots can be expressed in terms of* $U(z)$, *namely we have*

$$u_i(z) = U(\omega^{i-1}z^{1/h}) \quad \text{and} \quad v_i(z) = 1/U(\omega^{i-1}z^{1/h}), \quad i = 1, 2, \ldots, h,$$

where $\omega = e^{2\pi i/h}$ *is a primitive hth root of unity. The expansion of a power of the series* $U(z)$ *is explicitly given by*

$$U^m(z) = \sum_{n=m}^{\infty} \frac{m}{n} \binom{n/h}{n-m}_{2h+1} z^n.$$

Proof We want to solve $1 - zP(u) = 0$ for u. We may rewrite this equation as

$$z = \frac{u^h}{1 + u + \cdots + u^{2h}}.$$

Taking the hth root, we get

$$\omega^{i-1}z^{1/h} = \frac{u}{(1 + u + \cdots + u^{2h})^{1/h}},$$

for some i with $1 \leq i \leq h$.

Since an equation of the form $Z = u\phi(u)$, where $\phi(u)$ is a power series in u, has a unique power series solution $u(Z)$, the above equation has a unique solution $u_i(z)$, which turns out to have exactly the form described in the proposition. The equation for v_i follows from $u_i = 1/v_i$ as we have $P(u) = P(1/u)$.

The equation for U^m comes from Lagrange–Bürmann inversion:

$$\begin{aligned}
[z^n]U^m(z) &= \frac{1}{n}[z^{-1}](z^m)' P^{n/h}(z) \\
&= \frac{m}{n}[z^{-m}] \sum_k z^k \binom{n/h}{k+n}_{2h+1} \\
&= \frac{m}{n} \binom{n/h}{n-m}_{2h+1}.
\end{aligned}$$

Theorem 6.6 (Closed-form expression for walks with $\mathcal{S} = \{0, \pm 1, \ldots, \pm h\}$) *The numbers of positive walks and meanders from the origin to altitude k in n steps from* $\mathcal{S} = \{0, \pm 1, \ldots, \pm h\}$ *admit the closed-form expressions*

$$[z^n]G_{0,k}(z) = \sum_{n_1+\cdots+n_h=nh} \sum_{i_1+\cdots+i_h=k} \frac{i_1}{n_1}\binom{n_1/h}{n_1-i_1}_{2h+1} \cdots \frac{i_h}{n_h}\binom{n_h/h}{n_h-i_h}_{2h+1} \omega^{\sum_{j=1}^{h}(j-1)n_j},$$

$$[z^n]M_{>0}(z) = \sum_{n_1+\cdots+n_h=nh} \sum_{i_1,\ldots,i_h\geq 0} \frac{i_1}{n_1}\binom{n_1/h}{n_1-i_1}_{2h+1} \cdots \frac{i_h}{n_h}\binom{n_h/h}{n_h-i_h}_{2h+1} \omega^{\sum_{j=1}^{h}(j-1)n_j}.$$

Proof We use the expansions from Proposition 6.10 in the generating function formulas from Theorem 6.4.

Here are some sequences of numbers of positive walks with steps $\mathscr{S} = \{0, \pm 1, \ldots, \pm h\}$, starting at the origin, and ending at altitude 1, for different values of h:

$h = 1$ (A168049) : 0, 1, 1, 2, 4, 9, 21, 51, 127, 323, 835, ...
$h = 2$ (A104632) : 0, 1, 2, 6, 20, 73, 281, 1125, 4635, 19525, 83710, ...
$h = 3$ (A276902) : 0, 1, 3, 12, 56, 284, 1526, 8530, 49106, 289149, 1733347, ...
$h = 4$ (A277920) : 0, 1, 4, 20, 120, 780, 5382, 38638, 285762, 2162033, ...

Furthermore, here[7] are some sequences of numbers of positive walks with steps $\mathscr{S} = \{0, \pm 1, \ldots, \pm h\}$, starting at the origin, and ending at altitude 2, for small values of h:

$h = 1$ (A105695) : 0, 0, 1, 2, 5, 12, 30, 76, 196, 512, 1353, ...
$h = 2$ (A276903) : 0, 1, 2, 7, 25, 96, 382, 1567, 6575, 28096, 121847, 534953, ...
$h = 3$ (A276904) : 0, 1, 3, 14, 68, 358, 1966, 11172, 65104, 387029, 2337919, ...
$h = 4$ (A277921) : 0, 1, 4, 23, 142, 950, 6662, 48420, 361378, 2753687, ...

Here are the corresponding sequences for positive meanders:

$h = 1$ (A005773) : 1, 1, 2, 5, 13, 35, 96, 267, 750, 2123, 6046, 17303, ...
$h = 2$ (A278391) : 1, 2, 7, 29, 126, 565, 2583, 11971, 56038, 264345, ...
$h = 3$ (A278392) : 1, 3, 15, 87, 530, 3329, 21316, 138345, 906853, ...
$h = 4$ (A278393) : 1, 4, 26, 194, 1521, 12289, 101205, 844711, 7120398, ...

Here are the corresponding sequences for meanders (allowed to touch 0):

$h = 1$ (A005773) : 1, 2, 5, 13, 35, 96, 267, 750, 2123, 6046, 17303, 49721, ...
$h = 2$ (A180898) : 1, 3, 12, 51, 226, 1025, 4724, 22022, 103550, 490191, ...
$h = 3$ (A180899) : 1, 4, 22, 130, 803, 5085, 32747, 213419, 1403399, ...
$h = 4$ (A180900) : 1, 5, 35, 265, 2100, 17075, 141246, 1182719, 9994086, ...

[7] Axxxxxx refers to the corresponding sequence in the On-Line Encyclopedia of Integer Sequences, available electronically at https://oeis.org.

Here are the corresponding sequences for excursions:

$h = 1$ (A001006) : 1, 1, 2, 4, 9, 21, 51, 127, 323, 835, 2188, 5798, 15511, ...
$h = 2$ (A104184) : 1, 1, 3, 9, 32, 120, 473, 1925, 8034, 34188, 147787, ...
$h = 3$ (A204208) : 1, 1, 4, 16, 78, 404, 2208, 12492, 72589, 430569, 2596471, ...
$h = 4$ (A204209) : 1, 1, 5, 25, 155, 1025, 7167, 51945, 387000, 2944860, ...

Remark 6.2 Most of the above sequences for $h \geq 3$ were not contained in the On-Line Encyclopedia of Integer Sequences (OEIS) before we added them. In Sect. 5, we discuss the combinatorial structures related to the sequences which were already in the OEIS.

4.2 Counting Walks with Steps in $\mathcal{S} = \{\pm 1, \ldots, \pm h\}$

Here, we consider the same steps as in the previous one, except that we drop the 0-step.

Certainly, for any type of walks consisting of k steps with 0-step included, enumerated by f_k say, the number of walks of the same type consisting of n steps, all of which different from the 0-step, can be obtained by the inclusion–exclusion principle. The result is $\sum_{k=0}^{n}(-1)^{n-k}\binom{n}{k} f_k$.

Here, our way to derive the corresponding formulas is more ad hoc and relies on the shape of the considered steps in $\mathcal{S}$. This offers the advantage of leading to positive sum formulas, as opposed to the alternating sums produced by inclusion–exclusion. For convenience, we introduce the mock-m-nomial coefficients by

$$\binom{n}{k}_{2m}^{*} := [u^k](1 + \cdots + u^{m-1} + u^{m+1} + \cdots + u^{2m})^n .$$

Proposition 6.11 *The mock-m-nomial coefficients can be expressed in terms of the (ordinary) m-nomial coefficients in the form*[8]

$$\binom{n}{k}_{2m}^{*} = \sum_{i=0}^{n} \binom{n}{i} \binom{n}{k-(m+1)i}_{m} .$$

Proof Factoring the expression and extracting coefficients, we obtain

$$\begin{aligned}
\binom{n}{k}_{2m}^{*} &= [u^k](1 + \cdots + u^{m-1} + u^{m+1} + \cdots + u^{2m})^n \\
&= [u^k](1 + u^{m+1})^n (1 + u + \cdots + u^{m-1})^n \\
&= [u^k] \left(\sum_{i \geq 0} \binom{n}{i} u^{(m+1)i} \right) \left(\sum_{j \geq 0} \binom{n}{j}_{m} u^j \right) \\
&= \sum_{i=0}^{n} \binom{n}{i} \binom{n}{k-(m+1)i}_{m} .
\end{aligned}$$

[8]Here, the * is a mnemonic to remind us that we do not have the 0-step.

These mock-m-nomial coefficients have also a direct combinatorial interpretation in terms of lattice walk enumeration.

Theorem 6.7 (Unconstrained walk enumeration) *The mock-m-nomial coefficient* $\binom{n}{k+hn}^*_{2h}$ *is the number of unconstrained walks running from* 0 *to k in n steps taken from* $\{\pm 1, \pm 2, \ldots, \pm h\}$.

Proof We have

$$[u^k]P^n(u) = [u^k]\left(\sum_{i=-h}^{-1} u^i + \sum_{i=1}^{h} u^i\right)^n$$
$$= [u^k]u^{-hn}\left(\sum_{i=0}^{h-1} u^i + \sum_{i=h+1}^{2h} u^i\right)^n = \binom{n}{k+hn}^*_{2h}.$$

Proposition 6.12 (Explicit expansion of the roots u_i) *For lattice walks with step polynomial given by* $P(u) = u^{-h} + \cdots + u^{-1} + u^1 + \cdots + u^h$, *let* $U(z)$ *be the root of* $1 - z^h P(U) = 0$ *whose Taylor expansion at* 0 *starts* $U(z) = z + \cdots$. *Again,* $U(z)$ *is a power series, not a genuine Puiseux series. Then* $U(z)$ *satisfies*

$$U^m(z) = \sum_{n=1}^{\infty} \frac{m}{n}\binom{n/h}{n-m}^*_{2h} z^n,$$

and all small and large roots are expressed in terms of $U(z)$ *as*

$$u_i(z) = U(\omega^{i-1}z^{1/h}) \quad \text{and} \quad v_i(z) = 1/U(\omega^{i-1}z^{1/h}), \quad \text{for } i = 1, 2, \ldots, h,$$

where $\omega = e^{2\pi i/h}$ *is a primitive hth root of unity.*

Proof We apply Lagrange–Bürmann inversion to get

$$[z^n]U^m(z) = \frac{1}{n}[z^{-1}](z^m)'P^{n/h}(z) = \frac{m}{n}[z^{-m}]\sum_k u^k\binom{n/h}{k+n}^*_{2h} = \frac{m}{n}\binom{n/h}{n-m}^*_{2h}.$$

Theorem 6.8 (Closed-form expression for walks with $\mathcal{S} = \{\pm 1, \ldots, \pm h\}$) *The numbers of positive walks and meanders from the origin to altitude k in n steps from* $\mathcal{S} = \{\pm 1, \ldots, \pm h\}$ *admit the closed-form expressions*

$$[z^n]G_{0,k}(z) = \sum_{n_1+\cdots+n_h=nh}\ \sum_{i_1+\cdots+i_h=k} \frac{i_1}{n_1}\binom{n_1/h}{n_1-i_1}^*_{2h} \cdots \frac{i_h}{n_h}\binom{n_h/h}{n_h-i_h}^*_{2h} \omega^{\sum_{j=1}^h (j-1)n_j},$$
$$[z^n]M_{>0}(z) = \sum_{n_1+\cdots+n_h=nh}\ \sum_{i_1,\ldots,i_h\ge 0} \frac{i_1}{n_1}\binom{n_1/h}{n_1-i_1}^*_{2h} \cdots \frac{i_h}{n_h}\binom{n_h/h}{n_h-i_h}^*_{2h} \omega^{\sum_{j=1}^h (j-1)n_j}.$$

Proof We use the expansions from Proposition 6.12 in the generating function formulas from Theorem 6.4.

Here are some sequences of numbers of walks with steps in $\mathcal{S} = \{\pm 1, \pm 2, \ldots, \pm h\}$, starting at the origin, and ending at altitude 1, for different values of h:

$$\begin{aligned}
h &= 1 \quad (\texttt{A000108}): \; 0, 1, 0, 1, 0, 2, 0, 5, 0, 14, 0, \ldots \\
h &= 2 \quad (\texttt{A166135}): \; 0, 1, 1, 3, 7, 22, 65, 213, 693, 2352, 8034, \ldots \\
h &= 3 \quad (\texttt{A276852}): \; 0, 1, 2, 7, 28, 121, 560, 2677, 13230, 66742, 343092, \ldots \\
h &= 4 \quad (\texttt{A277922}): \; 0, 1, 3, 13, 71, 405, 2501, 15923, 104825, 704818, \ldots
\end{aligned}$$

Furthermore, here are some sequences of numbers of walks with steps in $\mathcal{S} = \{\pm 1, \pm 2, \ldots, \pm h\}$, starting at the origin, and ending at altitude 2, for different values of h:

$$\begin{aligned}
h &= 1 \quad (\texttt{A000108}): \; 0, 0, 1, 0, 2, 0, 5, 0, 14, 0, 42, \ldots \\
h &= 2 \quad (\texttt{A111160}): \; 0, 1, 1, 4, 9, 31, 91, 309, 1009, 3481, 11956, \ldots \\
h &= 3 \quad (\texttt{A276901}): \; 0, 1, 2, 9, 34, 159, 730, 3579, 17762, 90538, 467796, \ldots \\
h &= 4 \quad (\texttt{A277923}): \; 0, 1, 3, 16, 84, 505, 3121, 20180, 133604, 904512, \ldots
\end{aligned}$$

Here are the corresponding sequences for positive meanders:

$$\begin{aligned}
h &= 1 \quad (\texttt{A001405}): \; 1, 1, 1, 2, 3, 6, 10, 20, 35, 70, 126, 252, 462, 924, \ldots \\
h &= 2 \quad (\texttt{A278394}): \; 1, 2, 5, 17, 58, 209, 761, 2823, 10557, 39833, 151147 \ldots \\
h &= 3 \quad (\texttt{A278395}): \; 1, 3, 12, 60, 311, 1674, 9173, 51002, 286384, 1620776, \ldots \\
h &= 4 \quad (\texttt{A278396}): \; 1, 4, 22, 146, 1013, 7269, 53156, 394154, 2951950, \ldots
\end{aligned}$$

Here are the corresponding sequences for meanders (allowed to touch 0):

$$\begin{aligned}
h &= 1 \quad (\texttt{A001405}): \; 1, 1, 2, 3, 6, 10, 20, 35, 70, 126, 252, 462, 924, 1716, 3432, \ldots \\
h &= 2 \quad (\texttt{A047002}): \; 1, 2, 7, 23, 83, 299, 1107, 4122, 15523, 58769, 223848, \ldots \\
h &= 3 \quad (\texttt{A278398}): \; 1, 3, 15, 75, 400, 2169, 11989, 66985, 377718, 2144290, \ldots \\
h &= 4 \quad (\texttt{A278416}): \; 1, 4, 26, 174, 1231, 8899, 65492, 487646, 3664123, \ldots
\end{aligned}$$

Here are the corresponding sequences for excursions:

$$\begin{aligned}
h &= 1 \quad (\texttt{A126120}): \; 1, 0, 1, 0, 2, 0, 5, 0, 14, 0, 42, 0, 132, 0, 429, 0 \ldots \\
h &= 2 \quad (\texttt{A187430}): \; 1, 0, 2, 2, 11, 24, 93, 272, 971, 3194, 11293, 39148, 139687 \ldots \\
h &= 3 \quad (\texttt{A205336}): \; 1, 0, 3, 6, 35, 138, 689, 3272, 16522, 83792, 434749, \ldots \\
h &= 4 \quad (\texttt{A205337}): \; 1, 0, 4, 12, 82, 454, 2912, 18652, 124299, 841400, \ldots
\end{aligned}$$

Remark 6.3 The cases with $h = 1$ lead to famous sequences, having many links with the combinatorics of trees, via the Łukasiewicz correspondence (see Sect. 2). It is surprising that the cases with $h = 2$ also offer many links with trees, as we show in the next section.

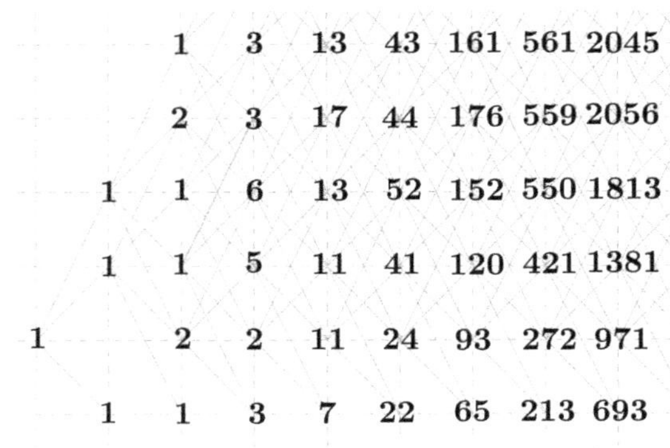

Fig. 7 Cutting a 4-nomial tree at one unit from its root gives the above picture, which thus naturally corresponds to the lattice supporting our lattice basketball walks. The numbers near each node indicate the number of walks from the root to this node

5 Some Links with Other Combinatorial Problems

In this section, we establish some links between our lattice walks and other combinatorial problems. Thereby, we prove several conjectures issued in the On-Line Encyclopedia of Integer Sequences.

5.1 Trees and Basketball Walks from 0 *to* 1

First, we prove that the sequence A166135 from the On-Line Encyclopedia of Integer Sequences, coming from the enumeration of certain tree structures used in financial mathematics, is in fact related to basketball walks and corresponds more precisely to the coefficients of $G_{0,1}(z)$.

The m-nomial tree is a lattice-based computational model used in financial mathematics to price options. It was developed by Phelim Boyle [21] in 1986. For example, for $m = 3$, the underlying stock price is modeled as a recombining tree, where, at each node, the price has three possible paths: an up, down, or stable path. The case $m = 2$ has a long history going back to one of the founding problems of financial mathematics and probability theory, the "ruin problem," analyzed in the eighteenth and nineteenth centuries by de Moivre, Laplace, Huygens, Ampère, Rouché, before to be revisited by combinatorialists like Catalan, Whitworth, Bertrand, André, Delannoy (see [9] for more on these aspects). Figure 7 illustrates a 4-nomial tree.

The following proposition gives the exact link between these trees and a generalization of basketball walks.

Proposition 6.13 (Link between lattice walks and m-nomial trees) *Consider the step sets*

$$\mathcal{S}_{2n} = \{-n, \ldots, -1, 1, \ldots, n\} \text{ and } \mathcal{S}_{2n+1} = \mathcal{S}_{2n} \cup \{0\}.$$

For each step set $\mathcal{S}_m$, define $T_m(z)$ to be the generating function for walks using steps from $\mathcal{S}_m$, starting at the origin and getting absorbed at -1. (By this, we mean that the walks may never touch $y = -1$ except, possibly, at the very last step.) Then, the coefficients of $T_2(z)$ are the Catalan numbers, the coefficients of $T_3(z)$ are the Motzkin numbers, while the coefficients of $T_4(z)$ count our basketball walks from 0 to 1 (walks with steps ± 2, ± 1, starting at the origin and ending at altitude 1, and never touching 0 in-between).

Table 3 By time reversal, $T_4(z) = G_{0,1}(z)$

T_4	$G_{0,1}$
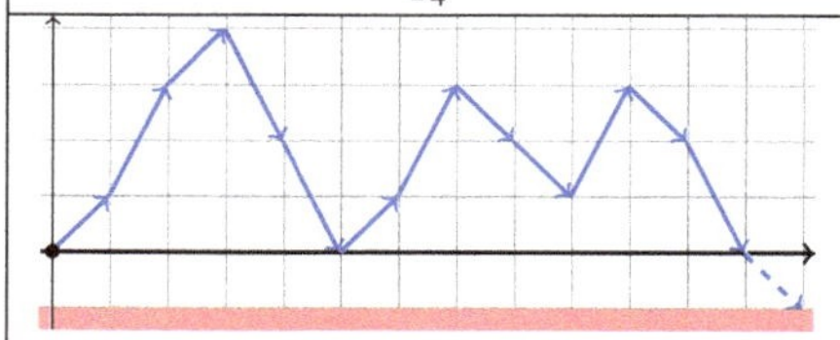 last step is a 1-step down	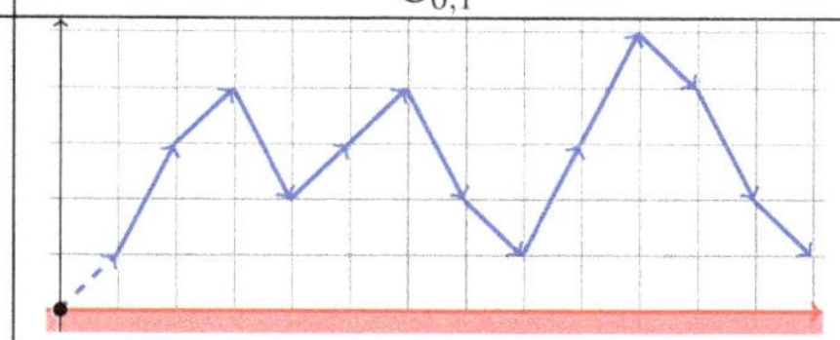 first step is a 1-step up
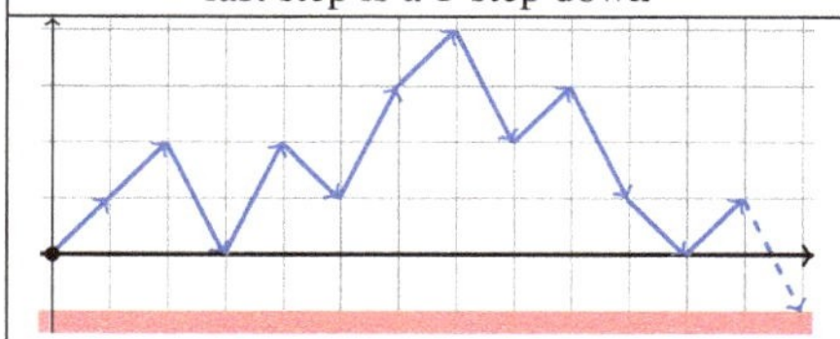 last step is a 2-step down	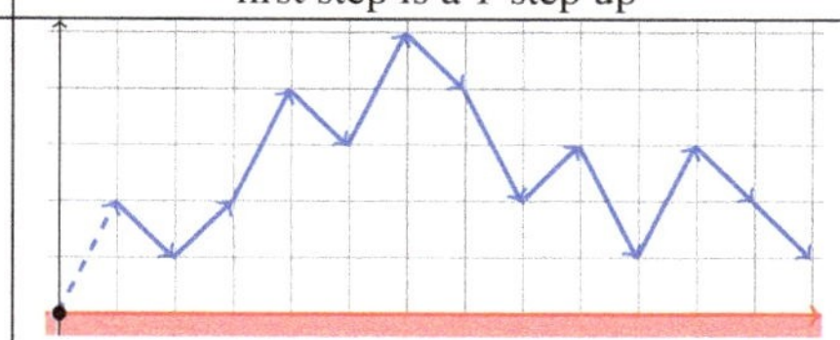 first step is a 2-step up

Proof While the correspondence is direct for $m \leq 3$, it follows for $m = 4$ from a time reversion, as each walk from T_4 can then be obtained from $G_{0,1}$ and vice versa (see Table 3). Thus, $T_4(z) = G_{0,1}(z)$.

5.2 *Increasing Trees and Basketball Walks*

A *unary-binary tree* is an ordered tree such that each node has 0, 1, or 2 children. An *increasing unary-binary tree on n vertices* is a unary-binary tree with n vertices labeled $1, 2, \ldots, n$ such that the labels along each walk from the root are increasing (cf. [48, p. 51]). Given an increasing unary-binary tree T, we associate with T the *permutation* σ_T constructed by reading the tree left to right, level by level, starting at the root. A permutation σ is said to *contain the pattern* π if there exists a subsequence of σ that has the same relative order as π. Otherwise, σ is said to *avoid the pattern* π. For example, the permutation $\sigma = 14235$ contains the pattern 213 because σ contains the subsequence 425, in which the numbers have the same relative order as in 213, while the permutation 12453 avoids 213.

Manda Riehl initiated studies of increasing trees for which the associated permutation avoids a given pattern (see also [39]). By a computer program, she obtained the first terms of the corresponding sequences for patterns of length 3. She observed that "the number of increasing unary-binary trees with associated permutation avoiding 213" seems to coincide with sequence `A166135`, which we proved to count basketballs walks from altitude 0 to altitude 1. Figure 8 shows a verification of this claim for $n = 5$: There are 39 increasing unary-binary trees on 5 vertices, and among them, 22 correspond to permutations avoiding the pattern 213. (The forbidden subsequences are highlighted in red. The trees in black all avoid 213. The trees are grouped according to their associated permutations. Tree labels are read left to right.)

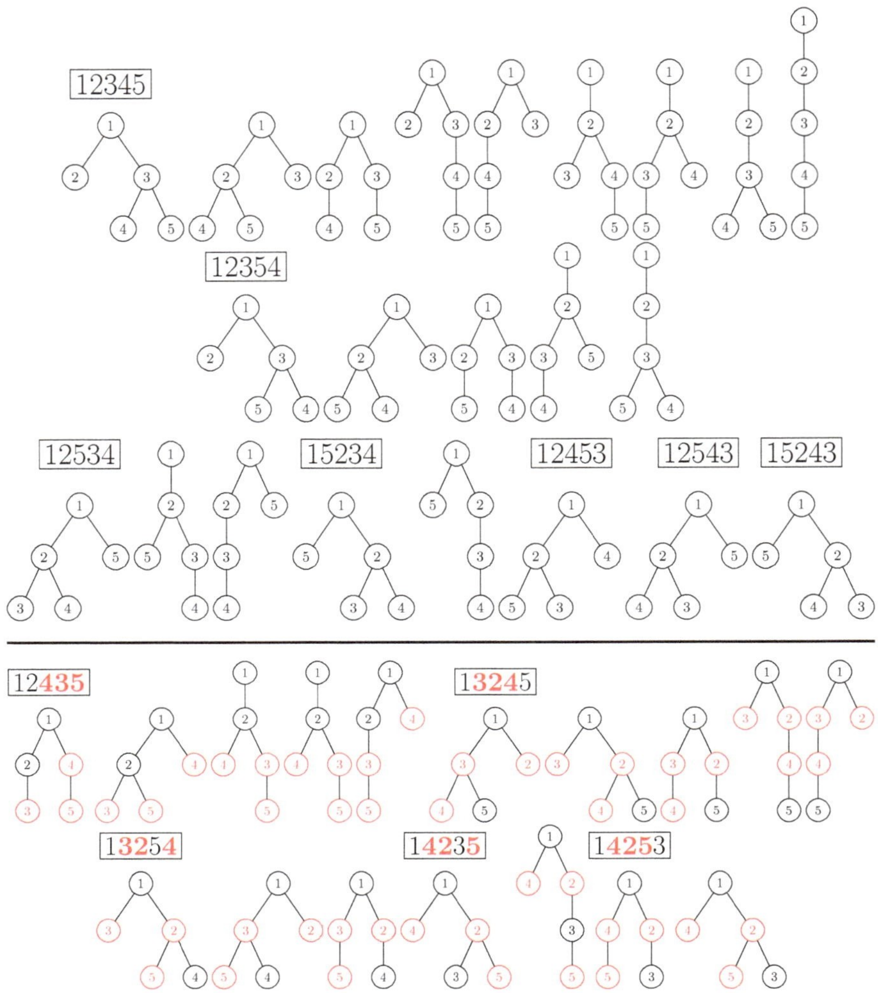

Fig. 8 All increasing unary-binary trees with 5 nodes, where patterns 213 are marked in red. There are 22 trees (drawn in black) for which the associated permutation avoids this pattern

Here is the reformulation of Riehl's conjecture which takes into account our findings.

Conjecture 6.1 The number of basketball walks of length n starting at the origin and ending at altitude 1 that never touch or pass below the x-axis equals the number of increasing unary-binary trees on n vertices with associated permutation avoiding 213.

After the first version of this article was circulated via the `arXiv`, Bettinelli, Fusy, Mailler, and Randazzo [14] found a nice bijective proof of this conjecture.

How strong is the constraint of avoiding the pattern 213? For this, we need to compute the probability that an increasing unary-binary tree avoids the pattern 213. Due to Conjecture 6.1, proved in [14], we know the number of increasing unary-binary trees which avoid 213. Hence, the question is to compute the total number t_n of increasing unary-binary trees, which can be done via the so-called boxed product.

The boxed product (written $^{\square}\!\!\times$) is the combinatorial construction corresponding to a labeled product, in which the minimal label is forced to be in the first component of this product (see [28]). This leads the following recursive decomposition for binary-ternary increasing trees $\mathcal{T}$:

$$\mathcal{T} = leaf + root \,^{\square}\!\!\times \mathcal{T} + root \,^{\square}\!\!\times \mathcal{T} \times \mathcal{T} \,,$$

which translates into the following functional equation for the corresponding exponential generating function:

$$T(z) = z + \int_0^z T(t)dt + \int_0^z T^2(t)dt \,.$$

By solving the associated differential equation $T'(z) = 1 + T(z) + T^2(z)$, we obtain

$$T(z) = \frac{\sqrt{3}}{2} \tan\left(\frac{\pi}{6} + \frac{\sqrt{3}}{2} z\right) - \frac{1}{2} \,.$$

The corresponding Taylor expansion is

$$T(z) = \sum_{n \geq 1} t_n \frac{z^n}{n!} = z + \frac{z^2}{2!} + 3\frac{z^3}{3!} + 9\frac{z^4}{4!} + 39\frac{z^5}{5!} + 189\frac{z^6}{6!} + 1107\frac{z^7}{7!} + O(z^8) \,.$$

Singularity analysis on the dominant poles of the tan function implies that

$$t_n \sim 3\sqrt{\frac{3}{2\pi}} \left(\frac{3^{3/2}}{2e\pi}\right)^n \sqrt{n}\, n^n \,.$$

In conclusion, increasing unary-binary trees grow like $n^{1/2} A^n n^n$, while the same trees avoiding the pattern 213 grow like $n^{-3/2} 4^n$. This observation suggests the following natural conjecture.

Conjecture 6.2 (*A Stanley–Wilf-like conjecture for pattern avoid- ance in increasing trees*) Let $\mathcal{T}$ be a class of increasing trees of prescribed arity encoded by a power series ϕ, i.e., one has $\mathcal{T}' = z\phi(\mathcal{T})$. Then, the number a_n of such trees avoiding a given pattern satisfies $a_n = O(C^n)$, for some C depending on the pattern and on ϕ.

This conjecture shares the spirit of the Stanley–Wilf conjecture (proven by a combination of [33, 40]), which asserted that any class of pattern-avoiding permutations has an exponential growth rate.

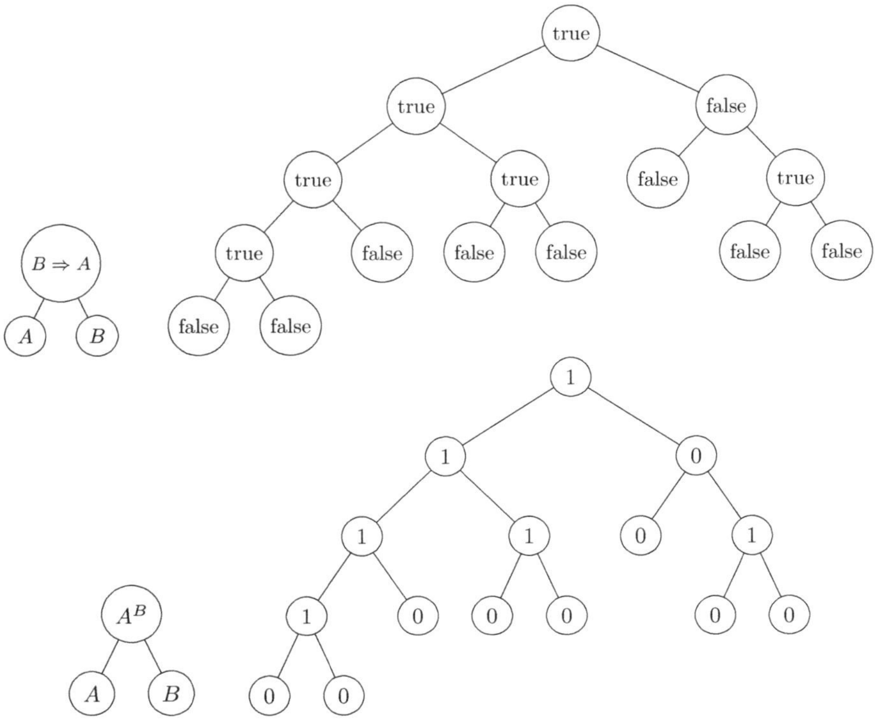

Fig. 9 Boolean trees (i.e., binary trees where each node is labeled either "false" or "true") such that a node having children with Boolean value A and B will have the Boolean value "$B \Rightarrow A$"

5.3 *Boolean Trees and Basketball Walks from* **0** *to* **2**

In [13], Bender and Williamson considered the problem of bracketing some binary operations (objects that are in bijection with the Boolean trees that we present in Fig. 9). It turns out that this problem is doubly related to our basketball walks (walks with steps ± 1, ± 2, always positive). This is what we address in the next two propositions.

Proposition 6.14 *Under the conventions* $1^1 = 1^0 = 0^0 = 1$ *and* $0^1 = 0$, *the number of bracketings of* $n+1$ *zeroes* $0\hat{}\cdots\hat{}0$ *giving result* 1 *is equal to the number of basketball walks from altitude* 0 *to altitude* 2 *of length* n.

Proof Let $W(z)$ (respectively $Z(z)$) be the generating function for the number of bracketings of n zeroes $0\hat{}0\hat{}\cdots\hat{}0$ producing result 1 (respectively 0). The objects that are counted by $W(z)$ are of the form ("1")^("1"), ("1")^("0"), or ("0")^("0"), where "1" stands for a bracketing producing the result 1, and "0" stands for a bracketing producing the result 0. This observation translates into the generating function equation

$$W(z) = W^2(z) + Z(z)W(z) + Z^2(z). \tag{6.23}$$

Similarly, a bracketing producing 0 may either be a single 0 or a bracketing of the form ("0")^("1"). This yields the equation

$$Z(z) = z + Z(z)W(z). \tag{6.24}$$

Let $C(z) := Z(z) + W(z)$. Equations (6.23) and (6.24) imply $C(z) = 1 + zC^2(z)$, i.e., $C(z) = \frac{1}{2z} - \frac{1}{2z}\sqrt{1-4z}$. This is not a surprise because $W + Z$ corresponds to well-parenthesized words, known to be counted by Catalan numbers.

We may "replace" $W(z)$ by $C(z)$ in Equation (6.24). This leads to

$$Z(z) = z + Z(z)(C(z) - Z(z)).$$

Solving for $Z(z)$, we obtain

$$\begin{aligned} Z(z) &= \frac{C(z) - 1 + \sqrt{(C(z)-1)^2 + 4z}}{2} \\ &= -\frac{1}{4} - \frac{1}{4}\sqrt{1-4z} + \frac{1}{4}\sqrt{2 + 12z + 2\sqrt{1-4z}}. \end{aligned} \tag{6.25}$$

Therefore, we get

$$W(z) = C(z) - Z(z) = \frac{3}{4} - \frac{1}{4}\sqrt{1-4z} - \frac{1}{4}\sqrt{2 + 12z + 2\sqrt{1-4z}}.$$

Comparison of this expression with Expression (6.17) for $G_{0,2}(z)$ shows that $W(z) = zG_{0,2}(z)$.

We leave it to the reader to find a bijective proof between bracketings of 0^ . . . ^0 having value 1 and basketball walks from altitude 0 to altitude 2.

Proposition 6.15 *The number of basketball walks of length n starting at the origin, ending at altitude* 1, *never running below the x-axis in-between, is equal to the number of bracketings of $n+2$ zeroes* 0^0^· · ·^0 *producing result* 0.

Proof The generating function $F_1(z)$ for walks ending at 1 is given by (6.8) in the form

$$F_1(z) = G_{1,2}(z) = \frac{u_1(z)u_2(z) + u_1(z) + u_2(z)}{z}.$$

The generating function $Z(z)$ for the number of bracketings of n zeroes 0^· · ·^0 having value 0 is given by (6.25). Substitution of the closed-form expressions for the small roots into $F_1(z)$ yields $z^2 F_1(z) = Z(z)$. This establishes the claim.

6 Conclusion

In this article, we show how to derive closed-form expressions for the enumeration of lattice walks satisfying various constraints (starting point, ending point, positivity, allowed steps, ...). The key is a proper use of the Lagrange–Bürmann inversion in combination with the expressions given by the kernel method. This technique admits many extensions, which will work in a similar way: It is possible to extend it to walks in which we want to keep track of some parameters (marking a specific step, pattern, altitude, ...), allowing an infinite set of steps, or unbounded steps (this would encode what is called catastrophes in queuing theory language). It is also possible to consider other constraints, such as to force the walk to live in some cone or to have some forbidden patterns. In all these cases, the kernel method will give a closed-form expression for the generating function, in terms of the roots of the kernel, and thus, our mix of kernel method and Lagrange–Bürmann inversion will lead in these situations also to some closed-form expression for the coefficients of the generating function (in terms of nested sums of binomials).

In several cases, these nested sums of binomials provide the nice challenge of finding bijective proofs. It is satisfying to find *some* formula for the enumeration of certain lattice paths which is efficient (in terms of algorithmic complexity), but the fact that many of these sums involve only positive terms is an indication that combinatorics has still its word to say on these formulas.

The holonomic approach, as well illustrated by the book of Petkovšek, Wilf, and Zeilberger [45], or Kauers and Paule [32], is a way to prove that different binomial expressions correspond in fact to the same sequence. It remains an open question to know which methods can lead to the most concise formula: The platypus algorithms and the Flajolet–Soria formula [7, 8], or the cycle lemma, and extraction of diagonals of rational functions seem to indicate that we could in fact need an arbitrarily large amount of nested sums. In some cases, one can reduce the number of nested sums with techniques from symbolic summation theory (e.g., by $\Sigma\Pi$ extension theory [47], or geometric simplifications in diagonal extractions of rational functions [17]), but it is still unknown if, for the directed lattice path models we considered, there is a miraculous simple formula (with just one or two nested sums).

Acknowledgements We thank the organizers of the 8th International Conference on Lattice Path Combinatorics & Applications, which provided the opportunity for this collaboration. Sri Gopal Mohanty played an important role in the birth of this sequence of conferences, and his book [42] was the first one (together with the book of his Ph.D. advisor Tadepalli Venkata Narayana [44]) to spur strong interest in lattice path enumeration. We are therefore pleased to dedicate our article to him.

References

1. André, D.: Mémoire sur les combinaisons régulières et leurs appliquations. Ann. Sci. École Norm. Sup. 2e série **5**, 155–198 (1876)
2. Andrews, G.E., Eriksson, H., Petrov, F., Romik, D.: Integrals, partitions and MacMahon's theorem. J. Combin. Theory Ser. A **114**, 545–554 (2007)
3. Andrews, G.E.: Euler's "exemplum memorabile inductionis fallacis" and q-trinomial coefficients. J. Am. Math. Soc. **3**, 653–669 (1990)
4. Andrews. G.E., Baxter, R.J.: Lattice gas generalization of the hard hexagon model. III. q-trinomial coefficients. J. Stat. Phys. **47**, 297–330 (1987)
5. Ayyer, A., Zeilberger, D.: The number of [old-time] basketball games with final score $n\,n$ where the home team was never losing but also never ahead by more than w points. Electron. J. Combin. **14**, Research Paper **19**, 8 (2007)

6. Banderier, B., Bousquet-Mélou, M., Denise, A., Flajolet, P., Gardy, D., Gouyou-Beauchamps, D.: Generating functions for generating trees. Discret. Math. **246**, 29–55 (2002)
7. Banderier, C., Drmota, M.: Formulae and asymptotics for coefficients of algebraic functions. Combin. Probab. Comput. **24**, 1–53 (2015)
8. Banderier, C., Flajolet, P.: Basic analytic combinatorics of directed lattice paths. Theor. Comput. Sci. **281**, 37–80 (2002)
9. Banderier, C., Schwer, S.: Why Delannoy numbers? J. Stat. Plan. Inference **135**, 40–54 (2005)
10. Banderier, C., Wallner, M.: The reflection-absorption model for directed lattice paths. In preparation
11. Banderier, C., Wallner, M.: The kernel method for lattice paths below a rational slope. This volume
12. Barcucci, E., Pinzani, R., Sprugnoli, R.: The Motzkin family. Pure Math. Appl. Ser. A **2**, 249–279 (1992)
13. Bender, E.A., Williamson, S.G.: Foundations of Combinatorics with Applications. Dover (2006)
14. Bettinelli, J., Fusy, É., Mailler, C., Randazzo, L.: A bijective study of basketball walks. Sém. Lothar. Combin. **77**, 24 (2017). Art. B77a
15. Blasiak, P., Dattoli, G., Horzela, A., Penson, K.A., Zhukovsky, K.: Motzkin numbers, central trinomial coefficients and hybrid polynomials. J. Integer Seq. **11**, 11 (2008). Article 08.1.1
16. Böhm, W.: Lattice path counting and the theory of queues. J. Stat. Plan. Inference **140**, 2168–2183 (2010)
17. Bostan, A., Lairez, P., Salvy, B.: Multiple binomial sums. J. Symb. Comput. **80**, 351–386 (2016)
18. Bousquet-Mélou, M.: Discrete excursions. Sém. Lothar. Combin. **57**, 23 (2008). Art. 57d
19. Bousquet-Mélou, M., Jehanne, A.: Polynomial equations with one catalytic variable, algebraic series and map enumeration. J. Combin. Theory Ser. B **96**, 623–672 (2006)
20. Bousquet-Mélou, M., Petkovšek, M.: Linear recurrences with constant coefficients: the multivariate case. Discret. Math. **225**, 51–75 (2000)
21. Boyle, P.: Option valuation using a three-jump process. Int. Options J. **3**, 7–12 (1986)
22. Bürmann, H.H.: Formules du développement, de retour et d'integration. Mém. Acad. Royale Sci. Belles-Lettres Berlin (1798). Submitted to the Institut National de France; the manuscript survives in the archives of the École Nationale des Ponts et Chaussées in Paris
23. Comtet, C.: Advanced Combinatorics. D. Reidel Publishing Co. (1974)
24. Dieudonné, J.: Infinitesimal Calculus. Houghton Mifflin Co., Hermann, Paris; Translated from the French (1971)
25. Egorychev, G.P.: Integral Representation and the Computation of Combinatorial Sums. Translations of Mathematical Monographs, vol. 59. American Mathematical Society, Providence (1984)
26. Eynard, B.: Counting Surfaces. Progress in Mathematical Physics, vol. 70. Birkhäuser/Springer, Cham (2016)
27. Fayolle, G., Iasnogorodski, R., Malyshev, V.: Random walks in the quarter-plane. Applications of Mathematics, vol. 40. Springer, Berlin (1999)
28. Flajolet, P., Sedgewick, R.: Analytic Combinatorics. Cambridge University Press, New York (2009)
29. Gross, D., Shortle, J.F., Thompson, J.M., Harris, C.M.: Fundamentals of Queueing Theory. Wiley Series in Probability and Statistics, 4th edn. Wiley, NJ (2008)
30. Jain, J.L., Mohanty, S.G., Böhm, W.: A Course on Queueing Models. Statistics: Textbooks and Monographs. Chapman and Hall/CRC, Boca Raton (2007)
31. Kauers, M., Johansson, F., Jaroschek, M.: Ore polynomials in sage. In: Computer Algebra and Polynomials. Lecture Notes in Computer Science, pp. 105–125. Springer, Berlin (2015)
32. Kauers, M., Paule, P.: The Concrete Tetrahedron. Texts and Monographs in Symbolic Computation. Springer, Berlin (2011)
33. Klazar, M.: The Füredi–Hajnal conjecture implies the Stanley–Wilf conjecture. In: Formal Power Series and Algebraic Combinatorics (Moscow, 2000), pp. 250–255. Springer, Berlin (2000)

34. Knuth, D.E.: The Art of Computer Programming. Vol. 2: Seminumerical Algorithms, 4th edn. Addison-Wesley (1998)
35. Krinik, A.C., Mohanty, S.G.: On batch queueing systems: a combinatorial approach. J. Stat. Plan. Inference **140**, 2271–2284 (2010)
36. Krinik, A.C., Shun, K.: Markov processes with constant transition rates of size one and two. J. Stat. Theory Pract. **5**, 475–495 (2011)
37. Lagrange, J.-L.: Nouvelle méthode pour résoudre les équations littérales par le moyen des séries. Mém. Acad. Royale Sci. Belles-Lettres Berlin 24, 251–326 (1770) Reprinted in Œuvres de Lagrange, tome 2. Paris: Gauthier-Villars, pp. 655–726 (1868)
38. Lagrange, J.-L., Legendre, A.-M.: Rapport sur deux mémoires d'analyse du professeur burmann. Mém. Inst. Nat. Sci. Arts: Sci Math. Phys. **2**, 13–17 (1799)
39. Levin, D., Pudwell, L.K., Riehl, M., Sandberg, A.: Pattern avoidance in k-ary heaps. Australas. J. Combin. **64**, 120–139 (2016)
40. Marcus, A., Tardos, G.: Excluded permutation matrices and the Stanley–Wilf conjecture. J. Combin. Theory Ser. A **107**, 153–160 (2004)
41. Margolius, B.: Asymptotic estimates for queueing systems with time-varying periodic transition rates. This volume
42. Mohanty, S.G.: Lattice Path Counting and Applications. Academic Press (1979)
43. Montel, P.: Sur les combinaisons avec répétitions limitées. Bull. Sci. Math. **2**(66), 86–103 (1942)
44. Narayana, T.V.: Lattice Path Combinatorics with Statistical Applications, Mathematical Expositions, vol. 23. University of Toronto Press (1979)
45. Petkovšek, M., Wilf, H.S., Zeilberger, D.: $A = B$. A.K. Peters (1996)
46. Salvy, B., Zimmermann, P.: Gfun: a Maple package for the manipulation of generating and holonomic functions in one variable. ACM Trans. Math. Softw. **20**, 163–177 (1994)
47. Schneider, C.: Symbolic summation assists combinatorics. Sém. Lothar. Combin. **56**, 36 (2007). Art. B56b
48. Stanley, R.P.: Enumerative combinatorics, vol. 1. 2nd edn. Cambridge Studies in Advanced Mathematics, vol. 49. Cambridge University Press, Cambridge (2012)
49. Stanley, R.P.: Enumerative combinatorics, vol. 2. Cambridge Studies in Advanced Mathematics, vol. 62. Cambridge University Press, Cambridge (1999)
50. Xin, G.: The ring of Malcev–Neumann series and the residue theorem. Ph.D. thesis, Brandeis University (2004)

The Kernel Method for Lattice Paths Below a Line of Rational Slope

Cyril Banderier and Michael Wallner

❋❋❋

We dedicate this article to the memory of Philippe Flajolet, who was and will remain a guide and a wonderful source of inspiration for so many of us.

❋❋❋

Abstract We analyze some enumerative and asymptotic properties of lattice paths below a line of rational slope. We illustrate our approach with Dyck paths under a line of slope $2/5$. This answers Knuth's problem #4 from his "Flajolet lecture" during the conference "Analysis of Algorithms" (AofA'2014) in Paris in June 2014. Our approach extends the work of Banderier and Flajolet for asymptotics and enumeration of directed lattice paths to the case of generating functions involving several dominant singularities and has applications to a full class of problems involving some "periodicities." A key ingredient in the proof is the generalization of an old trick by Knuth himself (for enumerating permutations sortable by a stack), promoted by Flajolet and others as the "kernel method." All the corresponding generating functions are algebraic, and they offer some new combinatorial identities, which can also be

---

This work extends our preliminary version "Lattice paths of slope 2/5" which appeared in the Proceedings of the ANALCO15 San Diego Conference.

C. Banderier
Laboratoire d'Informatique de Paris-Nord, CNRS/Université Paris 13, Villetaneuse, France
e-mail: Cyril.Banderier@lipn.univ-paris13.fr

M. Wallner (✉)
Institute of Discrete Mathematics and Geometry, TU Wien, Vienna, Austria
e-mail: Michael.Wallner@tuwien.ac.at

M. Wallner
LaBRI, Université de Bordeaux, Bordeaux, France

G. E. Andrews et al. (eds.), *Lattice Path Combinatorics and Applications*, Developments in Mathematics 58, https://doi.org/10.1007/978-3-030-11102-1_7

tackled in the $A = B$ spirit of Wilf–Zeilberger–Petkovšek. We show how to obtain similar results for any rational slope. An interesting case is, e.g., Dyck paths below the slope $2/3$ (this corresponds to the so-called Duchon's club model), for which we solve a conjecture related to the asymptotics of the area below such lattice paths. Our work also gives access to lattice paths below an irrational slope (e.g., Dyck paths below $y = x/\sqrt{2}$), a problem that we study in a companion article.

Keywords Lattice paths · Generating function · Analytic combinatorics · Singularity analysis · Kernel method · Generalized Dyck paths · Algebraic function · Rational Catalan combinatorics · Periodic support · Bizley formula · Grossman formula

2010 Mathematics Subject Classification Primary 05A15 · Secondary 05A16 68W40

1 Introduction

For the enumeration of simple lattice paths (allowing just the jumps -1, 0, and $+1$), many methods are often used, like the Lagrange inversion, determinant techniques, continued fractions, orthogonal polynomials, bijective proofs, and a lot is known in such cases [32, 45, 52, 54]. These nice methods do not apply to more complex cases of more generic jumps (or, if one adds a spacial boundary, like a line of rational slope). It is then possible to use some ad hoc factorization due to Gessel [35] or context-free grammars to enumerate such lattice paths [28, 47, 50]. One drawback of the grammar approach is that it leads to heavy case-by-case computations (resultants of equations of huge degree). In this article, we show how to proceed for the enumeration and the asymptotics in these harder cases: our techniques are relying on the "kernel method" which (contrary to the context-free grammar approach) offers access to the true simple *generic* structure of the final generating functions and the *universality* of their asymptotics via singularity analysis.

Let us start with the history of what Philippe Flajolet named the "kernel method": It has been part of the folklore of combinatorialists for some time and its simplest application deals with functional equations (with apparently more unknowns than equations!) of the form

$$K(z,u)F(z,u) = p(z,u) + q(z,u)G(z),$$

where the functions p, q, and K are given and F, G are the unknown generating functions we want to determine. $K(z,u)$ is a polynomial in u which we call the "kernel" as we "test" this functional equation on functions $u(z)$ cancelling this kernel.[1]

---

[1]The "kernel method" that we mention here for functional equations in combinatorics has nothing to do with what is known as the "kernel method" or "kernel trick" in statistics or machine learning. Also, there is no integral directly related to our kernel. For sure, in our case the word kernel was chosen as its zeros will play a key role, and also, in one sense, as this kernel has in its core the full description of the problem and its resolution.

The simplest case is when there is only one branch, $u_1(z)$, such that $K(z, u_1(z)) = 0$ and $u_1(0) = 0$; in that case, a single substitution gives a closed-form solution for G: namely $G(z) = -p(z, u_1(z))/q(z, u_1(z))$.

Such an approach was introduced in 1969 by Knuth to enumerate permutations sortable by a stack, see the detailed solution to Exercise 2.2.1–4 in *The Art of Computer Programming* ([43, pp. 536–537] and also Ex. 2.2.1.11 therein), which presents a "new method for solving the ballot problem," for which the kernel K is a quadratic polynomial (this specific case involves just one branch $u_1(z)$).

In combinatorics exist many applications of this method for solving variants of the above functional equation: one is known as the "quadratic method" in map enumeration, as initially developed in 1965 by Brown during his collaboration with Tutte (see Sect. 2.9.1 from [9, 24] for the analysis of about a dozen families of maps). During nearly 30 years, the kernel method was dealing only with "quadratic cases" like the ones of Brown for maps or of Knuth for a vast amount of examples involving trees, polyominoes, walks [57], or more exotic applications like the one mentioned by Odlyzko in his wonderful survey on asymptotic methods in enumeration [25]. Then, in 1998, the initial approach by Knuth was generalized by a group of four people, all of them being in contact and benefiting from mutual insights: Banderier in his memoir [5] solved some problems related to generating trees and walks, and this later led to the article with Flajolet [8] and to the solution of some conjectures due to Pinzani in the article with Bousquet-Mélou et al. [6]. At the same time, Petkovšek analyzed linear multivariate recurrences in [55], a work later extended in [23]. All these articles contributed to turn the original approach by Knuth into a method working when the equation has more unknowns (and the kernel has more roots). This solves equations of the type

$$K(z, u)F(z, u) = \sum_{i=1}^{m} p_i(z, u)G_i(z),$$

where K and the p_i's are known polynomials, and F and the G_i's are unknown functions.

A few years later, Bousquet-Mélou and Jehanne [21] solved the case of algebraic equations in F of arbitrary degree:

$$P(z, u, F(z, u), G_1(z), \ldots, G_m(z)) = 0.$$

The kernel method thus plays a key role in many combinatorial problems. A few examples are directed lattice paths and their asymptotics [8, 19], additive parameters like area [10, 61], generating trees [6], pattern avoiding permutations [49], prudent walks [4, 27], urn models [60], statistics in posets [20] and many other nice combinatorial structures...

Independently, in probability theory, in the 1970s, Malyshev invented an approach now sometimes called the "iterated kernel method." It can be used to analyze nearest neighbor random walks in queuing theory. In this context, these lead to the following type of equations:

$$K(t,x,y)F(t,x,y) = p_0(t,x,y) + p_1(t,x,y)F(x,0) + p_2(t,x,y)F(0,y),$$

where K and the p_i's are known polynomials, while F is the unknown function we are looking for. This approach culminated in the book [31], which was later revisited in the 2000s (e.g., in [46]), also with a more combinatorial point of view in [22]. It is still the subject of vivid activities, including the extension to higher dimensions [18]. Moreover, the kernel method also gives the transient solution of some birth–death queuing processes [37].

Also independently, in statistical mechanics, several authors developed other incarnations of the kernel method. For example, the WKB limit of the Bethe ansatz (also called thermodynamical Bethe ansatz) often leads to algebraic equations and to what is called the algebraic Bethe ansatz [34]. The kernel method is also used in the study of the Ising model of bicolored maps (see Theorem 8.4.5 in [30], and pushing further this method led Eynard to his "topological recurrence"), and in many articles on enumeration related to directed animals, polymers, walks [38–40].

After this short history of the kernel method, we want to show how to use it to derive explicit counting formulae and asymptotics for directed lattice paths below a line of rational slope. In the article by Banderier and Flajolet [8], the class of directed lattice paths in $\mathbb{Z}^2$ was investigated thoroughly by means of analytic combinatorics (see [33]). Our work is an extension of this article in mainly five ways:

1. Our work involves lattice paths having a "periodic support," and the comment in [8, Sect. 3.3] was incomplete for this more cumbersome case; indeed, there are then several dominant singularities, and we had to revisit in more detail the structural properties of the roots associated with the kernel method in order to understand the contribution of each of these singularities. It is pleasant that this new understanding gives a tool to deal with the asymptotics of many other lattice path enumeration problems.
2. We get new explicit formulae for the generating functions of walks with starting and ending at altitude other than 0, and links with complete symmetric homogeneous polynomials.
3. We give new closed forms for the coefficients of these generating functions.
4. We have an application to some harder parameters (like the area below a lattice path).
5. We extend the results to walks below a line of *arbitrary rational* slope, paving the way for our forthcoming article on walks below a line of arbitrary *irrational* slope [15].

Let us give a definition of the lattice paths we consider:

Definition 7.1 (*Jumps and lattice paths*) A *step set* $\mathcal{S} \subset \mathbb{Z}^2$ is a finite set of vectors $\{(x_1, y_1), \dots, (x_m, y_m)\}$. An n-step *lattice path* or *walk* is a sequence of vectors $(v_1, \dots, v_n)$, such that v_j is in $\mathcal{S}$. Geometrically, it may be interpreted as a sequence

Table 1 The four types of paths: walks, bridges, meanders, and excursions. We refer to these walks as the Banderier–Flajolet model, in contrast to the model in which we will consider lattice paths below a rational slope boundary

	ending anywhere	ending at 0
unconstrained (on $\mathbb{Z}$)	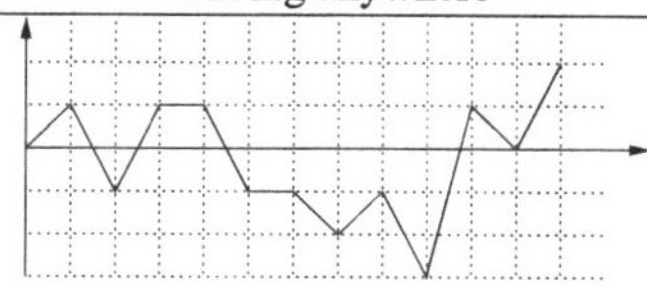 walk/path ($\mathcal{W}$)	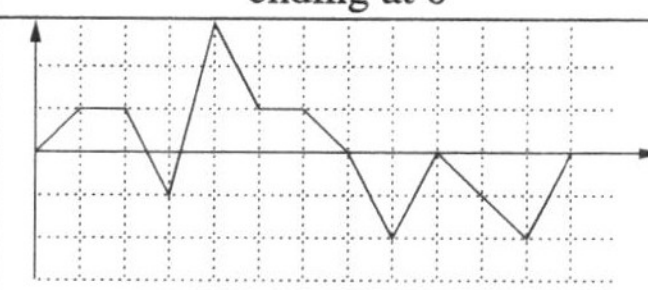 bridge ($\mathcal{B}$)
constrained (on $\mathbb{N}$)	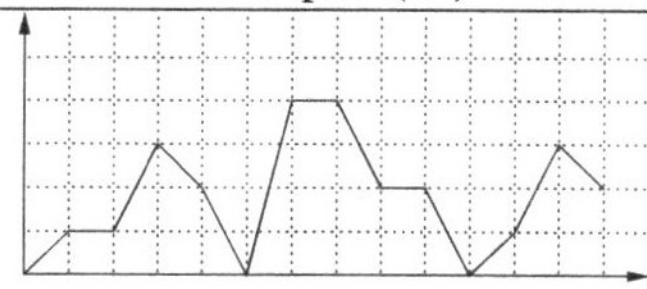 meander ($\mathcal{M}$)	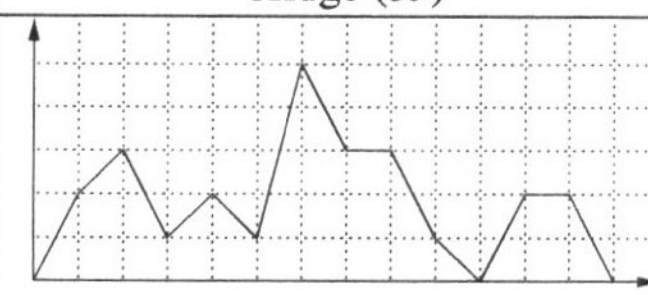 excursion ($\mathcal{E}$)

of points $\omega = (\omega_0, \omega_1, \ldots, \omega_n)$ where $\omega_i \in \mathbb{Z}^2$, $\omega_0 = (0, 0)$ (or another starting point) and $\omega_i - \omega_{i-1} = v_i$ for $i = 1, \ldots, n$. The elements of $\mathcal{S}$ are called *steps* or *jumps*. The *length* $|\omega|$ of a lattice path is its number n of jumps.

The lattice paths can have different additional constraints as shown in Table 1.

We restrict our attention to *directed paths* which are defined by the fact that, for each jump $(x, y) \in \mathcal{S}$, one must have $x \geq 0$. The next definition allows to merge the probabilistic point of view (random walks) and the combinatorial point of view (lattice paths):

Definition 7.2 (*Weighted lattice paths*) For a given step set $\mathcal{S} = \{s_1, \ldots, s_m\}$, we define the respective *system of weights* as $\{w_1, \ldots, w_m\}$ where $w_j > 0$ is the weight associated with step s_j for $j = 1, \ldots, m$. The *weight of a path* is defined as the product of the weights of its individual steps.

Plan of This Article

- First, in Sect. 2, we recall the fundamental results for lattice paths below a line of slope α (where α is an integer or the inverse of an integer) and the links with trees.
- Then, in Sect. 3, we give Knuth's open problem on lattice paths below a line of slope $2/5$.
- In Sect. 4, we give a bijection between lattice paths below any line of rational slope and lattice paths from the Banderier–Flajolet model.
- In Sect. 5, the needed bivariate generating function is defined and the governing functional equation is derived and solved: here the "kernel method" plays the most significant role in order to obtain the generating function (as typical for many combinatorial objects which are recursively defined with a "catalytic parameter").
- In Sect. 6, we tackle some questions on asymptotics, thus answering the question of Knuth.

- In Sect. 7, we comment on links with previous results of Nakamigawa and Tokushige, which motivated Knuth's problem, and we explain why some cases lead to particularly striking new closed-form formulae.
- In Sect. 8, we analyze what happens for the Duchon's club model (lattice paths below a line of slope 2/3), and we extend our formulae to general rational slopes.

2 Trees, Fractional Trees, Imaginary Trees

Due to their fundamental role in computer science trees were the subject of many investigations, and there exist many alternative representations of this key data structure. One of the most useful ones is an encoding by "traversing" the tree via a depth-first traversal (or via a breadth-first traversal). This directly gives a lattice path associated with the original tree. In fact, what are called "simple families of ordered trees" (rooted ordered trees in which each node has a degree prescribed to be in a given set) are in bijection with lattice paths. The reason is the famous *Łukasiewicz correspondence* between trees and lattice paths, see Fig. 1.

Basic manipulations on lattice paths also show that *Dyck paths* (paths with jumps North and East, see Fig. 2) below the line $y = \alpha x$ (α being here a positive integer), or below the line $y = x/\alpha$, are in bijection with trees (of arity α, i.e., every node has exactly 0 or α children).

The generating function $F(z) = \sum f_n z^n$, where f_n counts the number of trees with n nodes (internal and external ones), satisfies the functional equation $F(z) = z\phi(F(z))$, where ϕ encodes the allowed arities. Thus, we get binary trees: $\phi(F) = 1 + F^2$, unary-binary trees: $\phi(F) = 1 + F + F^2$, t-ary trees: $\phi(F) = 1 + F^t$, general trees: $\phi(F) = 1/(1 - F)$. See [33] for more on this approach, also extendible to unordered trees (i.e., the order of the children is not taken into account).

Because of the bijection with lattice paths, the enumeration of ordered trees solves the question of lattice paths below a line of integer slope. In the simplest case of classical Dyck paths, many tools were developed. In 1886, Delannoy was the first to

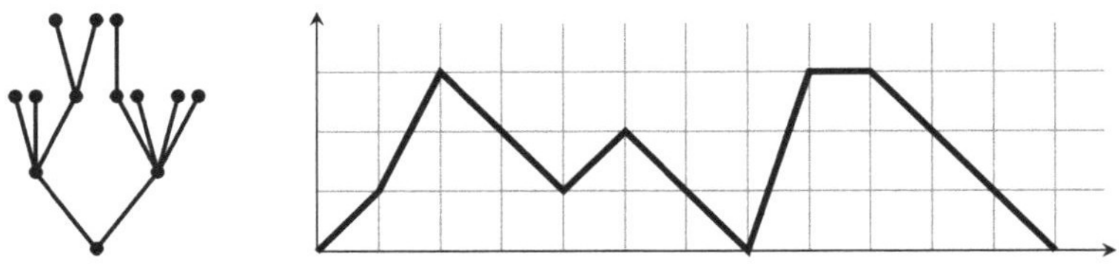

Fig. 1 The Łukasiewicz bijection between trees and lattice paths: A little fly is travelling along the full contour of the tree starting from the root. Whenever it meets a new node, one draws a new jump of size "arity of the node −1" in the lattice path. Without loss of generality, one can always remove the very last jump (as it will always be a "−1") and thus we get an excursion which is in bijection with the initial tree. It is straightforward to reverse this bijection. Additionally, note that any deterministic traversal of the tree offers such a bijection, so it could be a depth-first traversal, but also, e.g., a breadth-first traversal

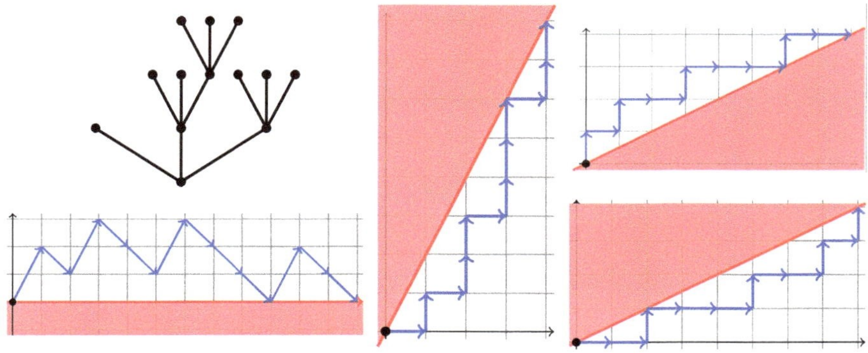

Fig. 2 Examples of combinatorial structures which are in bijection: ternary trees, excursions of directed lattice paths with jumps $+2$ and -1, Dyck paths of North-East steps below the line $y = 2x$, Dyck paths above the line $y = \frac{1}{2}x$, and Dyck paths below the line $y = \frac{1}{2}x$

promote a systematic way to enumerate lattice paths, using recurrences and an array representation (see [13] for more on this). Then, the Bertrand ballot problem [16] (already previously considered by Whitworth) and the ruin problem (as studied along centuries by Fermat, Pascal, the Bernoullis, Huygens, de Moivre, Lagrange, Laplace, Ampère and Rouché) were a strong motor for the birth of the combinatorics of lattice paths, one famous solution being the one by André [2] via a bijective proof involving "good minus bad" paths. Aebly [1] and Mirimanoff [51] gave a geometric variant of this bijective proof, which corresponds to what is nowadays known as the reflection principle. Later, the cycle lemma by Dvoretsky and Motzkin [29] proved useful for many similar problems. During the last century, all these tools were extended and applied to other cases than the classical Dyck paths, and we will use some of them in this article.

With respect to the closed form for the enumeration, another powerful tool is the Lagrange–Bürmann inversion formula (see, e.g., [33]). Applied on $T(z) = 1 + zT(z)^t$ (the equation for the generating function of t-ary trees where z marks internal nodes), it gives

$$T(z)^r = \sum_{k\geq 0} \binom{tk+r}{k} \frac{r}{tk+r} z^k = \sum_{k\geq 0} \binom{tk+(r-1)}{k} \frac{r}{(t-1)k+r} z^k . \quad (7.1)$$

Plugging rational values is not directly leading to a power series with integer coefficients, but it "miraculously" becomes the case after basic transformations (Fig. 3). For example, as observed by Knuth [44], for $t = 3/2$, one has the following neat non-trivial identity:

$$T(z)T(-z) = \left(\sum_{k\geq 0} \frac{\binom{3k/2}{k}}{k/2+1} z^k \right) \left(\sum_{k\geq 0} \frac{\binom{3k/2}{k}}{k/2+1} (-z)^k \right) = \sum_{n\geq 0} \frac{\binom{3n+1}{n}}{n+1} z^{2n} . \quad (7.2)$$

Fig. 3 It is possible to plug any value for t in $T(z)$, which is known to count trees and lattice paths when t is an integer. What happens when we consider generalized binomial series of order $3/2$ or of other fractional values? To recycle a nice pun by Don Knuth [44]: Nature is offering nice binary trees; will imaginary trees one day play a role in computer science?

What could be the meaning of such identities involving "half-trees"? The explanation behind this formula is better seen in terms of lattice paths, and we will shed light on it in the next sections via the kernel method. Another set of mysterious identities is, e.g., incarnated by:

$$\ln T(z) = \ln \sum_{n\geq 0} \frac{\binom{tn}{n}}{(t-1)n+1} z^n = \sum_{n\geq 1} \frac{\binom{tn}{n}}{tn} z^n \,. \tag{7.3}$$

In fact, this one is just another avatar of the cycle lemma, which is also the reason for the link between the generating function of bridges and the generating function of excursions (a fact also appearing in various disguises, e.g., in the Spitzer formula, in the Sparre Andersen formula), see [8] for explanations and proofs.

As we have seen, Dyck paths below an integer slope (or structures in bijection with them) were subject to many approaches, now considered as "folklore." The first result for lattice paths below a rational slope came much later and is best summarized by the following theorem:

Theorem 7.1 (Bizley's formula, Grossman's formula) *The number $f(an, bn)$ of Dyck paths from $(0,0)$ to (an, bn) staying weakly above $y = \frac{a}{b}x$ is given by the following expressions, where $c_j := \frac{1}{aj+bj}\binom{aj+bj}{aj}$:*

$$f(an, bn) = [t^n] \exp \sum_{j\geq 0}^{n} \frac{1}{(a+b)} \binom{(a+b)j}{a} t^j \,, \tag{7.4}$$

$$f(an, bn) = \sum_{\left\{\substack{\text{integer partitions of } n:\\ \sum_{j=1}^{k} j\, e_j = n}\right\}} \prod_{j=1}^{k} \frac{(c_j)^{e_j}}{e_j!} \,. \tag{7.5}$$

Formula (7.5) was first stated without proof by Grossman in 1950. A proof was then given by Bizley [17] in 1954. It starts with Formula (7.4), which is an avatar of the cycle lemma [29] expressed in terms of a generating function. Then, routine power series manipulation gives Formula (7.5). These formulae (or special cases of them) have since been rediscovered (and published ...) many times. One nice modern formulation of the method behind is found in the article by Gessel [35]. There exist alternative generic formulae as given by Banderier and Flajolet [8], Sato [59], which simplify for ad hoc cases [11, 28].

This formula admits many extensions as one could, for example, add parameters or take into account certain patterns. This would lead to "rational" Narayana numbers, "rational" q-analogues, "rational" Mahonian statistics (on lattice paths!), etc.

For each n, Grossman's formula (7.5) for $f(an, bn)$ involves $p(n)$ summands, where $p(n)$ is the integer partition sequence of Hardy–Ramanujan fame:

$$p(n) = [t^n] \prod_{n \geq 1} \frac{1}{1 - t^n} \sim \frac{1}{4n\sqrt{3}} \exp\left(\pi \sqrt{\frac{2n}{3}}\right).$$

Therefore, this nice closed-form formula of Grossman has many summands if n is large (computing it will have an exponential cost); it is thus useful to have an algorithmic alternative to it. Bizley's formula (7.4) allows to compute $f(an, bn)$ in quasi-linear time by a power series manipulation. This is also the advantage of other expressions like the ones given by [8] using the kernel method, on which we will come back in the next sections.

Formula (7.4) for $n = 1$ gives $f(a, b) = \frac{1}{a+b}\binom{a+b}{a}$, also known as the rational Catalan numbers $\mathrm{Cat}(a, b)$. In the last years, many properties of the Dyck paths and their "Catalan combinatorics" (i.e., the enumeration of the numerous combinatorial and algebraic structures related to them) were extended to Dyck paths below a line of rational slope. This new area of research is sometimes called "rational Catalan combinatorics" [3]. We expect that the recent developments of "rational Catalan combinatorics" have a generalization to $n > 1$, but with less simple formulae, as suggested by Table 2.

3 Knuth's AofA Problem #4

During the conference "Analysis of Algorithms" (AofA'2014) in Paris in June 2014, Knuth gave the first invited talk, dedicated to the memory of Philippe Flajolet (1948–2011). The title of his lecture was "Problems that Philippe would have loved" and he was pinpointing/developing five nice open problems with a good flavor of "analytic combinatorics" (his slides are available online.[2]) The fourth problem was on "Lattice

[2] http://www-cs-faculty.stanford.edu/~uno/flaj2014.pdf.

Table 2 The number $f(an, bn)$ of Dyck walks from $(0, 0)$ to (an, bn) staying weakly below $y = \frac{a}{b}x$. To shorten our expressions, we use the shorthand $c_j := \frac{1}{aj+bj}\binom{aj+bj}{aj}$ In the rest of the article, we will see further nice formulae for Dyck paths below a rational slope

	# Dyck walks from $(0,0)$ to (an,bn) staying weakly below $y = \frac{a}{b}x$
$n=1$	c_1
$n=2$	$c_2 + \frac{c_1^2}{2}$
$n=3$	$c_3 + c_1c_2 + \frac{c_1^3}{3!}$
$n=4$	$c_4 + \frac{c_2^2}{2} + c_1c_3 + \frac{c_1^2c_2}{2} + \frac{c_1^4}{4!}$
$n=5$	$c_5 + c_2c_3 + c_1c_4 + \frac{c_1c_2^2}{2} + \frac{c_1^2c_3}{2} + \frac{c_1^3}{3!}c_2 + \frac{c_1^5}{5!}$
$n=6$	$c_6 + c_5c_1 + c_4c_2 + \frac{c_1^2c_4}{2} + \frac{c_3^2}{2} + \frac{c_2^3}{3!} + \frac{c_2c_1^4}{4!} + \frac{{c_1}^3c_3}{3!} + \frac{{c_1}^2{c_2}^2}{4} + c_1c_2c_3 + \frac{c_1^6}{6!}$
$\vdots$	$\vdots$
n	$\sum\limits_{\left\{\substack{\text{integer partitions of } n:\\ \sum_{j=1}^k je_j=n}\right\}} \prod\limits_{j=1}^{k} \frac{(c_j)^{e_j}}{e_j!}$

paths of slope 2/5," in which Knuth investigated Dyck paths under a line of slope 2/5, following the work of [53]. This is best summarized by the two following original slides of Knuth:

$$A[i,j] = \begin{cases} 0, & \text{if } j \geq 2i/5 + 2/5, \\ A[i-1,j] + A[i,j-1], & \text{if } j < 2i/5 + 2/5; \end{cases}$$

$$B[i,j] = \begin{cases} 0, & \text{if } j \geq 2i/5 + 1/5, \\ B[i-1,j] + B[i,j-1], & \text{if } j < 2i/5 + 1/5; \end{cases}$$

$A[i,0] = B[i,0] = 1$. When $0 \leq i \leq 4$ and $0 \leq j \leq 10$ we have:

$$A = \begin{pmatrix} 1 & 1 & 1 & 1 & 1 & 1 & 1 & 1 & 1 & 1 & 1 \\ 0 & 0 & 1 & 2 & 3 & 4 & 5 & 6 & 7 & 8 & 9 \\ 0 & 0 & 0 & 0 & 0 & 4 & 9 & 15 & 22 & 30 & 39 \\ 0 & 0 & 0 & 0 & 0 & 0 & 0 & 15 & 37 & 67 & 106 \\ 0 & 0 & 0 & 0 & 0 & 0 & 0 & 0 & 0 & 0 & 106 \end{pmatrix}$$

$$B = \begin{pmatrix} 1 & 1 & 1 & 1 & 1 & 1 & 1 & 1 & 1 & 1 & 1 \\ 0 & 0 & 0 & 1 & 2 & 3 & 4 & 5 & 6 & 7 & 8 \\ 0 & 0 & 0 & 0 & 0 & 3 & 7 & 12 & 18 & 25 & 33 \\ 0 & 0 & 0 & 0 & 0 & 0 & 0 & 0 & 18 & 43 & 76 \\ 0 & 0 & 0 & 0 & 0 & 0 & 0 & 0 & 0 & 0 & 76 \end{pmatrix}$$

Thus $A[x,y]$ enumerates lattice paths from (0,0) that stay in the region $y < \frac{2}{5}x + \frac{2}{5}$, while $B[x,y]$ enumerates the paths that stay in the region $y < \frac{2}{5}x + \frac{1}{5}$.

Theorem (Nakamigawa, Tokushige, 2012):

$$A[5t-1,\ 2t-1] + B[5t-1,\ 2t-1] = \frac{2}{7t-1}\binom{7t-1}{2t}, \quad \text{for all } t \geq 1.$$

Empirical observation:

$$\frac{A[5t-1,\ 2t-1]}{B[5t-1,\ 2t-1]} = a - \frac{b}{t} + O(t^{-2}),$$

where $a \approx 1.63026$ and $b \approx 0.159$ (I think).

In the next sections, we prove that Knuth was indeed right! In order not to conflict with our notation, let us rename Knuth's constants a and b into κ_1 and κ_2.

4 A Bijection for Lattice Paths Below a Rational Slope

Consider paths in the $\mathbb{N}^2$ lattice,[3] starting in the origin, and whose allowed steps are of the type either East or North (i.e., steps $(1, 0)$ and $(0, 1)$, respectively). Let α, β

[3]We live in a world where $0 \in \mathbb{N}$.

be positive rational numbers. We restrict the walks to stay strictly below the barrier $L : y = \alpha x + \beta$. Hence, the allowed domain of our walks forms an obtuse cone with the x-axis, the y-axis and the barrier L as boundaries. The problem of counting walks in such a domain is equivalent to counting directed walks in the Banderier–Flajolet model [8], as seen via the following bijection:

Proposition 7.1 (Bijection: Lattice paths below a rational slope are directed lattice paths) *Let $\mathscr{D} : y < \alpha x + \beta$ be the domain strictly below the barrier L. From now on, we assume without loss of generality that $\alpha = a/c$ and $\beta = b/c$ where a, b, c are positive integers such that* $\gcd(a, b, c) = 1$ *(thus, it may be the case that a/c or b/c are reducible fractions). There exists a bijection between "walks starting from the origin with North and East steps" and "directed walks starting from* $(0, b)$ *with the step set* $\{(1, a), (1, -c)\}$*." What is more, the restriction of staying below the barrier L is mapped to the restriction of staying above the x-axis.*

Proof The following affine transformation gives the bijection (see Fig. 4):

$$\begin{pmatrix} x \\ y \end{pmatrix} \mapsto \begin{pmatrix} x + y \\ ax - cy + b \end{pmatrix}.$$

Indeed, the determinant of the involved linear mapping is $-(c + a) \neq 0$. What is more, the constraint of being below the barrier (i.e., one has $y < \alpha x + \beta$) is thus forcing the new abscissa to be positive: $ax - cy + b > 0$. The gcd conditions ensure an optimal choice (i.e., the thinnest lattice) for the lattice on which walks will live. Note that this affine transformation gives a bijection not only in the case of an initial step set North and East, but for any set of jumps.

The purpose of this bijection is to map walks of length n to meanders (i.e., walks that stay above the x-axis) which are constructed by n unit steps into the positive x direction.

Note that if one does not want the walk to touch the line $y = (a/c)x + b/c$, it corresponds to a model in which one allows to touch, but with a border at $y = (a/c)x + (b - 1)/c$. Time reversal is also giving a bijection between

- walks starting at altitude b with jumps $+a$, $-c$ and ending at 0,
- and walks starting at 0 and ending at altitude b with jumps $-a$, $+c$.

5 Functional Equation and Closed-Form Expressions for Lattice Paths of Slope 2/5

In this section, we show how to derive closed forms (i.e., explicit expressions) for the generating functions of lattice paths of slope 2/5 (and their coefficients). First, define the jump polynomial $P(u) := u^{-2} + u^5$. Note that the bijection in Proposition 7.1 gives jump sizes $+2$ and -5. However, a time reversal gives this equivalent model (jumps -2 and $+5$), which has the advantage of leading to more compact formulae

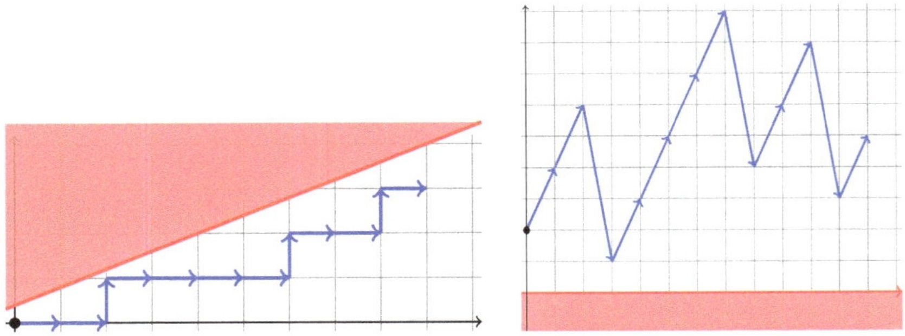

(a) Rational slope model (b) Banderier–Flajolet model

Fig. 4 Example showing the bijection from Proposition 7.1: Dyck paths below the line $y = (2/5)x + 2/5$ (or touching it) are in bijection with walks allowing jumps $+2$ and -5, starting at altitude 2, and staying above the line $y = 0$ (or touching it)

(see below). Let $f_{n,k}$ be the number of walks of length n which end at altitude k. The corresponding bivariate generating function is given by

$$F(z, u) = \sum_{n,k\geq 0} f_{n,k} z^n u^k = \sum_{n\geq 0} f_n(u) z^n = \sum_{k\geq 0} F_k(z) u^k,$$

where the $f_n(u)$ encode all walks of length n, and the $F_k(z)$ are the generating functions of walks ending at altitude k. A step-by-step approach yields the following linear recurrence

$$f_{n+1}(u) = \{u^{\geq 0}\} [P(u) f_n(u)] \qquad \text{for } n \geq 0,$$

with initial value $f_0(u)$ (i.e., the polynomial representing the walks of length 0), and where $\{u^{\geq 0}\}$ is a linear operator extracting all monomials in u with non-negative exponents. Summing the terms $z^{n+1} f_{n+1}(u)$ leads to the functional equation

$$(1 - zP(u))F(z, u) = f_0(u) - zu^{-2} F_0(z) - zu^{-1} F_1(z). \tag{7.6}$$

We apply the *kernel method* in order to transform this equation into a system of linear equations for F_0 and F_1. The factor $K(z, u) := 1 - zP(u)$ is called the *kernel* and the kernel equation is given by $K(z, u) = 0$. Solving this equation for u, we obtain 7 distinct solutions. These split into two groups, namely we get 2 small roots $u_1(z)$ and $u_2(z)$ (the ones going to 0 for $z \sim 0$) and 5 large roots which we call $v_i(z)$ for $i = 1, \ldots, 5$ (the ones going to infinity for $z \sim 0$). It is legitimate to insert the 2 small branches into (7.6) to obtain[4]

[4]In this article, whenever we thought it could ease the reading, without harming the understanding, we write u_1 for $u_1(z)$, or F for $F(z)$, etc.

$$zF_0 + zu_1F_1 = u_1^2 f_0(u_1),$$
$$zF_0 + zu_2F_1 = u_2^2 f_0(u_2).$$

This linear system is easily solved by Kramer's formula, which yields

$$F_0(z) = -\frac{u_1u_2\,(u_1 f_0(u_1) - u_2 f_0(u_2))}{z(u_1 - u_2)},$$
$$F_1(z) = \frac{u_1^2 f_0(u_1) - u_2^2 f_0(u_2)}{z(u_1 - u_2)}.$$

Now, let the functions $F(z,u)$ and $F_k(z)$ denote functions associated with $f_0(u) = u^3$ (i.e., there is one walk of length 0 at altitude 3) and let the functions $G(z,u)$ and $G_k(z)$ denote functions associated with $f_0(u) = u^4$. One thus gets the following theorem:

Theorem 7.2 (Closed forms for the generating functions) *Let us consider walks in $\mathbb{N}^2$ with jumps -2 and $+5$. The number of such walks starting at altitude 3 and ending at altitude 0 is given by $F_0(z)$, the number of such walks starting at altitude 4 and ending at altitude 1 is given by $G_1(z)$, and we have the following closed forms in terms of the small roots $u_1(z)$ and $u_2(z)$ of $1 - zP(u) = 0$ with $P(u) = u^{-2} + u^5$:*

$$F_0(z) = -\frac{u_1u_2\left(u_1^4 - u_2^4\right)}{z(u_1 - u_2)}, \tag{7.7}$$
$$G_1(z) = \frac{u_1^6 - u_2^6}{z(u_1 - u_2)}. \tag{7.8}$$

Thanks to the bijection given in Sect. 4 between walks in the rational slope model and directed lattice paths in the Banderier–Flajolet model (and by additionally reversing the time[5]), it is now possible to relate the quantities A and B of Knuth with F_0 and G_1:

$$A_n := A[5n-1, 2n-1] = [z^{7n-2}]G_1(z), \tag{7.9}$$
$$B_n := B[5n-1, 2n-1] = [z^{7n-2}]F_0(z). \tag{7.10}$$

Indeed, from the bijection of Proposition 7.1, the walks strictly below $y = \frac{a}{c}x + \frac{b}{c}$ (with $a = 2$, $c = 5$) and ending at $(x, y) = (5n-1, 2n-1)$ are mapped (in the Banderier–Flajolet model, not allowing to touch $y = 0$) to walks starting at $(0, b)$ and ending at $(x + y, ax - cy + b) = (7n - 2, 3 + b)$. Reversing the time and allowing to touch $y = 0$ (thus b becomes $b - 1$), we see that A_n counts walks starting at 4, ending at 1 (the generating function of this sequence is given by G_1!) and that B_n counts walks starting at 3, ending at 0 (the generating function of this sequence is

[5]Reversing the time allows us to express all generating functions in terms of just 2 roots. If one does not reverse time, everything works well but the expressions contain the 5 large roots, yielding more complicated closed forms.

given by F_0!). While there is no nice formula for A_n or B_n (see, however, [7] and page 136 for a formula involving nested sums of binomials), it is striking that there is a simple and nice formula for $A_n + B_n$:

Theorem 7.3 (Closed form for the sum of coefficients) *The sum of the number of Dyck paths (in our rational slope model) touching or staying below* $y = (2/5)x + 1/5$ *and* $y = (2/5)x$ *simplifies to the following expression:*

$$A_n + B_n = \frac{2}{7n-1}\binom{7n-1}{2n}. \tag{7.11}$$

Proof A first proof of this was given by [53] using a variant of the cycle lemma. (We comment more on this in Sect. 7.) We give here another proof; indeed, our Theorem 7.2 (Closed form for the generating functions) implies that

$$A_n + B_n = [z^{7n-1}]\left(u_1^5 + u_2^5\right). \tag{7.12}$$

This suggests to use holonomy theory to prove the theorem. First, a resultant equation gives the algebraic equation for $U := u_1^5$ (namely, $z^7 + (U-1)^5U^2 = 0$) and then, the Abel–Tannery–Cockle–Harley–Comtet theorem (see the comment after Proposition 4 in [7]) transforms it into a differential equation for the series $u_1^5(z^2)$. It is also the differential equation (up to distinct initial conditions) for $u_2^5(z^2)$ (as u_2 is defined by the same equation as u_1) and thus of $u_1^5(z^2) + u_2^5(z^2)$. Therefore, it directly gives the differential equation for the series $C(z) = \sum_n (A_n + B_n)z^n$, and it corresponds to the following recurrence for its coefficients:

$$C_{n+1} = \frac{7}{10}\frac{(7n+5)(7n+4)(7n+3)(7n+2)(7n+1)(7n-1)}{(5n+4)(5n+3)(5n+2)(5n+1)(2n+1)(n+1)}C_n\,,$$

which is exactly the hypergeometric recurrence for $\frac{2}{7n-1}\binom{7n-1}{2n}$ (with the same initial condition). This computation takes 1 second on an average computer, while, if not done in this way (e.g., if instead of the resultant shortcut above, one uses several `gfun[diffeq*diffeq]` or variants of it in Maple, see [58] for a presentation of the corresponding package), the computations for such a simple binomial formula surprisingly take hours.

Some additional investigations conducted by Manuel Kauers (private communication) show that this is the only linear combination of A_n and B_n which leads to a hypergeometric solution (to prove this, you can compute a recurrence for a formal linear combination $rA_n + sB_n$, and then check which conditions it implies on r and s if one wishes the associated recurrence to be of order 1, i.e., hypergeometric). It thus appears that $rA_n + sB_n$ is generically of order 5, with the exception of a sporadic $4A_n - B_n$ which is of order 4, and the miraculous $A_n + B_n$ which is of order 1 (hypergeometric).

However, there are many other hypergeometric expressions floating around: expressions of the type of the right-hand side of (7.12) have nice hypergeometric closed forms. This can also be explained in a combinatorial way; indeed, we observe that setting $k = -5$ in Formula (10) from [8] leads to $5W_{-5}(z) = \Theta(A(z) + B(z))$ (where Θ is the pointing operator). The "Knuth pointed walks" are thus in 1-to-5 correspondence with unconstrained walks (see our Table 1, top left) ending at altitude -5.

We want to end this section with exemplifying the miracles involved in the simplifications of (7.11). Using the Flajolet–Soria formula [7] for the coefficients of an algebraic function, we can extract the coefficient of z^{7n-2} of $G_1(z)$ and $F_0(z)$ in terms of nested sums. According to (7.9), this corresponds to A_n and B_n, which are thus given by formulae involving respectively 45 and 34 nested sums[6] (see Fig. 5).

Then, in the next section, we perform some analytic investigations in order to prove what Knuth conjectured:

$$\frac{A_n}{B_n} = \kappa_1 - \frac{\kappa_2}{n} + \mathcal{O}(n^{-2}),$$

with $\kappa_1 \approx 1.63026$ and $\kappa_2 \approx 0.159$.

6 Asymptotics

As usual, we need to locate the dominant singularities and to understand the local behavior there. The fact that there are several dominant singularities makes the game harder here, and this case was only sketched in [8]. Similarly to what happens in the rational world (Perron–Frobenius theory), or in the algebraic world (see [7]), a periodic behavior of the generating function leads to some more complicated proofs, because additional details have to be taken into account. With respect to walks, it is, e.g., crucial to understand how singularities spread among the roots of the kernel. To this aim, some quantities will play a key role: The structural constant τ is defined as the unique positive root of $P'(\tau)$, where

$$P(u) = u^{-2} + u^5$$

is encoding the jumps, and the structural radius ρ is given as $\rho = 1/P(\tau)$. For our problem, one thus has the explicit values (Fig. 6):

$$\tau = \sqrt[7]{\frac{2}{5}}, \qquad P(\tau) = \frac{7}{10}\sqrt[7]{2^5 5^2}, \qquad \rho = \frac{\sqrt[7]{2^2 5^5}}{7}.$$

From [8], we know that the small branches $u_1(z)$ and $u_2(z)$ are possibly singular only at the roots of $P'(u)$. Note that the jump polynomial has *periodic support* with

[6]Via the kernel method, as explained in [11], it is possible to express A_n and B_n with less nested sums than in Fig. 5 but the corresponding formulae are, however, still of the "ugly" type!

$$A_n = \sum_{m=0}^{7n-2} m! \sum_{\substack{m_1+\cdots+m_{44}=m+1 \\ b_1 m_1+\cdots+b_{44}m_{44}=7n-2 \\ c_1 m_2+\cdots+c_{44}m_{44}=m}} \Big(20^{m_1} 3^{m_2} (-190)^{m_3} (-39)^{m_4} 1140^{m_5} 239^{m_6} 4^{m_7} (-4845)^{m_8}$$
$$(-915)^{m_9} (-25)^{m_{10}} 15504^{m_{11}} 2443^{m_{12}} 68^{m_{13}} 1^{m_{14}} (-38760)^{m_{15}} (-4806)^{m_{16}} (-105)^{m_{17}}$$
$$77520^{m_{18}} 7173^{m_{19}} 100^{m_{20}} (-125970)^{m_{21}} (-8238)^{m_{22}} (-59)^{m_{23}} 167960^{m_{24}} 7305^{m_{25}} 20^{m_{26}}$$
$$(-184756)^{m_{27}} (-4971)^{m_{28}} (-3)^{m_{29}} 167960^{m_{30}} 2553^{m_{31}} (-125970)^{m_{32}} (-959)^{m_{33}} 77520^{m_{34}}$$
$$249^{m_{35}} (-38760)^{m_{36}} (-40)^{m_{37}} 15504^{m_{38}} 3^{m_{39}} (-4845)^{m_{40}} 1140^{m_{41}} (-190)^{m_{42}}$$
$$20^{m_{43}} (-1)^{m_{44}} \Pi_{k=1}^{44} \frac{1}{m_i!} \Big),$$

where $(b_n)_{n=1}^{44} = (2,5,4,7,6,9,12,8,11,14,10,13,16,19,12,15,18,14,17,20,16,19,22,18,21,24,20,23,26,22,25,24,27,26,29,28,31,30,33,32,34,36,38,40)$ and $(c_n)_{n=1}^{44} = (2,0,3,1,4,2,0,5,3,1,6,4,2,0,7,5,3,8,6,4,9,7,5,10,8,6,11,9,7,12,10,13,11,14,12,15,13,16,14,17,18,19,20,21)$.

$$B_n = \sum_{m=0}^{7n-2} m! \sum_{\substack{m_1+\cdots+m_{33}=m+1 \\ b_1 m_1+\cdots+b_{33}m_{33}=7n-2 \\ c_1 m_2+\cdots+c_{33}m_{33}=m}} \Big(20^{m_1} 2^{m_2} (-182)^{m_3} (-18)^{m_4} 1006^{m_5} 73^{m_6} (-1)^{m_7} (-3793)^{m_8}$$
$$(-176)^{m_9} 10349^{m_{10}} 279^{m_{11}} (-21084)^{m_{12}} (-294)^{m_{13}} 32521^{m_{14}} 190^{m_{15}} 1^{m_{16}} (-37980)^{m_{17}}$$
$$(-57)^{m_{18}} (-10)^{m_{19}} 33128^{m_{20}} 45^{m_{21}} (-20928)^{m_{22}} (-120)^{m_{23}} 9039^{m_{24}} 210^{m_{25}} (-2384)^{m_{26}}$$
$$(-252)^{m_{27}} 289^{m_{28}} 210^{m_{29}} (-120)^{m_{30}} 45^{m_{31}} (-10)^{m_{32}} 1^{m_{33}} \Pi_{k=1}^{33} \frac{1}{m_i!} \Big),$$

where $(b_n)_{n=1}^{33} = (2,5,4,7,6,9,12,8,11,10,13,12,15,14,17,13,16,19,15,18,17,20,19,22,21,24,23,26,25,27,29,31,33)$ and $(c_n)_{n=1}^{33} = (2,0,3,1,4,2,0,5,3,6,4,7,5,8,6,11,9,7,12,10,13,11,14,12,15,13,16,14,17,18,19,20,21)$.

$$A_n + B_n = \frac{2}{7n-1}\binom{7n-1}{2n}.$$

Fig. 5 The "ugly + ugly = nice" formula. A_n is counting Dyck paths touching or staying below the line $y = (2/5)x + 1/5$, and B_n is counting Dyck paths touching or staying below the line $y = (2/5)x$. They are given by complicated "ugly" nested sums, so the miracle is that the sum $A_n + B_n$ is nice. We give several explanations of this fact in this article

period $p = 7$ as $P(u) = u^{-2}H(u^7)$ with $H(u) = 1 + u$. Due to that, there are 7 possible singularities of the small branches

$$\zeta_k = \rho\omega^k, \qquad \text{with } \omega = e^{2\pi i/7}.$$

Definition 7.3 We call a function $F(z)$ *p-periodic* if there exists a function $H(z)$ such that $F(z) = H(z^p)$.

Additionally, we have the following local behaviors:

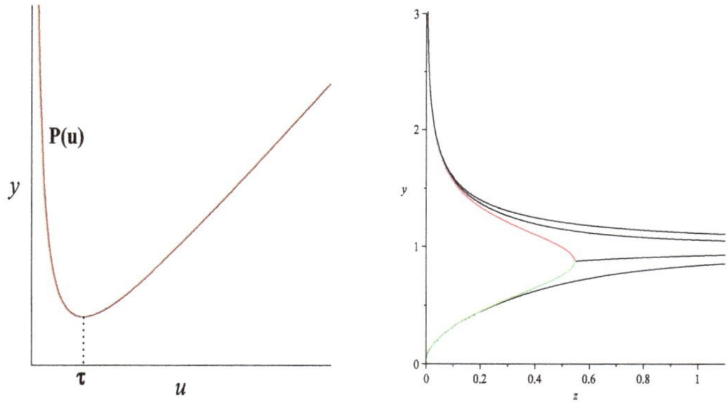

Fig. 6 $P(u)$ is the polynomial encoding the jumps, and its saddle point τ gives the singularity $\rho = 1/P(\tau)$ where the small root u_1 (in green) meets the large root v_1 (in red), with a square root behavior. (In black, we also plotted $|u_2|$, $|v_2| = |v_3|$, and $|v_4| = |v_5|$.) This is the key for all asymptotics of such lattice paths

Lemma 7.1 (Local behavior due to rotation law) *The limits of the small branches when $z \to \zeta_k$ exist and are equal to*

$$u_1(z) \underset{z \sim \zeta_k}{=} \begin{cases} \tau\omega^{-3k} + C_k\sqrt{1 - z/\zeta_k} + \mathcal{O}((1 - z/\zeta_k)^{3/2}), & \text{for } k = 2, 5, 7, \\ \tau_2\omega^{-3k} + D_k(1 - z/\zeta_k) + \mathcal{O}((1 - z/\zeta_k)^2), & \text{for } k = 1, 3, 4, 6, \end{cases}$$

$$u_2(z) \underset{z \sim \zeta_k}{=} \begin{cases} \tau_2\omega^{-3k} + D_k(1 - z/\zeta_k) + \mathcal{O}((1 - z/\zeta_k)^2), & \text{for } k = 2, 5, 7, \\ \tau\omega^{-3k} + C_k\sqrt{1 - z/\zeta_k} + \mathcal{O}((1 - z/\zeta_k)^{3/2}), & \text{for } k = 1, 3, 4, 6, \end{cases}$$

where $\tau_2 = u_2(\rho) \approx -.707723271$ is the unique real root of $500t^{35} + 3900t^{28} + 13540t^{21} + 27708t^{14} + 37500t^7 + 3125$, $C_k = -\frac{\tau}{\sqrt{5}}\omega^{-3k}$, *and* $D_k = \tau_2 \frac{\tau_2^7+1}{5\tau_2^7-2}\omega^{-3k}$.

Proof We will show the following *rotation law* for the small branches (for all $z \in \mathbb{C}$, with $|z| \leq \rho$ and $0 < \arg(z) < \pi - 2\pi/7$):

$$u_1(\omega z) = \omega^{-3} u_2(z),$$
$$u_2(\omega z) = \omega^{-3} u_1(z).$$

Let us consider the function $U(z) := \omega^3 u_i(wz)$ (with $i = 1$ or $i = 2$, as you prefer!) and the quantity X, defined by $X(z) := U^2 - z\phi(U)$ (where $\phi(u) := u^2 P(u)$). So we have $X(z) = (\omega^3 u_i(\omega z))^2 - z\phi(\omega^3 u_i(\omega z)) = \omega^6 u_i(\omega z)^2 - z\phi(u_i(\omega z))$ (because ϕ is 7-periodic) and thus $\omega X(z/\omega) = \omega(\omega^6 u_i(z)^2 - z/\omega\phi(u_i(z))) = u_i(z)^2 - z\phi(u_i(z))$, which is 0 because we recognize here the kernel equation. This implies that $X = U^2 - z\phi(U) = 0$ and thus U is a root of the kernel. Which one? It is one of the small roots, because it is converging to 0 at 0. What is more, this root U is not u_i, because it has a different Puiseux expansion (and Puiseux expansions are unique). So, by the analytic continuation principle (therefore, here, as far as we avoid the cut

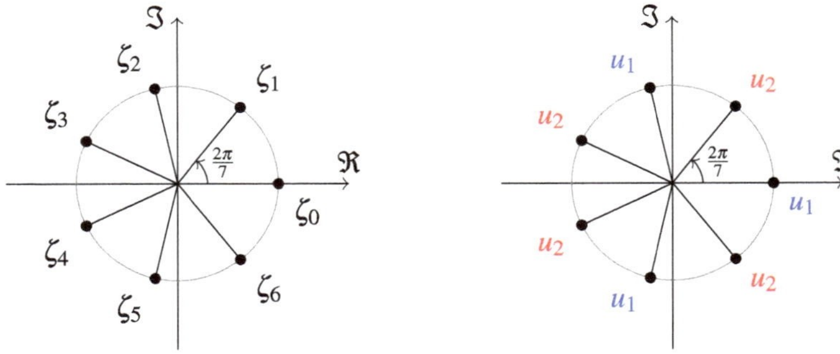

Fig. 7 The locations of the 7 possible singularities of the small branches (left); the small branch which is singular at that location (right)

line $\arg(z) = -\pi$), we just proved that $\omega^3 u_1(\omega z) = u_2(z)$ and $\omega^3 u_2(\omega z) = u_1(z)$ (and this also proves a similar rotation law for large branches, but we do not need it).

Accordingly, at every ζ_k, among the two small branches, only one branch becomes singular: This is u_1 for $k = 2, 5, 7$ and u_2 for $k = 1, 3, 4, 6$. This is illustrated in Fig. 7.

Hence, we directly see how the asymptotic expansion at the dominant singularities is correlated with the one of u_1 at $z = \rho = \zeta_7$, which we derive following the approach of [8]; this gives for $z \sim \rho$:

$$u_1(z) = \tau + C_7\sqrt{1 - z/\rho} + C_7'(1 - z/\rho)^{3/2} + \dots,$$

where $C_7 = -\sqrt{2\frac{P(\tau)}{P''(\tau)}}$. Note that in our case $P^{(3)}(\tau) = 0$ (this funny cancellation holds for any $P(u) = p_5 u^5 + p_0 + p_{-2}u^{-2}$), so even the formula for C_7' is quite simple: $C_7' = -\frac{1}{2}C_7$.

In the lemma, the formula for $\tau_2 = u_2(\rho)$ is obtained by a resultant computation.

For the local analysis of Knuth's generating functions $F_0(z)$ and $G_1(z)$ with periodic support, we introduce a shorthand notation.

Definition 7.4 (*Local asymptotics extractor* $[z^n]_{\zeta_k}$) Let $F(z)$ be an algebraic function with p dominant singularities ζ_k (for $k = 1, \dots, p$). Accordingly, for each ζ_k, $F(z)$ can be expressed as a Puiseux series; that is, there exist $r \in \mathbb{Q}$ and coefficients c_n (both depending on k) such that

$$F(z) = \sum_{j \geq 0} c_j (1 - z/\zeta_k)^{rj}, \qquad \text{for } z \sim \zeta_k.$$

Then, we define the local asymptotic extractor $[z^n]_{\zeta_k}$ as

$$[z^n]_{\zeta_k} F(z) := \sum_{j \geq 0} c_j [z^n](1 - z/\zeta_k)^{rj}.$$

This notation can be considered as "extracting the z^n-coefficient in the Puiseux expansion[7] of $F(z)$ at $z = \zeta_k$," and singularity analysis allows to write $[z^n]F(z) = \sum_k [z^n]_{\zeta_k} F(z) + o(C^{-n})$, for some constant $C > |\zeta_k|$.

Example 7.1 A sloppy but easy to remember formulation would be to say

$$[z^n]_{\zeta_k} F(z) := [z^n](\text{singular expansion of } F(z) \text{ at } z = \zeta_k).$$

This is well illustrated by the generating function $D(z)$ of Dyck paths defined by the functional equation $D(z) = 1 + z^2 D(z)^2$. In this case, we have $D(z) = \frac{1-\sqrt{1-4z^2}}{2z^2}$ with $p = 2$ and $\zeta_1 = 1/2$ and $\zeta_2 = -1/2$. Therefore, we get for any $\varepsilon > 0$

$$\begin{aligned}[z^n]D(z) &= [z^n]_{1/2}\, D(z) + [z^n]_{-1/2}\, D(z) + o\left((2-\varepsilon)^n\right) \\ &= [z^n](-2\sqrt{2})\sqrt{1-2z} + [z^n](-2\sqrt{2})\sqrt{1+2z} + O\left(\frac{2^n}{n^{5/2}}\right) + o\left((2-\varepsilon)^n\right).\end{aligned}$$

Proposition 7.2 (Periodic rule of thumb) *Let ρ be the positive real dominant singularity in the previous definition. If additionally the generating function $F(z)$ satisfies a rotation law $F(\omega z) = \omega^m F(z)$ (where $\omega = \exp(2i\pi/p)$, p maximal), then one has a neat simplification:*

$$[z^n]F(z) = p[z^n]_\rho F(z) + o(\rho^n),$$

if $n - m$ is a multiple of p. (The other coefficients are equal to 0.*)*

Proof As $F(z)$ is a generating function, it has real positive coefficients, and therefore, by Pringsheim's theorem [33, Theorem IV.6], one of the ζ_k's has to be real positive, called ρ. We relabel the ζ_k's such that $\zeta_k := \omega^k \rho$. Then

$$\begin{aligned}[z^n]F(z) - o(\rho^n) &= \sum_{k=1}^{p} [z^n]_{\zeta_k} F(z) = \sum_{k=1}^{p} [z^n]_{\zeta_k} (\omega^m)^k F(\omega^{-k} z) \\ &= \sum_{k=1}^{p} (\omega^m)^k (\omega^{-k})^n [z^n]_\rho F(z) \\ &= \left(\sum_{k=1}^{p} (\omega^k)^{m-n}\right) [z^n]_\rho F(z) = p[z^n]_\rho F(z),\end{aligned}$$

if $n - m$ is a multiple of p, and 0 elsewhere.

We can apply this proposition to $F_0(z)$ and $G_1(z)$, because the rotation law for the u_i's implies: $F_0(\omega z) = \omega^{-2} F_0(z)$ and $G_1(\omega z) = w^{-2} G_1(z)$. Thus, we just have to compute the asymptotics coming from the Puiseux expansion of $F_0(z)$ and $G_1(z)$

---

[7] In fact, this notation holds for singular expansions of alg-log functions [33], exp-log functions and more generally for expansions in Hardy fields [36] which are amenable to singularity analysis or saddle point methods.

at $z = \rho$, and multiply it by 7 (recall that it is classical to infer the asymptotics of the coefficients from the Puiseux expansion of the functions via the so-called transfer Theorem VI.3 from [33]); this gives:

Theorem 7.4 (Asymptotics of coefficients, answer to Knuth's problem) *The asymptotics for the number of excursions below $y = (2/5)x + 2/5$ and $y = (2/5)x + 1/5$ are given by:*

$$A_n = [z^{7n-2}]G_1(z) = \alpha_1 \frac{\rho^{-7n}}{\sqrt{\pi(7n-2)^3}} + \frac{3\alpha_2}{2} \frac{\rho^{-7n}}{\sqrt{\pi(7n-2)^5}} + \mathcal{O}(\rho^{-7n} n^{-7/2}),$$

$$B_n = [z^{7n-2}]F_0(z) = \beta_1 \frac{\rho^{-7n}}{\sqrt{\pi(7n-2)^3}} + \frac{3\beta_2}{2} \frac{\rho^{-7n}}{\sqrt{\pi(7n-2)^5}} + \mathcal{O}(\rho^{-7n} n^{-7/2}),$$

with the following constants, where we use the shorthand μ for τ_2/τ:

$$\alpha_1 = \frac{\mu^4 + 2\mu^3 + 3\mu^2 + 4\mu + 5}{\sqrt{5}}, \qquad \beta_1 = \sqrt{5} - \alpha_1, \qquad \beta_2 = -\frac{9}{10}\sqrt{5} - \alpha_2,$$

$$\alpha_2 = -\frac{1}{2} \frac{\tau_2^7 (13\mu^4 + 22\mu^3 + 29\mu^2 + 36\mu + 45)}{\sqrt{5}(5\tau_2^7 - 2)} - \frac{1}{5} \frac{15\mu^4 + 20\mu^3 + 13\mu^2 - 8\mu - 45}{\sqrt{5}(5\tau_2^7 - 2)}.$$

This theorem leads to the following asymptotics for $A_n + B_n$ (and this is for sure a good sanity test, coherent with a direct application of Stirling's formula to the closed-form formula (7.11) for $A_n + B_n$):

$$A_n + B_n = \sqrt{\frac{5}{7^3 \pi}} \frac{\rho^{-7n}}{\sqrt{n^3}} + \mathcal{O}(n^{-5/2}).$$

Finally, we directly get

$$\frac{A_n}{B_n} = \frac{\alpha_1 + \frac{3\alpha_2}{2(7n-2)}}{\beta_1 + \frac{3\beta_2}{2(7n-2)}} + \mathcal{O}(n^{-2}) = \frac{\alpha_1}{\beta_1} + \frac{3}{14}\left(\frac{\alpha_2\beta_1 - \alpha_1\beta_2}{\beta_1^2}\right)\frac{1}{n} + \mathcal{O}(n^{-2}),$$

which implies that Knuth's constants are

$$\kappa_1 = \frac{\alpha_1}{\beta_1} = -\frac{5}{\mu^4 + 2\mu^3 + 3\mu^2 + 4\mu} - 1$$
$$\approx 1.6302576629903501404248,$$

$$\kappa_2 = -\frac{3}{14}\left(\frac{\alpha_2\beta_1 - \alpha_1\beta_2}{\beta_1^2}\right) = \frac{3}{9800}(13 - 236\kappa_1 - 194\kappa_1^2 - 388\kappa_1^3 + 437\kappa_1^4)$$
$$\approx 0.1586682269720227755147.$$

Now a few resultant computations give the algebraic equations satisfied by τ_2, κ_1 and κ_2. We will illustrate their derivation with the required Maple commands. In what follows, these are always set in a typewriter font. First, we compute an annihilating polynomial for ρ:

```
>   R1:=resultant(numer(1-z*P),numer(diff(P,u)),u);
```

$$R1 := 823543\, z^7 - 12500$$

Then, we construct from it an annihilating polynomial for $u_i(\rho)$.

```
>   R2:=factor(resultant(numer(1-z*P),R1,z));
```

$$\left(500\, u^{35} + 3900\, u^{28} + 13540\, u^{21} + 27708\, u^{14} + 37500\, u^7 + 3125\right)\left(-2 + 5\, u^7\right)^2$$

This polynomial contains $u_1(\rho) = \tau$ and $u_2(\rho) = \tau_2$ as roots. It factorizes into smaller polynomials, and these two roots are in separate factors. Thus, we can go on with the right factor which we save in `Rtau2`. Then, we continue with the annihilating polynomial for μ.

```
>   resultant(x*t-t2,subs(u=t,diff(P,u)),t);
>   factor(resultant(%,subs(u=t2,Rtau2),t2));
```

We identify the algebraic relation for μ and save it in `Rmu`. Finally, we compute the minimal polynomial for κ_1:

```
>   Rmu:=2*u^5+4*u^4+6*u^3+8*u^2+10*u+5;
>   Rk1:=resultant((x+1)*(u^4+2*u^3
>   +3*u^2+4*u)+5,Rmu,u):
>   factor(Rk1/igcd(coeffs(Rk1)));
```

$$-23\, x^5 + 41\, x^4 - 10\, x^3 + 6\, x^2 + x + 1$$

In conclusion, κ_1 is the unique real root of the polynomial $23x^5 - 41x^4 + 10x^3 - 6x^2 - x - 1$, and similar computations show that $(7/3)\kappa_2$ is the unique real root of $11571875x^5 - 5363750x^4 + 628250x^3 - 97580x^2 + 5180x - 142$. The Galois group of each of these polynomials is S_5. This implies that there is no closed-form formula for the Knuth constants κ_1 and κ_2 in terms of basic operations on integers and roots of any degree.

In the next section, we want to establish a link with the results from Nakamigawa and Tokushige. We will show how Knuth derived his problem and how to establish more such nice identities.

7 Links with the Work of Nakamigawa and Tokushige

In this section, we show the connection between a result of Nakamigawa and Tokushige [53] and Knuth's statement. Furthermore, we derive extensions of this result.

Let α, β be positive rational numbers. The Nakamigawa–Tokushige model consists of a single boundary $L : y = \alpha x + \beta$ and a lattice point[8] $Q = (q_1, q_2) \in \mathbb{Z}^2$

---

[8] In the article [53], $Q = (m, n)$; we changed these coordinates in order to avoid a conflict with our other notations.

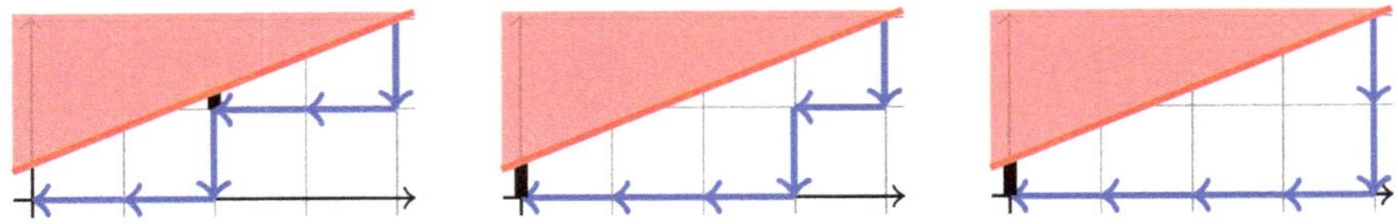

Fig. 8 The 3 walks of length 6 in the $(2/5)x + 2/5$ model with $\delta(w) > 0$. The vertical bars mark the minimal y-distance $\delta(w)$. The first walk has $\delta(w) = 1/5$, whereas the last two have $\delta(w) = 2/5$. All of them are members of $W_{1/5}$, but only the two last ones belong to $W_{2/5}$

on L, i.e., $q_2 = \alpha q_1 + \beta$. Furthermore, the walks go in the opposite direction; that is, they start in Q, use unit steps South and West (i.e., $(0, -1)$ and $(-1, 0)$, respectively) and end in the origin. Let V be the "vast" set of such walks without any restriction. The enumeration of V is a folklore result: $|V| = \binom{q_1+q_2}{q_1}$. Let $W \subset V$ be the set of walks which do not cross the line L and touch it only at Q.

Definition 7.5 (*Nearest distance to the boundary*) Let $w \in V$ be a walk from a point Q to the point $(0, 0)$. We define the *minimum y-distance* $\delta(w)$ as follows: If the walk w touches or crosses the boundary $y = \alpha x + \beta$ after the first step, then let $\delta(w) = 0$, otherwise let $\delta(w)$ be the minimum of $\alpha p_1 + \beta - p_2$, where (p_1, p_2) runs over all lattice points on w except Q, see Fig. 8.

Hence, we see that $\delta(w) = 0$ if and only if $w \in V \setminus W$, and so $\sum_{w \in V} \delta(w) = \sum_{w \in W} \delta(w)$. Note, if α and β are positive integers, then $\sum_{w \in V} \delta(w) = |W|$, because $\delta(w) = 1$ for all $w \in W$. This gives rise to the interpretation as a weighted sum corresponding to the number of walks.

For a real $t \geq 0$, let $W_t := \{w \in W \mid \delta(w) \geq t\}$; that is, the walks staying at least a y-distance of t away from the boundary. Due to the definition, $|W_t|$ is a left-continuous step function of t, and we get the representation

$$\int_0^1 |W_t|\, dt = \sum_{w \in V} \delta(w).$$

It is quite nice that this sum can be further simplified; this is what the next theorem states:

Theorem 7.5 (Nakamigawa–Tokushige lattice path integral) *Let q_1, q_2 be positive integers, and let α, β be positive reals with $q_2 = \alpha q_1 + \beta$. Let V be the set of walks from the origin to the point*[9] *(q_1, q_2). Then, we have*

$$\int_0^1 |W_t|\, dt = \sum_{w \in V} \delta(w) = \frac{\beta}{q_1 + q_2} \binom{q_1 + q_2}{q_1}. \tag{7.13}$$

[9]Nota bene: As proven in Lemma 7.2 (Possible starting points on the boundary), if α and β are irrational, then there is at most one such point. While if α and β are rational (with the right gcd condition), then there are infinitely many such points.

Proof This corresponds to [53, Theorem 1 and Corollary 1], where it is proven using a cycle lemma approach. We give a generalization of this formula in Sect. 8 hereafter, based on our kernel method approach and Lagrange inversion.

A geometric bijection. If α is a rational slope, i.e., $\alpha = a/c$ for some $a, c \in \mathbb{N} \setminus \{0\}$, then

$$\int_0^1 |W_t|\,dt = \frac{1}{c}\sum_{t\in T} |W_t|, \tag{7.14}$$

where $T = \{\delta(w) \mid w \in W\} = \{1/c, 2/c, \ldots, (c-1)/c\}$.

This gives rise to the following interpretation[10]: If $w \in W$, then the first step is a South step. Then, let $\tilde{w}$ be the walk obtained from w by omitting this step. Therefore, $\tilde{w}$ is a walk with $q_1 + q_2 - 1$ steps, starting from $Q - (0, 1) = (q_1, q_2 - 1)$ and ending in the origin. We see that all these walks which never cross or touch L are in bijection with all walks in W. Now, take a walk $w \in W_t$ and its corresponding walk $\tilde{w}$. As $\delta(w) \geq t$, we can translate the barrier L by $t - 1/c$ down and the walk $\tilde{w}$ still does not touch or cross this new barrier $\tilde{L}$. Hence, all walks in W_t are in bijection with walks from $(q_1, q_2 - 1)$ to the origin which stay strictly below the barrier $\tilde{L}$.

Example 7.2 This is the bijection that Knuth used in order to state his conjecture. In his case, we have $\alpha = \beta = 2/5$ and $q_1 = 5n - 1$, $q_2 = 2n$ for $n \in \mathbb{N} \setminus \{0\}$. We see that $q_2 = \alpha q_1 + \beta$. Hence, $a = 2$ and $c = 5$ which implies $T = \{1/5, 2/5, 3/5, 4/5\}$. In this case, the values $3/5$ and $4/5$ are playing no role, as $|W_{3/5}| = |W_{4/5}| = 0$ because $\beta = 2/5$ is the maximal value for $\delta(w)$ for all walks to the origin. Therefore, $\int_0^1 |W_t|\,dt$ can be represented by two summands involving $W_{1/5}$ and $W_{2/5}$. They correspond to the two models A and B with the barriers $L_1 : y < (2/5)x + 2/5$ and $L_2 : y < (2/5)x + 1/5$, respectively, where the paths start at $(5n - 1, 2n - 1)$ and move by South and West steps to the origin. Compare also Fig. 8. Note that in Knuth's case the walks move in the opposite direction, which is obviously equivalent.

In general, the number of summands $|W_t|$, which corresponds to the number of models in the equivalent formulation, is determined by the size of T minus the maximal y-distance at $(0, 0)$. Hence, we need to consider $\widetilde{T} = \{t \in T \mid t < \beta\} = \{1/c, \ldots, k/c\}$. This gives k models with walks from $(q_0, q_1 - 1)$ to the origin which stay strictly below the boundaries $L_i : y < \alpha x + (\beta - (i-1)/c)$ for $i = 1, \ldots, k$. Then, the above reasoning implies that the walks with boundary L_i correspond to the set $W_{i/c}$. Thus, counting the walks in these k models and summing them up give the binomial closed form appearing in the lattice path integral theorem (7.13) divided by c, compared with (7.14).

Up to now in this section, we explained which different counting models are connected with the Nakamigawa–Tokushige lattice path integral formula. Now, we discuss the possible starting points on the boundary and their interplay with the (ir)rationality of the slope.

[10]In the original work, a slightly different interpretation is given.

Lemma 7.2 (Possible starting points on the boundary) *Let α, β be positive reals. Then the equation $y = \alpha x + \beta$ possesses in the positive integers*

1. *infinitely many solutions (x, y), if $\alpha = a/c$, $\beta = b/c$ with $a, b, c \in \mathbb{N}$, and $\gcd(a, c)|b$;*

$$x = cs - r_a, \qquad\qquad y = as + r_c,$$

   with $s \geq S_0 := \max(\lceil r_a/c\rceil, \lceil -r_c/a\rceil)$, and r_a and r_c are integers such that $r_a a + r_c c = b$;
2. *exactly one solution $(x, y) = (q_1, q_2)$, if $\alpha \notin \mathbb{Q}$ and $\beta = q_2 - \alpha q_1 > 0$;*
3. *no solution, otherwise.*

Proof Let us start with rational slope $\alpha = a/c$, with $a, c \in \mathbb{N}$. In order to get integer solutions we need a rational $\beta = b/c$, with $b \in \mathbb{N}$. Then we need to find the solutions of the following linear Diophantine equation:

$$cy - ax = b. \tag{7.15}$$

These solutions exist if and only if $\gcd(a, c)|b$. By the extended Euclidean algorithm we get integers $r_a, r_c \in \mathbb{Z}$ such that

$$r_a a + r_c c = b.$$

This is done by first computing numbers r'_a, r'_c such that

$$r'_a a/\gcd(a, c) + r'_c/\gcd(a, c) = 1$$

and multiplying by b. All solutions are then given by the linear combination stated in the lemma. Due to the special form of (7.15) with a positive and a negative coefficient in front of the unknowns, it follows that for all $s \geq S_0$ the solutions are positive.

Finally, let α be irrational. Assume there exist two points $Q = (q_1, q_2)$ and $P = (p_1, p_2)$ fulfilling the assumptions. By taking the difference, we get $q_2 - p_2 = \alpha(q_1 - p_1)$ which implies that for $q_1 \neq q_2$ we get the contradiction $\alpha \in \mathbb{Q}$. But for $q_1 = q_2$ it also holds that $p_1 = p_2$ and therefore $Q = P$.

It is easy to see that this solution exists if and only if $\beta = q_2 - \alpha q_1$ for arbitrary $q_1, q_2 \in \mathbb{N}$ as long as $\beta > 0$.

The previous lemma also appeared in [42]; there, Kempner (of Kempner's series fame) also mentions that a similar claim holds for the number of algebraic rational (respectively algebraic) points on $y = \alpha x + \beta$ when α is algebraic (respectively transcendental) slope. The lemma gives us all possible integer solutions on a boundary with rational slope. With this knowledge, we can reformulate the lattice path integral from Theorem 7.5 in order to give a more explicit result for all possible starting points and for any slope.

Theorem 7.6 (Lattice path integral and explicit binomial expression) *Let a, b, c be positive integers such that $\gcd(a, c)|b$. Let r_a, r_c be integers such that $r_a a + r_c c = b$. Then, $q_1(s) := cs - r_a$ and $q_2(s) := as + r_c$ define all pairs $(q_1(s), q_2(s))$ of integers on the barrier $L : y = \frac{a}{c}x + \frac{b}{c}$. Furthermore, let V be the set of walks from $(q_1(s), q_2(s))$ to the origin strictly below the barrier L. Then, we have*

$$\int_0^1 |W_t|\, dt = \frac{b/c}{(a+c)s + (r_c - r_a)} \binom{(a+c)s + (r_c - r_a)}{as + r_c}, \tag{7.16}$$

for $s \geq S_0 := \max\left(\lceil r_a/c \rceil, \lceil -r_c/a \rceil\right)$.

For fixed s, the walks are ending after $q_1(s) + q_2(s) = (a+c)s + (r_c - r_a)$ steps, start at $(q_1(s), q_2(s))$ and go to the origin. In the equivalent formulation, the walks start at $(q_1(s), q_2(s) - 1)$ and go to the origin, but we consider $k = c\beta = b$ different boundaries, given by

$$L_1 : y < \frac{a}{c}x + \frac{b}{c}, \qquad L_2 : y < \frac{a}{c}x + \frac{b-1}{c}, \qquad \ldots, \qquad L_b : y < \frac{a}{c}x + \frac{1}{c}.$$

Example 7.3 Returning to Knuth's model, we have $y < \frac{2}{5}x + \frac{2}{5}$. Thus, the explicit values are $a = b = 2$ and $c = 5$ and the assumptions of Theorem 7.6 (Lattice path integral and explicit binomial expression) are satisfied, as $\gcd(a, c) = 1$. The Euclidean algorithm gives $r_a = -4$ and $r_c = 2$. From Lemma 7.2 on the possible starting point on the boundary, we deduce the possible integer coordinates on the barrier L:

$$q_1(s) = 5s + 4, \qquad\qquad q_2(s) = 2s + 2,$$

for $s \geq 0$ which represent the starting points of the walks. Finally, Theorem 7.6 directly gives the solution

$$\int_0^1 |W_t|\, dt = \frac{2/5}{7s+6}\binom{7s+6}{2s+2}.$$

This value can be equivalently interpreted as the number of walks in $k = 2$ models starting from $(5s + 4, 2s + 1)$ and moving to the origin below the barriers

$$L_1 : y < \frac{2}{5}x + \frac{2}{5}, \qquad\qquad L_2 : y < \frac{2}{5}x + \frac{1}{5}.$$

This is exactly Knuth's problem, where his index $t = s + 1$.

Formula (7.16) directly yields nice lattice path identities in the manner of Knuth's problem. Yet, there are even more formulae of this type that we will reveal in the next section. But let us start with an interesting (everyday) problem first.

8 Duchon's Club and Other Slopes

8.1 Duchon's Club: Slope 2/3 and Slope 3/2

A Duchon walk is a Dyck path starting from (0, 0), with East and North steps, and ending on the line $y = \frac{2}{3}x$ (see Fig. 9). This model was analyzed by Duchon [28] and further investigated by Banderier and Flajolet [8], who called it the "Duchon's club" model, as it can be seen as the number of possible "histories" of couples entering a club in the evening,[11] and exiting in groups of 3. What is the number of possible histories (knowing the club is closing empty)? Well, this is exactly the number E_n of excursions with n steps $+2, -3$, or (by reversal of the time) the number of excursions with n steps $-2, +3$. This gives the sequence $(E_{5n})_{n\in\mathbb{N}} = (1, 2, 23, 377, 7229, 151491, 3361598, \dots)$ (OEIS A060941). In fact, these numbers E_n appeared already in the article by Bizley [17] (who gave some binomial formulae, as we explained in Sect. 2). Duchon's club model should then be the Bizley–Duchon's club model; Stigler's law of eponymy strikes again.

One open problem in the article [28] was the following one: "The mean area is asymptotic to $Kn^{3/2}$, but the constant K can only be approximated to 3.43." Our method allows to identify this mysterious constant:

Theorem 7.7 (Area below Duchon lattice paths) *The average area below Duchon excursions of length n (lattice paths from* 0 *to* 0*, which jumps* -2 *and* $+3$*) is*

$$A_n \sim Kn^{3/2} \text{ where } K = \sqrt{15\pi}/2 \approx 3.432342124\,.$$

Proof The approach of [10] gives an expression for $A(z) = \sum A_n z^n$ in terms of the two small roots $u_1(z)$ and $u_2(z)$ of $1 - z(1/u^2 + u^3) = 0$. Then, using the rotation law gives the singular behavior of $A(z)$ and therefore the asymptotics of A_n with the explicit constant K (Fig. 11).

8.2 Arbitrary Rational Slope

The closed form for the coefficient (Theorem 7.3) generalizes to arbitrary rational slope:

Theorem 7.8 (General closed forms for any rational slope) *Let a, b, c be integers such that* $\gcd(a, c)|b$. *Let $A_s(k)$ be the number of Dyck walks below the line of slope $y = \frac{a}{c}x + \frac{k}{c}$, ending at (x_s, y_s) given by*

$$x_s = cs - r_a, \qquad\qquad y_s = as + r_c - 1,$$

[11] Caveat: There are no real life facts/anecdotes hidden behind this pun!

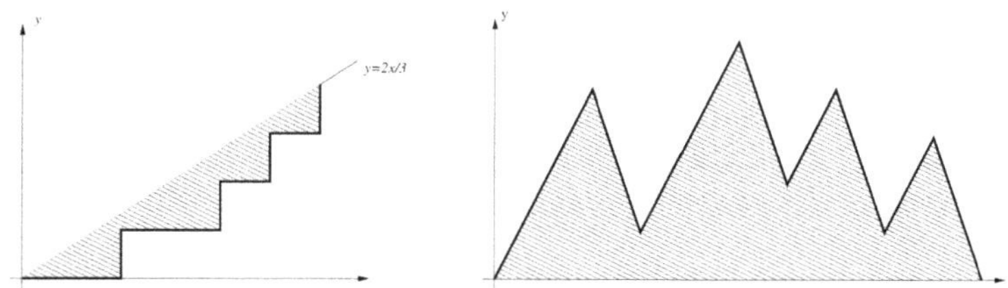

(a) North-East model: Dyck paths below the line of slope 2/3

(b) Banderier–Flajolet model: excursions with +2 and −3 jumps

Fig. 9 Dyck paths below the line of slope 2/3 and Duchon's club histories (i.e., excursions with jumps +2, −3) are in bijection. Duchon conjectured that the average area (in gray) after n jumps is asymptotically equal to $Kn^{3/2}$; our approach shows that $K = \sqrt{15\pi}/2$

where r_a and r_c are integers such that $r_a a + r_c c = b$. These numbers are non-negative for $s \geq S_0 := \max(\lceil r_a/c \rceil, \lceil -r_c/a \rceil)$. Then, we have

$$\sum_{k=1}^{b} A_s(k) = \frac{b}{(a+c)s + (r_c - r_a)} \binom{(a+c)s + (r_c - r_a)}{as + r_c}.$$

Proof This result is a direct consequence of Theorem 7.6 (lattice path integral and explicit binomial expression) and the geometric bijection (7.14).

The enumeration of lattice paths below the line $y = \frac{a}{c}x + \frac{b}{c}$ simplifies even more in the case $a = b$. Additionally, we are able to extend the nice counting formula in terms of binomial coefficients. In order to get these nice formulae, let us first state what becomes the equivalent of Theorem 7.2 (Closed form for the generating function) in the case of any rational slope.

Lemma 7.3 (Schur polynomial closed form for meanders ending at a given altitude) *Let us consider walks in $\mathbb{N}^2$ with jumps $-a$ and $+c$ starting at altitude $h \geq a$. Let $u_1(z), \ldots, u_a(z)$ be the small roots of the kernel equation $1 - zP(u) = 0$, with $P(u) = u^{-a} + u^c$. Let $F_0(z), \ldots, F_{a-1}(z)$ be the generating functions of meanders ending at altitude $0, \ldots, a-1$, respectively. They are given by*

$$F_i(z) = \frac{(-1)^{a-i-1}}{z} s_{(h+1, 1^{a-i-1}, 0^i)}(u_1(z), \ldots, u_a(z)), \tag{7.17}$$

where $s_\lambda(x_1, \ldots, x_a)$ is a Schur polynomial in a variables, and $\lambda = (\lambda_1, \ldots, \lambda_a)$ is an integer partition, i.e., $\lambda_1 \geq \lambda_2 \geq \cdots \geq \lambda_a \geq 0$. The notation 1^s denotes s repetitions of 1.

Proof Similar to (7.6) for the given step set, the functional equation is given by

$$(1 - zP(u))F(z,u) = f_0(u) - zu^{-a}F_0(z) - zu^{-a+1}F_1(z) - \ldots - zu^{-1}F_{a-1}(z).$$

Applying the kernel method, one may insert the a small branches into this equation. Then, one gets a independent linear equations for the a unknowns $F_0(z), \ldots, F_{a-1}(z)$. Expressing the solutions by Cramer's rule and rearranging the determinants, one uncovers the defining expressions for the claimed Schur polynomials (see, e.g., [62, Chap. 7.15] for an introduction to the relevant notions and notations).

Example 7.4 Let us consider the previous lemma for $a = 3$. We get the linear system

$$z \begin{pmatrix} 1 & u_1(z) & u_1(z)^2 \\ 1 & u_2(z) & u_2(z)^2 \\ 1 & u_3(z) & u_3(z)^2 \end{pmatrix} \begin{pmatrix} F_0(z) \\ F_1(z) \\ F_2(z) \end{pmatrix} = \begin{pmatrix} u_1(z)^{h+3} \\ u_2(z)^{h+3} \\ u_3(z)^{h+3} \end{pmatrix}.$$

Solving it with Cramer's rule and rearranging the determinants, we get

$$\begin{aligned} F_0(z) &= \frac{s_{(h+1,1,1)}(u_1, u_2, u_3)}{z}, \\ F_1(z) &= -\frac{s_{(h+1,1,0)}(u_1, u_2, u_3)}{z}, \\ F_2(z) &= \frac{s_{(h+1,0,0)}(u_1, u_2, u_3)}{z}, \end{aligned}$$

by the definition of Schur polynomials.

Now, we are able to extend the results of the closed form for the sum of coefficients (Theorem 7.3) even further. At its heart lies the nice expression (7.12): $u_1^5 + u_2^5$. We will see that such a phenomenon holds in full generality, involving a sum of u_i^h.

Theorem 7.9 (General closed forms for lattice paths below a rational slope $y = \frac{a}{c}x + \frac{b}{c}$, with b a multiple of a) *Let a, c be integers such that $a < c$, and let b be a multiple of a. Let $A_s(k)$ be the number of Dyck walks below the line of slope $y = \frac{a}{c}x + \frac{k}{c}$, $k \geq 1$, ending at (x_s, y_s) given by*

$$x_s = cs - 1, \qquad\qquad y_s = as - 1.$$

Then, it holds for $s \geq 1$ and $\ell \in \mathbb{N}$ such that $(\ell + 1)a < c$ that

$$\sum_{k=\ell a+1}^{(\ell+1)a} A_s(k) = \frac{\ell a + c}{(a+c)s + \ell - 1} \binom{(a+c)s + \ell - 1}{as - 1}.$$

Proof Consider walks starting at $(0, 0)$, ending at (x_s, y_s) and staying below the line $\frac{a}{c}x + \frac{1}{c}$. These are counted by $A_s(1)$. Let us transform such walks by adding a new horizontal jump at the end. Note that the first $\lfloor \frac{c}{a} \rfloor$ jumps must be horizontal jumps. Thus, we can interpret this walk as one starting from $(1, 0)$, ending at $(x_s + 1, y_s)$ staying below the given boundary. But as a horizontal jump increases the distance to the boundary by $\frac{a}{c}$, this is equivalent to counting walks starting at $(0, 0)$, ending at

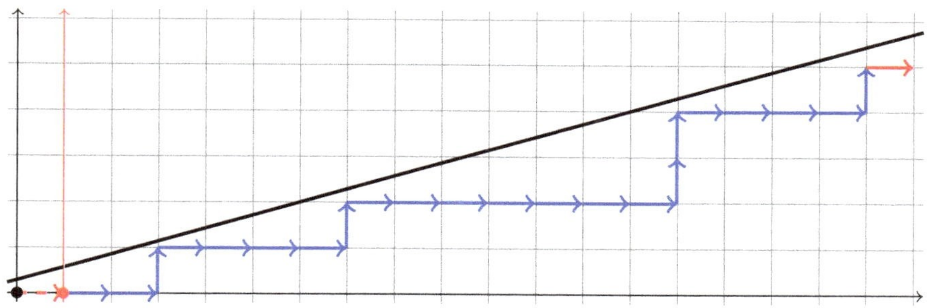

Fig. 10 Transforming walks by moving the first step to the end of the walk. The red dot at (1, 0) and the red y-axis mark the new origin

(x_s, y_s) and staying below the boundary $\frac{a}{c}x + \frac{a+1}{c}$. This process is shown in Fig. 10. Such walks are counted by $A_s(a+1)$.

Thus, the sequence $A_s(1), A_s(a+1), A_s(2a+1), \ldots$ can be interpreted as counting walks staying always below the boundary $\frac{a}{c}x + \frac{1}{c}$, starting at (0, 0) and ending at $(x_s, y_s), (x_s+1, y_s), (x_s+2, y_s), \ldots$, respectively. In particular, for $\ell \geq 0$ we define these new ending points as $(\tilde{x}_s, \tilde{y}_s)$ given by

$$\tilde{x}_s = x_s + \ell = cs + \ell - 1, \qquad\qquad \tilde{y}_s = y_s = as - 1.$$

Analogously, the same holds for $A_s(2), \ldots, A_s(a-1)$.

For the start, we then follow the line of thought from Theorem 7.3 (Closed form for the sum of coefficients). Let us first derive the respective generating functions. Therefore, we apply the bijection from Proposition 7.1, reverse the time and allow to touch $y = 0$. Then, the sum $\sum_{k=\ell a+1}^{(\ell+1)a} A_s(k)$ can be interpreted as walks of length $\tilde{x}_s + \tilde{y}_s = (a+c)s + \ell - 2$, starting at altitude $a\tilde{x}_s - c\tilde{y}_s + i = \ell a + (c-a) + i$ and ending at altitude i for $i = 0, \ldots, a-1$. To simplify notation, let us introduce the constant

$$h := \ell a + c\,.$$

Then, walks end at $h - a + i$. Therefore, we are now able to apply Lemma 7.3 (Schur polynomial closed form for meanders ending at a given altitude). Additionally, by reversing the summation order, we get:

$$\begin{aligned}\sum_{k=\ell a+1}^{(\ell+1)a} A_s(k) &= [z^{(a+c)s+\ell-2}] \sum_{j=0}^{a-1} \frac{(-1)^j}{z} s_{(h-j,1^j,0^{a-j-1})}(u_1(z), \ldots, u_a(z)) \\ &= [z^{(a+c)s+\ell-1}] \left(\sum_{i=1}^{a} u_i(z)^h \right). \end{aligned} \tag{7.18}$$

This surprisingly simple result is due to a nice representation theorem of power symmetric functions in terms of Schur polynomials: [62, Theorem 7.17.1]. One gets this equation by setting $\mu = \emptyset$ and restricting the case to a variables. Note that this is the analogue of (7.12). It is in one sense the reason for the nice closed forms in this article.

In contrast to Theorem 7.3 (Closed form for the sum of coefficients), we proceed now differently by Lagrange inversion [48]. From the kernel method, we know that the small branches $u_i(z)$ satisfy the kernel equation $1 - zP(u) = 0$, where $P(u) = u^{-a} + u^c$ for general slope a/c. The entire form of the kernel equation satisfies nearly a Lagrangian scheme

$$u_i(z)^a = z\left(1 + u_i(z)^{a+c}\right).$$

By taking the a-th root, one gets for an auxiliary power series $U(x)$:

$$U(x) = x\phi(U(x)), \qquad \text{with} \qquad \phi(u) = \left(1 + u^{a+c}\right)^{1/a}.$$

Let $\omega \neq 1$ be an a-th root of unity (i.e., $\omega^a = 1$). Then, we recover the $u_i(z)$, $i = 1, \ldots, a$, by

$$u_i(z) = U\left(\omega^{i-1} z^{1/a}\right).$$

Thus, coming back to (7.18) we are actually interested in

$$\sum_{i=1}^{a} u_i(z)^h = \sum_{i=1}^{a} U\left(\omega^{i-1} z^{1/a}\right)^h = \sum_{n\geq 0} U_n z^{n/a} \left(\sum_{i=1}^{a} \omega^{(i-1)n}\right) = a \sum_{n\geq 0} U_{an} z^n,$$

where $U(x)^h = \sum_{n\geq 0} U_n x^n$ (in fact, by construction many coefficients U_n are 0, because $U(z)$ has an $(a+c)$ periodic support, but this is not altering our reasoning hereafter). Considering (7.18) again, we need U_{an} for $n = (a+c)s + \ell - 1$. It is determined by the above Lagrangian scheme:

$$\begin{aligned}
U_{an} &= [x^{a((a+c)s+\ell-1)}]U(x)^h \\
&= \frac{\ell a + c}{a((a+c)s+\ell-1)} [u^{a((a+c)s+\ell-1)-1}] u^{\ell a + c - 1} \left(1 + u^{a+c}\right)^{(a+c)s+\ell-1} \\
&= \frac{\ell a + c}{a((a+c)s+\ell-1)} \binom{(a+c)s+\ell-1}{as-1}.
\end{aligned}$$

Rewriting the binomial coefficient by symmetry, the claim follows.

Example 7.5 Knuth's original problem was dealing with boundaries $y = \frac{2}{5}x + \frac{k}{5}$, $(k = 1, \ldots, 4)$. In particular, we may choose $\ell = 0$ and $\ell = 1$ to get:

$$\sum_{k=1}^{2} A_s(k) = \frac{5}{7s-1}\binom{7s-1}{2s-1} = \frac{2}{7s-1}\binom{7s-1}{2s},$$
$$\sum_{k=3}^{4} A_s(k) = \frac{1}{s}\binom{7s}{2s-1}.$$

The first one is the known result, whereas the second one is yet another surprising identity.

Now, we come back to the asymptotics of Sect. 6. Some key ingredients were Proposition 7.2 (Periodic rule of thumb) and the rotation law of the small branches. Happily, such a rotation law holds in general for any slope, and the derived techniques can also be applied. This is what we present now.

Let $P(u) = u^{-a} + u^c$ be the jump polynomial of directed walks. Thus, we have a small branches $u_i(z)$ satisfying the kernel equation $1 - zP(u_i(z)) = 0$. As before let τ be the unique positive root of $P'(\tau)$, and let ρ be defined as $\rho = 1/P(\tau)$. Recall that the small branches are possibly singular only at the roots of $P'(u)$. The jump polynomial has periodic support with period $p = a + c$ as $P(u) = u^{-a}H(u^p)$ with $H(u) = 1 + u$. Hence, there are p possible singularities of the small branches

$$\zeta_k = \rho\omega^k, \qquad \text{with} \qquad \omega = e^{2\pi i/p}.$$

The general version of Lemma 7.1 reads then as follows:

Lemma 7.4 (Rotation law of small branches) *Let* $\gcd(a, c) = 1$. *Then there exists a permutation* σ *of* $\{1, \ldots, p\}$ *without fix points and an integer* κ *(satisfying* $\kappa a + 1 \equiv 0 \mod p$*) such that*

$$u_i(\omega z) = \omega^{\kappa} u_{\sigma(i)}(z),$$

for all $z \in \mathbb{C}$ *with* $|z| \leq \rho$ *and* $0 < \arg(z) < \pi - 2\pi/p$.

Proof We proceed as in the proof of Lemma 7.1. Define $U(z) := \omega^{\kappa} u_i(\omega z)$ and a function $X(z) := U^a - z\phi(U)$ with $\phi(u) := u^a P(u)$. Then a straightforward computation shows that

$$X(z) = (\omega^{\kappa} u_i(\omega z))^a - z\phi\left(\omega^{\kappa} u_i(\omega z)\right) = \omega^{\kappa a} u_i(\omega z)^a - z\phi(u_i(\omega z)),$$

as $\phi(u)$ is p-periodic. Therefore, we get by the following transformation

$$\omega X(z/\omega) = \omega^{\kappa a+1} u_i(z)^a - z\phi(u_i(z)) = 0,$$

if $\kappa a + 1 \equiv 0 \mod p$, because of the kernel equation. Thus, $X = U^a - z\phi(U) = 0$ and therefore, $U(z)$ is a root of the kernel equation. It has to be a small root, as it is converging to 0 if z goes to 0. Furthermore, it has to be a different root, as it has a different Puiseux expansion. By the analytic continuation principle (as long as we avoid the cut line $\arg(z) = -\pi$), the result follows.

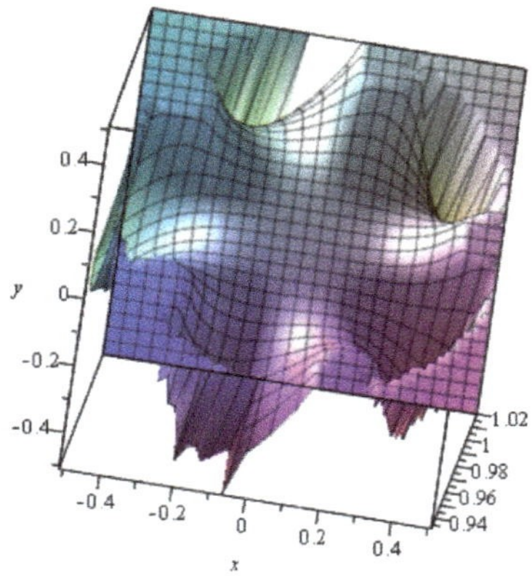

This is the landscape in the complex plane of $|F(z)|$, where F is here the generating function of Duchon's club excursions. One can see the five dominant singularities. It is enough to know the local behaviour near the real positive singularity, the rotation law implies the same behaviour at the other dominant singularities.

Fig. 11 Landscape in the complex plane of the generating function of lattice paths

The last lemma allows us to state the following "meta"-result:

Theorem 7.10 (Metatheorem/rule of thumb: enumeration and asymptotics of lattice paths) *Constrained* 1*-dimensional lattice paths have an algebraic generating function, expressible in terms of Schur functions (a symmetric function involving the small branches of the kernel). Singularity analysis gives its asymptotic behavior, which is equal to the asymptotics at the dominant real singularity (times the periodicity whenever the rotation law holds).*

We call this a metatheorem because it is rather informal in the description of the constraints allowed (it could be positivity, prescribed starting or ending points, to live in a cone, to stay below a line of rational slope, to have some additional Markovian behavior, to be multidimensional with one border, or in bijection with any of these constraints ...); in all these cases, the spirit of the kernel method and analytic combinatorics should give the enumeration and the asymptotics. Different incarnations of this rule of thumb appear in [7, 8, 10, 12, 19], and no doubt that many new lattice problems on the one hand, and many new combinatorial problems involving some type of periodicity on the other hand, will offer additional incarnations of this metatheorem (Fig. 11).

9 Conclusion

In this article, we analyzed some models of directed lattice paths below a line of rational slope. As a guiding thread, we first illustrated our method on Dyck paths below the line of slope $2/5$. Beside the (pleasant) satisfaction of answering a problem of Don Knuth, this sheds light on properties of constrained lattice paths, including the delicate case (for analysis) of a periodic behavior.

We can shortly recall the main methods used in this article to attack lattice path problems:

Firstly, the method of choice of Nakamigawa and Tokushige was the *cycle lemma*. It is a classical result for lattice paths which uses the geometry of the problem. However, its applications are limited to certain cases.

Secondly, a more general result is given in Theorem 7.9 (General closed forms for lattice paths below a rational slope $y = \frac{a}{c}x + \frac{b}{c}$), via the *Lagrange inversion*. This directly gives the sought closed form. However, it does not give access to the asymptotics.

Thus, thirdly, we used the *kernel method* to express the generating functions explicitly in terms of (known) algebraic functions. This gave us access to the asymptotics and is an alternative way to access the closed forms. Our Proposition 7.2 (Periodic rule of thumb) explains in which way the asymptotic expansions are modified in the case of a periodic behavior (via some local asymptotics extractor and the rotation law); we expect this approach to be reused in many other problems.

Also, the method of *holonomy theory* used in Theorem 7.3 (Closed form for the sum of coefficients) shows the possible usage of computer algebra to *prove* such *conjectured* identities. This is probably the fastest technique for checking given identities and can be automatized to a great extent. The interested reader is referred to the nicely written introductions [41, 56].

Our approach extends to any lattice path (with any set of jumps of positive coordinates) below a line of (ir)rational slope (see [15]). This leads to some nice universal results for the enumeration and asymptotics. As an open question, it could be natural to look for similar results for lattice paths (with any set of jumps with positive and negative coordinates, and not just jumps to the nearest neighbors) in a cone given by two lines of rational slope. This is equivalent to the enumeration of non-directed lattice paths in dimension 2. Despite the nice approach from the probabilistic school [26, 31] and from the combinatorial school [22] via the iterated kernel method, this remains a terribly simple problem (to state!), but a challenge for the mathematics of this century.

Acknowledgements This work is the result of a collaboration founded by the SFB project F50 “Algorithmic and Enumerative Combinatorics” and the Franco-Austrian PHC “Amadeus.” Michael Wallner is supported by the Austrian Science Fund (FWF) grant SFB F50-03 and by ÖAD, grant F04/2012. A preliminary version of this work [14] was presented at the conference ANALCO’15 (San Diego, January 2015) and at the 8th International Conference on Lattice Path Combinatorics and Applications (Pomona, August 2015). Last but not least, we thank Don Knuth, Ernst Schulte-Geers and Manuel Kauers for exchanging references on this problem, and the referee for the detailed feedback!

References

1. Aebly, J.: Démonstration du problème du scrutin par des considérations géométriques. Enseign. Math. **23**, 185–186 (1923)
2. André, D.: Solution directe du problème résolu par M. Bertrand. C. R. Acad. Sci. Paris **105**, 436–437 (1887)
3. Armstrong, D., Rhoades, B., Williams, N.: Rational Catalan combinatorics: the associahedron. Discret. Math. Theor. Comput. Sci. Proc. AS 933–944 (2013)
4. Bacher, A., Beaton, N.: Weakly prudent self-avoiding bridges. Discret. Math. Theor. Comput. Sci. Proc. AT 827–838 (2014)
5. Banderier, C.: Combinatoire analytique : application aux marches aléatoires. Mémoire de DEA, INRIA Rocquencourt/Univ. Paris 6 (1998)
6. Banderier, C., Bousquet-Mélou, M., Denise, A., Flajolet, P., Gardy, D., Gouyou-Beauchamps, D.: Generating functions for generating trees. Discret. Math. **246**(1–3), 29–55 (2002)
7. Banderier, C., Drmota, M.: Formulae and asymptotics for coefficients of algebraic functions. Combin. Probab. Comput. **24**(1), 1–53 (2015)
8. Banderier, C., Flajolet, P.: Basic analytic combinatorics of directed lattice paths. Theor. Comput. Sci. **281**(1–2), 37–80 (2002)
9. Banderier, C., Flajolet, P., Schaeffer, G., Soria, M.: Random maps, coalescing saddles, singularity analysis, and Airy phenomena. Random Struct. Algorithms **19**(3–4), 194–246 (2001)
10. Banderier, C., Gittenberger, B.: Analytic combinatorics of lattice paths: enumeration and asymptotics for the area. Discret. Math. Theor. Comput. Sci. Proc. AG 345–355 (2006)
11. Banderier, C., Krattenthaler, C., Krinik, A., Kruchinin, D., Kruchinin, V., Nguyen, D., Wallner, M.: Explicit formulas for enumeration of lattice paths: basketball and the kernel method. This volume
12. Banderier, C., Nicodème, P.: Bounded discrete walks. Discret. Math. Theor. Comput. Sci. Proc. AM 35–48 (2010)
13. Banderier, C., Schwer, S.: Why Delannoy numbers? J. Stat. Plan. Inference **135**(1), 40–54 (2005)
14. Banderier, C., Wallner, M.: Lattice paths of slope $2/5$. In: Proceedings of the Twelfth Workshop on Analytic Algorithmics and Combinatorics (ANALCO), pp. 105–113 (2015)
15. Banderier, C., Wallner, M.: Lattice paths below a line of irrational slope. in preparation (2016)
16. Bertrand, J.: Solution d'un problème. C. R. Acad. Sci. Paris **105**, 369 (1887)
17. Bizley, M.T.L.: Derivation of a new formula for the number of minimal lattice paths from (0, 0) to (km, kn) having just t contacts with the line $my = nx$ and having no points above this line; and a proof of Grossman's formula for the number of paths which may touch but do not rise above this line. J. Inst. Actuar. **80**, 55–62 (1954)
18. Bostan, A., Bousquet-Mélou, M., Kauers, M., Melczer, S.: On 3-dimensional lattice walks confined to the positive octant. Ann. Combin. **20**, 661–704 (2016)
19. Bousquet-Mélou, M.: Discrete excursions. Sém. Lothar. Combin. **57**, 23 (2008). Article B57d
20. Bousquet-Mélou, M., Fusy, É., Préville-Ratelle, L.-F.: The number of intervals in the m-Tamari lattices. Electron. J. Combin. **18**(2), 26 (2011). Article #R31
21. Bousquet-Mélou, M., Jehanne, A.: Polynomial equations with one catalytic variable, algebraic series and map enumeration. J. Combin. Theory Ser. B **96**(5), 623–672 (2006)
22. Bousquet-Mélou, M., Mishna, M.: Walks with small steps in the quarter plane. Algorithmic Probab. Combinatorics. Contemp. Math. **520**, 1–39 (2010)
23. Bousquet-Mélou, M., Petkovšek, M.: Linear recurrences with constant coefficients: the multivariate case. Discret. Math. **225**(1–3), 51–75 (2000)
24. Brown, W.G.: On the existence of square roots in certain rings of power series. Math. Ann. **158**, 82–89 (1965)
25. Chung, F., Graham, R., Morrison, J., Odlyzko, A.: Pebbling a chessboard. Am. Math. Mon. **102**(2), 113–123 (1995)
26. Denisov, D., Wachtel, V.: Random walks in cones. Ann. Probab. **43**(3), 992–1044 (2015)

27. Duchi, E.: On some classes of prudent walks. In: Proceedings of FPSAC'2005, Taormina, Italy (2015)
28. Duchon, P.: On the enumeration and generation of generalized Dyck words. Discret. Math. **225**(1–3), 121–135 (2000)
29. Dvoretzky, A., Motzkin, T.: The asymptotic density of certain sets of real numbers. Duke Math. J. **14**, 315–321 (1947)
30. Eynard, B.: Counting Surfaces. Progress in mathematical physics, vol. 70. Birkhäuser/Springer [Cham] (2016)
31. Fayolle, G., Iasnogorodski, R., Malyshev, V.: Random Walks in the Quarter-plane. Applications of mathematics, vol. 40. Springer, Berlin (1999)
32. Flajolet, P.: Combinatorial aspects of continued fractions. Discret. Math. **32**(2), 125–161 (1980)
33. Flajolet, P., Sedgewick, R.: Analytic Combinatorics. Cambridge University Press, Cambridge (2009)
34. Gaudin, M.: The Bethe Wavefunction. Cambridge University Press, New York (2014)
35. Gessel, I.M.: A factorization for formal Laurent series and lattice path enumeration. J. Combin. Theory Ser. A **28**(3), 321–337 (1980)
36. Hardy, G.H., Ramanujan, S.: Asymptotic formulae for the distribution of integers of various types. Proc. Lond. Math. Soc. **16**(2), 112–132 (1917). Collected papers of Srinivasa Ramanujan, pp. 245–261. AMS Chelsea Publication, Providence (2000)
37. Jain, J.L., Mohanty, S.G., Böhm, W.: A Course on Queueing Models. Statistics: textbooks and monographs. Chapman and Hall/CRC, FL (2007)
38. Janse van Rensburg, E.J.: Square lattice directed paths adsorbing on the line $y = qx$. J. Stat. Mech.: Theory Exp. **2005**(09), P09010 (2005)
39. Janse van Rensburg, E.J., Prellberg, T., Rechnitzer, A.: Partially directed paths in a wedge. J. Combin. Theory Ser. A **115**(4), 623–650 (2008)
40. Janse van Rensburg, E.J., Rechnitzer, A.: Adsorbing and collapsing directed animals. J. Stat. Phys. **105**(1–2), 49–91 (2001)
41. Kauers, M., Paule, P.: The Concrete Tetrahedron. Texts and monographs in symbolic computation. Springer, Berlin (2011)
42. Kempner, A.J.: A theorem on lattice-points. Ann. Math. **19**(2), 127–136 (1917)
43. Knuth, D.E.: The Art of Computer Programming, vol. 1: Fundamental Algorithms. Addison–Wesley (1968)
44. Knuth, D.E.: 20th Annual Christmas Tree Lecture: (3/2)-ary Trees. Stanford University (2014). https://www.youtube.com/watch?v=P4AaGQIo0HY
45. Krattenthaler, C.: Lattice Path Enumeration. In: Bóna, M. (ed.) Handbook of Enumerative Combinatorics. Discrete Mathematics and its Applications, pp. 589–678. CRC Press (2015)
46. Kurkova, I., Raschel, K.: On the functions counting walks with small steps in the quarter plane. Publ. Math. Inst. Hautes Études Sci. **116**, 69–114 (2012)
47. Labelle, J., Yeh, Y.-N.: Generalized Dyck paths. Discret. Math. **82**(1), 1–6 (1990)
48. Lagrange, J.-L.: Nouvelle méthode pour résoudre les équations littérales par le moyen des séries. Mémoires de l'Académie Royale des Sciences et Belles-Lettres de Berlin 24, 251–326 (1770). Reprinted in Œuvres de Lagrange, tome 2, pp. 655–726. Gauthier-Villars, Paris (1868)
49. Mansour, T., Shattuck, M.: Pattern avoiding partitions, sequence A054391 and the kernel method. Appl. Appl. Math. **6**(12), 397–411 (2011)
50. Merlini, D., Sprugnoli, R., Verri, M.C.: The area determined by underdiagonal lattice paths. In: Proceedings of CAAP'96. Lecture Notes in Computer Science, vol. 1059, pp. 59–71 (1996)
51. Mirimanoff, D.: À propos de l'interprétation géométrique du problème du scrutin. Enseign. Math. **23**, 187–189 (1923)
52. Mohanty, S.G.: Lattice Path Counting and Applications. Academic Press, New York (1979)
53. Nakamigawa, T., Tokushige, N.: Counting lattice paths via a new cycle lemma. SIAM J. Discret. Math. **26**(2), 745–754 (2012)
54. Narayana, T.V.: Lattice Path Combinatorics with Statistical Applications. Mathematical Expositions, vol. 23. University of Toronto Press (1979)
55. Petkovšek, M.: The irrational chess knight. In: Proceedings of FPSAC'98, pp. 513–522 (1998)

56. Petkovšek, M., Wilf, H.S., Zeilberger, D.: A = B. AK Peters (1996)
57. Prodinger, H.: The kernel method: a collection of examples. Sém. Lothar. Combin. **50**, 19 (2003/2004). Article B50f
58. Salvy, B., Zimmermann, P.: Gfun: a Maple package for the manipulation of generating and holonomic functions in one variable. ACM Trans. Math. Softw. **20**(2), 163–177 (1994)
59. Sato, M.: Generating functions for the number of lattice paths between two parallel lines with a rational incline. Math. Japon. **34**(1), 123–137 (1989)
60. Schulte-Geers, E., Stadje, W.: Maximal percentages in Pólya's urn. J. Appl. Probab. **52**(1), 180–190 (2015)
61. Schwerdtfeger, U.: Linear functional equations with a catalytic variable and area limit laws for lattice paths and polygons. Eur. J. Combin. **36**, 608–640 (2014)
62. Stanley, R.P.: Enumerative Combinatorics, vol. 2. Cambridge Studies in Advanced Mathematics, vol. 62. Cambridge University Press, Cambridge (1999)

Enumeration of Colored Dyck Paths Via Partial Bell Polynomials

Daniel Birmajer, Juan B. Gil, Peter R. W. McNamara and Michael D. Weiner

Abstract We consider a class of lattice paths with certain restrictions on their ascents and down-steps and use them as building blocks to construct various families of Dyck paths. We let every building block P_j take on c_j colors and count all of the resulting colored Dyck paths of a given semilength. Our approach is to prove a recurrence relation of convolution type, which yields a representation in terms of partial Bell polynomials that simplifies the handling of different colorings. This allows us to recover multiple known formulas for Dyck paths and related lattice paths in a unified manner.

Keywords Colored Dyck paths · Colored Dyck words · Colored Motzkin paths · Partial Bell polynomials

2010 Mathematics Subject Classification Primary: 05A15 · Secondary: 05A19

1 Introduction

A *Dyck path* of semilength n is a lattice path in the first quadrant, which begins at the origin $(0, 0)$, ends at $(2n, 0)$, and consists of steps $(1, 1)$ and $(1, -1)$. It is customary to encode an up-step $(1, 1)$ with the letter u and a down-step $(1, -1)$ with the letter

D. Birmajer
Department of Mathematics, Nazareth College, 4245 East Ave., Rochester, NY 14618, USA
e-mail: abirmaj6@naz.edu

J. B. Gil (✉) · M. D. Weiner
Penn State Altoona, 3000 Ivyside Park, Altoona, PA 16601, USA
e-mail: jbg10@psu.edu

M. D. Weiner
e-mail: mdw8@psu.edu

P. R. W. McNamara
Department of Mathematics, Bucknell University, 1 Dent Drive, Lewisburg, PA 17837, USA
e-mail: peter.mcnamara@bucknell.edu

G. E. Andrews et al. (eds.), *Lattice Path Combinatorics and Applications*,
Developments in Mathematics 58, https://doi.org/10.1007/978-3-030-11102-1_8

d. Thus, every Dyck path can be encoded by a corresponding *Dyck word* of u's and d's. We will freely pass from paths to words and vice versa.

Much is known about Dyck paths and their connection to other combinatorial structures like rooted trees, noncrossing partitions, polygon dissections, Young tableaux, and other lattice paths. While there is a vast literature on the enumeration of Dyck paths and related combinatorial objects according to various statistics, for the scope of the present work, we only refer to the closely related papers [1, 7, 10]. For more information, the reader is referred to the general overview on lattice path enumeration written by Krattenthaler in [4, Chap. 10].

For $a, b \in \mathbb{N}_0 = \mathbb{N} \cup \{0\}$ with $a + b \neq 0$ and $\mathbf{c} = (c_1, c_2, \dots)$ with $c_j \in \mathbb{N}_0$, we define

> $\mathfrak{D}_n^{\mathbf{c}}(a, b)$ as the set of Dyck words of semilength $(a + b)n$ created from strings of the form $P_0 =$ "d" and $P_j =$ "$u^{(a+b)j} d^{b(j-1)+1}$" for $j = 1, \dots, n$, such that each maximal $(a + b)j$-ascent substring $u^{(a+b)j}$ may be colored in c_j different ways. We use $c_j = 0$ if $(a + b)j$-ascents are to be avoided. We will refer to the elements of $\mathfrak{D}_n^{\mathbf{c}}(a, b)$ as *colored Dyck paths* or *colored Dyck words*.

Note that if $a = 1$, $b = 0$, and $\mathbf{c}$ is the sequence of ones $\mathbf{c} = \mathbb{1} = (1, 1, \dots)$, then the building blocks take the form $P_0 =$ "d", $P_j =$ "$u^j d$" for $j = 1, \dots, n$, and $\mathfrak{D}_n^{\mathbb{1}}(1, 0)$ is just the set of regular Dyck words of semilength n.

In this paper, we are interested in counting the number of elements in $\mathfrak{D}_n^{\mathbf{c}}(a, b)$. For the sequence given by $y_n = \left|\mathfrak{D}_n^{\mathbf{c}}(a, b)\right|$, we prove a recurrence relation of convolution type (see Theorem 1) and give a representation of y_n in terms of partial Bell polynomials in the elements of the sequence $\mathbf{c} = (c_1, c_2, \dots)$ (see Theorem 2).

We conclude with several examples that illustrate the use of our formulas for various values of the parameters a and b as well as some interesting coloring choices.

2 Enumeration of Colored Dyck Words

Our technique for enumerating $\mathfrak{D}_n^{\mathbf{c}}(a, b)$ will be to show in Theorem 1 and Proposition 1 that the sequence $y_n = \left|\mathfrak{D}_n^{\mathbf{c}}(a, b)\right|$ satisfies the same initial condition and recurrence relation as a sequence (z_n) involving Bell polynomials. As a direct consequence, we get the promised enumeration of $\mathfrak{D}_n^{\mathbf{c}}(a, b)$ in terms of partial Bell polynomials (Theorem 2).

Theorem 1 *For $a, b \in \mathbb{N}_0$ with $a + b \neq 0$ and $\mathbf{c} = (c_1, c_2, \dots)$ with $c_j \in \mathbb{N}_0$, let (y_n) be the sequence defined by $y_0 = 1$ and $y_n = \left|\mathfrak{D}_n^{\mathbf{c}}(a, b)\right|$ for $n \geq 1$. Then, y_n satisfies the recurrence*

$$y_n = \sum_{\ell=1}^{n} c_\ell \sum_{i_1+\cdots+i_{a\ell+b}=n-\ell} y_{i_1} \cdots y_{i_{a\ell+b}}, \tag{8.1}$$

where each i_j is a nonnegative integer.

Proof We will prove (8.1) by showing that there is a bijection between the sets of objects counted by each side of the equation. The left-hand side counts colored Dyck words of semilength $(a+b)n$. The right-hand side counts tuples of the form

$$(\ell, C; D_1, D_2, \dots, D_{a\ell+b}),$$

where

- $1 \le \ell \le n$,
- C is a color from a choice of c_ℓ colors,
- D_j is a member of $\mathfrak{D}^{\mathfrak{c}}_{i_j}(a,b)$, i.e., a colored Dyck word of semilength $(a+b)i_j$, and
- $i_1 + \cdots + i_{a\ell+b} = n - \ell$.

From this tuple, we will construct a colored Dyck word w of semilength $(a+b)n$ in the following fashion.

Due to the ℓ and C appearing in the tuple, we begin with the string $v = u^{(a+b)\ell} d^{b(\ell-1)+1}$ with the substring $u^{(a+b)\ell}$ colored C. We then append D_1, D_2, ..., $D_{a\ell+b}$ to v, separating each adjacent pair (D_i, D_{i+1}) by an additional copy of the letter d. In this way, a string w is obtained. We need to check that this map is well defined, meaning that w is a colored Dyck word of semilength $(a+b)n$.

Let us first check that w contains equal numbers of the letters u and d. Since the D_i already satisfy this condition, we need

$$(a+b)\ell = (b(\ell-1)+1) + (a\ell + b - 1),$$

which is true. Similar reasoning shows the "Dyck" property, i.e., that any prefix of w has at least as many appearances of u as of d. To determine the semilength of w, we count the number of appearances of u as

$$(a+b)\ell + (a+b)(n-\ell) = (a+b)n,$$

as desired. By construction, each maximal $(a+b)j$-ascent has an appropriate color and we conclude that w is a colored Dyck word of semilength $(a+b)n$.

To show that this map f from the tuple to w is a bijection, we argue that it has a well-defined inverse g. Thus, let w be a colored Dyck path of semilength $(a+b)n$, and recall that a and b are fixed. The length L and color of the ascent sequence at the beginning of w determine $\ell = \frac{L}{a+b}$ and C at the start of the tuple $g(w)$. Let w^1 denote the word obtained from w by removing this prefix $u^{(a+b)\ell} d^{b(\ell-1)+1}$ from w. See Fig. 1 for a schematic example. We next wish to determine $D_1, \dots, D_{a\ell+b}$ from w^1. Let us say that w^1 has *starting height* $a\ell + b - 1$, meaning that it has this many more copies of d than of u. Notice that this starting height is nonnegative.

To determine D_1, we proceed by finding the smallest r such that the suffix of w^1 corresponding to those letters strictly after position r has starting height one less than that of w^1. Then, we let w^2 be that suffix of w and we let D_1 be the prefix of w^1 corresponding to those letters strictly before position r. In particular, if the first

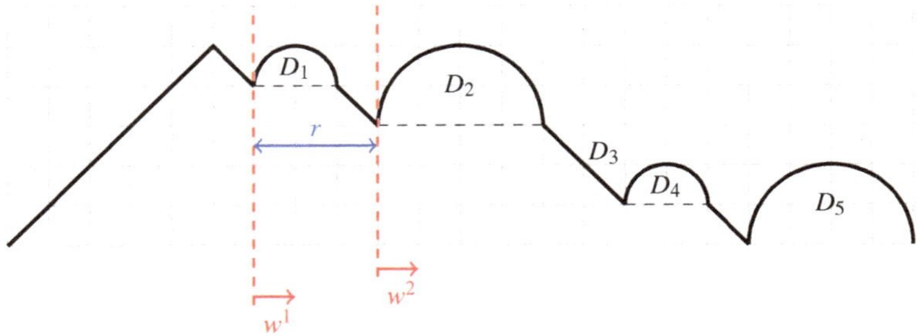

Fig. 1 A schematic example of determining $(\ell, C; D_1, D_2, \ldots, D_{a\ell+b})$ from w as in the proof of Theorem 1, where the semicircles represent colored Dyck paths. We have $L = 5$, $a = 5$, $b = 0$, $\ell = 1$, and D_3 is an empty word. For $i = 1, 2$, we see that w^i is the portion of w to the right of the corresponding dashed line

letter of w^1 is d, then we get that D_1 is the empty word, and we let w^2 be the word obtained from w^1 by deleting this 1-letter prefix. If the first letter of w^1 is u, then D_1 will be nonempty. If no such position r exists, then it must be the case that w^1 has excess 0 and we let $D_1 = w^1$.

By the definition of r, D_1 has equal numbers of u's and d's, and it satisfies the Dyck property. Moreover, every maximal $(a+b)j$-ascent sequence is immediately followed by a $(b(j-1)+1)$-descent sequence because w has this property, and because these ascent lengths $(a+b)j$ are at least as large as their partnering descent lengths $(b(j-1)+1)$. In other words, D_1 is a colored Dyck word. We continue in this exact manner to determine the full sequences $w^3, \ldots, w^s$ and $D_1, \ldots, D_s$ for some s. Since each w^i has starting height one less than w^{i-1}, we deduce that $s = a\ell + b$, as desired.

Notice that the resulting tuple $g(w) = (\ell, C; D_1, D_2, \ldots, D_{a\ell+b})$ satisfies the properties in the four bullet points given at the beginning of this proof. In particular, since w has semilength $(a+b)n$, the total number of u's in $D_1, D_2, \ldots, D_{a\ell+b}$ equals $(a+b)n - \ell(a+b)$, and so $i_1 + \cdots + i_{a\ell+b} = n - \ell$. We conclude that g maps colored Dyck words of semilength $(a+b)n$ to tuples of the desired type. Finally, one can readily observe that $g \circ f$ and $f \circ g$ both equal the identity map.

3 Representation in Terms of Partial Bell Polynomials

Our goal for this section is to use the result of Theorem 1 to give a formula for $y_n = \left|\mathfrak{D}_n^{\mathbf{c}}(a,b)\right|$ in terms of partial Bell polynomials.

For $a, b \in \mathbb{R}$ (not both $= 0$) and $\mathbf{c} = (c_1, c_2, \ldots)$, consider the sequence (z_n) defined by

$$z_0 = 1, \quad z_n = \sum_{k=1}^{n} \binom{an+bk}{k-1} \frac{(k-1)!}{n!} B_{n,k}(1!c_1, 2!c_2, \ldots) \text{ for } n \geq 1, \tag{8.2}$$

where $B_{n,k}$ denotes the (n,k)th partial Bell polynomial defined as

$$B_{n,k}(x_1,\dots,x_{n-k+1}) = \sum_{\alpha\in\pi(n,k)} \frac{n!}{\alpha_1!\cdots\alpha_{n-k+1}!}\left(\frac{x_1}{1!}\right)^{\alpha_1}\cdots\left(\frac{x_{n-k+1}}{(n-k+1)!}\right)^{\alpha_{n-k+1}}$$

with $\pi(n,k)$ denoting the set of multi-indices $\alpha\in\mathbb{N}_0^{n-k+1}$ such that $\alpha_1+\cdots+\alpha_{n-k+1}=k$ and $\alpha_1+2\alpha_2+\cdots+(n-k+1)\alpha_{n-k+1}=n$. For more information on partial Bell polynomials, see [6, Sect. 3.3].

The sequence (8.2) satisfies the following convolution formula.

Lemma 1 (cf. [2, Theorem 2.1]) *For $r,n\ge 1$, we have*

$$z_n^{(r)} \overset{\text{def}}{=} \sum_{m_1+\cdots+m_r=n} z_{m_1}\cdots z_{m_r} = r\sum_{k=1}^{n}\binom{an+bk+r-1}{k-1}\frac{(k-1)!}{n!}B_{n,k}(1!c_1,2!c_2,\dots).$$

Proposition 1 *Suppose $a,b\in\mathbb{N}_0$. For $n\ge 1$, the sequence (z_n) defined by* (8.2) *satisfies the recurrence*

$$z_n = \sum_{\ell=1}^{n} c_\ell \sum_{i_1+\cdots+i_{a\ell+b}=n-\ell} z_{i_1}\dots z_{i_{a\ell+b}} = \sum_{\ell=1}^{n} c_\ell\, z_{n-\ell}^{(a\ell+b)}, \tag{8.3}$$

where each i_j is a nonnegative integer and $z_0^{(an+b)}=1$. In other words, the sequence (z_n) satisfies the same recurrence as the sequence (y_n).

Proof By the previous lemma, omitting the argument of the Bell polynomials,

$$\begin{aligned}
\sum_{\ell=1}^{n-1} c_\ell\, z_{n-\ell}^{(a\ell+b)} &= \sum_{\ell=1}^{n-1} c_\ell(a\ell+b)\sum_{k=1}^{n-\ell}\tbinom{an+b(k+1)-1}{k-1}\tfrac{(k-1)!}{(n-\ell)!}B_{n-\ell,k}\\
&= \sum_{\ell=1}^{n-1} c_{n-\ell}\big(a(n-\ell)+b\big)\sum_{k=1}^{\ell}\tbinom{an+b(k+1)-1}{k-1}\tfrac{(k-1)!}{\ell!}B_{\ell,k}\\
&= \sum_{k=1}^{n-1}\tbinom{an+b(k+1)-1}{k-1}(k-1)!\sum_{\ell=k}^{n-1}\tfrac{c_{n-\ell}(a(n-\ell)+b)}{\ell!}B_{\ell,k}\\
&= \sum_{k=2}^{n}\tbinom{an+bk-1}{k-2}\tfrac{(k-2)!}{n!}\sum_{\ell=k-1}^{n-1}\tfrac{n!}{\ell!}c_{n-\ell}(a(n-\ell)+b)B_{\ell,k-1}\\
&= \sum_{k=2}^{n}\tbinom{an+bk-1}{k-2}\tfrac{(k-2)!}{n!}\sum_{\ell=k-1}^{n-1}\Big(an\tbinom{n-1}{\ell}+b\tbinom{n}{\ell}\Big)(n-\ell)!c_{n-\ell}B_{\ell,k-1}.
\end{aligned}$$

Now, using equations (11.11) and (11.12) in [5, Theorem 11.2], one can easily verify the identities

$$\sum_{\ell=k-1}^{n-1} an\binom{n-1}{\ell}(n-\ell)!c_{n-\ell}B_{\ell,k-1} = anB_{n,k},$$

$$\sum_{\ell=k-1}^{n-1} b\binom{n}{\ell}(n-\ell)!c_{n-\ell}B_{\ell,k-1} = bkB_{n,k},$$

which imply

$$\sum_{\ell=1}^{n-1} c_\ell z_{n-\ell}^{(a\ell+b)} = \sum_{k=2}^{n} \binom{an+bk-1}{k-2}\frac{(k-2)!}{n!}(an+bk)B_{n,k} = \sum_{k=2}^{n} \binom{an+bk}{k-1}\frac{(k-1)!}{n!}B_{n,k}.$$

Finally, by adding c_n to each of these sums, we arrive at (8.3).

We now arrive at our main result.

Theorem 2 *For $a, b \in \mathbb{N}_0$ with $a+b \neq 0$ and $\mathbf{c} = (c_1, c_2, \dots)$, the sequence $y_n = \left|\mathfrak{D}_n^{\mathbf{c}}(a,b)\right|$ can be written as*

$$y_n = \sum_{k=1}^{n} \binom{an+bk}{k-1}\frac{(k-1)!}{n!}B_{n,k}(1!c_1, 2!c_2, \dots) \text{ for } n \geq 1. \tag{8.4}$$

Moreover, the quantity $\binom{an+bk}{k-1}\frac{(k-1)!}{n!}B_{n,k}(1!c_1, 2!c_2, \dots)$ counts the number of Dyck paths in $\mathfrak{D}_n^{\mathbf{c}}(a,b)$ having exactly k peaks.

Proof Equation (8.4) is a direct consequence of Theorem 1 and Proposition 1. The second assertion follows by considering both sides of (8.4) as polynomials in the c_i's and by equating the terms of degree k. Indeed, note that $B_{n,k}(1!c_1, 2!c_2, \dots)$ contains as many monomials as there are partitions of n into k parts, and each such monomial has degree k in the c_i's. On the other hand, each appearance of a c_i in a monomial of y_n corresponds to a coloring of a maximal ascent substring and therefore to a peak.

4 Examples

In this section, we proceed to illustrate the use and versatility of the representation (8.4). The goal is to take advantage of the partial Bell polynomials to derive combinatorial formulas for the given enumerating sequence.

First of all, as we mentioned in the introduction, $\mathfrak{D}_n^{\mathbb{1}}(1,0)$ is nothing but the set of Dyck paths of semilength n. Recall that we are using the symbol $\mathbb{1}$ to denote the sequence of ones $\mathbf{c} = (1, 1, \dots)$.

Example 1 (Narayana numbers) By [6, Sec. 3.3, eqn. (3h)] for example,

$$B_{n,k}(1!, 2!, 3!, \dots) = \frac{n!}{k!}\binom{n-1}{k-1} = \frac{(n-1)!}{(k-1)!}\binom{n}{k} \text{ for } n, k \geq 1,$$

so Theorem 2 gives the known fact that the number of Dyck paths of semilength n with exactly k peaks is given by

$$\binom{n}{k-1}\frac{(k-1)!}{n!}B_{n,k}(1!, 2!, \dots) = \frac{1}{n}\binom{n}{k-1}\binom{n}{k},$$

the Narayana number $N(n, k)$.

In general, for any given parameters a and b, and coloring sequence $\mathbf{c}$, the expressions

$$N_{a,b}^{\mathbf{c}}(n,k) = \binom{an+bk}{k-1}\frac{(k-1)!}{n!}B_{n,k}(1!c_1, 2!c_2, \dots)$$

provide the appropriate analog of the Narayana numbers.

Example 2 (Colored Motzkin paths) It is known that the number of Motzkin paths of length n is the same as the number of Dyck words of semilength n that avoid uuu (via the bijection $u^2d \to u$, $d \to d$, and $ud \to h$, where h denotes a horizontal step (1,0)). Thus, for $n \geq 1$, the number of Motzkin n-paths whose horizontal steps admit c_1 colors and whose up-steps admit c_2 colors is given by

$$\begin{aligned} y_n &= \sum_{k=1}^{n}\binom{n}{k-1}\frac{(k-1)!}{n!}B_{n,k}(1!c_1, 2!c_2, 0, \dots) \\ &= \sum_{k=\lceil\frac{n}{2}\rceil}^{n}\binom{n}{k-1}\frac{(k-1)!}{n!}\frac{n!}{k!}\binom{k}{n-k}c_1^{2k-n}c_2^{n-k} \\ &= \sum_{k=\lceil\frac{n}{2}\rceil}^{n}\frac{1}{n+1}\binom{n+1}{k}\binom{k}{n-k}c_1^{2k-n}c_2^{n-k} \\ &= \sum_{k=0}^{\lfloor\frac{n}{2}\rfloor}\frac{1}{n+1}\binom{n+1}{n-k}\binom{n-k}{k}c_1^{n-2k}c_2^{k} \\ &= \sum_{k=0}^{\lfloor\frac{n}{2}\rfloor}\binom{n}{2k}C_k\, c_1^{n-2k}c_2^{k}, \end{aligned}$$

where C_k denotes the Catalan number $\frac{1}{k+1}\binom{2k}{k}$. Letting $c_1 = c_2 = 1$ gives one of the better-known expressions for the Motzkin numbers.

Example 3 (Schröder numbers) The numbers in the sequence [11, A001003] are called little Schröder numbers and are known to count (among other things) Dyck paths in which the interior vertices of the ascents admit two colors, that is, Dyck paths in which a maximal j-ascent may be colored in 2^{j-1} different ways. The number y_n of such colored paths of semilength n can be obtained from (8.4) with $a = 1, b = 0$, and $\mathbf{c} = (1, 2, 2^2, \dots)$. Thus

$$\begin{aligned} y_n &= \sum_{k=1}^{n} \binom{n}{k-1} \frac{(k-1)!}{n!} B_{n,k}(1! \cdot 1,\ 2! \cdot 2,\ 3! \cdot 2^2, \dots) \\ &= \sum_{k=1}^{n} \binom{n}{k-1} \frac{(k-1)!}{n!} 2^{n-k} B_{n,k}(1!, 2!, \dots) \\ &= \sum_{k=1}^{n} \frac{1}{n} \binom{n}{k-1} \binom{n}{k} 2^{n-k} = \sum_{k=1}^{n} N(n,k)\, 2^{n-k}. \end{aligned}$$

Example 4 (m-ary paths) For $m \in \mathbb{N}$, we consider the set $\mathfrak{D}_n^{\mathbb{1}}(m, 0)$ of Dyck words of semilength mn created from strings of the form $P_0 = d$ and $P_j = u^{mj} d$ for $j = 1, \dots, n$.

The elements of $\mathfrak{D}_n^{\mathbb{1}}(m, 0)$ are in one-to-one correspondence with the elements of the set $\mathfrak{L}_n(m)$ of m-ary paths of length $(m+1)n$, i.e., lattice paths in the first quadrant from $(0, 0)$ to $((m+1)n, 0)$ with steps $(1, m)$ or $(1, -1)$. Here is an example for $m = 2$:

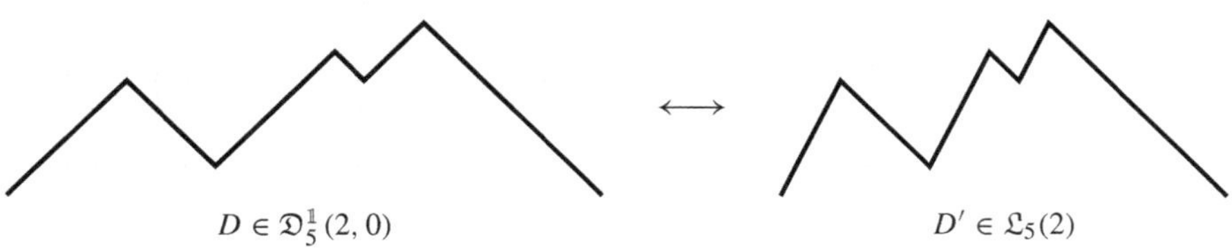

By equation (8.4), the sequence $y_n = |\mathfrak{L}_n(m)| = \left|\mathfrak{D}_n^{\mathbb{1}}(m, 0)\right|$ is given by

$$\begin{aligned} y_n &= \sum_{k=1}^{n} \binom{mn}{k-1} \frac{(k-1)!}{n!} B_{n,k}(1!, 2!, \dots) \\ &= \sum_{k=1}^{n} \frac{1}{k} \binom{mn}{k-1} \binom{n-1}{k-1} = \sum_{k=1}^{n} \frac{1}{mn+1} \binom{mn+1}{k} \binom{n-1}{n-k}, \end{aligned}$$

which by Vandermonde's identity becomes

$$y_n = \frac{1}{mn+1} \binom{(m+1)n}{n}.$$

Moreover, the number of such paths with exactly k peaks is given by the expression

$$N_{m,0}^{\mathbb{1}}(n,k) = \frac{1}{k}\binom{mn}{k-1}\binom{n-1}{k-1} = \frac{1}{n}\binom{mn}{k-1}\binom{n}{k}.$$

These formulas are consistent with [8, Corollary 4.12]. Clearly, Theorem 2 also provides formulas for other choices of the coloring sequence $\mathbf{c}$.

The next three examples illustrate simple connections with other types of lattice paths.

Example 5 ([11, A052709]) If $a = 0, b = 2$, and $\mathbf{c} = (1, 1, 0, 0, \dots)$, the set $\mathfrak{D}_n^{\mathbf{c}}(0, 2)$ consists of Dyck words of semilength $2n$ created from strings of the form $P_0 = d$, $P_1 = u^2d$, and $P_2 = u^4d^3$. With the simple map $d \to (1, -1)$, $u^2d \to (1, 1)$, and $u^4d^3 \to (3, 1)$, we get a one-to-one correspondence between $\mathfrak{D}_n^{\mathbf{c}}(0, 2)$ and the set $\mathfrak{L}_n(0, 2)$ of lattice paths in the first quadrant from $(0, 0)$ to $(2n, 0)$ with steps $(1, 1)$, $(1, -1)$, or $(3, 1)$.

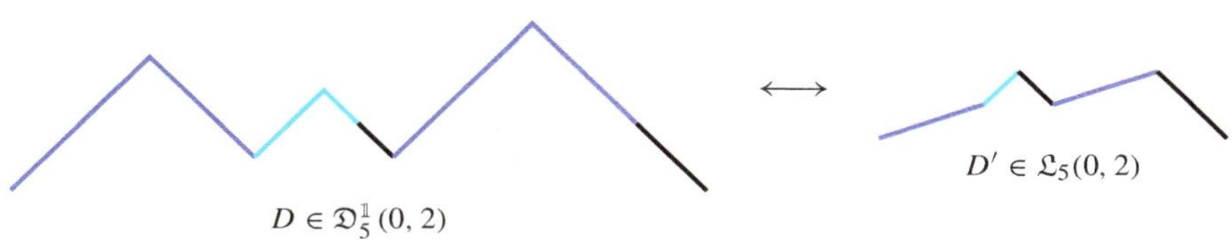

By means of (8.4), we then get that $y_n = |\mathfrak{L}_n(0, 2)| = \left|\mathfrak{D}_n^{\mathbf{c}}(0, 2)\right|$ satisfies

$$y_n = \sum_{k=1}^{n}\binom{2k}{k-1}\frac{(k-1)!}{n!}B_{n,k}(1!, 2!, 0, \dots) = \sum_{k=\lceil\frac{n}{2}\rceil}^{n}\frac{1}{k}\binom{2k}{k-1}\binom{k}{n-k}.$$

Example 6 ([11, A186997]) If $a = 1, b = 2$, and $\mathbf{c} = (1, 1, 0, 0, \dots)$, the set $\mathfrak{D}_n^{\mathbf{c}}(1, 2)$ consists of Dyck words of semilength $3n$ created from strings of the form $P_0 = d$, $P_1 = u^3d$, and $P_2 = u^6d^3$. With the simple map $d \to (1, -1)$, $u^3d \to (1, 2)$, and $u^6d^3 \to (3, 3)$, we get a one-to-one correspondence between $\mathfrak{D}_n^{\mathbf{c}}(1, 2)$ and the set $\mathfrak{L}_n(1, 2)$ of lattice paths in the first quadrant from $(0, 0)$ to $(3n, 0)$ with steps $(1, 2)$, $(1, -1)$, or $(3, 3)$.

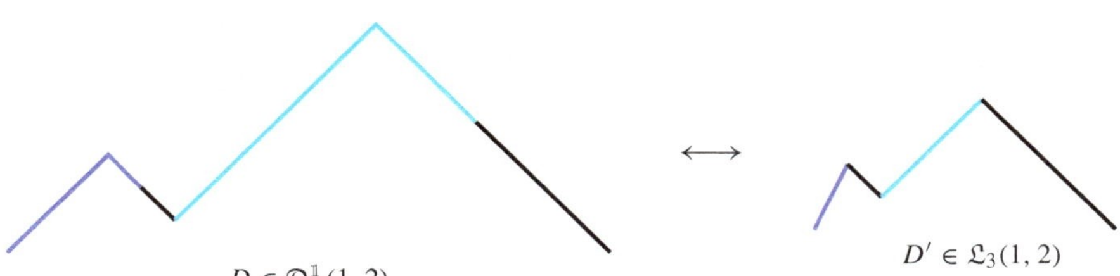

Again, by means of (8.4), we get that $y_n = |\mathfrak{L}_n(1,2)| = \left|\mathfrak{D}_n^{\mathbf{c}}(1,2)\right|$ satisfies

$$y_n = \sum_{k=1}^{n} \binom{n+2k}{k-1} \frac{(k-1)!}{n!} B_{n,k}(1!, 2!, 0, \dots) = \sum_{k=\lceil \frac{n}{2} \rceil}^{n} \frac{1}{k} \binom{n+2k}{k-1} \binom{k}{n-k}.$$

Example 7 ($\frac{3}{2}$-Dyck paths)

In the context of generalized Dyck languages with only two letters, Duchon [9] studied rational Dyck paths and suggests the need for colored Dyck words. In particular, he considered the set of Dyck words with slope $\frac{3}{2}$ and length $5n$, which can be visualized as generalized Dyck paths starting at $(0,0)$ and ending at $(2n, 3n)$, without crossing the line $y = \frac{3}{2}x$. For example, for $n = 2$,

ababbaabbb $\longleftrightarrow$ 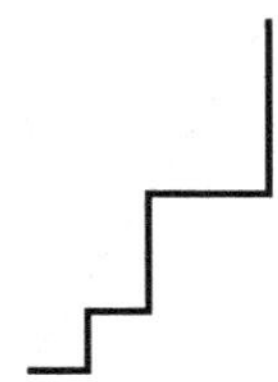

We denote this set by $\mathscr{D}_{3/2}(5n)$. In op. cit. Duchon proved that the number of factor-free elements of $\mathscr{D}_{3/2}(5n)$ is given by $C_{n-1} + C_n$, where C_n is the nth Catalan number.[1] Moreover, for $d_n = \left|\mathscr{D}_{3/2}(5n)\right|$, he gives the formula

$$d_n = \sum_{j=0}^{n} \frac{1}{5n+j+1} \binom{5n+1}{n-j} \binom{5n+2j}{j}.$$

This is sequence A060941 in [11].

It turns out that these numbers may also be generated by counting the elements of $\mathfrak{D}_n^{\mathbf{c}}(5,0)$ with coloring sequence $\mathbf{c} = (C_{j-1} + C_j)_{j \ge 1}$. In other words, there is a bijection between $\mathscr{D}_{3/2}(5n)$ and the set of Dyck words of semilength $5n$ created from strings of the form $P_0 =$ "d" and $P_j =$ "$u^{5j}d$" for $j = 1, \dots, n$, such that each maximal ascent u^{5j} is colored by a factor-free Dyck word with slope $\frac{3}{2}$ and length $5j$.[2]

Consequently, since $d_n = y_n = \left|\mathfrak{D}_n^{\mathbf{c}}(5,0)\right|$, Theorem 2 gives the alternative formula

$$d_n = \sum_{k=1}^{n} \binom{5n}{k-1} \frac{(k-1)!}{n!} B_{n,k}(1!(C_0 + C_1), 2!(C_1 + C_2), \dots).$$

Finally, since $j!(C_{j-1} + C_j) = (2j-2)_{j-1} + (2j)_{j-1}$, we can use the second identity in [12, Example 3.2] with $a = 2$, $b = -1$, and $c = 2$ to obtain

[1] A word in a language L is said to be *factor-free* if it has no proper factor in L.

[2] For more on this bijection for general rational Dyck paths, we refer to [3].

$$
\begin{aligned}
d_n &= \sum_{k=1}^{n} \binom{5n}{k-1} \sum_{j=0}^{k} \frac{(-1)^{k-j}}{k} \binom{k}{j} (2j-k) \frac{(2j-k+2n-1)_{n-1}}{n!} \\
&= \sum_{k=1}^{n} \binom{5n}{k-1} \sum_{j=0}^{k} \frac{(-1)^{k-j}}{nk} \binom{k}{j} (2j-k) \binom{2j-k+2n-1}{n-1} \\
&= \sum_{k=1}^{n} \binom{5n}{k-1} \sum_{j=0}^{k} \frac{(-1)^{j}}{n} \left[\binom{k-1}{j} - \binom{k-1}{j-1} \right] \binom{2n+k-2j-1}{n-1}.
\end{aligned}
$$

References

1. Asinowski, A., Mansour, T.: Dyck paths with coloured ascents. Eur. J. Comb. **29**, 1262–1279 (2008)
2. Birmajer, D., Gil, J.B., Weiner, M.D.: Convolutions of tribonacci, Fuss-Catalan, and Motzkin sequences. Fibonacci Quart. **52**, 54–60 (2014)
3. Birmajer, D., Gil, J.B., Weiner, M.D.: On rational Dyck paths and the enumeration of factor-free Dyck words. Discret. Appl. Math. **244**, 36–43 (2018)
4. Bóna, M.: Handbook of Enumerative Combinatorics. Discrete Mathematics and Its Applications. CRC Press, Boca Raton (2015)
5. Charalambides, C.A.: Enumerative Combinatorics. Chapman and Hall, Boca Raton (2002)
6. Comtet, L.: Advanced Combinatorics: The Art of Finite and Infinite Expansions. D. Reidel Publishing Co., Dordrecht (1974)
7. Deutsch, E.: Dyck path enumeration. Discret. Math. **204**, 167–202 (1999)
8. Du, R.R., Nie, Y., Sun, X.: Enumerations of humps and peaks in (k, a)-paths and (n, m)-Dyck paths via bijective proofs. Discret. Appl. Math. **190**(191), 42–49 (2015)
9. Duchon, P.: On the enumeration and generation of generalized Dyck words. Discret. Math. **225**, 121–135 (2000)
10. Mansour, T., Sun, Y.: Dyck paths and partial Bell polynomials. Australas. J. Comb. **42**, 285–297 (2008)
11. Sloane, N.J.A.: The on-line encyclopedia of integer sequences. http://oeis.org/
12. Wang, W., Wang, T.: General identities on Bell polynomials. Comput. Math. Appl. **58**, 104–118 (2009)

A Review of the Basic Discrete q-Distributions

Ch. A. Charalambides

Abstract Consider a sequence of independent Bernoulli trials and assume that the probability (or odds) of success (or the probability (or odds) of failure) at a trial varies (increases or decreases) geometrically, with rate q, either with the number of trials or with the number of successes. Let X_n be the number of successes up the nth trial and W_n (or T_k) be the number of failures (or trials) until the occurrence of the nth (or kth) success. The distributions of these random variables turned out to be q-analogues of the binomial and negative binomial (or Pascal) distributions. The Heine and Euler distributions, which are q-analogues of the Poisson distribution, are obtained as limiting distributions of q-binomial distributions (or negative q-binomial distributions), as the number of trials (or the number of successes) tends to infinity. Also, introducing the notion of a q-drawing of a ball from an urn containing balls of various kinds, a q-analogue of the Pólya urn model is constructed and q-Pólya and inverse q-Pólya distributions are examined. Finally, considering a stochastic model that is developing in time or space, in which events (successes) may occur at continuous points, a Heine and an Euler stochastic processes are presented.

Keywords Euler distribution · Euler process · Heine distribution · Heine process · Negative q-binomial distribution · q-binomial distribution · q-Poisson distribution · q-Poisson process.

2010 Mathematics Subject Classification Primary 60C05 · Secondary 05A30 33D90.

1 Introduction

The classical binomial and negative binomial (or Pascal) distributions are defined in the stochastic model of a sequence of independent and identically distributed Bernoulli trials. The Poisson distribution may be considered as a limiting case of the

Ch. A. Charalambides (✉)
Department of Mathematics, University of Athens, 15784 Athens, Greece
e-mail: ccharal@math.uoa.gr

G. E. Andrews et al. (eds.), *Lattice Path Combinatorics and Applications*, Developments in Mathematics 58, https://doi.org/10.1007/978-3-030-11102-1_9

binomial (or negative binomial) distribution as the number of trials (or failures) tends to infinity. Also, considering a stochastic model that is developing in time or space, in which events (successes) may occur at continuous points, a Poisson stochastic process, is introduced as a stochastic process with independent and homogeneous increments and with success probability in a small time interval analogous to its length.

Poisson (1837) generalized the binomial distribution (and implicitly the negative binomial distribution) by assuming that the probability of success at a trial varies with the number of previous trials. The negative binomial distribution (and implicitly the binomial distribution) can be generalized to a different direction by assuming that the probability of success at a trial varies with the number of successes occurring in the previous trials. The probability function of the number of successes up to a given number of trials was derived by Woodbury [23]. The Pólya urn model, which was introduced by Eggenberger and Pólya [11], may be considered as a sequence of independent Bernoulli trials, with the probability of success at a trial varying with both the number of trials and the number of successes.

It should be noticed that a stochastic model of a sequence of independent Bernoulli trials, in which the probability of success at a trial is assumed to vary with the number of trials and/or the number of successes, is advantageous in the sense that it permits incorporating the experience gained from previous trials and/or successes. If the probability of success at a trial is a very general function of the number of trials and/or the number successes, very little can be inferred from it about the distributions of the various random variables that may be defined on this model. The assumption that the probability of success (or failure) at a trial varies geometrically, with rate (proportion) q, leads to the introduction of discrete q-distributions. The study of these distributions is greatly facilitated by the wealth of existing q-sequences and q-functions, in q-combinatorics, and the theory of q-hypergeometric series.

In Sect. 2, after introducing the notions of a q-power, a q-factorial, and a q-binomial coefficient of a real number, two q-factorial convolution formulae are given, without proof, and, as a corollary of them, two q-binomial convolution formulae are deduced. Also, the q-binomial and the negative q-binomial formulae are obtained. In addition, a general q-binomial formula is given and, as limiting forms of it, q-exponential functions are deduced. Moreover, the q-factorial moments, which apart from the interest in their own, are used as an intermediate step in the calculation of the usual factorial moments of a discrete q-distribution are introduced. Also, a formula expressing the usual factorial moments in terms of the q-factorial moments is given.

Section 3 deals with discrete q-distributions defined in the stochastic model of a sequence of independent Bernoulli trials, with success probability at a trial varying geometrically with the number of previous trials. Specifically, assuming that the odds of success at a trial are a geometrically varying (increasing or decreasing) sequence, a q-binomial distribution of the first kind and a negative q-binomial distribution of the first kind are introduced and studied. In addition, the Heine distribution, which is a q-Poisson distribution of the first kind, is obtained as a limiting distribution of the

q-binomial distribution (or the negative q-binomial distribution) of the first kind, as the number of trials (or the number of successes) tends to infinity.

Section 4 is devoted to the study of discrete q-distributions defined in the stochastic model of a sequence of independent Bernoulli trials, with success probability varying geometrically with the number of previous successes. Introducing the notion of a geometric sequence of (Bernoulli) trials as a sequence of independent Bernoulli trials, with constant probability of success, which is terminated with the occurrence of the first success, an equivalent stochastic model is constructed as follows. A sequence of independent geometric sequences of trials with success probability at a geometric sequence of trials varying (increasing or decreasing) geometrically with the number of previous sequences (successes), is considered. In this model, a negative q-binomial distribution of the second kind and a q-binomial distribution of the second kind are introduced and examined. In addition, the Euler distribution, which is a q-Poisson distribution of the second kind, is obtained as a limiting distribution of the q-binomial distribution (or the negative q-binomial distribution) of the second kind, as the number of trials (or the number of successes) tends to infinity.

In Sect. 5, after introducing the notion of a q-drawing of a ball from an urn containing balls of various kinds, a q-Pólya urn model is presented and q-Pólya and inverse q-Pólya distributions and examined.

Section 6 is devoted to q-Poisson (Heine and Euler) stochastic processes. This article is based on the tutorial lecture on the basic discrete q-distributions, prepared for the 8th International Conference on Lattice Path Combinatorics and Applications.

2 Basic q-Combinatorics and q-Hypergeometric Series

Let x and q be real numbers, with $q \neq 1$, and k be an integer. The number

$$[x]_q = \frac{1-q^x}{1-q}$$

is called *q-number* and in particular $[k]_q$ is called q-integer. The base (parameter) q, in the theory of discrete q-distributions, varies in the interval $0 < q < 1$ or in the interval $1 < q < \infty$.

The kth order factorial of the q-number $[x]_q$, which is defined by

$$[x]_{k,q} = [x]_q[x-1]_q \cdots [x-k+1]_q, \quad k = 0, 1, \ldots,$$

is called *q-factorial* of x of order k. In particular $[k]_q! = [1]_q[2]_q \cdots [k]_q$ is called q-factorial. The *q-binomial coefficient* (or *Gaußian polynomial*) is defined by

$$\begin{bmatrix} x \\ k \end{bmatrix}_q = \frac{[x]_{k,q}}{[k]_q!}, \quad k = 0, 1, \ldots .$$

Note that

$$\lim_{q\to 1}[x]_q = x, \quad \lim_{q\to 1}[x]_{k,q} = (x)_k, \quad \lim_{q\to 1}\begin{bmatrix} x \\ k \end{bmatrix}_q = \binom{x}{k}, \quad k = 0, 1, \ldots .$$

Furthermore, the lack of uniqueness of q-analogues of expressions and formulae in q-combinatorics and q-hypergeometric series should be pointed out. It is due to the presence of powers of q in pseudo-isomorphisms as

$$[x + y]_q = [x]_q + q^x[y]_q \text{ and } [x + y]_q = q^y[x]_q + [y]_q,$$

where $0 < q < 1$ or $1 < q < \infty$. It should also be noticed that the two formulae may be considered as *equivalent* in the sense that any of these implies the other by replacing the base q by q^{-1}. In this framework, the existence of two versions of the q-analogue of the sum, which may be considered as equivalent, is attributed to the lack of uniqueness.

The particular cases of the q-binomial coefficients $\begin{bmatrix} n \\ k \end{bmatrix}_q$ and $\begin{bmatrix} n+k-1 \\ k \end{bmatrix}_q$, with n and k positive integers, admit q-combinatorial interpretations, which are given, without proof, in the following theorem.

Theorem 9.1 *The q-binomial coefficient $\begin{bmatrix} n \\ k \end{bmatrix}_q$, for n and k positive integers, equals the k-combinations of the set $\{1, 2, \ldots, n\}$, $\{m_1, m_2, \ldots, m_k\}$, weighted by $q^{m_1+m_2+\cdots+m_k-\binom{k+1}{2}}$,*

$$\sum_{1\le m_1<m_2<\cdots<m_k\le n} q^{m_1+m_2+\cdots+m_k-\binom{k+1}{2}} = \begin{bmatrix} n \\ k \end{bmatrix}_q .$$

Also, the q-binomial coefficient $\begin{bmatrix} n+k-1 \\ k \end{bmatrix}_q$, for n and k positive integers, equals the k-combinations of the set $\{1, 2, \ldots, n\}$ with repetition, $\{r_1, r_2, \ldots, r_k\}$, weighted by $q^{r_1+r_2+\cdots+r_k-k}$,

$$\sum_{1\le r_1\le r_2\le\cdots\le r_k\le n} q^{r_1+r_2+\cdots+r_k-k} = \begin{bmatrix} n+k-1 \\ k \end{bmatrix}_q .$$

A discrete q-uniform distribution is derived in the following probabilistic number theoretic example.

Example 9.1 Discrete q-uniform distribution.Consider a sequence of independent Bernoulli trials, with constant failure probability q, and let X be the number of failures until the occurrence of the first success. Clearly, the random variable X follows a geometric distribution with probability function

$$P(X = x) = (1 - q)q^x, \quad x = 0, 1, \ldots,$$

where $0 < q < 1$. Also, consider a fixed positive integer n and let

$$C_x = \{x + kn : k = 0, 1, \ldots\}, \quad x = 0, 1, \ldots, n-1.$$

Clearly, each of the possible values of the random variable X, $\{0, 1, \ldots\}$, belongs in one of these n congruence classes (pairwise disjoint sets), modulo n, $\{C_0, C_1, \ldots, C_{n-1}\}$. Furthermore, consider a sequence of independent Bernoulli trials, with constant failure probability q, and let X_n be the index of the congruence class in which the number of failures, until the occurrence of the first success, belongs. Since the random variable X_n assumes the value x if and only if X belongs to C_x, its probability function,

$$P(X_n = x) = P(X \in C_x) = \sum_{k=0}^{\infty}(1-q)q^{x+kn} = \frac{(1-q)q^x}{1-q^n},$$

may be expressed as

$$P(X_n = x) = \frac{q^x}{[n]_q}, \quad x = 0, 1, \ldots, n-1, \quad 0 < q < 1.$$

Note that the limiting probability function, as $q \to 1$,

$$\lim_{q \to 1} P(X_n = x) = \frac{1}{n}, \quad x = 0, 1, \ldots, n-1,$$

is the discrete uniform probability function on the set $\{0, 1, \ldots, n-1\}$. For this reason, the distribution of X_n is called *discrete q-uniform distribution.*

It is worth noting that the probability function $P(X = x | X \leq n-1)$, of a right truncated geometric distribution, since

$$P(X = x | X \leq n-1) = \frac{P(X = x, X \leq n-1)}{P(X \leq n-1)} = \frac{P(X = x)}{P(X \leq n-1)},$$

for $x = 0, 1, \ldots, n-1$, and

$$P(X \leq n-1) = \sum_{x=0}^{n-1}(1-q)q^x = 1 - q^n,$$

is readily deduced as

$$P(X = x | X \leq n-1) = \frac{q^x}{[n]_q}, \quad x = 0, 1, \ldots, n-1, \quad 0 < q < 1.$$

Two versions of a q-Vandermonde (q-factorial convolution) formula are presented in the next theorem.

Theorem 9.2 *Let n be a positive integer and let x, y, and q be real numbers, with $q \neq 1$. Then,*

$$[x+y]_{n,q} = \sum_{k=0}^{n} \begin{bmatrix} n \\ k \end{bmatrix}_q q^{(n-k)(x-k)} [x]_{k,q} [y]_{n-k,q}.$$

Alternatively,

$$[x+y]_{n,q} = \sum_{k=0}^{n} \begin{bmatrix} n \\ k \end{bmatrix}_q q^{k(y-n+k)} [x]_{k,q} [y]_{n-k,q}.$$

Two versions of a q-Cauchy (q-binomial convolution) formula, which by virtue of

$$\begin{bmatrix} x \\ k \end{bmatrix}_q = \frac{[x]_{k,q}}{[k]_q!}, \quad k = 0, 1, \ldots,$$

may be considered as reformulations of the corresponding two versions of the q-Vandermonde (q-factorial convolution) formula, are stated in the following corollary of Theorem 9.2.

Corollary 9.1 *Let n be a positive integer and let x, y, and q be real numbers, with $q \neq 1$. Then,*

$$\begin{bmatrix} x+y \\ n \end{bmatrix}_q = \sum_{k=0}^{n} q^{(n-k)(x-k)} \begin{bmatrix} x \\ k \end{bmatrix}_q \begin{bmatrix} y \\ n-k \end{bmatrix}_q.$$

Alternatively,

$$\begin{bmatrix} x+y \\ n \end{bmatrix}_q = \sum_{k=0}^{n} q^{k(y-n+k)} \begin{bmatrix} x \\ k \end{bmatrix}_q \begin{bmatrix} y \\ n-k \end{bmatrix}_q.$$

q-Newton (q-binomial and negative q-binomial) formulae are given in the following theorems.

Theorem 9.3 *Let n be a positive integer and let t and q be real numbers, with $q \neq 1$. Then,*

$$\prod_{i=1}^{n} (1 + tq^{i-1}) = \sum_{k=0}^{n} q^{\binom{k}{2}} \begin{bmatrix} n \\ k \end{bmatrix}_q t^k.$$

Theorem 9.4 *Let n be a positive integer and let t and q be real numbers, with $|t| < 1$ and $|q| < 1$ or $|t| < |q|^{-(n-1)}$ and $|q| > 1$. Then,*

$$\prod_{j=1}^{n} \left(1 - tq^{j-1}\right)^{-1} = \sum_{k=0}^{\infty} \begin{bmatrix} n+k-1 \\ k \end{bmatrix}_q t^k.$$

Notice that the two versions of the q-Vandermonde formula, given in Theorem 9.2, as well as the two versions of the q-Cauchy formula, deduced in Corollary 9.1, are equivalent in the sense that any of two implies the other by replacing the base q by q^{-1}. It should be pointed out that the replacement of q by q^{-1} in the q-Newton (q-binomial and negative q-binomial) formulae, stated in Theorems 9.3 and 9.4, lead, respectively, to the same expression.

The *q-binomial* and the *negative q-binomial formulae* may be extended to $n = x$, a real number, as

$$\prod_{i=1}^{\infty} \frac{1+tq^{i-1}}{1+tq^{x+i-1}} = \sum_{k=0}^{\infty} q^{\binom{k}{2}} \begin{bmatrix} x \\ k \end{bmatrix}_q t^k, \quad |t| < 1, \quad 0 < q < 1, \tag{9.1}$$

and

$$\prod_{i=1}^{\infty} \frac{1-tq^{x+i-1}}{1-tq^{i-1}} = \sum_{k=0}^{\infty} \begin{bmatrix} x+k-1 \\ k \end{bmatrix}_q t^k, \quad |t| < 1, \quad 0 < q < 1, \tag{9.2}$$

respectively. A q-analogue of the exponential function can be obtained from (9.1) by replacing t by $(1-q)t$ and then taking the limit as $x \to \infty$. Since, for $|q| < 1$,

$$\lim_{x\to\infty} (1-q)[x-j]_q = \lim_{x\to\infty} (1-q^{x-j}) = 1, \quad \lim_{x\to\infty} (1-q)^k [x]_{k,q} = 1,$$

a *q-exponential function* is deduced as

$$E_q(t) = \prod_{i=1}^{\infty} (1 + t(1-q)q^{i-1}) = \sum_{k=0}^{\infty} q^{\binom{k}{2}} \frac{t^k}{[k]_q!}, \quad -\infty < t < \infty.$$

Another *q-exponential function* can be similarly obtained from (9.2) as

$$e_q(t) = \prod_{i=1}^{\infty} (1 - t(1-q)q^{i-1})^{-1} = \sum_{k=0}^{\infty} \frac{t^k}{[k]_q!}, \quad |t| < 1/(1-q).$$

Clearly, $E_q(t)e_q(-t) = 1$ and $E_{q^{-1}}(t) = e_q(t)$.

The q-factorial moments of a discrete q-distribution are introduced as follows.

Let X be a nonnegative integer valued random variable, with probability function $f(x) = P(X = x)$, $x = 0, 1, \ldots$. The expected value

$$E\big([X]_{m,q}\big) = \sum_{x=m}^{\infty} [x]_{m,q} f(x), \quad m = 1, 2, \ldots,$$

provided the series is convergent, is called the m th order q -factorial moment of the random variable X.

The usual factorial moments are expressed in terms of the q-factorial moments by

$$E\big[(X)_j\big] = j!\sum_{m=j}^{\infty}(-1)^{m-j}(1-q)^{m-j}s_q(m,j)\frac{E\big([X]_{m,q}\big)}{[m]_q!},$$

for $j = 1, 2, \ldots,$ where $s_q(m, j)$ is the q-Stirling number of the first kind.

Remark 9.1 q-Deformed distributions in quantum physics. Consider a nonnegative integer valued random variable X with probability mass function $f_X(x) = P(X = x)$, $x = 0, 1, \ldots$. Furthermore, consider the q-number transformation $Y = [X]_q$, which in the language of quantum physics is known as a q-deformation. The distribution of the random variable Y, with probability function

$$f_Y([x]_q) = P(Y = [x]_q) = P(X = x) = f_X(x), \quad x = 0, 1, \ldots,$$

is called *q-deformed distribution*. The mean and the variance of the q-deformed distribution of Y are the q-mean and the q-variance of the distribution of X.

3 Success Probability Varying with the Number of Trials

Consider a sequence of independent Bernoulli trials and assume that the odds of success at the ith trial is given by

$$\theta_i = \theta q^{i-1}, \quad i = 1, 2, \ldots, \quad 0 < \theta < \infty, \quad 0 < q < 1 \text{ or } 1 < q < \infty,$$

which is a geometrically varying sequence, with rate q. Note that the case $q = 1$ corresponds to the classical case of constant odds (or constant probability) of success. Also, for $0 < q < 1$, the sequence θ_i, $i = 1, 2, \ldots,$ is geometrically decreasing, while for $1 < q < \infty$, is geometrically increasing. Since $p_i = \theta_i/(1+\theta_i)$, it follows that the probability of success at the ith trial is given by

$$p_i = \frac{\theta q^{i-1}}{1+\theta q^{i-1}}, \quad i = 1, 2, \ldots, \quad 0 < \theta < \infty, \; 0 < q < 1 \text{ or } 1 < q < \infty. \tag{9.3}$$

Let X_n be the number of successes in a sequence of n independent Bernoulli trials, with probability of success at the ith trial given by (9.3). The distribution of the random variable X_n is called *q-binomial distribution of the first kind*, with parameters n, θ, and q.

Theorem 9.5 *The probability function of the q-binomial distribution of the first kind, with parameters n, θ, and q, is given by*

$$P(X_n = x) = \begin{bmatrix} n \\ x \end{bmatrix}_q \frac{\theta^x q^{\binom{x}{2}}}{\prod_{i=1}^{n}(1+\theta q^{i-1})}, \quad x = 0, 1, \ldots, n, \tag{9.4}$$

for $0 < \theta < \infty$ *and* $0 < q < 1$ *or* $1 < q < \infty$. *Its* q*-factorial moments are given by*

$$E([X_n]_{m,q}) = \frac{[n]_{m,q}\,\theta^m q^{\binom{m}{2}}}{\prod_{i=1}^{m}(1+\theta q^{i-1})}, \quad m = 1, 2, \ldots, n,$$

and $E([X_n]_{m,q}) = 0$, *for* $m = n+1, n+2, \ldots$. *Furthermore, its (usual) factorial moments are given by*

$$E[(X_n)_j] = j! \sum_{m=j}^{n} (-1)^{m-j} \begin{bmatrix} n \\ m \end{bmatrix}_q \frac{\theta^m q^{\binom{m}{2}} (1-q)^{m-j} s_q(m, j)}{\prod_{i=1}^{m}(1+\theta q^{i-1})},$$

for $j = 1, 2, \ldots, n$, *and* $E[(X_n)_j] = 0$, *for* $j = n+1, n+2, \ldots$, *where* $s_q(m, j)$ *is the* q*-Stirling number of the first kind. In particular, its mean and variance are given by*

$$E(X_n) = \sum_{i=1}^{n} \frac{\theta q^{i-1}}{1+\theta q^{i-1}}, \quad V(X_n) = \sum_{i=1}^{n} \frac{\theta q^{i-1}}{(1+\theta q^{i-1})^2}.$$

Example 9.2 Weldon's classical dice data. Walter F. R. Weldon obtained the data from $m = 26,306$ throws of $n = 12$ dice. Among the $mn = 315,672$ recorded numbers from the set of the six faces of a die, $\{1, 2, 3, 4, 5, 6\}$, the number of dice showing face 5 or face 6 was $s = 106,602$.

A discrete probability distribution that fits to these data may be defined on a sequence of independent Bernoulli trials. Specifically, a throw of a die is considered as a Bernoulli trial, with success the event of showing face 5 or face 6. Then, each throw of the 12 dice constitutes a sequence of $n = 12$ independent Bernoulli trials.

Kemp and Kemp [18] examined first the fair dice assumption, which leads to the usual binomial distribution, with $n = 12$ and $p = 1/3$. It was found out that this distribution does not fit to these data. After this conclusion, they replaced the success probability $p = 1/3$ by its moment estimate

$$\hat{p} = \frac{s}{mn} = \frac{106,602}{315,672} = 0.3377.$$

Although this equally unbalanced dice hypothesis may give a satisfactory fit to these data, the assumption that all $n = 12$ dice are identically unbalanced seems inherently implausible.

More realistic hypotheses for unfair dice, p_i, $i = 1, 2, \ldots, n$, were examined by Kemp and Kemp [18]. They supposed that there is a spectrum of unfairness among the dice. A log-linear odds assumption,

$$\log \theta_i = \log \theta + (i-1) \log q, \quad i = 1, 2, \ldots, n,$$

implies (9.3). Then, the number X_n of successes in n trials obeys the q-binomial distribution of the first kind with probability function (9.4).

Also, let W_n be the number of failures until the occurrence of the nth success in a sequence of independent Bernoulli trials, with probability of success at the ith trial given by (9.3). The distribution of the random variable W_n is called *negative q-binomial distribution of the first kind*, with parameters n, θ, and q.

Theorem 9.6 *The probability function of the negative q-binomial distribution of the first kind, with parameters n, θ, and q, is given by*

$$P(W_n = w) = \begin{bmatrix} n+w-1 \\ w \end{bmatrix}_q \frac{\theta^n q^{\binom{n}{2}+w}}{\prod_{i=1}^{n+w}(1+\theta q^{i-1})}, \quad w = 0, 1, \ldots, \tag{9.5}$$

for $0 < \theta < \infty$ and $0 < q < 1$ or $1 < q < \infty$. Its q-factorial moments are given by

$$E\big([W_n]_{m,q}\big) = \frac{[n+m-1]_{m,q}}{\theta^m q^{\binom{m}{2}+(n-1)m}}, \quad m = 1, 2, \ldots .$$

Furthermore, its (usual) factorial moments are given by

$$E[(W_n)_j] = j! \sum_{m=j}^{\infty} (-1)^{m-j} \begin{bmatrix} n+m-1 \\ m \end{bmatrix}_q \frac{(1-q)^{m-j} s_q(m,j)}{\theta^m q^{\binom{m}{2}+(n-1)m}},$$

for $j = 1, 2, \ldots,$ where $s_q(m, j)$ is the q-Stirling number of the first kind.

Remark 9.2 Another negative q-binomial distribution of the first kind. The probability function of the number U_n of successes until the occurrence of the nth failure is closely connected to (9.5). Specifically,

$$P(U_n = u) = P(X_{n+u-1} = u)(1 - p_{n+u}) = P(W_{u+1} = n-1)\frac{1 - p_{n+u}}{p_{n+u}}$$

and so

$$P(U_n = u) = \begin{bmatrix} n+u-1 \\ u \end{bmatrix}_q \frac{\theta^u q^{\binom{u}{2}}}{\prod_{i=1}^{n+u}(1+\theta q^{i-1})}, \quad u = 0, 1, \ldots, \tag{9.6}$$

for $0 < \theta < \infty$ and $0 < q < 1$ or $1 < q < \infty$.

Finally, let X be a discrete random variable with probability function

$$f(x) = P(X = x) = e_q(-\lambda)\frac{q^{\binom{x}{2}}\lambda^x}{[x]_q!}, \quad x = 0, 1, \ldots, \tag{9.7}$$

where $0 < \lambda < \infty$, $0 < q < 1$ and $e_q(t) = \prod_{i=1}^{\infty}(1 - t(1-q)q^{i-1})^{-1}$ is a q-exponential function. The distribution of the random variable X is called *Heine distribution*, with parameters λ and q. The Heine distribution is a q-Poisson distribution since its probability function, for $q \to 1$, approaches the probability function of the Poisson distribution.

Theorem 9.7 *The q-factorial moments of the Heine distribution are given by*

$$E([X]_{m,q}) = \frac{q^{\binom{m}{2}}\lambda^m}{\prod_{i=1}^{m}(1+\lambda(1-q)q^{i-1})}, \quad m = 1, 2, \ldots .$$

Moreover, its factorial moments are given by

$$E[(X)_j] = j!\sum_{m=j}^{\infty}(-1)^{m-j}\frac{q^{\binom{m}{2}}\lambda^m}{[m]_q!} \cdot \frac{(1-q)^{m-j}s_q(m,j)}{\prod_{i=1}^{m}(1+\lambda(1-q)q^{i-1})},$$

for $j = 1, 2, \ldots,$ where $s_q(m, j)$ is the q-Stirling number of the first kind.

The q-binomial and the negative q-binomial distributions of the first kind can be approximated by the Heine distribution, according to the following theorem.

Theorem 9.8 *The limit of the probability function of the q-binomial distribution of the first kind,* (9.4), *as $n \to \infty$, is the probability function of the Heine distribution,*

$$\lim_{n\to\infty}\begin{bmatrix} n \\ x \end{bmatrix}_q \frac{q^{\binom{x}{2}}\theta^x}{\prod_{i=1}^{n}(1+\theta q^{i-1})} = e_q(-\lambda)\frac{q^{\binom{x}{2}}\lambda^x}{[x]_q!}, \quad x = 0, 1, \ldots,$$

for $0 < \lambda < \infty$ and $0 < q < 1$, with $\lambda = \theta/(1-q)$.

Also, the limit of the probability function of the negative q-binomial distribution of the first kind, (9.6), *as $n \to \infty$, is the probability function of the Heine distribution,*

$$\lim_{n\to\infty}\begin{bmatrix} n+u-1 \\ u \end{bmatrix}_q \frac{q^{\binom{u}{2}}\theta^u}{\prod_{i=1}^{n+u}(1+\theta q^{i-1})} = e_q(-\lambda)\frac{q^{\binom{u}{2}}\lambda^u}{[u]_q!}, \quad u = 0, 1, \ldots,$$

for $0 < \lambda < \infty$ and $0 < q < 1$, with $\lambda = \theta/(1-q)$.

4 Success Probability Varying with the Number of Successes

Consider a random experiment with sample space $\Omega = \{f, s\}$, where the sample points (events) f and s are characterized as *failure* and *success*, respectively. An experiment with such a sample space is called *Bernoulli trial*. Furthermore, a sequence of independent Bernoulli trials, with constant success probability, which is terminated with the occurrence of the first success, $g = (f, f, \ldots, f, s)$, is called *geometric sequence of trials*. Let us first consider, the case with probability of success at the jth geometric sequence of trials given by

$$p_j = 1 - \theta q^{j-1}, \quad j = 1, 2, \ldots, \quad 0 < \theta < 1, \quad 0 < q < 1 \text{ or } 1 < q < \infty, \tag{9.8}$$

where, for $0 < \theta < 1$ and $1 < q < \infty$, the number j of geometric sequences of trials is restricted by $\theta q^{j-1} < 1$, ensuring that $0 < p_j < 1$. This restriction imposes on j an upper bound, $j = 1, 2, \ldots, [r]$, with $[r]$ denoting the integral part of $r = -\log\theta/\log q > 0$. Notice that the probability p_j is essentially the conditional probability of success at any Bernoulli trial, given that $j - 1$ successes occur in the previous trials.

Let W_n be the number of failures until the occurrence of the nth success, in a sequence of independent geometric sequences of trials, with probability of success at the jth geometric sequence of trials given by (9.8). The distribution of the random variable W_n is called *negative q-binomial distribution of the second kind*, with parameters n, θ, and q.

Theorem 9.9 *The probability function of the negative q-binomial distribution of the second kind, with parameters n, θ, and q, is given by*

$$P(W_n = w) = \begin{bmatrix} n + w - 1 \\ w \end{bmatrix}_q \theta^w \prod_{j=1}^{n} (1 - \theta q^{j-1}), \quad w = 0, 1, \ldots, \tag{9.9}$$

for $0 < \theta < 1$ and $0 < q < 1$ or $1 < q < \infty$ with $\theta q^{n-1} < 1$. Its q-factorial moments are given by

$$E([W_n]_{m,q}) = \frac{[n + m - 1]_{m,q}\theta^m}{\prod_{j=1}^{m}(1 - \theta q^{n+j-1})}, \quad m = 1, 2, \ldots .$$

Furthermore, its factorial moments are given by

$$E[(W_n)_j] = j! \sum_{m=j}^{\infty} (-1)^{m-j} \begin{bmatrix} n + m - 1 \\ m \end{bmatrix}_q \frac{\theta^m (1 - q)^{m-j} s_q(m, j)}{\prod_{i=1}^{m}(1 - \theta q^{n+i-1})},$$

for $j = 1, 2, \ldots$, where $s_q(m, j)$ is the q-Stirling number of the first kind. In particular, its mean and variance are given by

$$E(W_n) = \sum_{j=1}^{n} \frac{\theta q^{j-1}}{1 - \theta q^{j-1}}, \quad V(W_n) = \sum_{j=1}^{n} \frac{\theta q^{j-1}}{(1 - \theta q^{j-1})^2}.$$

Remark 9.3 A q-geometric distribution of the second kind. The probability function of the number Z_1 of successes until the occurrence of the first failure is of interest and may be obtained as follows. Considering the event A_j of success at the jth trial, for

$j = 1, 2, \ldots$, the probability $P_z = P(A_1 A_2 \cdots A_z A'_{z+1})$, that z successes precede the occurrence of the first failure, on using the multiplication formula, is deduced as

$$P_z = P(A_1)P(A_2|A_1) \cdots P(A_z|A_1 A_2 \cdots A_{z-1})P(A'_{z+1}|A_1 A_2 \cdots A_z)$$
$$= \prod_{j=1}^{z} (1 - \theta q^{j-1})\theta q^z,$$

for $z = 0, 1, \ldots$. Also, the probability Q of the occurrence of at least one failure in an infinite number of Bernoulli trials is readily deduced as

$$Q = \lim_{n\to\infty} P(A'_1 \cup A'_2 \cup \cdots \cup A'_n) = 1 - \lim_{n\to\infty} P(A_1 A_2 \cdots A_n)$$
$$= 1 - \lim_{n\to\infty} \prod_{j=1}^{n} \big(1 - \theta q^{j-1}\big) = 1 - E_q\big(-\theta/(1-q)\big),$$

where $E_q(t) = \prod_{j=1}^{\infty} \big(1 + t(1-q)q^{j-1}\big)$ is a q-exponential function. Clearly, the probability function of the random variable Z_1 is the conditional probability that z successes precede the occurrence of the first failure, given the occurrence of at least one failure in an infinite number of Bernoulli trials, $P(Z_1 = z) = P_z/Q$, and so

$$P(Z_1 = z) = \big(1 - E_q\big(-\theta/(1-q)\big)\big)^{-1} \prod_{j=1}^{z} \big(1 - \theta q^{j-1}\big)\theta q^z,$$

for $z = 0, 1, \ldots$, where $0 < \theta < 1$ and $0 < q < 1$.

Also, let X_n be the number of failures in a sequence of n independent Bernoulli trials, with probability of success at the jth geometric sequence of trials given by (9.8). The distribution of the random variable X_n is called *q-binomial distribution of the second kind*, with parameters n, θ, and q.

Theorem 9.10 *The probability function of the q-binomial distribution of the second kind, with parameters n, θ, and q, is given by*

$$P(X_n = x) = \begin{bmatrix} n \\ x \end{bmatrix}_q \theta^x \prod_{j=1}^{n-x} (1 - \theta q^{j-1}), \quad x = 0, 1, \ldots, n, \tag{9.10}$$

for $0 < \theta < 1$ and $0 < q < 1$ or $1 < q < \infty$, with $\theta q^{n-1} < 1$. Its q-factorial moments are given by

$$E\big([X_n]_{m,q}\big) = [n]_{m,q}\theta^m, \quad m = 1, 2, \ldots, n,$$

and $E\big([X_n]_{m,q}\big) = 0$, for $m = n+1, n+2, \ldots$. Moreover, its factorial moments are given by

$$E[(X_n)_j] = j! \sum_{m=j}^{n} (-1)^{m-j} \begin{bmatrix} n \\ m \end{bmatrix}_q \theta^m (1-q)^{m-j} s_q(m, j),$$

for $j = 1, 2, \ldots, n$, *and* $E[(X_n)_j] = 0$, *for* $j = n+1, n+2, \ldots$, *where* $s_q(m, j)$ *is the* q*-Stirling number of the first kind. In particular, its mean and variance are given by*

$$E(X_n) = \sum_{m=1}^{n} \frac{[n]_{m,q}(1-q)^{m-1}\theta^m}{[m]_q}$$

and

$$V(X_n) = 2\sum_{m=2}^{n} \frac{[n]_{m,q}(1-q)^{m-2}\theta^m \zeta_{m-1,q}}{[m]_q} + E(X_n) - [E(X_n)]^2,$$

where $\zeta_{m,q} = \sum_{j=1}^{m} 1/[j]_q$.

Remark 9.4 Absorption and inverse absorption distributions. The success probability (9.8), in the case $1 < q < \infty$, may be preferably expressed as follows. Replacing the parameter q by q^{-1}, with $0 < q < 1$, and then setting $\theta = q^r$, we get

$$p_j = 1 - q^{r-j+1}, \quad j = 1, 2, \ldots, [r], \quad 0 < r < \infty, \quad 0 < q < 1, \tag{9.11}$$

which is a geometrically decreasing sequence of a finite number of terms. The probability function of the number $Y_n = n - X_n$ of successes in n independent Bernoulli trials, with probability of success at the jth geometric sequence of trials given by (9.8), on using the relation $P(Y_n = y) = P(X_n = n - y)$, $y = 0, 1, \ldots, n$, and the expression (9.10), is obtained as

$$P(Y_n = y) = \begin{bmatrix} n \\ y \end{bmatrix}_q q^{(n-y)(r-y)}(1-q)^y [r]_{y,q}, \quad y = 0, 1, \ldots, n, \tag{9.12}$$

for $0 < r < \infty$, $0 < q < 1$, and $n \leq [r]$. This q-binomial distribution of the second kind is particularly known as *absorption distribution*. Also, in the same stochastic model, the probability function of the number of failures until the occurrence of the nth success, on using the expression (9.9), is deduced as

$$P(W_n = w) = \begin{bmatrix} n+w-1 \\ w \end{bmatrix}_q q^{(r-n+1)w}(1-q)^n [r]_{n,q}, \quad w=0, 1, \ldots, \tag{9.13}$$

for $0 < r < \infty$, $0 < q < 1$, and $n \leq [r]$. This negative q-binomial distribution of the second kind is particularly known as *inverse absorption distribution*.

Example 9.3 Proofreading a manuscript. Assume that a proofreader reads a manuscript, which has a fixed number of errors m and when he/she finds an error corrects it and starts reading the manuscript from the beginning. Also the proofreader starts reading the manuscript from the beginning when he/she reaches its end. A scan (reading) of the manuscript is successful if the proofreader finds (and corrects) an error and is a failure otherwise. Thus, a scan of the manuscript constitutes a Bernoulli trial. Assume that the probability of finding any particular error is $p = 1 - q$. Then, the conditional probability that a scan (trial) is successful, given that $j - 1$ scans (trials) were successful in the previous scans, is

$$p_j = 1 - q^{m-j+1}, \quad j = 1, 2, \ldots, m, \quad 0 < q < 1,$$

which is of the form (9.11), with $r = m$ a positive integer. Consequently, the distribution of the number Y_n of errors found (and corrected) in n scans (readings) of the manuscript is an absorption distribution, with probability function (9.12). Also, the distribution of the number W_n of unsuccessful scans until the occurrence of the nth successful scan, with $n \leq m$, is the inverse absorption distribution, with probability function (9.13).

Finally, let X be a discrete random variable with probability function

$$f(x) = P(X = x) = E_q(-\lambda)\frac{\lambda^x}{[x]_q!}, \quad x = 0, 1, \ldots, \tag{9.14}$$

where $0 < \lambda < 1/(1-q)$, $0 < q < 1$, and $E_q(t) = \prod_{i=1}^{\infty}(1 + t(1-q)q^{i-1})$ is a q-exponential function. The distribution of the random variable X is called *Euler distribution*, with parameters λ and q. The Euler distribution is a q-Poisson distribution since the probability function (9.14), for $q \to 1$, converges to the probability function of the Poisson distribution.

Theorem 9.11 *The q-factorial moments of the Euler distribution are given by*

$$E([X]_{m,q}) = \lambda^m, \quad m = 1, 2, \ldots .$$

Moreover, its factorial moments are given by

$$E[(X)_j] = j! \sum_{m=j}^{\infty}(-1)^{m-j}\frac{\lambda^m}{[m]_q!}(1-q)^{m-j}s_q(m, j), \quad j = 1, 2, \ldots,$$

where $s_q(m, j)$ is the q-Stirling number of the first kind.

The q-binomial and the negative q-binomial distributions of the second kind can be approximated by the Euler distribution, according to the following theorem.

Theorem 9.12 *The limit of the probability function* (9.10) *of the q-binomial distribution of the second kind, as $n \to \infty$, is the probability function of the Euler distribution,*

$$\lim_{n\to\infty}\begin{bmatrix} n \\ x \end{bmatrix}_q \theta^x \prod_{i=1}^{n-x}(1-\theta q^{i-1}) = E_q(-\lambda)\frac{\lambda^x}{[x]_q!}, \quad x = 0, 1, \ldots,$$

for $0 < \lambda < 1/(1-q)$ *and* $0 < q < 1$, *with* $\lambda = \theta/(1-q)$.

Also, the limit of the probability function (9.9) *of the negative q-binomial distribution of the second kind, as $n \to \infty$, is the probability function of the Euler distribution,*

$$\lim_{n\to\infty}\begin{bmatrix} n+w-1 \\ w \end{bmatrix}_q \theta^w \prod_{i=1}^{n}(1-\theta q^{i-1}) = E_q(-\lambda)\frac{\lambda^w}{[w]_q!}, \quad w = 0, 1, \ldots,$$

for $0 < \lambda < 1/(1-q)$ *and* $0 < q < 1$, *with* $\lambda = \theta/(1-q)$.

An interesting application of the Heine and Euler distributions, as feasible prior in a Bayesian model for oil exploration, was presented by Benkherouf and Bather [2].

5 Success Probability Varying with the Number of Trials and the Number of Successes

Consider a sequence of independent Bernoulli trials, with the conditional probability of success at the ith trial, given that $j-1$ successes occur in the $i-1$ previous trials, given by

$$p_{i,j} = \frac{a_j}{b_i}, \quad j = 1, 2, \ldots, i, \quad i = 1, 2, \ldots,$$

where $0 < a_j \le b_i$, for $j = 1, 2, \ldots, i$ and $i = 1, 2, \ldots$. The Pólya urn model, which belongs in this family of stochastic models, may be extended to a q-Pólya urn model by introducing an appropriate q-analogue of a random drawing of a ball from an urn.

Let us consider an urn containing r white balls, $\{b_1, b_2, \ldots, b_r\}$, and s black balls, $\{b_{r+1}, b_{r+2}, \ldots, b_{r+s}\}$. A *random q-drawing* of a ball from the urn is carried out as follows. Assume that the balls in the urn are forced to pass through a random mechanism, one by one, in the order $(b_1, b_2, \ldots, b_{r+s})$ or in the reverse order $(b_{r+s}, b_{r+s-1}, \ldots, b_1)$. Also, suppose that each passing ball may or may not be caught by the mechanism, with probabilities $p = 1-q$ and q, respectively. The first caught ball is drawn out of the urn. In the case all $r+s$ balls pass through the mechanism and no ball is caught, the ball passing procedure is repeated, with the same order. Clearly, the probability that ball b_x is drawn from the urn is given by

$$\sum_{k=0}^{\infty}(1-q)q^{(x-1)+k(r+s)} = (1-q)q^{x-1}\sum_{k=0}^{\infty}q^{(r+s)k} = \frac{q^{x-1}}{[r+s]_q},$$

or by

$$\sum_{k=0}^{\infty}(1-q)q^{(r+s-x)+k(r+s)} = \frac{q^{r+s-x}}{[r+s]_q} = \frac{q^{-(x-1)}}{[r+s]_{q^{-1}}},$$

where $0 < q < 1$, according to whether the ball passing order is $(b_1, b_2, \dots, b_{r+s})$ or $(b_{r+s}, b_{r+s-1}, \dots, b_1)$. These probabilities may be expressed as

$$p_{r+s}(x; q) = P(X_{r+s} = x) = \frac{q^{x-1}}{[r+s]_q}, \quad x = 1, 2, \dots, r+s,$$

where $0 < q < 1$ or $1 < q < \infty$. Note that this is the probability function of the discrete q-uniform distribution on the set $\{1, 2, \dots, r+s\}$. Also, the probability $P_{r+s}(r; q)$, that a white ball is drawn from the urn is given by

$$P_{r+s}(r; q) = P(X_{r+s} \leq r) = \frac{[r]_q}{[r+s]_q} = \frac{(q^{-1})^s [r]_{q^{-1}}}{[r+s]_{q^{-1}}},$$

where $0 < q < 1$ or $1 < q < \infty$. It is worth noticing that the probability $Q_{r+s}(s; q)$ that a black ball is drawn from the urn is given by

$$Q_{r+s}(s; q) = P(r < X_{r+s} \leq r+s) = \frac{q^r [s]_q}{[r+s]_q} = \frac{[s]_{q^{-1}}}{[r+s]_{q^{-1}}},$$

where $0 < q < 1$ or $1 < q < \infty$, which conforms with the relation

$$P_{r+s}(r; q) + Q_{r+s}(s; q) = 1.$$

Finally, notice that a random q-drawing of a ball, for $q \to 1$ and since

$$\lim_{q \to 1} P_{r+s}(r; q) = \frac{r}{r+s}, \quad \lim_{q \to 1} Q_{r+s}(s; q) = \frac{s}{r+s},$$

reduces to the usual random drawing of a ball from the urn.

Furthermore, assume that random q-drawings of balls are sequentially carried out, one after the other, from an urn, initially containing r white and s black balls, according to the following scheme. After each q-drawing, the drawn ball is placed back in the urn together with k balls of the same color. Then, the conditional probability of drawing a white ball at the ith q-drawing, given that $j-1$ white balls are drawn in the previous $i-1$ q-drawings, is given by

$$p_{i,j} = \frac{1 - q^{r+k(j-1)}}{1 - q^{r+s+k(i-1)}} = \frac{[\alpha - j + 1]_{q^{-k}}}{[\alpha + \beta - i + 1]_{q^{-k}}}, \tag{9.15}$$

for $j = 1, 2, \ldots, i$ and $i = 1, 2, \ldots$, where $0 < q < 1$ or $1 < q < \infty$ and $\alpha = -r/k$ and $\beta = -s/k$, with r and s positive integers and k an integer. This model, which for $q \to 1$ and since

$$\lim_{q\to 1} p_{i,j} = \frac{r + k(j-1)}{r + s + k(i-1)} = \frac{\alpha - j + 1}{\alpha + \beta - i + 1},$$

for $j = 1, 2, \ldots, i$ and $i = 1, 2, \ldots$, reduces to the (classical) Pólya urn model, may be called *q-Pólya urn model*. Characterizing the q-drawing of a white ball as success and the q-drawing of a black ball as failure, the q-Pólya urn model reduces to the stochastic model of a sequence of independent Bernoulli trials, with probability of success at a trial varying with the number of trials and the number of previous successes, according to (9.15).

Let X_n be the number of white balls drawn in n q-drawings in a q-Pólya urn model, with the conditional probability of drawing a white ball at the ith q-drawing, given that $j-1$ white balls are drawn in the previous $i-1$ q-drawings, given by (9.15). The distribution of the random variable X_n is called *q-Pólya distribution*, with parameters n, α, β, k, and q.

Theorem 9.13 *The probability function of the q-Pólya distribution, with parameters n, α, β, k, and q, is given by*

$$\begin{aligned} P(X_n = x) &= \begin{bmatrix} n \\ x \end{bmatrix}_{q^{-k}} q^{-k(n-x)(\alpha-x)} \frac{[\alpha]_{x,q^{-k}}[\beta]_{n-x,q^{-k}}}{[\alpha+\beta]_{n,q^{-k}}} \\ &= q^{-k(n-x)(\alpha-x)} \begin{bmatrix} \alpha \\ x \end{bmatrix}_{q^{-k}} \begin{bmatrix} \beta \\ n-x \end{bmatrix}_{q^{-k}} \Big/ \begin{bmatrix} \alpha+\beta \\ n \end{bmatrix}_{q^{-k}}, \end{aligned} \tag{9.16}$$

for $x = 0, 1, \ldots, n$, where $0 < q < 1$ or $1 < q < \infty$, and $\alpha = -r/k$, $\beta = -s/k$, with r and s positive integers and k an integer. Its q-factorial moments are given by

$$E\big([X_n]_{i,q^{-k}}\big) = \frac{[n]_{i,q^{-k}}[\alpha]_{i,q^{-k}}}{[\alpha+\beta]_{i,q^{-k}}},$$

for $i = 1, 2, \ldots, n$ and $E\big([X_n]_{i,q^{-k}}\big) = 0$, for $i = n+1, n+2, \ldots$. Furthermore, its factorial moments are given by

$$E[(X_n)_j] = j! \sum_{i=j}^{n} (-1)^{i-j} \begin{bmatrix} n \\ i \end{bmatrix}_{q^{-k}} \frac{s_{q^{-k}}(i,j)(1-q^{-k})^{i-j}[\alpha]_{i,q^{-k}}}{[\alpha+\beta]_{i,q^{-k}}},$$

for $j = 1, 2, \ldots, n$, where $s_q(i,j)$ is the q-Stirling number of the first kind, and $E[(X_n)_j] = 0$, for $j = n+1, n+2, \ldots$.

The q-Pólya distribution, for large $r+s$, can be approximated by a q-binomial distribution of the second kind.

Theorem 9.14 *Consider the q-Pólya distribution with probability function* (9.16). *For $0 < q < 1$, assume that*

$$\lim_{r+s\to\infty} \frac{[s]_{q^{-1}}}{[r+s]_{q^{-1}}} = \lim_{r+s\to\infty} \frac{q^{-s}-1}{q^{-(r+s)}-1} = \theta \tag{9.17}$$

and in the case of a negative integer k assume, in addition, that $\theta < q^{-k(m-1)}$, for some positive integer m. Then,

$$\lim_{r+s\to\infty} P(X_n = x) = \begin{bmatrix} n \\ x \end{bmatrix}_{q^k} \theta^{n-x} \prod_{i=1}^{x} (1 - \theta q^{k(i-1)}),$$

for $x = 0, 1, \ldots, n$, where $0 < q < 1$ and $0 < \theta < 1$, in the case k is a positive integer, or $0 < \theta < q^{-k(m-1)}$, for some positive integer $m \geq n$, in the case k is a negative integer.

Also, for $1 < q < \infty$, assume that

$$\lim_{r+s\to\infty} \frac{[r]_q}{[r+s]_q} = \lim_{r+s\to\infty} \frac{q^r - 1}{q^{r+s}-1} = \lambda \tag{9.18}$$

and in the case of a negative integer k assume, in addition, that $\lambda < q^{k(m-1)}$, for some positive integer m. Then,

$$\lim_{r+s\to\infty} P(X_n = x) = \begin{bmatrix} n \\ x \end{bmatrix}_{q^{-k}} \lambda^x \prod_{i=1}^{n-x} (1 - \lambda q^{-k(i-1)}),$$

for $x = 0, 1, \ldots, n$, where $1 < q < \infty$ and $0 < \lambda < 1$, in the case k is a positive integer, or $0 < \lambda < q^{-k(m-1)}$, for some positive integer $m \geq n$, in the case k is a negative integer.

An interesting application of the q-hypergeometric distribution, which is a particular case of the q-Pólya distribution for $k = -1$, in estimating the errors in a manuscript, is discussed in Charalambides [6].

Now, let W_n be the number of black balls drawn until the nth white ball is drawn in a q-Pólya urn model, with the conditional probability of drawing a white ball at the ith q-drawing, given that $j-1$ white balls are drawn in the previous $i-1$ q-drawings, given by (9.15). The distribution of the random variable W_n is called *inverse q-Pólya distribution*, with parameters n, α, β, k, and q.

Theorem 9.15 *The probability function of the inverse q-Pólya distribution, with parameters n, α, β, k, and q, is given by*

$$P(W_n = w) = \begin{bmatrix} n+w-1 \\ w \end{bmatrix}_{q^{-k}} q^{-wk(\alpha-n+1)} \frac{[\alpha]_{n,q^{-k}}[\beta]_{w,q^{-k}}}{[\alpha+\beta]_{n+w,q^{-k}}}, \tag{9.19}$$

for $w = 0, 1, \ldots,$ *where* $0 < q < 1$ *or* $1 < q < \infty$, $\alpha = -r/k$ *and* $\beta = -s/k$, *with* r *and* s *positive integers and* k *an integer. Its* q*-factorial moments are given by*

$$E\big([W_n]_{i,q^{-k}}\big) = \frac{[n-i+1]_{i,q^{-k}}[\beta]_{i,q^{-k}}}{q^{ik(\alpha-n+1)}[\alpha+i]_{i,q^{-k}}},$$

for $i = 1, 2, \ldots,$ *provided* $\alpha + i \neq 0$. *Furthermore, its factorial moments are given by*

$$E[(W_n)_j] = j!\sum_{i=j}^{\infty}(-1)^{i-j}\begin{bmatrix} n+i-1 \\ i \end{bmatrix}_{q^{-k}} \frac{s_{q^{-k}}(i,j)(1-q^{-k})^{i-j}[\beta]_{i,q^{-k}}}{q^{ik(\alpha-n+1)}[\alpha+i]_{i,q^{-k}}},$$

for $j = 1, 2, \ldots,$ *provided* $\alpha + j \neq 0$, *where* $s_q(i, j)$ *is the* q*-Stirling number of the first kind.*

The inverse q-Pólya distribution, for large $r + s$, can be approximated by a negative q-binomial distribution of the second kind.

Theorem 9.16 *Consider the inverse* q*-Pólya distribution with probability function* (9.19).

For $0 < q < 1$, *assume that the limiting expression* (9.17) *holds true. Then,*

$$\lim_{r+s\to\infty} P(W_n = w) = \begin{bmatrix} n+w-1 \\ w \end{bmatrix}_{q^k} \theta^w \prod_{i=1}^{n}(1-\theta q^{k(i-1)}),$$

for $w = 0, 1, \ldots,$ *where* $0 < q < 1$ *and* $0 < \theta < 1$, *in the case* k *is a positive integer, or* $0 < \theta < q^{-k(m-1)}$, *for some positive integer* $m \geq n$, *in the case* k *is a negative integer.*

Also, for $1 < q < \infty$ *assume that the limiting expression* (9.18) *holds true. Then,*

$$\lim_{r+s\to\infty} P(W_n = w) = \begin{bmatrix} n+w-1 \\ w \end{bmatrix}_{q^{-k}} q^{-kw}\lambda^n \prod_{i=1}^{w}(1-\lambda q^{-k(i-1)}),$$

for $w = 0, 1, \ldots,$ *where* $1 < q < \infty$ *and* $0 < \lambda < 1$, *in the case* k *is a positive integer, or* $\lambda = q^{km}$, *for some positive integer* m, *in the case* k *is a negative integer.*

6 Heine and Euler Stochastic Processes

A nonnegative integer valued stochastic process X_t, $t \geq 0$, with independent and homogeneous increments is called Poisson process, if in a small time interval, $(t, t+\delta t]$, either a success, $A = \{s\}$, occurs, with probability analogous to the length of the interval, $\lambda\delta t$, or a failure, $A' = \{f\}$. A Heine process, which was studied by Kyriakoussis and Vamvakari [20] and constitutes a q-analogue of a Poisson process, may be introduced as follows.

Consider a stochastic model that is developing in time or space and let $X_t, t \geq 0$, be the number of successes (occurrences of event A) in the interval $(0,\ t]$. Assume that X_t, $t \geq 0$, is a stochastic process, with independent and homogeneous increments, which starts at time $t = 0$ from state 0, $P(X_0 = 0) = 1$, and, in the q-small time interval $(qt,\ t]$, of length $\delta t = (1-q)t$, satisfies the condition

$$p_j(\delta t) = P(X_t - X_{qt} = j) = \begin{cases} \dfrac{1}{1+\lambda(1-q)t}, & j = 0, \\ \dfrac{\lambda(1-q)t}{1+\lambda(1-q)t}, & j = 1, \\ 0, & j > 1, \end{cases} \tag{9.20}$$

with $0 < \lambda < \infty$ and $0 < q < 1$. Then, X_t, $t \geq 0$, is called *Heine process*, with parameters λ and q.

Theorem 9.17 *The probability function of the Heine process X_t, $t \geq 0$, with parameters λ and q, is given by*

$$p_x(t) = P(X_t = x) = e_q(-\lambda t)\frac{q^{\binom{x}{2}}(\lambda t)^x}{[x]_q!}, \quad x = 0, 1, \ldots, \tag{9.21}$$

where $0 < \lambda < \infty$, $0 < q < 1$ and $e_q(u) = \prod_{i=1}^{\infty}(1 - u(1-q)q^{i-1})^{-1}$ is a q-exponential function.

Furthermore, let W_n be the waiting time until the occurrence of the nth success. The distribution of W_n is called *q-Erlang distribution of the first kind*, with parameters n, λ, and q. In particular, the distribution of the waiting time until the occurrence of the first success, $W \equiv W_1$, is called *q-exponential distribution of the first kind*, with parameters λ and q.

Theorem 9.18 *The distribution function $F_n(w) = P(W_n \leq w)$, $-\infty < w < \infty$, of the q-Erlang distribution of the first kind, with parameters n, λ, and q, is given by*

$$F_n(w) = 1 - \sum_{x=0}^{n-1} e_q(-\lambda w)\frac{q^{\binom{x}{2}}(\lambda w)^x}{[x]_q!}, \quad 0 < w < \infty, \tag{9.22}$$

and $F_n(w) = 0$, $-\infty < w < 0$, where n is a positive integer, $0 < \lambda < \infty$, and $0 < q < 1$. Its q-density function $f_n(w) = d_q F_n(w)/d_q w$ is given by

$$f_n(w) = \frac{q^{\binom{n}{2}}\lambda^n}{[n-1]_q!}w^{n-1}e_q(-\lambda w), \quad 0 < w < \infty. \tag{9.23}$$

Also, its jth q-moment is given by

$$\mu'_{j,q} = E(W_n^j) = \frac{[n+j-1]_{j,q}}{\lambda^j q^{\binom{j}{2}+nj}}, \quad j = 1, 2, \dots .$$

Kyriakoussis and Vamvakari [20], on using condition (9.20), derived a system of q-differential equations and deduced the probability function (9.21) of the Heine process. Also, using the relation $P(W_n > w) = P(X_w < n)$ and expression (9.21), they derived the distribution function of the q-Erlang distribution of the first kind in the form (9.22).

An Euler process, which is another q-analogue of a Poisson process, may be introduced, by considering the geometrically decreasing sequence of time differences

$$\delta t_i = (1-q)q^{i-1}t, \quad i = 1, 2, \dots, \quad 0 < q < 1,$$

with $\sum_{i=1}^{\infty} \delta t_i = t$, to partition the time interval $(0, t]$. Specifically, consider a stochastic model that is developing in time or space and let X_t, $t \geq 0$, be the number of successes (occurrences of event A) in the interval $(0, t]$. Assume that X_t, $t \geq 0$, is a stochastic process, with dependent and homogeneous increments, which starts at $t = 0$ from state 0, $P(X_0 = 0) = 1$, and, in the q-small time interval $(q^i t, q^{i-1}t]$, of length $\delta t_i = (1-q)q^{i-1}t$, for $i = 1, 2, \dots$, satisfies the condition

$$p_{j,k}(\delta t_i) = P(X_{q^{i-1}t} = k | X_{q^i t} = j) = \begin{cases} 1 - \lambda(1-q)q^{i-j-1}t, & k = j, \\ \lambda(1-q)q^{i-j-1}t, & k = j+1, \\ 0, & k > j+1, \end{cases} \tag{9.24}$$

for $j = 0, 1, \dots, i-1$ and $i = 1, 2, \dots$, with $0 < \lambda t < 1/(1-q)$ and $0 < q < 1$. Then, X_t, $t \geq 0$, is called *Euler process*, with parameters λ and q.

It is worth noticing that, in contrast to a Poisson process, an Euler process does not have independent increments. Also, the condition of the occurrence of at most one success in a small time interval is expressed in terms of a series of small time intervals of varying (q-decreasing) lengths.

Theorem 9.19 *The probability function of the Euler process X_t, $t \geq 0$, with parameters λ and q, is given by*

$$p_x(t) = P(X_t = x) = E_q(-\lambda t)\frac{(\lambda t)^x}{[x]_q!}, \quad x = 0, 1, \dots, \tag{9.25}$$

where $0 < \lambda t < 1/(1-q)$, $0 < q < 1$, and $E_q(u) = \prod_{i=1}^{\infty}(1 + u(1-q)q^{i-1})$ is a q-exponential function.

Proof The probability function $p_x(q^{i-1}t)$ of the Euler process, by the total probability theorem,

$$p_x(q^{i-1}t) = p_x(q^i t + \delta t_i) = \sum_{k=0}^{x} p_{x-k}(q^i t) p_{x-k,x}(\delta t_i),$$

for $x = 0, 1, \ldots, i-1$, and condition (9.24), satisfies the system of equations

$$p_0(q^{i-1}t) = (1 - \lambda(1-q)q^{i-1}t)p_0(q^i t),$$

$$p_x(q^{i-1}t) = (1 - \lambda(1-q)q^{i-x-1}t)p_x(q^i t) + \lambda(1-q)q^{i-x}tp_{x-1}(q^i t),$$

for $x = 1, 2, \ldots, i-1$. Setting $u = q^{i-1}t$, this system of equations may be rewritten as

$$p_0(u) = (1 - \lambda(1-q)u)p_0(qu),$$

$$p_x(u) = (1 - \lambda(1-q)q^{-x}u)p_x(qu) + \lambda(1-q)q^{-(x-1)}up_{x-1}(u),$$

for $x = 1, 2, \ldots,$ or as

$$\frac{p_0(u) - p_0(qu)}{(1-q)u} = -\lambda p_0(qu),$$

$$\frac{p_x(u) - p_x(qu)}{(1-q)u} = -\lambda q^{-x}p_x(qu) + \lambda q^{-(x-1)}p_{x-1}(qu),$$

for $x = 1, 2, \ldots$. Introducing the q-derivative operator $\mathscr{D}_q$, with respect to u, we deduce the system of q-differential equations

$$\mathscr{D}_q p_0(u) = -\lambda p_0(qu),$$

$$\mathscr{D}_q p_x(u) = -\lambda q^{-x}p_x(qu) + \lambda q^{-(x-1)}p_{x-1}(qu), \quad x = 1, 2, \ldots.$$

Introducing the function $g(u)$ by

$$p_x(u) = g(u)\frac{(\lambda u)^x}{[x]_q!}, \quad x = 0, 1, \ldots, \tag{9.26}$$

and since

$$\mathscr{D}_q p_x(u) = \frac{(\lambda u)^x}{[x]_q!}\mathscr{D}_q g(u) + \lambda\frac{(\lambda u)^{x-1}}{[x-1]_q!}g(qu),$$

the system of q-differential equations reduces to the q-differential equation

$$\mathscr{D}_q g(u) = -\lambda g(qu),$$

with initial condition $g(0) = p_0(0) = 1$. Its solution is readily obtained as $g(u) = E_q(-\lambda u)$, and so, by (9.26), expression (9.25) is established, with u instead of t.

Let W_n be the waiting time until the occurrence of the nth success, with the successes occurring according to an Euler process. The distribution of W_n is called *q-Erlang distribution of the second kind*, with parameters n, λ, and q. In particular, the distribution of the waiting time until the occurrence of the first success, $W \equiv W_1$, is called *q-exponential distribution of the second kind*, with parameters λ and q.

Theorem 9.20 *The distribution function $F_n(w) = P(W_n \leq w)$, $-\infty < w < \infty$, of the q-Erlang distribution of the second kind, with parameters n, λ, and q, is given by*

$$F_n(w) = 1 - \sum_{x=0}^{n-1} E_q(-\lambda w) \frac{(\lambda w)^x}{[x]_q!}, \quad 0 < w < \infty, \tag{9.27}$$

and $F_n(w) = 0$, $-\infty < w < 0$, where n is a positive integer, $0 < \lambda < \infty$ and $0 < q < 1$. Its q-density function $f_n(w) = d_q F_n(w)/d_q w$ is given by

$$f_n(w) = \frac{\lambda^n}{[n-1]_q!} w^{n-1} E_q(-\lambda q w), \quad 0 < w < \infty. \tag{9.28}$$

Also, its jth q-moment is given by

$$\mu'_{j,q} = E(W_n^j) = \frac{[n+j-1]_{j,q}}{\lambda^j}, \quad j = 1, 2, \ldots. \tag{9.29}$$

Proof The event $\{W_n > w\}$, that the nth success occurs after time w, is equivalent to the event $\{X_w < n\}$, that the number of successes up to time w is less than n and so

$$P(W_n > w) = P(X_w < n) = \sum_{x=0}^{n-1} P(X_w = x).$$

Thus, the distribution function of the random variable W_n, on using the relation $F_n(w) = P(W_n \leq w) = 1 - P(W_n > w)$ and expression (9.25), is deduced as (9.27).

The q-density function of W_n, on taking the q-derivative of (9.27), by using the q-Leibniz formula, is obtained in the form

$$f_n(w) = \lambda E_q(-\lambda q w) \sum_{x=0}^{n-1} \frac{(\lambda w)^x}{[x]_q!} - \lambda E_q(-\lambda q w) \sum_{x=1}^{n-1} \frac{(\lambda w)^{x-1}}{[x-1]_q!},$$

which reduces to (9.28). The jth q-moment of W_n,

$$\mu'_{j,q} = E(W_n^j) = \frac{\lambda^n}{[n-1]_q!} \int_0^\infty w^{n+j-1} E_q(-\lambda q w) d_q w,$$

using the transformation $u = \lambda w$ and expression

$$\int_0^\infty u^{n-1} E_q(-qu) d_q u = [n-1]_q!,$$

is obtained as

$$\mu'_{j,q} = \frac{\lambda^n}{[n-1]_q!\lambda^{n+j}} \int_0^\infty u^{n+j-1} E_q(-qu) d_q u = \frac{[n+j-1]_q!}{[n-1]_q!\lambda^j}.$$

Since $[n+j-1]_q! = [n-1]_q![n+j-1]_{j,q}$, the last relation implies the desired expression.

Remark 9.5 q-Poisson stochastic processes. As has already been noted, the Euler and Heine stochastic processes constitute q-analogues of the Poisson stochastic process. Their probability functions may be expressed by the same functional formula, with different parametric spaces. Specifically, the probability function (9.21), of the Heine stochastic process,

$$p_x(t) = e_q(-\lambda t) \frac{q^{\binom{x}{2}} (\lambda t)^x}{[x]_q!}, \quad x = 0, 1, \ldots,$$

with $0 < \lambda t < \infty$ and $0 < q < 1$, on replacing q by the q^{-1}, with $1 < q < \infty$, and using the relations $[x]_{q^{-1}}! = q^{-\binom{x}{2}} [x]_q$ and $e_{q^{-1}}(-\lambda t) = E_q(-\lambda t)$, may be expressed as

$$p_x(t) = E_q(-\lambda t) \frac{(\lambda t)^x}{[x]_q!}, \quad x = 0, 1, \ldots,$$

with $0 < \lambda t < \infty$ and $1 < q < \infty$. Note that this is the same expression as that of the probability function, (9.25), of the Euler stochastic process, with a different parametric space. It should also be remarked the significant difference in the definitions of the two q-Poisson stochastic processes; the increments of a Heine process are independent, while those of an Euler process are dependent.

Remark 9.6 Elementary derivation of the probability functions of the Heine and Euler processes. Consider a stochastic model in which successes or failures (events A or A') may occur at continuous time (or space) points. Also, consider a time interval $(0, t]$ and its partition in n subintervals

$$\left(\frac{[i-1]_q t}{[n]_q}, \frac{[i]_q t}{[n]_q} \right], \quad i = 1, 2, \ldots, n, \quad 0 < q < 1,$$

with lengths $\delta_{n,i}(t) = tq^{i-1}/[n]_q$, $i = 1, 2, \ldots, n$, and suppose that in each subinterval either a success or a failure may occur.

Kyriakoussis and Vamvakari [20] assumed that the odds of success is analogous to the length of the subinterval, $\theta_{n,i}(t) = \lambda \delta_{n,i}(t) = \lambda t q^{i-1}/[n]_q$, $i = 1, 2, \ldots, n$, with $0 < \lambda < \infty$. Clearly, by Theorem 9.5, the number of successes $X_{t,n}$ in the n

subintervals of $(0, t]$ obeys the q-binomial distribution of the first kind with probability function

$$P(X_{t,n} = x) = \begin{bmatrix} n \\ x \end{bmatrix}_q \frac{q^{\binom{x}{2}}(\lambda t/[n]_q)^x}{\prod_{i=1}^{n}\left(1 + \lambda t q^{i-1}/[n]_q\right)}, \quad x = 0, 1, \ldots, n,$$

where $0 < \lambda < \infty, 0 < t < \infty$ and $0 < q < 1$. Taking the limit as $n \to \infty$ and using Theorem 9.8, the limiting probability function $P(X_t = x) = \lim_{n\to\infty} P(X_{t,n} = x)$, $x = 0, 1, \ldots$, is readily deduced as (9.21).

Furthermore, assume that the conditional probability of success at any subinterval, given that $j - 1$ successes occur in the previous subintervals, is given by $p_{n,j}(t) = 1 - \lambda t q^{j-1}/[n]_q$, for $j = 1, 2, \ldots, n$, with $0 < \lambda t < [n]_q$. Clearly, by Theorem 9.10, the number of failures $X_{t,n}$ in the n subintervals of $(0, t)$ obeys the q-binomial distribution of the second kind, with probability function

$$P(X_{t,n} = x) = \begin{bmatrix} n \\ x \end{bmatrix}_q \left(\frac{\lambda t}{[n]_q}\right)^x \prod_{j=1}^{n-x}\left(1 - \frac{\lambda t}{[n]_q} q^{j-1}\right), \quad x = 0, 1, \ldots, n,$$

where $0 < \lambda t < [n]_q$ and $0 < q < 1$. Taking the limit as $n \to \infty$ and using Theorem 9.12, the limiting probability function $P(X_t = x) = \lim_{n\to\infty} P(X_{t,n} = x)$, $x = 0, 1, \ldots$, is obtained as (9.25).

7 Bibliographic Notes

The introduction of the q-number and its notation stems from Jackson [16], who published important and influential papers on the subject. A list of his publications is included in the obituary note by Chaudry [9]. Gauß [14] introduced the q-binomial coefficients (or Gaußian polynomials) and presented q-analogues of Pascal's triangular recurrence relation. The discrete q-uniform distribution of the number theoretic random variable examined in Example 9.1 was discussed by Rawlings [21]. Cauchy [3] derived the q-factorial and q-binomial convolution formulae, which were stated in Theorem 9.2 and its Corollary 9.1. The origin of the general q-binomial formulae is quite uncertain; Hardy [15] attributed these formulae to Euler. The derivation of the power series expressions of the two q-exponential functions are, indeed, due to Euler [12]. Several other interesting q-series expansions are presented in the classical book of Andrews [1]. An authoritative and comprehensive account of the basic q-hypergeometric series is given by Gasper and Rahman [13]. The q-factorial moments and their connection to the usual factorial moments were discussed in Charalambides and Papadatos [8].

The q-binomial distribution of the first kind was examined by Kemp and Kemp [18] in their study of Weldon's classical dice data. The Heine and Euler distributions were derived by Benkherouf and Bather [2] as feasible priors in a simple Bayesian model for oil exploration. The derivation of the Heine distribution as a limiting distribution of the q-binomial distribution of the first kind was given by Kemp and Newton [19]. The q-binomial and the negative q-binomial distributions of the second kind were presented in Charalambides [4]. The absorption and the inverse absorption distributions were studied by Dunkl [10] and Kemp [17]. The negative q-binomial distribution of the second kind and its limit to the Euler distribution was derived by Rawlings [22]. Charalambides [6] introduced the q-Pólya urn model and studied in detail the q-Pólya and the inverse q-Pólya distributions. Kupershmidt (2000) introduced a q-hypergeometric distribution and a q-contagious distribution (q-Pólya distribution) and represented the corresponding random variable as a sum of two-valued dependent random variables. The Heine process was recently discussed by Kyriakoussis and Vamvakari [20], while the presentation of the Euler process has not been published elsewhere.

Additional q-discrete distributions, defined on Bernoulli trials with geometrically varying success probability, are presented in the review article of Charalambides [5]. A comprehensive presentation of discrete q-distributions is given in the book of Charalambides [7].

References

1. Andrews, G.E.: The Theory of Partitions. Addison-Wesley, MA (1976)
2. Benkherouf, L., Bather, J.A.: Oil exploration: sequential decisions in the face of uncertainty. J. Appl. Probab. **25**, 529–543 (1988)
3. Cauchy, A.-L.: Mémoire sur les fonctions dont plusieurs valeurs sont liées entre elles par une équation linéaire, et sur diverses transformations de produits composés d'un nombre indéfini de facteurs. C. R. Acad. Sci. Paris XVII, p. 523 (1843). Œuvres de Cauchy. série 1, tome 8, pp. 42–50. Gauthier-Villars, Paris (1893)
4. Charalambides, Ch.A.: The q-Bernstein basis as a q-binomial distribution. J. Stat. Plan. Inference **140**, 2184–2190 (2010)
5. Charalambides, Ch.A.: Discrete q-distributions on Bernoulli trials with geometrically varying success probability. J. Stat. Plan. Inference **140**, 2355–2383 (2010)
6. Charalambides, Ch.A.: A q-Pólya urn model and the q-Pólya and inverse q-Pólya distributions. J. Stat. Plan. Inference **142**, 276–288 (2012)
7. Charalambides, Ch.A.: Discrete q-Distributions. Wiley, NJ (2016)
8. Charalambides, Ch.A., Papadatos, N.: The q-factorial moments of discrete q-distributions and a characterization of the Euler distribution. In: Balakrishnan, N., Bairamov, I.G., Gebizlioglu, O.L. (eds.) Advances on Models, Characterizations and Applications, pp. 57–71. Chapman and Hall/CRC Press, Boca Raton, FL (2005)
9. Chaudry, T.W.: Frank Hilton Jackson. J. Lond. Math. Soc. **37**, 126–128 (1962)
10. Dunkl, C.F.: The absorption distribution and the q-binomial theorem. Commun. Stat. Theory Methods **10**, 1915–1920 (1981)
11. Eggenberger, F., Pólya, G.: Über die Statistik verketteter Vorgänge. Z. Angew. Math. Mechanik **3**, 279–289 (1923)
12. Euler, L.: Introductio in Analysin Infinitorum. Marcum-Michaelem Bousquet, Lausannae (1748)
13. Gasper, G., Rahman, M.: Basic Hypergeometric Series, 2nd edn. Cambridge University Press, Cambridge (2004)
14. Gauß, C.F.: Werke, vol. 2. Königliche Gesellschaft der Wissenschaften, Göttingen (1863)

15. Hardy, G.H.: Ramanujan: Twelve Lectures on his Life and Work. Cambridge University Press, Cambridge (1940); reprinted by Chelsea, New York (1978)
16. Jackson, F.H.: q-Difference equations. Am. J. Math. **32**, 305–314 (1910)
17. Kemp, A.: Absorption sampling and the absorption distribution. J. Appl. Probab. **35**, 489–494 (1998)
18. Kemp, A., Kemp, C.D.: Weldon's dice data revisited. Am. Stat. **45**, 216–222 (1991)
19. Kemp, A., Newton, J.: Certain state-dependent processes for dichotomised parasite populations. J. Appl. Probab. **27**, 251–258 (1990)
20. Kyriakoussis, A., Vamvakari, M.: Heine process as a q-analog of the Poisson process-waiting and interarrival times. Commun. Stat. Theory Methods **46**, 4088–4102 (2017)
21. Rawlings, D.: Bernoulli trials and number theory. Am. Math. Mon. **101**, 948–952 (1994)
22. Rawlings, D.: A probabilistic approach to some of Euler's number theoretic identities. Trans. Am. Math. Soc. **350**, 2939–2951 (1998)
23. Woodbury, M.A.: On a probability distribution. Ann. Math. Stat. **20**, 311–313 (1949)

Families of Parking Functions Counted by the Schröder and Baxter Numbers

Robert Cori, Enrica Duchi, Veronica Guerrini and Simone Rinaldi

Abstract We define two new families of parking functions: one counted by Schröder numbers and the other by Baxter numbers. These families both include the well-known class of non-decreasing parking functions, which is counted by Catalan numbers and easily represented by Dyck paths, and they both are included in the class of underdiagonal sequences, which are bijective to permutations. We investigate their combinatorial properties exhibiting bijections between these two families and classes of lattice paths (Schröder paths and triples of non-intersecting lattice paths) and discovering a link between them and some classes of pattern avoiding permutations. Then, we provide a quite natural generalization for each of these families that results in some enumeration problems tackled by ECO method.

Keywords Parking functions · Lattice paths · Pattern avoiding permutations · Schröder numbers · Baxter numbers · ECO method

2010 Mathematics Subject Classification Primary: 05A15 · Secondary: 05A19

R. Cori
Labri, Université Bordeaux, Bordeaux, France
e-mail: robert.cori@labri.fr

E. Duchi
LIAFA, Université Paris 7, Paris, France
e-mail: enrica.duchi@liafa.univ-paris-diderot.fr

V. Guerrini · S. Rinaldi (✉)
Dipartimento di Ingegneria dell'Informazione e Scienze Matematiche,
University of Siena, Siena, Italy
e-mail: rinaldi@unisi.it

V. Guerrini
e-mail: guerrini55@student.unisi.it

G. E. Andrews et al. (eds.), *Lattice Path Combinatorics and Applications*,
Developments in Mathematics 58, https://doi.org/10.1007/978-3-030-11102-1_10

1 Introduction

The notion of parking functions is recurring in discrete mathematics and arises naturally in the so-called *parking problem*, which can be stated as follows: there are n cars $C_1, \dots, C_n$ that want to park on a one-way street with ordered parking spaces $0, 1, \dots, n-1$. Each car C_i has a preferred space a_i, and the cars enter the street one at a time in the order $C_1, \dots, C_n$. A car tries to park in its preferred space. If that space is occupied, then it parks in the next available space. If there is no space, then the car leaves the street. The sequence $a_1 \dots a_n$ is called a *parking function* of length n if all the cars can park, i.e., no car leaves the street. It is easy to see that a sequence $a_1 \dots a_n$ is a parking function if and only if it has at least i terms less than i, for each $1 \leq i \leq n$. Another equivalent definition is that a sequence $a_1 \dots a_n$ is a parking function if and only if there is a permutation σ of length n such that $0 \leq a_{\sigma_i} < i$, for each $1 \leq i \leq n$.

It is also worth recalling that parking functions have a simple representation in terms of lattice paths, precisely as *labelled Dyck paths* [13]. We recall that a *Dyck path* of length $2n$ is a path made of *up* steps $U = (1, 1)$, of *down* steps $D = (1, -1)$, running from $(0, 0)$ to $(2n, 0)$ and remaining weakly above the x-axis, and a *labelled Dyck path* is a Dyck path of length $2n$ whose up steps are labelled by integers from 1 to n, provided that the labels of consecutive up steps are increasing.

The number of parking functions of length n was analytically proved to be equal to $(n+1)^{n-1}$ in [15], but then several combinatorial explanations of this formula were provided (see, for instance, [16, 21]). Among them, there are also many bijections between parking functions of length n and labelled trees on $n+1$ vertices which by Cayley's formula are counted by $(n+1)^{n-1}$ as well.

Such bijective proofs show remarkable connections between parking functions and other combinatorial structures and lead to various generalizations and applications in different fields, notably in algebra, interpolation theory, probability and statistics, representation theory and geometry.

We are mainly interested in exploring the connections between parking functions, lattice paths and permutations, so we underline that two of the most well-known families of parking functions have simple representations both as lattice paths and as permutations, and precisely, they are as follows:

(a) *non-decreasing parking functions*: sequences $u_1 \dots u_n$ such that

$$u_i < i \text{ and } u_i \leq u_{i+1}, \quad \text{for } 1 \leq i \leq n-1\,. \tag{10.1}$$

These sequences are particularly relevant since it is proved that every parking function can be obtained as a rearrangement of a non-decreasing parking function [23].

The number of non-decreasing parking functions of length n is given by the nth *Catalan number*. This number sequence, reviewed in Sect. 2.2, is almost ubiquitous in combinatorics, as it counts several classes of combinatorial objects, most of which are listed in [23]: one of them is given by (ordinary) *Dyck paths* of length $2n$; another is that of permutations avoiding a pattern of length 3.

(b) *underdiagonal sequences*: sequences $u_1 \dots u_n$ such that

$$u_i < i, \quad \text{for } 1 \leq i \leq n\,. \tag{10.2}$$

These sequences of length n are counted by $n!$ as they are trivially bijective to permutations of length n. They are particularly relevant, since, in the context of parking functions, they represent all configurations in which cars park in the same order as they enter the street. Also these sequences have a simple representation as lattice paths, precisely as *underdiagonal paths*.

The researches carried on in this paper aim at exploring connections between parking sequences and families of combinatorial objects (especially lattice paths and permutations) which are counted by the *Schröder* and the *Baxter* numbers (for brevity, *Schröder* and *Baxter structures*). These two sequences, as well as their most remarkable combinatorial interpretations, are reviewed in Sects. 2.3 and 2.4.

Like Catalan and factorial structures, also Schröder and Baxter numbers have quite popular combinatorial representations in terms of lattice paths (Schröder paths [11, 22] and Baxter triples [12]), and in terms of pattern avoiding permutations (separable permutations [25] and Baxter permutations [9]).

The general purpose of this paper is to introduce new families of parking functions which are as follows:

- point-wise larger than the Catalan sequence and contain the family of non-decreasing parking functions;
- point-wise smaller than the factorial sequence and contained in the family of inversion tables of permutations.

To reach our goal, such families of parking functions are defined by imposing constraints on the entries of each sequence, weaker than (10.1) but stronger than (10.2). The most important of these families are those of *Schröder* and *Baxter parking functions*. Our aim is that of studying combinatorial properties of each of these families and, in particular, how the combinatorial properties of the two families (a) and (b) can be translated to them. Precisely: provide a description of these families of parking functions in terms of lattice paths and permutations.

Another aspect which unifies our investigation of these families of parking functions is the application of the ECO method and generating trees to handle all enumeration problems concerning these objects. ECO method and the related notions of generating tree and succession rule prove to be powerful tools for describing the recursive growth of all the considered families of parking functions, and in fact, for each of these classes we are able to provide an associated succession rule. These concepts are recalled in Sect. 2.1, but we address the reader to [1, 2, 25] for further details.

In Sect. 2, we comment upon the numerical sequences which are studied in the paper and their most important combinatorial interpretations in terms of lattice paths and permutations: we consider the Catalan, Schröder, Baxter and factorial numbers. For each of these sequences, we provide a generating tree describing its recursive growth, according to the ECO method.

The main character of the first part of the paper is the family of the *Schröder parking functions*. These sequences are defined in Sect. 3 by slightly modifying the definition (10.1) of non-decreasing parking functions. First, we prove that Schröder parking functions of length n are actually counted by the Schröder numbers by providing a recursive construction of them by means of the ECO method and then by providing a bijection with Schröder paths of length n. To reach this goal, we use an encoding of Schröder parking functions as words of a context-free language. Moreover, we show how this class of sequences results closely related to a pattern avoiding permutation class. In Sect. 4, we provide quite a natural extension of the notion of Schröder parking function, by introducing *generalized Schröder parking functions* of degree m, such that with $m = 0$ and $m = 1$ we have non-decreasing and Schröder parking functions. We extend the results of the previous section for a generic $m \geq 2$ and give a representation of generalized Schröder parking functions as *labelled Dyck paths*. Then, we also show how to extend to generalized Schröder parking functions the characterization given for Schröder parking functions in terms of an algebraic language and a pattern avoiding permutation class.

The second part of the paper is dedicated to the study of *Baxter parking functions* and their generalizations. Analogously to what we have done for Schröder parking functions, firstly we provide an ECO operator for the recursive construction of this class and then determine a bijection with Baxter triples, proving that Baxter parking functions are enumerated by Baxter numbers. Then, we study two families of *generalized Baxter parking functions*, defined by relaxing the definition of Baxter parking function. Also for these families, we determine a recursive growth according to the ECO method and the associated succession rule, leaving open some problems concerning the nature of their generating functions and their representation in terms of pattern avoiding permutations.

2 Basic Definitions

In this section, we recall the basic definitions and properties of the combinatorial objects that are studied in the paper.

2.1 ECO Method and Generating Trees

Enumeration of combinatorial objects (ECO) is a method for the enumeration and the recursive construction of a class $\mathcal{O}$ of combinatorial objects by means of an operator ϑ which performs "local expansions" on the objects of $\mathcal{O}$. Let p be a parameter on

$\mathcal{O}$, such that $|\mathcal{O}_n| = |\{O \in \mathcal{O} : p(O) = n\}|$ is finite, and let $2^{\mathcal{O}_n}$ denote the set of subsets of $\mathcal{O}_n$.

Definition 10.1 An ECO operator ϑ on the class $\mathcal{O}$ is a function from $\mathcal{O}_n$ to $2^{\mathcal{O}_{n+1}}$ such that:

1. for each $O' \in \mathcal{O}_{n+1}$, there exists $O \in \mathcal{O}_n$ such that $O' \in \vartheta(O)$,
2. for each $O, O' \in \mathcal{O}_n$ such that $O \neq O'$, then $\vartheta(O) \cap \vartheta(O') = \emptyset$,

Clearly, if ϑ is an ECO operator, then the family of sets $\mathcal{F}_{n+1} = \{\vartheta(O) : O \in \mathcal{O}_n\}$ is a partition of $\mathcal{O}_{n+1}$.

ECO method was successfully applied to the enumeration of various classes of walks, permutations, and polyominoes. We refer to [2] for further details and results.

The recursive construction determined by ϑ can be suitably described through a *generating tree*, i.e. a rooted tree whose vertices are objects of $\mathcal{O}$. The objects having the same value of the parameter p lie at the same level, and the sons of an object are the objects it produces through ϑ. We point out that generating trees have first been introduced by West in [25].

If the construction determined by the ECO operator ϑ is regular enough, it is then possible to describe it by means of a *succession rule* (sometimes called generating tree, as well) of the form:

$$\begin{cases} (b) \\ \\ (h) \rightsquigarrow (c_1)(c_2)\dots(c_h), \end{cases}$$

where $b, h, c_i \in \mathbb{N}$, meaning that the root object has b sons, and the h objects $O'_1, \dots, O'_h$, produced by an object O are such that $\left|\vartheta(O'_i)\right| = c_i$, $1 \leq i \leq h$. A succession rule defines a sequence $\{f_n\}_{n\geq 0}$ of positive integers, where f_n is the number of nodes at level n of the generating tree. In the years, succession rules have shown their applicability to several combinatorial problems and have become a versatile tool to solve enumeration problems as shown in [1].

More recently, in order to solve enumeration problems, the notion of succession rule has been extended, allowing the nodes of the generating tree to have two or more labels which take into account different parameters of the object. Some examples of these succession rules (or generating trees) with two labels have been studied in [4].

2.2 Catalan Structures

The sequence of *Catalan numbers* (sequence A000108 in [17]) is one of the most well-know combinatorial sequences. They are defined by the closed formula:

$$C_n = \frac{1}{n+1}\binom{2n}{n}.$$

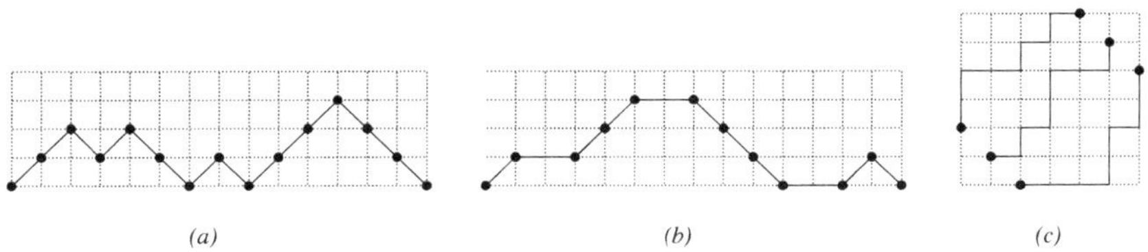

Fig. 1 **a** A Dyck path; **b** a Schröder path; **c** a Baxter triple

A considerably exhaustive list of combinatorial structures enumerated by Catalan numbers, that we are going to call *Catalan structures*, is provided in [23]. Among those, we will use the following:

- *Dyck paths*. In the discrete plane, a *Dyck path* of semi-length n is a path made of *up* steps $U = (1, 1)$, of *down* steps $D = (1, -1)$, running from $(0, 0)$ to $(2n, 0)$ and remaining weakly above the x-axis (see Fig. 1a). The number of Dyck paths of semi-length n is given by the nth Catalan number C_n.
- *Non-decreasing parking functions*. Sequences $u = u_1 \dots u_n$ such that $u_i < i$ and $u_i \leq u_{i+1}$, for any $i = 1, \dots, n-1$; these are a special case of parking functions, and in particular, it holds that parking functions can be obtained as all possible rearrangements of non-decreasing parking functions [23].
- *τ-avoiding permutations, for any permutation τ of size* 3. Recall that a permutation $\sigma = \sigma_1\sigma_2 \dots \sigma_n$ contains $\tau = \tau_1\tau_2 \dots \tau_k$ if there exists $i_1 < i_2 < \dots < i_k$ such that $\sigma_{i_a} < \sigma_{i_b}$ if and only if $\tau_a < \tau_b$. Otherwise, σ avoids τ [23].

Probably, the most well-known succession rule for Catalan numbers is the following (see [2]):

$$\Omega_{Cat} : \quad \begin{cases} (2) \\ (k) \rightsquigarrow (2) \dots (k)(k+1) \, . \end{cases}$$

The first levels of the generating tree of Ω_{Cat} are shown in Fig. 2.

2.3 Schröder Structures

Schröder numbers (sequence A006318 in [17]) are defined by the formula:

$$S_n = \frac{1}{n} \sum_{k=1}^{n} 2^k \binom{n}{k} \binom{n}{k-1} .$$

There are several combinatorial structures enumerated by this sequence, and in this paper, we will consider:

- *Schröder paths*. In the discrete plane, a *Schröder path* of length $2n$ is a path made of *up* steps $U = (1, 1)$, of *down* steps $D = (1, -1)$ and *horizontal steps* $H = (2, 0)$, running from $(0, 0)$ to $(2n, 0)$ and remaining weakly above the x-axis (see Fig. 1b). The number of Schröder paths of semi-length n is given by the nth Schröder number S_n.
- *Separable permutations*. These are introduced and enumerated in [22]. These permutations can be easily described by the avoidance of the two patterns 2413 and 3142. There are, however, other classes of permutations avoiding two patterns of length 4 counted by Schröder numbers. The complete list of them is provided in [24]. Among them, we have permutations avoiding 1423 and 1432, which will be reconsidered in Sect. 3.3.

Probably, the most well-known succession rule for Schröder numbers is the following, determined in [25], and describing a recursive growth both for Schröder paths and separable permutations:

$$\Omega_{Sch}: \begin{cases} (2) \\ (k) \rightsquigarrow (3)\ldots(k)(k+1)^2 . \end{cases}$$

Strictly related to our sequence are the *small Schröder numbers* s_n, sequence A001003 in [17]. The term s_n is precisely half the nth Schröder number S_n, for $n \geq 1$, whereas $s_0 = 1$; in particular, they count Schröder paths with no horizontal step at level 0 (see [11]).

2.4 Baxter Structures

Baxter numbers (sequence A001181 in [17]) were first introduced in [9], where it is shown that they count Baxter permutations. Precisely, Baxter numbers are given by $B_n = \sum_{k=0}^{n-1} \theta_{k,n-k-1}$, where

$$\theta_{k,l} = \frac{\binom{n+1}{k}\binom{n+1}{k+1}\binom{n+1}{k+2}}{\binom{n+1}{1}\binom{n+1}{2}}, \tag{10.3}$$

where $n = k + l + 1$. Baxter numbers also enumerate numerous families of combinatorial objects, and their study has attracted significant attention, see, for instance, [3, 12]. Among these structures, we will use:

- *triples of non-intersecting lattice paths* (briefly, *Baxter triples*) of length $n - 1$ in the discrete plane (see Fig. 1c). Precisely, the number of Baxter triples running from $A_1 = (0, 2)$, $A_2 = (1, 1)$ and $A_3 = (2, 0)$ to $(k, l + 2)$, $(k + 1, l + 1)$ and $(k + 2, l)$, using north $(1, 0)$ and east $(0, 1)$ unit steps, is given by $\theta_{k,l}$ in (10.3).
- *Baxter permutations* can be defined by the avoidance of the two vincular patterns $2\underline{41}3$ and $3\underline{14}2$, meaning that in a Baxter permutation σ no subsequence

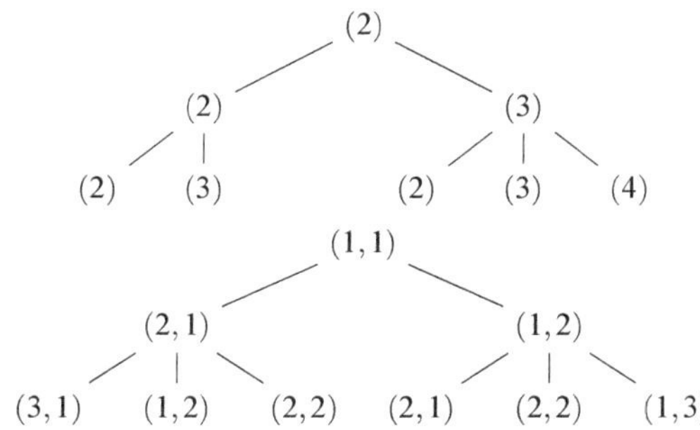

Fig. 2 The first levels of the generating trees for rules Ω_{Cat} and Ω_{Bax}

$\sigma_i\sigma_j\sigma_{j+1}\sigma_k$ of σ satisfies $\sigma_{j+1} < \sigma_i < \sigma_k < \sigma_j$ (resp. $\sigma_j < \sigma_k < \sigma_i < \sigma_{j+1}$). Note that we do not represent vincular patterns with dashes, as it was done originally. We prefer the more modern and more coherent notation that indicates by a symbol $\llcorner\!\lrcorner$ the elements of the pattern that are required to be adjacent in an occurrence. Concerning Baxter permutations, the term $\theta_{k,l}$ in (10.3) counts Baxter permutations of length n with k ascents and l descents.

A succession rule with two labels, describing the recursive growth of Baxter permutations, in [4], is the following:

$$\Omega_{Bax}: \quad \begin{cases} (1,1) \\ (p,q) \rightarrow (p+1,1)(p+1,2)\ldots(p+1,q) \\ \qquad\qquad\quad (1,q+1)(2,q+1)\ldots(p,q+1)\,. \end{cases}$$

The first levels of the generating tree of Ω_{Bax} are shown in Fig. 2.

Remark 10.1 The three sequences we have presented are strictly related, since Schröder numbers form a sequence point-wise larger than the Catalan sequence, and it is additionally point-wise smaller than the Baxter sequence. As a matter of fact, many Baxter families can be immediately seen to contain a Catalan or a Schröder subfamily: for instance, the set of triples of non-intersecting lattice paths contains all pairs of non-intersecting lattice paths (that are in essence parallelogram polyominoes); Baxter permutations (defined by the avoidance of the vincular patterns $2\,\underline{41}\,3$ and $3\,\underline{14}\,2$) include separable permutations (avoiding 2413 and 3142), which, on their turn, include τ-avoiding permutations, for any permutation τ of length 3.

2.5 Factorial Structures

Factorial numbers (sequence A000142 in [17]) are a well-known sequence that enumerates permutations of length n and some related structures such as underdiagonal sequences.

Given a permutation π of length n, we say that the pair of indices (p, q) forms an *inversion* if $p < q$ and $\pi_p > \pi_q$. The array $I(\pi) = (i_1, \ldots, i_n)$, where

$$i_p = |\{q : p < q, \text{ such that } (p, q) \text{ is an inversion}\}|$$

is called the *left inversion table* of π. It can be easily proved that I yields a bijective mapping ϕ_I between permutations of length n and *underdiagonal sequences* of length n, i.e. sequences $u_1 \ldots u_n$ such that, for any $i = 1, \ldots, n$, we have $u_i < i$. The mapping ϕ_I is defined by setting $\phi_I(\pi)$ equal to the reverse word of $I(\pi)$.

For instance, with $\pi = 4\,3\,6\,1\,5\,2$, $I(\pi) = (3, 2, 3, 0, 1, 0)$, and the corresponding underdiagonal sequence is $\phi_I(\pi) = 0\,1\,0\,3\,2\,3$. The family of underdiagonal sequences of length n will be denoted by $\mathscr{U}_n$. A succession rule describing the recursive growth of these numbers is [1]:

$$\Omega_{Fac} : \quad \begin{cases} (2) \\ \\ (k) \rightsquigarrow (k+1)^k \,. \end{cases}$$

3 Schröder Parking Functions

In this section, we define a family of parking functions counted by the Schröder numbers, and then, we study the relations between these objects and some other Schröder structures.

Definition 10.2 A *Schröder parking function* s is a sequence $s_1\, s_2 \ldots s_n$ such that

- $s_i < i$, for all i;
- If $i < j$ then $s_i - s_j \leq 1$.

The family of Schröder parking functions of length n will be denoted by $\mathscr{S}(n)$. All the elements of a Schröder parking function $s_1 \ldots s_n$ can be classified into two groups, as follows:

- s_1 is a fall;
- a generic element s_j which is equal to 0 or there is a $i < j$ such that $s_i > s_j$ is a fall. All the other elements are not falls.

Example 10.1 The sequence $s = 0\,0\,0\,1\,4\,3\,3\,6\,7\,6$ is a Schröder parking function of length 10, and its falls are $s_1 = 0$, $s_2 = 0$, $s_3 = 0$, $s_6 = 3$, $s_7 = 3$, $s_{10} = 6$. On the other side, the sequence $s = 0\,0\,2\,1\,0$ is not a Schröder parking function, since $s_3 - s_5 = 2 > 1$.

Proposition 10.1 *Schröder parking functions are counted by Schröder numbers.*

Proof To prove that Schröder parking functions are counted by Schröder numbers, we describe an ECO operator θ_{Sch} for the recursive construction of these objects and show that the generating tree associated with θ_{Sch} is precisely Ω_{Sch}, defined in the previous section. Given $s_1 \dots s_n$ a Schröder parking function of length n, the application of θ_{Sch} produces a certain number of Schröder parking functions of length $n+1$, depending on the last value s_n, as follows:

(a) If s_n is not a fall, then θ_{Sch} adds s_{n+1} to the sequence $s_1 \dots s_n$, where s_{n+1} is any value among $n, n-1, \dots, s_n, s_n - 1$.
(b) If s_n is a fall, then θ_{Sch} adds s_{n+1} to $s_1 \dots s_n$, where s_{n+1} is any value among $n, n-1, \dots, s_n + 1, s_n$.

Note that operation performed in case (a) (resp. (b)) produces $n + 2 - s_n$ (resp. $n + 1 - s_n$) sequences of length $n + 1$ that satisfy Definition 10.2 and among them only one is such that s_{n+1} is a fall, namely $s_{n+1} = s_n - 1$ (resp. $s_{n+1} = s_n$). Easily, one can verify that θ_{Sch} satisfies conditions 1 and 2 in Definition 10.1.

Now, it is simple to determine the generating tree associated with θ_{Sch}. To a sequence $s = s_1 \dots s_n$ satisfying condition at point (a) we assign label (k), where $k = n - s_n + 2$. Then, the sequence $s' = s_1 \dots s_n\, s_{n+1}$, where $s_{n+1} = n$ (resp. $n - 1, \dots, s_n + 1, s_n, s_n - 1$) has label (3) (resp. (4), $\dots$, (k), $(k+1)$, $(k+1)$). While to a sequence $s = s_1 \dots s_n$ satisfying condition at point (b) we assign label (k), where $k = n - s_n + 1$. And sequence $s' = s_1 \dots s_n\, s_{n+1}$, where $s_{n+1} = n$ (resp. $n - 1, \dots, s_n + 2, s_n + 1, s_n$), has label (3) (resp. (4), $\dots$, (k), $(k+1)$, $(k+1)$). These simple computations allow to prove that Ω_{Sch} is the generating tree associated with this construction.

Example 10.2 The sequence $s = 0\,1\,0\,3$, where s_4 is not a fall, has label (3). Hence, the application of operator θ_{Sch} produces the sequences $0\,1\,0\,3\,4$, $0\,1\,0\,3\,3$, and $0\,1\,0\,3\,2$, with labels (3), (4) and (4).

The sequence $s = 0\,0\,1\,0$, where s_4 is a fall, has label (5). Hence, the application of operator θ_{Sch} produces the sequences $0\,0\,1\,0\,4$, $0\,0\,1\,0\,3$, $0\,0\,1\,0\,2$, $0\,0\,1\,0\,1$ and $0\,0\,1\,0\,0$, with labels (3), (4), (5), (6) and (6).

We would like to observe that it is easy to find out a subclass of Schröder parking functions counted by the small Schröder numbers.

Definition 10.3 A *small Schröder parking function* is a Schröder parking function s such that $s = 0$ or s begins with the factor $0\,1$.

By simple symmetry arguments it follows that:

Corollary 10.1 *Small Schröder parking functions are counted by the small Schröder numbers.*

3.1 Encoding Schröder Parking Functions as Words of an Algebraic Language

A Schröder parking function s can be represented uniquely as a word $w = w(s)$ in the alphabet $\{a, b, c\}$ as follows.

1. $w(s_1)$ is the empty word ε.
2. Let $w' = w(s_1 \dots s_{n-1})$, with $n > 2$. If s_n is a fall, then $w(s_1 \dots s_n) = w'a$. Otherwise $w(s_1 \dots s_n) = w'c^k b$ and $k \geq 0$ is determined either by the difference between the values s_n and s_r, where s_r is the rightmost non-fall element of $s_1 \dots s_{n-1}$ if there is any, or by $k = s_n - 1$.

In particular, note that the length n of a Schröder parking function s is given by the number of occurrences of a and b in the word $w(s)$ plus 1, i.e. $|w|_a + |w|_b = n - 1$.

Example 10.3 The Schröder parking function 0 0 0 1 4 3 3 6 7 6 given at the beginning of this section is codified by the word

$$a\,a\,b\,cccb\,a\,a\,ccb\,cb\,a.$$

From the definition above, we can provide a combinatorial description of the set:

$$\mathscr{L}_S(n) = \{w(s) : s \in \mathscr{S}(n+1)\}.$$

Proposition 10.2 *A word w in the alphabet $\Sigma = \{a, b, c\}$ belongs to $\mathscr{L}_S(n)$ if and only if*

- *for each prefix v of w, $|v|_c \leq |v|_a + |v|_b$,*
- *w does not contain the factor ca,*
- *the last letter is not c,*
- *$|w|_a + |w|_b = n$.*

To our knowledge, the language $\mathscr{L}_S$ provides a new occurrence of Schröder numbers.

3.2 Schröder Parking Functions and Schröder Paths

Now, we describe a bijective way to pass from a word of $\mathscr{L}_S(n)$ to a Schröder path of length $2n$.

Let $\mathscr{C}(n)$ be the set of prefixes of Dyck paths of length n with up steps labelled a or b. Then, each word w of $\mathscr{L}_S(n)$ can be represented as a path in $\mathscr{C}(n)$, ending at $(n, n - |w|_c)$, simply by coding each a (resp. b) as an *up* step U labelled a (resp. b), and each c as a *down* step D. Observe that not all paths in $\mathscr{C}(n)$ correspond to a word in $\mathscr{L}_S(n)$: it is the case of the path $abcab$, which contains the factor ca. From

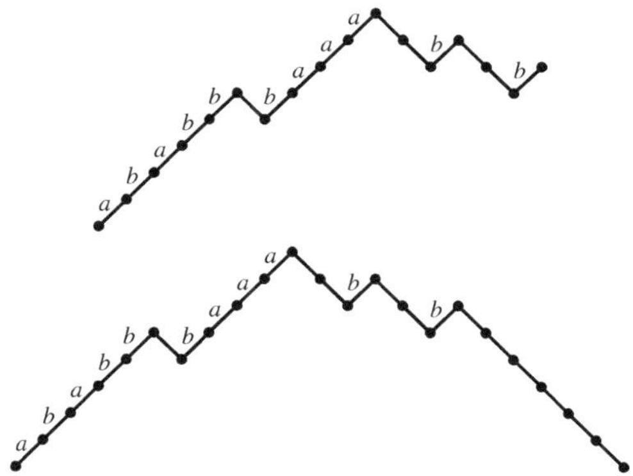

Fig. 3 The word $w = a\,a\,b\,cccb\,a\,a\,ccb\,cb\,a$ and its closure

now on, we will use words of $\mathscr{L}_S(n)$ and their graphical representations in terms of Dyck prefixes, indifferently.

Given a path w of $\mathscr{L}_S(n)$, we define its *closure* $\overline{w}$ as the smallest Dyck path containing w as a prefix, i.e. $\overline{w} = w\,c^h$, where $h = n - |w|_c \geq 1$. Therefore,

$$\overline{\mathscr{L}_S}(n) = \{\overline{w} : w \in \mathscr{L}_S(n)\}\,.$$

Clearly, $\mathscr{L}_S(n)$ and $\overline{\mathscr{L}_S}(n)$ are bijective. Also observe that, to a generic Dyck path of length $2n$ correspond 2^q paths in $\overline{\mathscr{L}_S}(n)$, where q is the number of up steps not directly preceded by a down step.

Example 10.4 The graphical representation of the word $w = a\,a\,b\,cccb\,a\,a\,ccb\,cb\,a$ of Example 10.3 and its closure $\overline{w}$ are shown in Fig. 3. The Schröder path $\Psi(w)$ is depicted in Fig. 4.

Given an up step labelled a (resp. b) in a path $\overline{w} = u_1\,a\,u_2$ (resp. $v_1\,b\,v_2$) of $\overline{\mathscr{L}_S}(n)$, there exists a unique down step c such that $\overline{w} = u_1\,a\,u_3\,c\,u_4$ (resp. $v_1\,b\,v_3\,c\,v_4$) and u_3 (resp. v_3) is a Dyck path. We say that the pair (a, c) (resp. (b, c)) forms a *matching*, and a (resp. b) *matches* c.

We define the function $\Psi : \overline{\mathscr{L}_S}(n) \to \mathscr{S}P(n)$, where $\mathscr{S}P(n)$ denotes Schröder paths of length $2n$. There is a unique decomposition for paths w in $\overline{\mathscr{L}_S}(n)$

(1) w is the empty path, or	
(2) $w = a\,v'\,c\,v''$	(a, c) is a matching, or
(3) $w = b\,v'\,c\,v''$	(b, c) is a matching.

Observe that in cases (2) and (3) owing to the definition of words in $\overline{\mathscr{L}_S}(n)$, we have that v'' is the empty path or

$$v'' = b\,g_1\,c\,\ldots\,b\,g_k\,c,$$

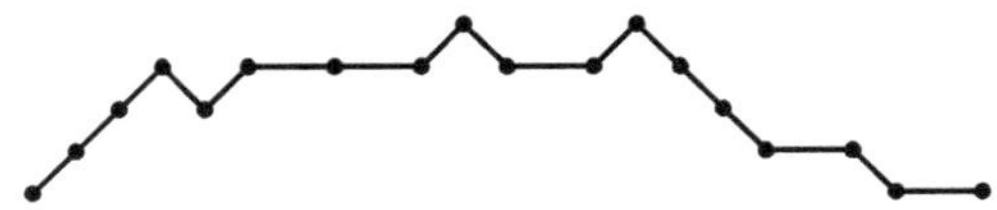

Fig. 4 The path $\Psi(w)$

where g_i is any Dyck path with up steps labelled a or b, and $k \geq 1$. According to this decomposition, the function Ψ is defined as follows:

$$\begin{cases} \Psi(\varepsilon) = \varepsilon\,; \\ \Psi(a\,v'\,c\,v'') = \Psi(v')\,H\,\Psi(v'')\,; \\ \Psi(b\,v'\,c\,v'') = U\,\Psi(v')\,D\,\Psi(v'')\,. \end{cases}$$

where, as usual, ε is the empty path, U (resp. D) denotes up (resp. down) steps, while H denotes horizontal steps of length 2.

The Schröder path $\Psi(w)$ obtained from the word w considered in Example 10.4 is depicted in Fig. 4.

Proposition 10.3 *The function* $\Psi : \overline{\mathscr{L}_S}(n) \to \mathscr{S}P(n)$ *is a bijection.*

Proof To prove the main statement, it is sufficient to define the function Φ : $\mathscr{S}P(n) \to \overline{\mathscr{L}_S}(n)$ and prove that, for all words $w \in \overline{\mathscr{L}_S}(n)$ we have $\Phi(\Psi(w)) = w$. So, let P be a Schröder path of length $2n$, the function Φ is defined as follows

$$\Phi(P) = \begin{cases} \varepsilon & \text{if } P = \varepsilon \\ a\,\Phi(S)\,c & \text{if } P = S\,H \\ \Phi(S)\,b\,\Phi(S')c & \text{if } P = S\,U\,S'\,D\,, \end{cases}$$

where S and S' are Schröder paths.

Let us now prove that $\Phi(\Psi(w)) = w$, by induction on the length of w.

- Clearly if $w = \varepsilon$, $\Phi(\Psi(w)) = w$.
- Let $w = b\,v'\,c\,v''$ (resp. $w = a\,v'\,c\,v''$), where b (resp. a) matches c.
  Suppose $v'' = \varepsilon$, then

$$\Phi(\Psi(w)) = \Phi(\Psi(b\,v'\,c)) = \Phi(U\,\Psi(v')\,D) = b\,\Phi(\Psi(v'))\,c = b\,v'\,c$$

(resp. $\Phi(\Psi(a\,v'\,c)) = \Phi(\Psi(v')\,H) = a\,\Phi(\Psi(v'))\,c = a\,v'\,c$).
Else if $v'' = b\,g_1\,c\,\ldots\,b\,g_k\,c$, with $k \geq 1$, then by applying Ψ recursively it holds

$$\Psi(v'') = U\,\Psi(g_1)\,D \ldots U\,\Psi(g_k)\,D.$$

Therefore, if $w = b\,v'\,c\,v''$, then

$$\begin{aligned}\Phi(\Psi(w)) &= \Phi(U\,\Psi(v')\,D\,U\,\Psi(g_1)\,D\ldots U\,\Psi(g_k)\,D)\\ &= \Phi(U\,\Psi(v')\,D\,U\,\Psi(g_1)\,D\ldots\,U\Psi(g_{k-1})\,D)\,b\,\Phi(\Psi(g_k))\,c\\ &= \ldots\quad\ldots\\ &= b\,\Phi(\Psi(v'))\,c\,b\,\Phi(\Psi(g_1))\,c\ldots b\,\Phi(\Psi(g_k))\,c\\ &= b\,v'\,c\,b\,g_1\,c\,\ldots\,b\,g_k\,c.\end{aligned}$$

While, if $w = a\,v'\,c\,v''$, then $\Phi(\Psi(w)) = \Phi(\Psi(v')\,H\,U\,\Psi(g_1)\,D\ldots U\,\Psi(g_k)\,D) = \ldots = \Phi(\Psi(v')\,H)\,b\,\Phi(\Psi(g_1))\,c\ldots b\,\Phi(\Psi(g_k))\,c = a\,v'\,c\,b\,g_1\,c\,\ldots\,b\,g_k\,c$.

We observe that, in the language $\mathscr{L}_S$, small Schröder parking functions are precisely words beginning with b. So we have the following:

Corollary 10.2 *The function Ψ yields a bijection between small Schröder parking functions and Schröder paths having no horizontal steps on the x axis.*

3.3 Schröder Parking Functions and Pattern Avoiding Permutations

Since Schröder parking functions are underdiagonal sequences it is natural to investigate if they correspond, via the bijection ϕ_I described in Sect. 2.5, to some known family of permutations. Below, we are going to prove that they correspond precisely to the set of inversion tables of a class of pattern avoiding permutations counted by the Schröder numbers.

Proposition 10.4 *Schröder parking functions are bijective to permutations avoiding* 1432 *and* 1423.

Proof We claim that a permutation $\pi \in \mathscr{AV}_n(1432, 1423)$ if and only if the reverse word of its left inversion table is a Schröder parking function of length n.

Let $s_1\,s_2\ldots s_n$ be the reverse word of the left inversion table of any permutation π. The property $s_i < i$ holds for all π and all $i \leq n$. Using this fact, we can reduce the negative form of our statement to prove that a permutation $\pi \notin \mathscr{AV}_n(1432, 1423)$ if and only if there exist two indices i, j, with $i < j$, such that $s_i - s_j \geq 2$.

A permutation $\pi \notin \mathscr{AV}_n(1432, 1423)$ if and only if π contains 1423 or 1432; in other words, if and only if there exist four indices $u < v < w < t$ such that $\pi_u < \pi_v, \pi_w, \pi_t$ and $\pi_v > \pi_w, \pi_t$.

We can suppose without loss of generality that there is no index $u < z < w$ such that $\pi_z < \pi_u$. Let $t_1\,t_2\ldots t_n$ be the left inversion table of such π. Since $\pi_u < \pi_v$ and no elements smaller than π_u are between them, $t_v \geq t_u + 2$ holds. Considering $s_1\,s_2\ldots s_n$ as the reverse word of $t_1\,t_2\ldots t_n$. Set $i = n + 1 - v$ and $j = n + 1 - u$, it results $i < j$ because of $u < v$ and $s_i = t_v \geq t_u + 2 = s_j + 2$.

Conversely, let $s_1\,s_2\ldots s_n$ be the reverse word of the left inversion table $t_1\,t_2\ldots t_n$ of a permutation π. If there exist i and j, with $i < j$, such that $s_i - s_j \geq 2$, then

$t_{n+1-i} - t_{n+1-j} \geq 2$; namely, there exist two indices $u = n+1-j < v = n+1-i$ such that $t_u \leq t_v - 2$. Therefore, there must be $\pi_u < \pi_v$, otherwise $t_u \geq t_v$, and moreover, since $t_v - t_u \geq 2$, there must be two indices w and t, with $w < t \leq n$, such that $\pi_v > \pi_w, \pi_t$ and $\pi_u < \pi_w, \pi_t$. This conclude the proof.

From the statement above, it follows that there is a simple subclass of permutations of $\mathscr{A}\mathscr{V}_n(1432, 1423)$ counted by the small Schröder numbers.

Corollary 10.3 *Small Schröder parking functions of length n are bijective to permutations of* $\mathscr{A}\mathscr{V}_n(1432, 1423)$ *such that* $(n-1, n)$ *is an inversion.*

4 Generalized Schröder Parking Functions

In this section, we provide quite a natural extension of the notion of Schröder parking function, by introducing *generalized Schröder parking functions* of degree m. Then, we study combinatorial properties of these sequences.

Definition 10.4 A generalized Schröder parking function of degree $m \geq 0$ is a sequence $p_1 p_2 \ldots p_n$ such that

- $p_i < i$, for all i;
- If $i < j$ then $p_i - p_j \leq m$.

Clearly, with $m = 0$ we have non-decreasing parking functions and with $m = 1$, we have Schröder parking functions, which are studied in the previous section.

Our aim is to extend the results of the previous section for a generic $m \geq 2$, and in particular, we start providing a representation of generalized Schröder parking functions as *labelled Dyck paths* in order to obtain a generating tree for them, hence their enumeration. As a special case, we retrieve a new combinatorial definition of Schröder parking functions in terms of labelled Dyck paths. Then, we also show how to extend to generalized Schröder parking functions the characterization given for Schröder parking functions in terms of an algebraic language and a pattern avoiding permutation class.

4.1 Generalized Schröder Parking Functions and Labelled Dyck Paths

It is known that generic parking functions can be represented as labelled Dyck paths [13]. We recall that, to each up step U of a Dyck path, we can assign a nonnegative integer $\ell(U)$ (called *level* of U) such that U lies on the line $y = x + \ell(U)$ (see Fig. 5).

Definition 10.5 A *labelled Dyck path P* of length $2n$ is a Dyck path of length $2n$ in which every up step is labelled with a integer $i \in \{1, \ldots, n\}$ and such that the labels of all the up steps at the same level are increasing.

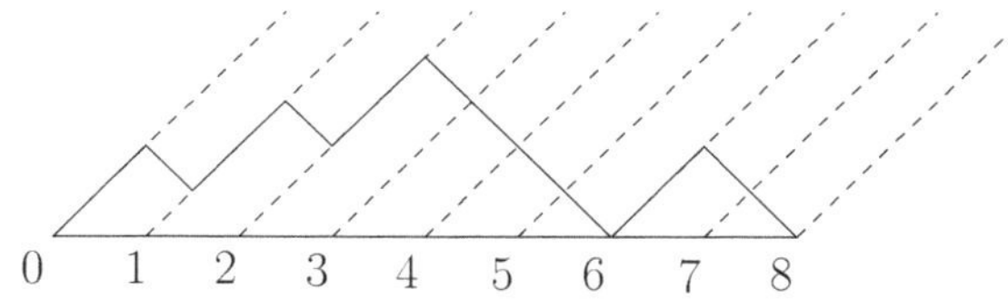

Fig. 5 A Dyck path and the levels of its up steps

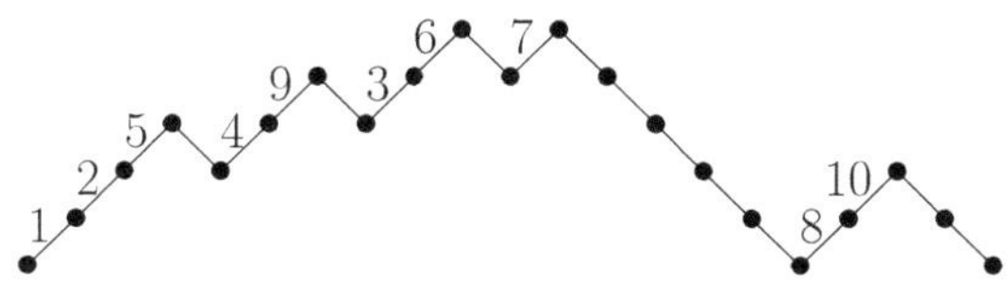

Fig. 6 The generalized labelled Dyck path associated with the parking function 0021023818 according to the bijection ϕ

In a labelled Dyck path P, let $i \in \{1, \ldots, n\}$ be the label of an up step $\overline{U}$, we denote $\ell_i := \ell(\overline{U})$. By [13], the bijection ϕ between labelled Dyck paths and parking functions is given by $\phi(P) = \ell_1, \ldots, \ell_n$. So, for instance, the parking function corresponding to the labelled Dyck path depicted in Fig. 6 is 0021023818.

Our aim is to give a characterization to the family of labelled Dyck paths associated with generalized Schröder parking functions.

Definition 10.6 A *generalized labelled Dyck path P of degree $m \geq 0$* and length $2n$ is a labelled Dyck path such that:

(1) $\ell_i < i$, for any $i \in \{1, \ldots, n\}$;
(2) if $i < j$, with $i, j \in \{1, \ldots, n\}$, then $\ell_i - \ell_j \leq m$.

Let us denote by $\mathscr{D}^m(n)$ the class of generalized labelled Dyck paths of degree m and length $2n$. Observe that, for $m = 0$, we obtain Dyck paths whose up steps are labelled increasingly from 1 to n, which are trivially bijective to Dyck paths and their corresponding parking functions are precisely the non-decreasing parking functions.

Proposition 10.5 *The mapping ϕ yields a bijection between generalized labelled Dyck paths of degree m and generalized Schröder parking functions of degree m.*

Now, we provide a recursive construction for the class $\mathscr{D}^m$ according to the ECO method. In order to do it, we introduce the notion of *color* of a path. Let $P \in \mathscr{D}^m$, then $c(P)$ (the color of P) is defined as $\min(l, m)$, where l is the level of the rightmost up step of P. Moreover, the *last descent* of P is the last sequence of down steps of P.

Let ϑ be an operator performing the following operations on P:

(a) For each point in the last descent of P, ϑ adds an up step labelled n at this point and a down step at the end of the path. This operation produces a number of paths of length $2(n+1)$ equal to the number of points in the last descent of P.
(b) Let us consider the $c(P)$ down steps preceding the last descent of P. Then, ϑ adds an up step labelled by n before each of these steps, and a down step at the end of the path, thus producing $c(P)$ paths of length $2(n+1)$.

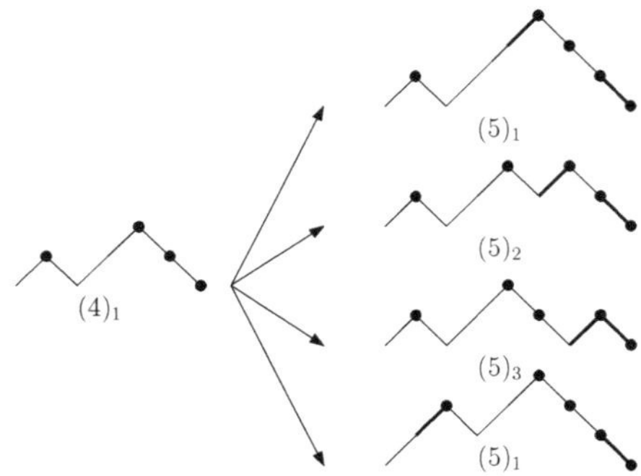

Fig. 7 The production of a path with $c = 1, m = 4, k = 4$

One can easily verify that ϑ satisfies conditions 1 and 2 in Definition 10.1, thus it is an ECO operator. Now, we aim at formalizing the recursive construction of the operator ϑ by a generating tree. To do this, let us denote by $k(P)$ the number of objects produced by P through the application of θ, and distinguish the following three cases depending on m and $k(P)$:

- $0 \leq c < k \leq m$. This means that $c(P) = l$, where l is the level of the rightmost up step of P. Operation (a) adds an up step at height $h = 0, \ldots, k - c - 1$ of the last descent of P, thus producing paths with color $c' = \min(l', m) = k - 1 - h = k - 1, \ldots, c$ and parameter $k' = h + 2 + c' = k + 1$. Operation (b) can be performed at level $\ell = 0, \ldots, c - 1$, producing paths with color $c' = c$ and $k' = k + 1$. See Fig. 7 for an example.
- $0 \leq c < m < k$. This means again that $c(P) = l$, but in this case the result of the application of Operation (a) depends on the height of the point of the last descent to which the up step is added:

  (i) if Operation (a) is applied to a point at height $h = 0, \ldots, k - m - 1$ of the last descent of P, we obtain $k - m$ paths with color $c' = m$ and $k' = h + 2 + c' = m + 2, \ldots, k + 1$;
  (ii) if Operation (a) is applied to a point at height $h = k - m, \ldots, k - c - 1$, we obtain paths with color $c' = m - 1, \ldots, c$, and $k' = k + 1$.

  Operation (b) can be performed at level $\ell = 0, \ldots, c - 1$, producing paths with color $c' = c$ and parameter $k' = k + 1$. See Fig. 8, for an example.
- Otherwise, if $c = m$ then Operation (a), performed at height $h = 0, \ldots, k - c - 1$ of the last descent of P, produces paths with color $c' = m$ and $k' = h + 2 + c' = m + 2, \ldots, k + 1$. Operation (b), performed at level $\ell = l - m, \ldots, l - 1$, produces paths with color $c' = m$ and $k' = k + 1$. See Fig. 9, for an example.

We are now able to give the succession rule for generalized Schröder parking functions, obtained from θ operator.

Proposition 10.6 *The succession rule associated with the operator ϑ for generalized Schröder parking functions of degree $m \geq 0$ is:*

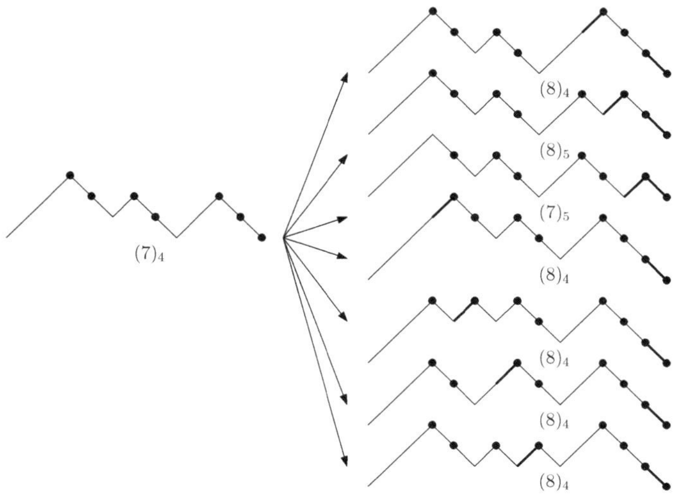

Fig. 8 The production of a path with $c = 4$, $m = 5$ and $k = 7$

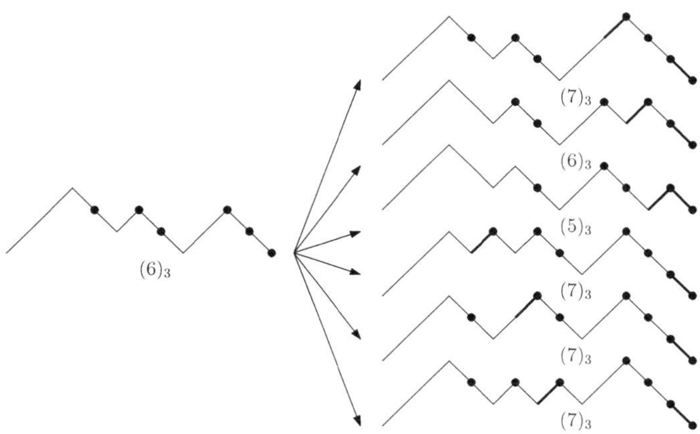

Fig. 9 The production of a path with $c = 3$, $m = 3$ and $k = 6$

$$\Omega^m_{Sch} \quad \begin{cases} (2) \\ (k) \rightarrow (k+1)^k & \textit{if } k < m+2 \\ (k) \rightarrow (m+2)(m+3)\dots(k)(k+1)^{m+1} & \textit{otherwise.} \end{cases}$$

Proof Let $(k)_c$ be the label of $P \in \mathscr{D}^m(n)$, where $k = k(P)$ and $c = c(P)$. From the description of θ operator, we have the following succession rule

$$
\Omega_\theta^m \quad \begin{cases} (2)_0 \\ (k)_c \rightarrow (k+1)_{k-1} \dots (k+1)_{c+1} (k+1)_c^{c+1}, & 0 \le c < k \le m \\ (k)_c \rightarrow (m+2)_m (m+3)_m \dots (k)_m (k+1)_m (k+1)_{m-1} \\ \qquad \dots (k+1)_{c+1} (k+1)_c^{c+1}, & 0 \le c < m < k \\ (k)_m \rightarrow (m+2)_m (m+3)_m \dots (k)_m (k+1)_m^{m+1}, & \text{otherwise.} \end{cases}
$$

Hence, it follows rule Ω_{Sch}^m considering the following cases for the parameter k:

- If $k \le m$, then by the first case of rule Ω_θ^m, for any c, a label $(k)_c$ produces $(k+1)^k$.
- If $k = m+1$ and $c < m$ (resp. $c = m$), then by the second (resp. third) case of rule Ω_θ^m, for any c, a label $(k)_c$ produces $(k+1)^k$.
- If $k \ge m+2$ and $c < m$ (resp. $c = m$), then by the second (resp. third) case of rule Ω_θ^m, for any c, a label $(k)_c$ produces $(m+2)(m+3)\dots(k)(k+1)^{m+1}$.

Observe that, if m goes to infinity, the succession rule becomes:

$$
\Omega_{Fac} \quad \begin{cases} (2) \\ (k) \rightarrow (k+1)^k \end{cases}
$$

which is the usual succession rule for factorial numbers.

Proposition 10.7 *Let us denote by $F_{Sch}^m(x)$ the generating function of the generalized Schröder parking functions of degree m, then*

$$
F_{Sch}^m(x) = \sum_{i=1}^{m} i!x^i + x^m(m+1)! \frac{(1 - x(m+2) - \sqrt{(mx+1)^2 - 4(m+1)x}}{2(m+1)x}
$$

Indeed, from the succession rule Ω_{Sch}^m we have that $F_{Sch}^m(x) = \sum_{i=1}^m i!x^i + x^m G^m(x)$, where $G^m(x)$ is the generating function associated with the succession rule starting at level $(m+1)$ of the generating tree $\mathscr{T}$:

$$
\Omega^m \quad \begin{cases} (m+2) \\ (k) \rightarrow (m+2)(m+3)\dots(k)(k+1)^{m+1} \end{cases}
$$

Let us take a node $o \in \mathscr{T}$, we denote by $k(o)$ its label and by $n(o)$ its level in the generating tree. Note that at level $m+1$ of the generating tree of Ω_{Sch}^m there are $(m+1)!$ objects all having label $(m+2)$. Then:

$$\begin{aligned}G^m(x,y) &= \sum_{o\in\mathcal{T}} y^{k(o)}x^{n(o)} \\ &= (m+1)!y^{m+2}x + \sum_{o\in\mathcal{T}}((y^{m+2}+\ldots+y^{k(o)}+(m+1)y^{k(o)+1})x^{n(o)+1} \\ &= (m+1)!y^{m+2}x + \frac{xy}{1-y}(y^{m+1}G^m(x,1)-G^m(x,y))+(m+1)xyG^m(x,y)\end{aligned}$$

We apply the kernel method [1], obtaining the following equation for the kernel $Y(x)$:

$$1-Y(x)+xY(x)-(m+1)xY(x)(1-Y(x))$$

from which we obtain:

$$Y(x) = \frac{(mx+1)-\sqrt{(mx+1)^2-4(m+1)x}}{2(m+1)x}$$

Since

$$G^m(x) = G^m(x,1) = (m+1)!(Y(x)-1)$$

then

$$F^m_{Sch}(x) = \sum_{i=1}^{m} i!x^i + x^m(m+1)!\frac{(1-x(m+2)-\sqrt{(mx+1)^2-4(m+1)x}}{2(m+1)x}$$

We point out that the rule Ω^m_{Sch} was already studied in [10] in the enumeration of permutations avoiding the set of patterns $\{\sigma\ m\ m+1\ :\ \sigma\in S_{m-1}\}$. We will reconsider and comment this combinatorial interpretation in Sect. 4.3.

4.2 An Algebraic Language for Generalized Schröder Parking Functions of Degree m

The algebraicity of the generating function can be explained by providing a coding of generalized Schröder parking functions of degree m as words of an algebraic language. As for Schröder parking functions, also generalized Schröder parking functions of degree m can be described by means of words of an algebraic language $\mathcal{L}^m$ in the alphabet $\{a_0, a_1, \ldots, a_m, c\}$ as follows. Observe that the case of Schröder parking functions is readily obtained with $m=1$ and by setting $a_0=a$, $a_1=b$.

So, let $p=p_1\ldots p_n$ be a generalized Schröder parking function of degree m. For $i\neq 0$, an entry p_i is a *left-to-right maximum* of p if $p_i > p_j$ for all $j<i$. For instance, in $p = 0\,0\,\underline{1}\,\underline{2}\,1\,2\,0\,1\,\underline{3}\,2\,2\,1\,3\,\underline{6}\,\underline{7}\,6$ every left-to-right maximum is underlined. Therefore, we define $w(p)\in\mathcal{L}^m$ recursively as follows:

Basis. Let $p = p_1 \dots p_n$ be such that $p_i < m+1$, for all $i \leq n$. We set $w(p) = a_{p_2} \dots a_{p_n}$. In particular, if $p = p_1$, $w(p) = \varepsilon$, the empty word.

Inductive step. Let $w' = w(p_1 \dots p_{n-1})$ and suppose there exists a $p_i \geq m+1$, with $i \leq n-1$. We consider two cases:

1. if p_n is a left-to-right maxima, let p_j be the left-to-right maxima preceding p_n (if there is any, otherwise let $p_j = 0$), and set $h = p_n - p_j > 0$. So, we have $w(p) = w'c^h a_m$.
2. if p_n is not a left-to-right maxima, and p_j is the left-to-right maxima preceding p_n, let $h = p_j - p_n < m+1$, we set $w(p) = w'a_h$.

So, for instance, given the parking function of degree 2, $p = 0012120132213676$, we have

$$w(p) = a_0a_1a_2a_1a_2a_0a_1ca_2a_1a_1a_0a_2ccca_2ca_2a_1.$$

The characterization of the set

$$\mathscr{L}^m(n) = \{w(p) : p \text{ is a Schröder parking function of degree } m \text{ and length } n+1\}$$

is then straightforward.

Proposition 10.8 *A word $w = w_1 \dots w_r$ in the alphabet $\{a_0, \dots, a_m, c\}$ belongs to $\mathscr{L}^m(n)$ if and only if*

- *if $w_i = a_j$ then $j \leq i$,*
- *for each prefix v of w, $|v|_c \leq |v|_{a_0} + \dots + |v|_{a_m}$,*
- *an entry c can be followed only by c or a_m,*
- *the last letter is not c,*
- *$|w|_{a_0} + \dots + |w|_{a_m} = n$.*

4.3 Generalized Schröder Parking Functions of Degree m and Pattern Avoiding Permutations

Proposition 10.4 shows that Schröder parking functions are precisely the reverse words of left inversion tables of permutations avoiding the patterns 1423 and 1432. Now we extend this result to generalized Schröder parking functions of degree m, proving that they are bijective to the permutation class described by the avoidance of the set of patterns:

$$\Delta_m = \{1 \ m+3 \ \sigma \ : \ \sigma \text{ is a permutation of length } m+1\}.$$

Proposition 10.9 *Generalized Schröder parking functions of degree m are in bijection with permutations avoiding the set of patterns Δ_m.*

Proof Analogously to Proposition 10.4 it can be shown that generalized Schröder parking function of degree m is precisely the set of reverse words of the left inversion tables of permutations of $\mathcal{AV}(\Delta_m)$.

As a matter of fact, with $m = 0$ we have $\mathcal{AV}(132)$ counted by the Catalan numbers [23], and with $m = 1$ we have $\mathcal{AV}(1432, 1423)$.

We point out that in [10] it was proved that, for any $m \geq 0$, the rule Ω^m_{Sch} describes the recursive growth of an ECO operator for permutations avoiding the set of patterns

$$\Gamma_m = \{\sigma \; m+2 \; m+3 \, : \, \sigma \text{ is a permutation of length } m+1\},$$

namely $\mathcal{AV}(\Gamma_m)$. Due to what we have proved in the previous section and to Proposition 10.9, we conclude that there is a (non trivial) bijection between $\mathcal{AV}(\Gamma_m)$ and $\mathcal{AV}(\Delta_m)$.

5 Baxter Parking Functions

In this section, we define a new family of parking functions, and we prove that it is enumerated by the Baxter numbers, first by describing a recursive growth of them according to the generating tree Ω_{Bax}, then determining a bijection with Baxter triples, defined in Sect. 2.4.

For any sequence of integers $u_1u_2 \ldots u_i$, denote by $Max_2(u_1u_2 \ldots u_i)$ the maximal value among the u_j appearing twice in the sequence. We use the convention that $Max_2(u_1u_2 \ldots u_i) = -1$ if all the entries u_j's are different, so that we can write formally:

$$Max_2(u_1u_2 \ldots u_i) = Max\{u_{j_1} | \exists j_2 \neq j_1, u_{j_1} = u_{j_2}\}$$

For the sake of simplicity, from now on, we write $Max(u_1u_2 \ldots u_i)$ in place of $Max\{u_1, u_2, \ldots, u_i\}$.

Definition 10.7 A *Baxter parking function* is a sequence $u_1u_2 \ldots u_n$ such that, for any $i > 1$, $u_i < i$ and one of the two constraints are satisfied:

$$u_i > Max(u_1u_2 \ldots u_{i-1}) \tag{10.4}$$

$$\exists j < i \quad \text{such that} \quad u_j = u_i \quad \text{and} \tag{10.5.1}$$

$$u_i \geq Max_2(u_1u_2 \ldots u_{i-1}) \tag{10.5.2}$$

Since every entry u_i of a Baxter parking function u (apart from $u_1 = 0$) satisfies either (10.4) or (10.5.1) and (10.5.2), we classify all the entries $u_2 \ldots u_n$ into two groups and we call the entries that satisfy (10.4) *left-to-right maxima* of u.

Example 10.5 An example of Baxter parking function is given by the sequence $u = 0\,1\,1\,3\,4\,3\,3\,4\,6\,7$, whose left-to-right maxima are $u_2, u_4, u_5, u_9, u_{10}$.

Clearly $u_i < i$ for all i, and condition (10.4) is satisfied for $i \neq 3, 6, 7, 8$. For these indices, conditions (10.5.1) and (10.5.2) are satisfied:

$$u_3 = u_2 = 1 > Max_2(0\,1) = -1 \quad u_6 = u_4 = 3 > Max_2(0\,1\,1\,3\,4) = 1$$

$$u_7 = u_4 = 3 \geq Max_2(0\,1\,1\,3\,4\,3) = 3 \quad u_8 = u_5 = 4 \geq Max_2(0\,1\,1\,3\,4\,3\,3) = 3\,.$$

Example 10.6 Examples of non-Baxter parking functions are given by the sequences $v = 0\,0\,1\,1\,4\,3\,3\,4$ and $w = 0\,0\,1\,1\,4\,0\,3\,4$.

Indeed he have $v_6 = 3 < Max(0\,0\,1\,1\,4)$ and $v_6 \neq v_i$ for all $i < 6$, also $w_6 = 0 < Max(0\,0\,1\,1\,4)$ and $w_6 = 0 < Max_2(0\,0\,1\,1\,4) = 1$.

Proposition 10.10 *Baxter parking functions are counted by Baxter numbers.*

Proof To prove that Baxter parking functions are counted by Baxter numbers, we define an ECO operator θ_{Bax} for this class and show that the generating tree associated with θ_{Bax} is equal to Ω_{Bax}, defined in Sect. 2.4.
To any Baxter parking function $u = u_1\,u_2\,\ldots\,u_n$ we associate the label (p, q) where:

- $p(u) = n - Max(u_1 u_2 \ldots u_n)$,
- $q(u)$ is the cardinality of $Q(u)$, the set of all the entries of u greater than $Max_2(u_1 \ldots u_n)$ included,

$$Q(u) = \left\{x \mid \exists\, j \text{ s.t. } u_j = x \quad \text{and} \quad x \geq Max_2(u_1, \ldots, u_n)\right\}.$$

The operator θ_{Bax} adds the entry u_{n+1} to $u_1 \ldots u_n$ such that:

(a) $u_{n+1} = Max(u) + j$, for any $1 \leq j \leq p(u)$;
(b) $u_{n+1} = x$, for any $x \in Q(u)$.

Observe that the sequence $u' = u_1\,u_2\,\ldots\,u_n\,u_{n+1}$ obtained by applying operation (a) (resp. (b)) satisfies condition (10.4) (resp. conditions (10.5.1) and (10.5.2)), thus u' is a Baxter parking function, and θ_{Bax} is an ECO operator for this class. Moreover, there are exactly $p + q$ possible values for u_{n+1}, i.e. the application of θ_{Bax} to u produces $p + q$ elements of size $n + 1$.

Now, we prove that the generating tree associated with θ_{Bax} is equal to Ω_{Bax}. Indeed, let (p, q) be the label of a Baxter parking function u such that $p = p(u)$ and $q = q(u)$:

(a) if $u_{n+1} = Max(u) + j$, for $1 \leq j \leq p$, then $p(u') = (n + 1) - Max(u') = (n + 1) - u_{n+1} = p + 1 - j$ and $Q(u') = Q(u) \cup \{u_{n+1}\}$. Then, $q(u') = q + 1$ and $1 \leq p(u') \leq p$ leading to the productions $(i, q + 1)$, for $1 \leq i \leq p$, of Ω_{Bax}.
(b) if $u_{n+1} = x$, with $x \in Q(u)$, then $p(u') = p + 1$ and $Q(u')$ is a subset of $Q(u)$, whose cardinality $q(u')$ depends on x and varies from 1 (if $x = Max(u)$) to q (if $x = Max_2(u)$). This leads to have productions $(p + 1, i)$, for $1 \leq i \leq q$, of Ω_{Bax}.

Example 10.7 For instance, let $u = 0\,1\,1\,3\,4\,3\,3\,4\,6\,7$; here $Max_2(u) = 4$, $Q(u) = \{4, 6, 7\}$. Then, u has label $(3, 3)$, and the application of θ_{Bax} to u produces the Baxter parking functions:

$$\begin{array}{ll} 0\,1\,1\,3\,4\,3\,3\,4\,6\,7\,10 & \text{with label } (1,4) \\ 0\,1\,1\,3\,4\,3\,3\,4\,6\,7\,9 & \text{with label } (2,4) \\ 0\,1\,1\,3\,4\,3\,3\,4\,6\,7\,8 & \text{with label } (3,4) \\ 0\,1\,1\,3\,4\,3\,3\,4\,6\,7\,7 & \text{with label } (4,1) \\ 0\,1\,1\,3\,4\,3\,3\,4\,6\,7\,6 & \text{with label } (4,2) \\ 0\,1\,1\,3\,4\,3\,3\,4\,6\,7\,4 & \text{with label } (4,3)\,. \end{array}$$

5.1 *Baxter Parking Functions and Baxter Triples*

In this section, we provide a bijective mapping between Baxter parking functions and Baxter triples defined in Sect. 2.4.

We define a mapping Θ from the class of Baxter triples to the family of underdiagonal sequences, which sends a triple (p^u, p^m, p^d) of non-intersecting lattice paths of length $n \geq 0$ into an underdiagonal sequence $u = u_1 u_2 \ldots u_{n+1}$. As usual, p^u, p^m, p^d denote the upper, medium and lower paths. Then, we prove that the image through Θ of any (p^u, p^m, p^d) is a Baxter parking function.

First of all, we set that the Θ image of the empty Baxter triple is $u = 0$.

Then, given a triple (p^u, p^m, p^d) of non-intersecting lattice paths of length n, with $n > 0$, we define its Φ image $u = u_1 \ldots u_{n+1}$ setting $u_1 = 0$ and each u_i, with $i > 1$, depending on the distance between some points of those paths. More precisely, for any $i > 1$, let (a_i, b_i) be the ending point of the $i-1$th step of p^m, usually denoted by p^m_{i-1}. If $p^m_{i-1} = E$ (resp. $p^m_{i-1} = N$), let us consider the distance $j := j_{i-1}$ between the point $(a_i - 1, b_i + 1)$ (resp. $(a_i + 1, b_i - 1)$) and the ending point of the east (resp. north) step of path p^d (resp. p^u) ending at abscissa $a_i + 1$ (resp. ordinate $b_i + 1$). It is well worth noticing that the value j strictly depends on i, but in the following we write j instead of j_{i-1} since no misunderstanding occur.

Therefore, we set $u_i = (i-1) - j$, if $p^m_{i-1} = E$. This assignment is well defined since $i - 1 \geq j$.

Else if $p^m_{i-1} = N$, let us consider the set

$$Q_i(u) = \{u_h \mid \exists h' \text{ s. t. } u_{h'} = u_h,\ u_h \geq Max_2(u_1 \ldots u_i) \text{ and } h \leq i\},$$

for any $i \leq n+1$, as defined in the proof of Proposition 10.10. Let $q_0 \ldots q_l$, for some l, be the sequence of the elements of $Q_i(u)$ ordered decreasingly. Then, we set $u_i = q_j$. Also in this case, Φ is well defined as for any $i > 1$, $Q_i(u)$ is non-empty and $j \leq l$.

Observe that by Θ the entry u_i is a left-to-right of u if and only if $p^m_{i-1} = E$.

For instance, the Baxter triple depicted in Fig. 10 is mapped in the Baxter parking function $0\,1\,0\,2\,0\,2\,3\,6\,3$.

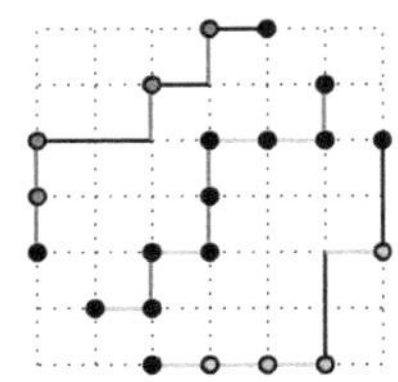

Fig. 10 A Baxter triple mapped by Θ into 0 1 0 2 0 2 3 6 3

Proposition 10.11 *The mapping Θ is a bijection between Baxter triples of length n, having k east steps, and Baxter parking functions of length $n+1$, having k left-to-right maxima.*

Proof Given any triple (p^u, p^m, p^d) of non-intersecting lattice paths of length n, with $n > 0$, we prove that its Θ image u is a Baxter parking function and Θ is a one-to-one mapping, so that it is a bijection.

To prove that $u = \Theta(p^u, p^m, p^d)$ satisfies Definition 10.7, we want to show that for any $i > 1$ such that $p^m_{i-1} = E$, the value $(i-1) - j$ is greater than $Max(u_1 \ldots u_i)$. Hence, u_i satisfies either condition (10.4), if $p^m_{i-1} = E$, or conditions (10.5.1) and (10.5.2), otherwise. Follows from this fact that Θ is one-to-one.

By the recursive definition of Θ proving $(i-1) - j > Max(u_1 \ldots u_i)$, for any i, is the same as proving that $u_s < u_t$, for s and t such that $p^m_{s-1} = p^m_{t-1} = E$. Without loss of generality, we suppose that p^m_{s-1} and p^m_{t-1} are two consecutive instances of east steps in p^m, i.e. p^m contains the factor $p^m_{s-1}\, N^{t-s-1}\, p^m_{t-1}$. Let j_1 (resp. j_2) be the distance between (a_s+1, b_s-1) (resp. (a_s+2, b_t-1)) and the ending point (a_s+1, y_1) (resp. (a_s+2, y_2)) of the corresponding east step of p^d. It holds that

$$j_2 \leq t - s - 1 + j_1,$$

hence $s - 1 - j_1 < t - 1 - j_2$.

As a consequence of Proposition 10.10 we have that the term $\theta_{k,l}$ in (10.3) counts Baxter parking functions of length n and having k left-to-right maxima. Indeed, there are $\theta_{1,1} = 4$ Baxter parking functions of length 3 having only one left-to-right maximum, namely 001, 010, 011, 002.

6 Generalized Baxter Parking Functions

In this section, we study two families of parking functions, which are still contained in the family of underdiagonal sequences and are defined by relaxing the definition of Baxter parking function. Precisely, we obtain GB_1-parking functions (resp. GB_2-parking functions) by removing condition (10.5.1) (resp. (10.5.2)) in Definition 10.7.

6.1 GB_1-Parking Functions

Definition 10.8 A GB_1-parking function is an underdiagonal sequence $u_1 u_2 \dots u_n$ such that, for any $i > 1$, u_i satisfies conditions (10.4) or (10.5.1) of Definition 10.7, precisely:

$$u_i > Max(u_1 u_2 \dots u_{i-1}) \tag{4}$$

$$\exists j < i \quad \text{such that} \quad u_j = u_i \tag{5.1}$$

The first terms of the sequence enumerating GB_1-parking functions are as follows:

$$1, 2, 6, 23, 106, 566, 3415, 22872, 167796, 1334596, \dots$$

We point out that this sequence does not appear in the Encyclopedia of Integer sequences [17]; nevertheless, we are able to write a rule that enumerates it:

Proposition 10.12 *GB_1-parking functions grow according to the succession rule:*

$$\Omega_{GB_1} : \begin{cases} (1, 1) \\ (p, q) \to (1, q+1)(2, q+1) \cdots (p, q+1)(p+1, q)^q \end{cases}$$

Proof The proof is analogous to that of Proposition 10.10 for the recursive growth of Baxter parking functions. We just have to observe that the label (p, q) of a GB_1-parking function $u = u_1 \dots u_n$ is obtained by setting $p = p(u) = n - Max(u)$, and $q = q(u)$ is the cardinality of the set

$$Q_1(u) = \{x \mid \exists j \text{ s.t. } u_j = x\}.$$

Then, if $u_{n+1} = x$, for any $x \in Q_1(u)$, the sequence $u' = u_1 \dots u_n u_{n+1}$ has label $(p + 1, q)$.

For any integers n, p, q such that $p, q \leq n$, let $a_{n,p,q}$ be the number of GB_1-parking functions u of length n such that $p(u) = p$ and $q(u) = q$. From Proposition 10.12 follows a recursive formula satisfied by these numbers.

Corollary 10.4 *The numbers $a_{n,p,q}$ satisfy the following recursive formula:*

$$\begin{cases} a_{1,1,1} = 1, \\ a_{n,p,q} = q\, a_{n-1,p-1,q} + \sum_{j=p}^{n} a_{n-1,j,q-1} \quad \textit{for } n > 1\,. \end{cases} \tag{10.6}$$

Note that we always have $p + q \leq n + 1$. In particular, a GB_1-parking function such that $p = n + 1 - q$ is a sequence in which the set of entries is $\{0, 1, \dots, Max(u)\}$, and then, $Max(u) = q - 1$. For such particular sequences, we are able to prove the following striking result:

Proposition 10.13 *The numbers $a_{n,n+1-q,q}$ are equal to the Stirling numbers of the second kind* $\left\{ n \atop q \right\}$.

Proof It is well known that the Stirling number of the second kind $\left\{ n \atop q \right\}$ counts the number of set partitions of $[n] = \{1, 2, \ldots n\}$ into q parts. Using a very simple bijection, we prove that GB_1-parking functions of length n such that $p = n + 1 - q$ are $\left\{ n \atop q \right\}$.

Given a GB_1-parking function u, define

$$X_{i+1} = \{j \mid u_j = i\}, \quad \text{for } 0 \le i \le q - 1.$$

Then, for any i, X_{i+1} is non empty and $\bigcup_i X_{i+1} = [n]$. Therefore, we have a set partition of $[n]$ into q parts.

Conversely, let $\{X_i\}_{i=1}^{q}$ be a set partition of $[n]$ and assume without loss of generality that $\min(X_i) < \min(X_{i+1})$, for any i. Construct a GB_1-parking function of length n such that $p + q = n + 1$ simply setting for all $j \in X_i$,

$$u_j = i - 1.$$

Let us now consider the generating function $F(t; x, y)$ of GB_1-parking functions according to the length, the parameter p, and the parameter q of a sequence:

$$F(t; x, y) = \sum_{n \ge 1} \left(\sum_{p=1}^{n} \sum_{q=1}^{n} a_{n,p,q}\, x^p y^q \right) t^n. \tag{10.7}$$

Using standard techniques, we can translate the succession rule Ω_{GB_1} into a functional equation satisfied by $F(t; x, y)$:

$$F(t; x, y) = xyt + xyt \left[\frac{F(t; 1, y) - F(t; x, y)}{1 - x} + \frac{\partial F(t; x, y)}{\partial y} \right] \tag{10.8}$$

We observe that (10.8) resembles the functional equation (3), Proposition 8 in [6]. A more general family of functional equations, including the one in [6], was studied in [8]. Unfortunately, after a discussion with Guillaume Chapuy, we are led to conclude that the methods described in [6, 8] cannot be easily extended to our case, so we have not been able to calculate the generating function $F(t; 1, 1)$ of GB_1-parking functions.

On the other side, there is some experimental evidence that the *anisotropic generating function* $F(t; x, y)$ is not differentiably-finite (briefly, D-finite).

As pointed out by Guillaume Chapuy in a personal communication, writing $F(t; x, y) = \sum_{k \ge 1} x^k q_k(t, y)$, it holds that, for each k, $q_k(t, y)$ is a rational function in (t, y): this can be demonstrated by induction since Equation (10.8) is equivalent to

$$(1 - tyk)q_k(t, y) = \delta_{k,1}yt + \frac{yt}{(y-1)}(q_{k-1}(t, y) - q_{k-1}(t, 1)) .$$

Using induction, we can state something more about these rational terms $q_k(t; y)$. In fact, there exists a polynomial $p_k(t, y)$ such that:

$$q_k(t, y) = \frac{p_k(t, y)}{\prod_{i=1}^{k}(iyt - 1)\prod_{i=1}^{k-1}(it - 1)^{k-i}} ,$$

and $p_k(t, y)$ has degree $k(k + 1)/2 - 1$ in t and degree $(k - 1)$ in y (except the case $k = 1$). The first cases are:

$$q_1(t, y) = -\tfrac{yt}{ty-1}$$

$$q_2(t, y) = -\tfrac{yt^2}{(t-1)(ty-1)(2ty-1)}$$

$$q_3(t, y) = -\tfrac{y(2t^2y-1)t^3}{(t-1)^2(2t-1)(ty-1)(2ty-1)(3ty-1)}$$

$$q_4(t, y) = -\tfrac{y(12t^5y^2-6t^3y^2-18t^3y+13t^2y+2t^2-1)t^4}{(t-1)^3(2t-1)^2(3t-1)(ty-1)(2ty-1)(3ty-1)(4ty-1)} .$$

Recently, Tony Guttmann [14] suggested a numerical procedure for testing the solvability of lattice models based on the study of the singularities of their *anisotropic generating functions*. T. Guttmann observed that for a large number of *unsolved models* (leading to non D-finite generating functions) the number of different factors in the denominators increases with n, and suggested that this property could be used as a test of *solvability*. This test has been used successfully by A. Rechnitzer for conjecturing (and then proving) the non D-finiteness of self-avoiding polygons [19], of directed bond animals [20], and of bargraphs according to the site perimeter [5]. Motivated by Guttmann's test we make the following conjecture:

Conjecture 1 *The anisotropic generating function of GB_1-parking functions is not D-finite.*

To prove such a conjecture, we would just need to prove that the number of poles of $q_k(t, y)$ increases, as n grows, so that the function $F(t; x, y)$ has an infinite number of poles. Due to our previous observations, it would be sufficient to prove that the denominators do not simplify with the numerators. Such proofs are in general quite complex, since the expression for the numerators $p_k(t, y)$ may be very difficult to obtain. So, we believe that, to obtain such a proof, it might be convenient to use the so-called *haruspicy techniques* developed in [18–20].

6.2 GB_2-Parking Functions

Definition 10.9 A GB_2-parking function is an underdiagonal sequence $u_1u_2\dots u_n$ such that, for any $i > 1$, u_i satisfies conditions (10.4) or (10.5.2) of Definition 10.7, precisely:

$$u_i > Max(u_1u_2\dots u_{i-1}) \tag{4}$$

$$u_i \geq Max_2(u_1u_2\dots u_{i-1}) \tag{5.2}$$

The first terms of the sequence g_n enumerating GB_2-parking functions are as follows:

$$1, 2, 6, 23, 105, 549, 3207, 20577, 143239, 1071704, \dots$$

We start defining a recurrence relation satisfied by the number $g_{n,k}$ of GB_2-parking functions of length n and having k occurrences of 0, whose first values are shown in the table below.

k	1	2	3	4	5	6
n=1	1					
n=2	1	1				
n=3	2	3	1			
n=4	6	10	6	1		
n=5	23	40	31	10	1	
n=6	105	187	166	75	15	1

For instance, $g_{4,2} = 10$ corresponds to the following 10 GB_2-parking functions of length 4 with 2 occurrences of 0.

$$0\,0\,1\,1, 0\,0\,1\,2, 0\,0\,1\,3, 0\,0\,2\,1, 0\,0\,2\,2, 0\,0\,2\,3, 0\,1\,0\,1, 0\,1\,0\,2, 0\,1\,0\,3, 0\,1\,2\,0$$

We can obtain a recurrence relation satisfied by the numbers $g_{n,k}$ by providing a mapping ψ from GB_2-parking functions of length $n > 1$ onto GB_2-parking functions of length $n - 1$.

Definition 10.10 To any GB_2-parking function $u = u_1u_2\dots u_n$ of length n, we associate the sequence $\psi(u) = u' = u'_1u'_2\dots u'_{n-1}$, such that

$$u'_i = \begin{cases} 0 & \text{if } u_{i+1} = 0 \\ u_{i+1} - 1 & \text{otherwise} \end{cases} \tag{10.9}$$

It is not difficult to see that $\psi(u)$ is a GB_2-parking function of length $n - 1$. Moreover, let us denote by $\mathcal{G}B_2(n, k)$ the set of GB_2-parking functions of length n such that there are exactly k occurrences of 0. Then, we have:

Proposition 10.14 *For any* $u' \in \mathcal{G}B_2(n-1,k)$ *the number of* $u \in \mathcal{G}B_2(n,j)$ *such that* $\psi(u) = u'$ *is equal to*

$$\begin{cases} 0 \text{ if } j > k+1 \\ 1 \text{ if } j = k+1 \\ j \text{ if } j \leq k \end{cases} \tag{10.10}$$

Proof Since each sequence u in $\mathcal{G}B_2(n,j)$ has j occurrences of 0, its image through ψ has at least $j-1$ occurrences of 0 (note that the first value $u_1 = 0$ is not used to compute $\psi(u)$). Hence, $\psi(u)$ is an element of $\mathcal{G}B_2(n-1,j')$, with $j' \geq j-1$, proving the first case of (10.10).

The second case of (10.10) holds, since the unique u in $\mathcal{G}B_2(n,k+1)$ satisfying $\psi(u) = u'$ is obtained by taking $u_{i+1} = u'_i + 1$ for any $u'_i > 0$ and setting u_1 and all the other entries equal to 0.

Concerning the third case, let $u' \in \mathcal{G}B_2(n-1,k)$. There are j ways to obtain a sequence u in $\mathcal{G}B_2(n,j)$ such that $u' = \psi(u)$: we have to replace each $u'_i > 0$ by $u'_i + 1$, add an occurrence of 0 at the beginning of u, keep unchanged $j-1$ occurrences of 0 and replace the other ones by 1. Since u has to satisfy conditions (10.4) and (10.5.2) defining GB_2-parking functions, there is at most one occurrence of 1 on the left of the rightmost occurrence of 0, so that restricting the sequence u to the occurrences of 0's and 1's we must have the situation below.

$$0, \underbrace{\overbrace{0, \cdots, 0, 1, 0, \cdots, 0}^{j}, 1, \cdots, 1}_{k}$$

This proves that there are exactly j possible positions for this unique occurrence of 1 followed by some occurrences of 0.

Proposition 10.14 has the following consequence:

Theorem 10.1 *The numbers* $g_{n,k} = |\mathcal{G}B_2(n,k)|$ *satisfy the following recurrence formula:*

$$\begin{cases} g_{1,1} = 1, \\ g_{n,k} = g_{n-1,k-1} + k \sum_{i=k}^{n-1} g_{n-1,i} \, . \end{cases} \tag{10.11}$$

Proof In order to prove this corollary, we use Proposition 10.14 considering the subsets $\mathcal{G}B_2(n-1,i)$, $i \geq k-1$, and $\mathcal{G}B_2(n,k)$.

From the proof of Proposition 10.14, we obtain a simple generating tree for GB_2-parking sequences, with succession rule given by:

$$\Omega_{GB_2}: \qquad \begin{cases} (1) \\ (k) \rightsquigarrow (1)(2)^2 \ldots (k)^k (k+1)\,, \end{cases}$$

where any GB_2-parking sequence with exactly k occurrences of 0 has label (k).

The sequence $\{g_n\}_{n\geq 0}$, counting GB_2-parking functions, is registered in the Encyclopedia of Integer Sequences as $A113227$ and counts also the number of permutations avoiding the generalized pattern $1\,\underline{23}\,4$. More precisely, in [7] David Callan shows that permutations avoiding $1\,\underline{23}\,4$ are enumerated by a sequence $\{r_n\}_{n\geq 0}$, where $r_n = \sum_k r_{n,k}$, and the terms $r_{n,k}$ satisfy the same recurrence relation as the terms $g_{n,k}$ in (10.11). In [7], David Callan studies and determines the exponential generating function of $1\,\underline{23}\,4$-avoiding permutations according to several parameters. Unfortunately, we have not been able to find parameters on GB_2-parking functions which have the same distribution.

Since the number of GB_2-parking functions of length n is equal to the number of permutations of length n avoiding the pattern $1\,\underline{23}\,4$, to our opinion it would be interesting to find a (direct) bijection between Baxter parking functions of type 2, namely $GB_2(n)$, and permutations of $\mathscr{AV}_n(1\,\underline{23}\,4)$. We believe that this can be done by determining a recursive growth of $\mathscr{AV}_n(1\,\underline{23}\,4)$ according to Ω_{GB_2}.

Acknowledgements The authors would like to thank Guillaume Chapuy for his valuable help on the study of the functional equation for GB_1-parking functions.

References

1. Banderier, C., Bousquet-Mélou, M., Denise, A., Flajolet, P., Gardy, D., Gouyou-Beauchamps, D.: Generating functions for generating trees. Discrete Math. **246**, 29–55 (2002)
2. Barcucci, E., Del Lungo, A., Pergola, E., Pinzani, R.: ECO: a methodology for the enumeration of combinatorial objects. J. Diff. Eq. Appl. **5**, 435–490 (1999)
3. Bonichon, N., Bousquet-Mélou, M., Fusy, É.: Baxter permutations and plane bipolar orientations. Sém. Lothar. Combin. **61A** (2008). Article B61Ah
4. Bousquet-Mélou, M.: Four classes of pattern-avoiding permutations under one roof: generating trees with two labels. Electron. J. Comb. **9**(2) (2003). Article #R19
5. Bousquet-Mélou, M., Rechnitzer, A.: The site-perimeter of bargraphs. Adv. Appl. Math. **31**, 86–112 (2003)
6. Bousquet-Mélou, M., Chapuy, G., Préville-Ratelle, L.-F.: The representation of the symmetric group on m-Tamari intervals. Adv. Math. **247**, 309–342 (2013)
7. Callan, D.: A bijection to count $(1 - 23 - 4)$-avoiding permutations. Unpublished manuscript (2010). arXiv:1008.2375
8. Chapuy, G., Fang, W.: Generating functions of bipartite maps on orientable surfaces. Electron. J. Comb. **23** (2016). Paper 3.31
9. Chung, F.R.K., Graham, R., Hoggatt, V., Kleiman, M.: The number of Baxter permutations. J. Comb. Theory Ser. A **24**(3), 382–394 (1978)
10. Del Lungo, A., Barcucci, E., Pergola, E., Pinzani, R.: Permutations avoiding an increasing number of length-increasing forbidden subsequences. Discrete Math. Theor. Comput. Sci. **4**, 31–44 (2000)
11. Deutsch, E.: A bijective proof of the equation linking the Schröder numbers, large and small. Discrete Math. **241**, 235–240 (2001)
12. Felsner, S., Fusy, É., Noy, M., Orden, D.: Bijections for Baxter families and related objects. J. Comb. Theory Ser. A **118**(3), 993–1020 (2011)
13. Garsia, A.M., Haiman, M.: A remarkable q, t-Catalan sequence and q-Lagrange Inversion. J. Algebr. Comb. **5**(3), 191–244 (1996)
14. Guttmann, A.J.: Indicators of solvability for lattice models. Discrete Math. **217**, 167–189 (2000)

15. Konheim, A.G., Weiss, B.: An occupancy discipline and applications. SIAM J. Appl. Math. **14**, 1266–1274 (1966)
16. Loehr, N.A.: Combinatorics of q, t-parking functions. Adv. Appl. Math. **34**, 408–425 (2005)
17. OEIS Foundation Inc.: The on-line encyclopedia of integer sequences (2011). http://oeis.org
18. Rechnitzer, A.: Haruspicy and anisotropic generating functions. Adv. Appl. Math. **30**, 228–257 (2003)
19. Rechnitzer, A.: Haruspicy 2: The self-avoiding polygon generating function is not D-finite. J. Comb. Theory Ser. A **113**(3), 520–546 (2006)
20. Rechnitzer, A.: Haruspicy 3: the anisotropic generating function of directed bond-animals is not D-finite. J. Comb. Theory Ser. A **113**(6), 1031–1049 (2006)
21. Riordan, J.: Ballots and trees. J. Comb. Theory **6**(4), 408–411 (1969)
22. Shapiro, L., Stephens, A.B.: Bootstrap percolation, the Schröder numbers, and the N-kings problem. SIAM J. Discrete Math. **4**, 275–280 (1991)
23. Stanley, R.P.: Enumerative Combinatorics, vol. 2. Cambridge University Press, Cambridge (1999)
24. Tenner, B.: Database of permutation pattern avoidance. http://math.depaul.edu/bridget/patterns.html
25. West, J.: Generating trees and the Catalan and Schröder numbers. Discrete Math. **146**, 247–262 (1995)

Some Tilings, Colorings and Lattice Paths via Stern Polynomials

Karl Dilcher and Larry Ericksen

Abstract We use certain subsequences of two different but related types of generalized Stern polynomials to characterize all lattice paths, with specific restrictions, that go from the origin to the line $x + y = n$ in the first quadrant of the xy-plane. The first kind of lattice paths can also be interpreted as tilings with squares and dominoes in one case and "black and white" colorings in another case. The second kind of lattice paths is certain weighted Delannoy paths; from our analysis, we obtain results on weighted Delannoy numbers and extensions with polynomial weights. Finally, we establish some connections with Jacobi polynomials.

Keywords Lattice paths, Stern polynomials, Generating functions, Delannoy numbers, Jacobi polynomials

2010 Mathematics Subject Classification Primary: 05A15 · Secondary: 11B75 11B83

1 Introduction

It is a well-known fact that the number of tilings of an $n \times 1$ rectangle (also known as an n-board) by 1×1 squares and 2×1 dominoes is F_{n+1}, where F_k is the kth Fibonacci number defined recursively by $F_0 = 0$, $F_1 = 1$ and $F_k = F_{k-1} + F_{k-2}$

Research supported in part by the Natural Sciences and Engineering Research Council of Canada, Grant # 145628481.

K. Dilcher
Department of Mathematics and Statistics, Dalhousie University,
Halifax, Nova Scotia B3H 4R2, Canada
e-mail: dilcher@mathstat.dal.ca

L. Ericksen (✉)
P.O. Box 172, Millville, NJ 08332-0172, USA
e-mail: LE22@cornell.edu

G. E. Andrews et al. (eds.), *Lattice Path Combinatorics and Applications*,
Developments in Mathematics 58, https://doi.org/10.1007/978-3-030-11102-1_11

($k \geq 2$). It is also known that the number of such tilings with exactly j dominoes ($0 \leq j \leq \lfloor \frac{n}{2} \rfloor$) is $\binom{n-j}{j}$; see, e.g., [4, Ch. 1] for these and other related facts.

We now consider the equivalent problem of counting the following lattice paths from the origin to a point on the line $x + y = n$ that intersects the lattice $\mathbb{Z} \times \mathbb{Z}$, with $x \geq 0$, $y \geq 0$. How many such paths are there if the allowable moves are two units in the vertical (up) and one unit in the horizontal direction (to the right)?

Clearly, this is equivalent to the tiling mentioned above if we identify a vertical move with a domino and a horizontal move with a square. This means that there are F_{n+1} such lattice paths, and the number of paths with exactly j vertical moves is $\binom{n-j}{j}$. This is, of course, also related to the well-known identity

$$F_{n+1} = \sum_{j=0}^{\lfloor n/2 \rfloor} \binom{n-j}{j}. \tag{11.1}$$

This identity, in turn, can be seen as a special case of the explicit expansion

$$F_{n+1}(x, y) = \sum_{j=0}^{\lfloor n/2 \rfloor} \binom{n-j}{j} x^{n-2j} y^j \tag{11.2}$$

for the (bivariate) Fibonacci polynomials defined by the recurrence relation $F_0(x, y) = 0$, $F_1(x, y) = 1$, and

$$F_k(x, y) = x F_{k-1}(x, y) + y F_{k-2}(x, y). \tag{11.3}$$

These polynomials, which are related to the Chebyshev polynomials, have a long history, going back to Lucas [20], and are still being applied and extended; see, e.g., [1] or [7]. Either one of the univariate polynomials $F_{n+1}(x, 1)$ or $F_{n+1}(1, y)$ would serve to encode not only the number of lattice paths discussed above, but also the additional information of the number of paths with exactly j vertical moves.

It is the main purpose of this paper to go a step further and introduce a one-parameter extension of the polynomial sequence $\{F_{n+1}(1, y)\}_n$, which will allow us to "read off," for each fixed $n \geq 1$, all the individual F_{n+1} paths. We achieve this by generalizing one of two recently introduced polynomial extensions of the Stern (diatomic) sequence.

We define this generalization in Sect. 2, along with an analogous generalization that has previously been introduced and applied. In Sect. 3, we then apply the first generalization to the lattice paths (and tilings) discussed above, and to a certain coloring of the n-board. In Sect. 4, we use the second type of generalized Stern polynomials to deal with a Delannoy-type lattice path problem and also present a method of identifying, for a given n, each individual path from the origin to a lattice point on the line $x + y = n$. The proofs follow from a more general treatment, using polynomial weights; this is done in Sect. 5. We conclude this paper with some further remarks in Sect. 6, including connections with Jacobi polynomials.

2 Generalized Stern Polynomials

The Stern sequence, also known as Stern's (diatomic) sequence, is one of the most remarkable integer sequences in number theory and combinatorics. Using the notation $\{a(n)\}_{n\geq 0}$, it can be defined by $a(0) = 0$, $a(1) = 1$, and for $n \geq 1$ by

$$a(2n) = a(n), \qquad a(2n+1) = a(n) + a(n+1). \tag{11.4}$$

Numerous properties and references can be found, e.g., in [5], [22, A002487], or [24]. This sequence was independently extended to two different concepts of Stern *polynomials* in [12, 18]; see also [8, 13], resp. [25, 26, 28, 29], for further properties.

We are now going to define one-parameter extensions of these two polynomial sequences and derive some important properties.

Definition 11.1 Let t be a fixed real or complex number.
(*a*) The type-1 generalized Stern polynomials $a_{1,t}(n; z)$ are polynomials in z defined by $a_{1,t}(0; z) = 0$, $a_{1,t}(1; z) = 1$, and for $n \geq 1$ by

$$a_{1,t}(2n; z) = z\, a_{1,t}(n; z^t), \tag{11.5}$$
$$a_{1,t}(2n+1; z) = a_{1,t}(n+1; z^t) + a_{1,t}(n; z^t). \tag{11.6}$$

(*b*) The type-2 generalized Stern polynomials $a_{2,t}(n; z)$ are polynomials in z defined by $a_{2,t}(0; z) = 0$, $a_{2,t}(1; z) = 1$, and for $n \geq 1$ by

$$a_{2,t}(2n; z) = a_{2,t}(n; z^t), \tag{11.7}$$
$$a_{2,t}(2n+1; z) = a_{2,t}(n+1; z^t) + z\, a_{2,t}(n; z^t). \tag{11.8}$$

See Table 1 for the first 21 of each of these polynomials. When $z = 1$, both sequences reduce to the Stern (diatomic) sequence, by comparing with (11.4). Furthermore, $\{a_{1,1}(n; z)\}$ is the sequence of Stern polynomials introduced in [18], and $\{a_{2,2}(n; z)\}$ is the one introduced in [12]. The generalized sequence $\{a_{2,t}(n; z)\}$ was recently introduced by the present authors in [9], where it was used in a detailed study of hyperbinary expansions; further properties were derived in [10]. The sequence $\{a_{1,t}(n; z)\}$ is new.

Table 1 indicates that both sequences of polynomials have a special structure. While it is easily seen that for $t = 1$ the exponents in a given polynomial can coincide, for $t \geq 2$ the situation is quite different.

Proposition 11.1 *For integers $t \geq 2$ and $n \geq 0$, the coefficients of $a_{1,t}(n; z)$ and $a_{2,t}(n; z)$ are only* 0 *or* 1. *Furthermore, all exponents of z are polynomials in t with only* 0 *or* 1 *as coefficients.*

Proof For $a_{2,t}(n; z)$ this was proved in [9]. To deal with $a_{1,t}(n; z)$, we first note that by (11.5) all *exponent* polynomials for even n have constant coefficient 1, while by (11.6) all exponent polynomials for odd n have constant coefficients 0. We are done

Table 1 $a_{1,t}(n; z)$ and $a_{2,t}(n; z)$, $1 \le n \le 21$

n	$a_{1,t}(n; z)$	$a_{2,t}(n; z)$
1	1	1
2	z	1
3	$1 + z^t$	$1 + z$
4	z^{t+1}	1
5	$1 + z^t + z^{t^2}$	$1 + z + z^t$
6	$z + z^{t^2+1}$	$1 + z^t$
7	$1 + z^{t^2} + z^{t^2+t}$	$1 + z + z^{t+1}$
8	z^{t^2+t+1}	1
9	$1 + z^{t^2} + z^{t^2+t} + z^{t^3}$	$1 + z + z^t + z^{t^2}$
10	$z + z^{t^2+1} + z^{t^3+1}$	$1 + z^t + z^{t^2}$
11	$1 + z^t + z^{t^2} + z^{t^3} + z^{t^3+t}$	$1 + z + z^{t+1} + z^{t^2} + z^{t^2+1}$
12	$z^{t+1} + z^{t^3+t+1}$	$1 + z^{t^2}$
13	$1 + z^t + z^{t^3} + z^{t^3+t} + z^{t^3+t^2}$	$1 + z + z^t + z^{t^2+1} + z^{t^2+t}$
14	$z + z^{t^3+1} + z^{t^3+t^2+1}$	$1 + z^t + z^{t^2+t}$
15	$1 + z^{t^3} + z^{t^3+t^2} + z^{t^3+t^2+t}$	$1 + z + z^{t+1} + z^{t^2+t+1}$
16	$z^{t^3+t^2+t+1}$	1
17	$1 + z^{t^3} + z^{t^3+t^2} + z^{t^3+t^2+t} + z^{t^4}$	$1 + z + z^t + z^{t^2} + z^{t^3}$
18	$z + z^{t^3+1} + z^{t^3+t^2+1} + z^{t^4+1}$	$1 + z^t + z^{t^2} + z^{t^3}$
19	$1 + z^t + z^{t^3} + z^{t^3+t} + z^{t^3+t^2}$ $+ z^{t^4} + z^{t^4+t}$	$1 + z + z^{t+1} + z^{t^2} + z^{t^2+1}$ $+ z^{t^3} + z^{t^3+1}$
20	$z^{t+1} + z^{t^3+t+1} + z^{t^4+t+1}$	$1 + z^{t^2} + z^{t^3}$
21	$1 + z^t + z^{t^2} + z^{t^3} + z^{t^3+t} + z^{t^4}$ $+ z^{t^4+t} + z^{t^4+t^2}$	$1 + z + z^t + z^{t^2+1} + z^{t^2+t} + z^{t^3}$ $+ z^{t^3+1} + z^{t^3+t}$

if we can show that for a given n the exponent polynomials are all distinct and have coefficients 0 and 1 only. We do this by induction on n.

The statement is clearly true for $n = 1$ and 2 (see Table 1). Now suppose it is true up to some $n - 1 \ge 2$. If n is even, then by (11.5) it will also be true for $a_{1,t}(n; z)$. Suppose then that n is odd, say $n = 2k + 1$. By induction hypothesis, the property in question holds for $a_{1,t}(k; z)$ and $a_{1,t}(k + 1; z)$. But one of k, $k + 1$ is even and the other one is odd; therefore, by the previous paragraph, no exponent polynomial in $a_{1,t}(k; z)$ will match one in $a_{1,t}(k + 1; z)$, and all have coefficients 0 and 1. The same is true when z is replaced by z^t, and therefore, by (11.6), $a_{1,t}(2k + 1; z)$ has the desired property. This completes the proof.

Although generating functions will not be used for the applications in this paper, we state them as important properties of the sequences in question.

Proposition 11.2 *For integers $t \ge 1$, the generalized Stern polynomials have the following generating functions:*

Table 2 α_n, β_n, $1 \le n \le 10$

n	1	2	3	4	5	6	7	8	9	10
α_n	1	1	3	5	11	21	43	85	171	341
β_n		2	3	7	13	27	53	107	213	427

$$x\prod_{j=0}^{\infty}\left(1+x^{2^j}z^{t^j}+x^{2^{j+1}}\right)=\sum_{n=1}^{\infty}a_{1,t}(n;z)x^n, \tag{11.9}$$

$$x\prod_{j=0}^{\infty}\left(1+x^{2^j}+x^{2^{j+1}}z^{t^j}\right)=\sum_{n=1}^{\infty}a_{2,t}(n;z)x^n. \tag{11.10}$$

The generating function (11.9) was proved in [11], with the special case $t = 1$ obtained by Ulas in [28]. The identity (11.10) was proved in [9], with the cases $t = 1$ and $t = 2$ earlier obtained in [3, 12], respectively. The corresponding generating function for Stern's diatomic sequence was first obtained by Carlitz [6].

The remainder of this section will be devoted to two special subsequences for each of the two types of generalized Stern polynomials. Through their relationship with Fibonacci numbers we will ultimately establish the desired connection with lattice paths.

An important and interesting property of Stern's diatomic sequence defined by (11.4) is the fact that in each interval $2^{n-2} \le m \le 2^{n-1}$ the maximum value of $a(m)$ is the Fibonacci number F_n. It was apparently first shown by Lehmer [19] that this maximum occurs at

$$\alpha_n := \frac{1}{3}\left(2^n-(-1)^n\right) \quad \text{and} \quad \beta_n := \frac{1}{3}\left(5\cdot 2^{n-2}+(-1)^n\right) \quad (n \ge 2), \tag{11.11}$$

where α_n is also defined for $n = 0, 1$. The two sequences can be found in [22] as A001045 and A048573, respectively, with numerous properties and references. The first few values of both are listed in Table 2.

Of the various properties of these sequences we require the recurrence relations

$$\alpha_{n+1} = 2\alpha_n + (-1)^n, \qquad \beta_{n+1} = 2\beta_n - (-1)^n, \tag{11.12}$$

which immediately follow from (11.11). In particular, these identities show that with the exception of $\beta_2 = 2$, all α_n and β_n, $n \ge 2$, are odd.

In [13] the two subsequences of the Stern polynomials $a_{2,2}(k; z)$ given by $k = \alpha_n$ and $k = \beta_n$ were introduced and studied in some detail. In analogy we consider the sequences

$$a_{1,t}(\alpha_n; z), \qquad a_{1,t}(\beta_n; z), \qquad a_{2,t}(\alpha_n; z), \qquad a_{2,t}(\beta_n; z).$$

Since for $z = 1$ the sequences of generalized Stern polynomials reduce to Stern's diatomic sequence, by the remarks preceding (11.11) we have

$$a_{1,t}(\alpha_n; 1) = a_{1,t}(\beta_n; 1) = a_{2,t}(\alpha_n; 1) = a_{2,t}(\beta_n; 1) = F_n \quad (n \geq 2), \tag{11.13}$$

independent of t. By Proposition 11.1 this means that the number of terms in $a_{\varepsilon,t}(\alpha_n; z)$ and $a_{\varepsilon,t}(\beta_n; z)$ is F_n for both $\varepsilon = 1$ and 2.

The relation (11.13) also shows that the following recurrence relations can be seen as analogues of the basic recurrence of the Fibonacci numbers.

Proposition 11.3 *For a fixed integer $t \geq 1$ we have*

$$a_{1,t}(\alpha_{n+1}; z) = a_{1,t}(\alpha_n; z) + z^{t^{n-1}} a_{1,t}(\alpha_{n-1}; z) \quad (n \geq 2), \tag{11.14}$$

$$a_{1,t}(\beta_{n+1}; z) = a_{1,t}(\beta_n; z^t) + z^t a_{1,t}(\beta_{n-1}; z^{t^2}) \quad (n \geq 3), \tag{11.15}$$

$$a_{2,t}(\alpha_{2n+1}; z) = z\, a_{2,t}(\alpha_{2n}; z^t) + a_{2,t}(\alpha_{2n-1}; z^{t^2}) \quad (n \geq 1), \tag{11.16}$$

$$a_{2,t}(\alpha_{2n}; z) = a_{2,t}(\alpha_{2n-1}; z^t) + z\, a_{2,t}(\alpha_{2n-2}; z^{t^2}) \quad (n \geq 2), \tag{11.17}$$

$$a_{2,t}(\beta_{2n+1}; z) = a_{2,t}(\beta_{2n}; z^t) + z\, a_{2,t}(\beta_{2n-1}; z^{t^2}) \quad (n \geq 2), \tag{11.18}$$

$$a_{2,t}(\beta_{2n}; z) = z\, a_{2,t}(\beta_{2n-1}; z^t) + a_{2,t}(\beta_{2n-2}; z^{t^2}) \quad (n \geq 2). \tag{11.19}$$

The proofs of these identities are routine and come from the recurrence relations (11.5)–(11.8), using the two identities in (11.12).

3 Tilings and Colorings of the n-Board

3.1 Tilings

As mentioned in the introduction, there is an easy 1-1 correspondence between certain lattice paths and tilings of the n-board with squares and dominoes. We therefore restrict our attention in this section to the latter and begin with the main result of this section. We first fix some terminology. Given an n-board, we number its squares, from the left, by $1, 2, \ldots, n$, and we say that a domino is in position k, $1 \leq k \leq n-1$, if it covers the squares numbered k and $k+1$.

Theorem 11.1 *Given an integer $n \geq 1$, let $\mathscr{E}_{n+1}$ be the set of exponents of z in $a_{1,t}(\alpha_{n+1}; z)$, i.e.,*

$$a_{1,t}(\alpha_{n+1}; z) = \sum_{p \in \mathscr{E}_{n+1}} z^{p(t)}. \tag{11.20}$$

Then each tiling of the n-board corresponds to exactly one polynomial in $\mathscr{E}_{n+1}$ as follows:

The zero polynomial corresponds to the tiling without dominoes. Otherwise, if

$$p(t) = t^{a_1} + \cdots + t^{a_r} \in \mathscr{E}_{n+1}, \quad 1 \le a_1 < \cdots < a_r, \quad r \ge 1, \tag{11.21}$$

then the corresponding tiling has r dominoes, each in position a_j, $j = 1, \dots, r$.

Before proving this, we consider an example.

Example 11.1 For tiling the 5-board we consider $a_{1,t}(\alpha_6; z) = a_{1,t}(21; z)$, and in Table 1 we see that

$$\mathscr{E}_6 = \{0, t, t^2, t^3, t^3 + t, t^4, t^4 + t, t^4 + t^2\}.$$

Accordingly, all the tilings are given by dominoes in positions 1, 2, 3, 1&3, 4, 1&4, and 2&4, respectively, in addition to the tiling with only squares.

We also note that there is $\binom{5}{0} = 1$ tiling with no domino, and there are $\binom{4}{1} = 4$ tilings with one domino and $\binom{3}{2} = 3$ tilings with two dominoes, consistent with the remarks at the beginning of the introduction.

Proof (*of Theorem* 11.1) We use induction on n. For $n = 1$ and 2 the tilings clearly correspond to $a_{1,t}(\alpha_2; z) = 1$ and $a_{1,t}(\alpha_3; z) = 1 + z^t$, respectively. Suppose now that the result holds up to $n - 1$ ($n \ge 3$). We get all tilings of the n-board by combining the following:

(1) The tilings of the $(n-1)$-board, with one square added to the end. By induction hypothesis these are given by the exponents in $a_{1,t}(\alpha_n; z)$.

(2) The tilings of the $(n-2)$-board, with one domino attached to the end, i.e., in position $n-1$. This amounts to adding t^{n-1} to each of the coefficient polynomials in $a_{1,t}(\alpha_{n-1}; z)$ or, equivalently, multiplying $a_{1,t}(\alpha_{n-1}; z)$ by $z^{t^{n-1}}$.

Finally, adding the polynomials obtained in (1) and (2) and using (11.14), we find that all tilings of the n-board are given by $a_{1,t}(\alpha_{n+1}; z)$. This completes the proof.

Remarks (1) It follows from Theorem 11.1 that in any $p(t) \in \mathscr{E}_n$ no two exponents of t can be adjacent. This property is special to $a_{1,t}(\alpha_n; z)$; it does not hold for Stern polynomials in general, as can be seen in Table 1.

(2) Theorem 11.1 also implies that, by setting $t = 1$, the coefficients of z^j, $j \ge 0$, in $a_{1,1}(\alpha_{n+1}; z)$ count the number of tilings of the n-board with exactly j dominoes. By the remark at the beginning of the introduction, this means that we have the explicit expansion

$$a_{1,1}(\alpha_{n+1}; z) = \sum_{j=0}^{\lfloor \frac{n}{2} \rfloor} \binom{n-j}{j} z^j. \tag{11.22}$$

Since the left-hand side is a Stern polynomial as defined in [18] and the right-hand side can be written in terms of Chebyshev polynomials, one can obtain results on zeros and irreducibility of these polynomials; see [14].

3.2 Colorings

Given the similarities between the sequences α_n and β_n, it is natural to ask whether the Stern polynomials $a_{1,t}(\beta_n; z)$ have a similar combinatorial interpretation as the polynomials $a_{1,t}(\alpha_n; z)$ do in Theorem 11.1. This is indeed the case, as we shall now see.

Given an integer $n \geq 1$, we color the squares of the n-board in any of the following ways:

- If $n = 1$, the square can be white (W) or black (B).
- If $n \geq 2$, then
  (i) the positions $n-1, n$ can only be WW, WB and BB (not BW), and
  (ii) apart from positions $n-1$ and n, no two adjacent B are allowed.

Example 11.2 For $n = 2$, the allowable colorings are WW, WB, BB, and for $n = 3$ they are WWW, WWB, WBB, BWW, BWB. Note that their numbers are $F_4 = 3$ and $F_5 = 5$, respectively.

Theorem 11.2 *Given an integer $n \geq 1$, let $\mathscr{E}'_{n+2}$ be the set of exponents of z in $a_{1,t}(\beta_{n+2}; z)$, i.e.,*

$$a_{1,t}(\beta_{n+2}; z) = \sum_{p \in \mathscr{E}'_{n+2}} z^{p(t)}. \tag{11.23}$$

Then each of the above colorings of the n-board corresponds to exactly one polynomial in $\mathscr{E}'_{n+2}$ as follows:

The zero polynomial corresponds to all squares being W. *Otherwise, if*

$$p(t) = t^{b_1} + \cdots + t^{b_r} \in \mathscr{E}'_{n+2}, \quad 1 \leq b_1 < \cdots < b_r, \quad r \geq 1, \tag{11.24}$$

then exactly r squares are B, *each in position b_j, $j = 1, \ldots, r$.*

Once again, before proving this result we give an example.

Example 11.3 For coloring the 4-board we consider

$$\begin{aligned} a_{1,t}(\beta_6; z) &= a_{1,t}(27; z) \\ &= 1 + z^t + z^{t^2} + z^{t^4} + z^{t^4+t} + z^{t^4+t^2} + z^{t^4+t^3+t} + z^{t^4+t^3}, \end{aligned}$$

which can be obtained from Table 1 with (11.6). Hence

$$\mathscr{E}'_6 = \{0, t, t^2, t^4, t + t^4, t^2 + t^4, t + t^3 + t^4, t^3 + t^4\}.$$

Accordingly, the colorings are

WWWW, BWWW, WBWW, WWWB, BWWB, WBWB, BWBB, WWBB.

Proof (*of Theorem* 11.2) We use again induction on n. For $n = 1$ and $n = 2$ the colorings correspond to

$$a_{1,t}(\beta_3; z) = 1 + z^t \quad \text{and} \quad a_{1,t}(\beta_4; z) = 1 + z^{t^2} + z^{t^2+t},$$

respectively. Suppose now that the result holds up to $n - 1$ ($n \geq 3$). We get all colorings of the n-board by combining the following:

(1) A W in position 1, followed by the colorings of the $(n - 1)$-board in positions $2, \ldots, n$. This shift by one position is achieved if we replace z by z^t in the polynomial $a_{1,t}(\beta_{n+1}; z)$ which comes from the induction hypothesis. Hence the contribution is $a_{1,t}(\beta_{n+1}; z^t)$.

(2) The pair BW in positions 1, 2, followed by the coloring of the $(n - 2)$-board in positions $3, \ldots, n$. This shift by two positions is achieved if we replace z by z^{t^2} in the polynomial $a_{1,t}(\beta_n; z)$ which comes from the induction hypothesis, while having a B in position 1 is achieved by multiplying this polynomial by z. Hence the contribution in this case is $z\, a_{1,t}(\beta_n; z^{t^2})$.

Finally, adding the polynomials obtained in (1) and (2) and using (11.15), we find that all colorings of the n-board are given by $a_{1,t}(\beta_{n+2}; z)$, which completes the proof.

Remark In analogy to Remark (2) above, we can set $t = 1$ in (11.23). Then Theorem 11.2 implies that the coefficients of z^j, $j \geq 0$, in $a_{1,1}(\beta_{n+2}; z)$ count the number of colorings of the n-board with exactly j black squares. Also, if we set $z = 1$, then (11.13) shows that the total number of such colorings of the n-board is F_{n+2}. Now, it can be shown by induction, using (11.15) with $t = 1$, that

$$a_{1,1}(\beta_{n+2}; z) = 1 + \sum_{j=1}^{\lfloor \frac{n}{2} \rfloor + 1} \left(\binom{n+1-j}{j-1} + \binom{n-1-j}{j} \right) z^j, \tag{11.25}$$

with the convention that $\binom{k}{j} = 0$ whenever $k < j$. We have thus obtained the following consequence of Theorem 11.2.

Corollary 11.1 *For $n \geq 1$, the number of colorings of the n-board, as described above, with exactly j black squares is $\binom{n+1-j}{j-1} + \binom{n-1-j}{j}$.*

3.3 Lattice Paths

In the introduction we already discussed the one-to-one correspondence between the tilings from Sect. 3.1 and a certain class of lattice paths. Thus, with the obvious translation "domino $\leftrightarrow$ two steps up" and "square $\leftrightarrow$ one step to the right" (or with the slight variation of interchanging "up" and "right"), Theorem 11.1 can also be seen as a result on these lattice paths.

Similarly, the colorings of Sect. 3.2 can also be translated to a class of lattice paths, somewhat similar to the ones above. We find it convenient to consider the colored

n-board "from back to front," and we translate a B into an up (U) move by one unit, and a W into a right (R) move by one unit. Thus we consider the following class of lattice paths:

Given an integer $n \geq 1$, an allowable path goes from the origin to a lattice point (x, y), $x \geq 0$ and $y \geq 0$, on the line $x + y = n$ in a sequence of U and R steps under the following conditions:

- Starting with U or UU, or with any number of R;
- the initial steps RU are not allowed;
- apart from the beginning, no two consecutive U steps are allowed.

We have seen in the previous subsection that the total number of such lattice paths is F_{n+2}, and the number of those with exactly j U moves is $\binom{n+1-j}{j-1} + \binom{n-1-j}{j}$. Furthermore, Theorem 11.2, appropriately interpreted, gives each lattice path explicitly. Of course, one could also consider the obvious variants with U and R interchanged, or "reading the colorings from front to back."

4 Delannoy Paths

4.1 Some Basics

An interesting and well-known class of lattice paths are the Delannoy paths, defined as follows. Let $(r, s) \in \mathbb{Z} \times \mathbb{Z}$ be a lattice point with $r \geq 0$ and $s \geq 0$. Then a *Delannoy path* is a lattice path from the origin to (r, s), consisting of a sequence of up $(0, 1)$, right $(1, 0)$, and diagonal $(1, 1)$ steps. The number of Delannoy paths to (r, s) is called the *Delannoy number* $D(r, s)$, and it is known that

$$D(r, s) = \sum_{j=0}^{r} \binom{r}{j}\binom{s}{j} 2^j. \qquad (11.26)$$

For this and other properties see, e.g., [2] or [15], with historical and biographical remarks in [2]; see also [22, A008288] for further properties and references.

A generalization of Delannoy numbers is given by *weighted Delannoy numbers*, where the up, right, and diagonal steps are assigned real or complex weights α, β, γ, respectively. Then the weight of each path is the product of the weights of each of its component steps, and the weighted Delannoy number of (r, s) is the sum of all the weighted Delannoy paths to (r, s). For further details, including the analogue to (11.26), namely,

$$D_{\alpha,\beta,\gamma}(r, s) = \sum_{j=0}^{r} \binom{r}{j}\binom{s}{j} \alpha^{s-j}\beta^{r-j}(\alpha\beta + \gamma)^j, \qquad (11.27)$$

see [16]. Asymptotic expansions for these numbers were recently obtained in [21], and a further generalization was introduced and investigated in [15].

In this section we will consider the special weights $\alpha = 2$ (for `up`), $\beta = 1$ (for `right`), and $\gamma = -1$ (for `diagonal`). With (11.27) we immediately get

$$D_{2,1,-1}(r,s) = \sum_{j=0}^{r} \binom{r}{j}\binom{s}{j} 2^{s-j}. \tag{11.28}$$

In general this is different from (11.26), but when $r = s$, we get by changing the order of summation and comparing with (11.26) that

$$D_{2,1,-1}(r,r) = D(r,r). \tag{11.29}$$

These are the central Delannoy numbers $1, 3, 13, 63, 321, \ldots$ (for $r = 0, 1, \ldots$), which have been particularly well studied (see, e.g., [22, A001850]). For further remarks related to (11.28) and (11.29), see Sect. 6.3 below.

4.2 Connections with Stern Polynomials

In this section we will establish a connection between the special weighted Delannoy numbers (and paths) and the second type of generalized Stern polynomials, $a_{2,t}(m; z)$. In particular, we will be taking $m = \alpha_n$, as defined in (11.11).

We begin with the special case $t = 1$ and note that the polynomials $a_{2,1}(m; z)$, in a different notation, were introduced and studied independently in [3, 27] in connection with hyperbinary expansions; see also [9] for an extension. For $m = \alpha_n$, the case of interest here, these polynomials are easy to compute by alternately using the recurrence relations (11.16) and (11.17). The first few are listed in Table 3.

While in general the degree of $a_{2,t}(m; z)$ is not obvious (see [9, Sect. 4]), here it is easy to show by induction, using (11.16) and (11.17), that

Table 3 $a_{2,1}(\alpha_n; z)$, $1 \le n \le 12$

n	$a_{2,1}(\alpha_n; z)$	n	$a_{2,1}(\alpha_n; z)$
1	1	7	$1 + 3z + 5z^2 + 4z^3$
2	1	8	$1 + 4z + 8z^2 + 8z^3$
3	$1 + z$	9	$1 + 4z + 9z^2 + 12z^3 + 8z^4$
4	$1 + 2z$	10	$1 + 5z + 13z^2 + 20z^3 + 16z^4$
5	$1 + 2z + 2z^2$	11	$1 + 5z + 14z^2 + 25z^3 + 28z^4 + 16z^5$
6	$1 + 3z + 4z^2$	12	$1 + 6z + 19z^2 + 38z^3 + 48z^4 + 32z^5$

$$\deg a_{2,1}(\alpha_n; z) = \left\lfloor \tfrac{n-1}{2} \right\rfloor. \tag{11.30}$$

The following theorem shows the close connection between the special weighted Delannoy numbers and the Stern polynomials $a_{2,1}(\alpha_n; z)$. To fix some notation, we write

$$a_{2,1}(\alpha_n; z) = \sum_{j=0}^{\lfloor \frac{n-1}{2} \rfloor} b_{n,j} z^j. \tag{11.31}$$

Theorem 11.3 *For $n \geq 0$, the special weighted Delannoy numbers for the lattice line $r + s = n$ are given by*

$$D_{2,1,-1}(n-s, s) = b_{2n+2,s}, \qquad 0 \leq s \leq n. \tag{11.32}$$

This result is in fact a special case of a more general result which we will state and prove in the following section. Before we do this, we give an example and then derive some easy consequences.

Example 11.4 Let $n = 3$. Then, taking the weights into account, we have
$D_{2,1,-1}(3,0) = 1 \cdot 1 \cdot 1 = 1,$
$D_{2,1,-1}(2,1) = 1 \cdot 1 \cdot 2 + 1 \cdot 2 \cdot 1 + 2 \cdot 1 \cdot 1 + 1 \cdot (-1) + (-1) \cdot 1 = 4,$
$D_{2,1,-1}(1,2) = 2 \cdot 2 \cdot 1 + 2 \cdot 1 \cdot 2 + 1 \cdot 2 \cdot 2 + 2 \cdot (-1) + (-1) \cdot 2 = 8,$
$D_{2,1,-1}(0,3) = 2 \cdot 2 \cdot 2 = 8.$
This is consistent with $a_{2,1}(\alpha_8; z) = 1 + 4z + 8z^2 + 8z^3$.

As a first consequence of Theorem 11.3 we get the following result by comparing (11.32) with (11.28).

Corollary 11.2 *The coefficients of the polynomial $a_{2,1}(\alpha_{2n}; z)$ are given by*

$$b_{2n,s} = \sum_{j=0}^{n-1-s} \binom{n-1-s}{j} \binom{s}{j} 2^{s-j}, \qquad 0 \leq s \leq n-1. \tag{11.33}$$

For a different explicit expansion, see Corollary 11.8 in Sect. 6.3. Next we use the connection with Fibonacci numbers given by (11.13). Setting $z = 1$ in (11.31), we get the following relation.

Corollary 11.3 *For any $n \geq 0$ we have*

$$\sum_{s=0}^{n} D_{2,1,-1}(n-s, s) = F_{2n+2}. \tag{11.34}$$

We note that the analogous sum for the usual Delannoy numbers gives the sequence of Pell numbers $1, 2, 5, 12, 29, 70, \ldots$, which satisfy the recurrence relation $P_1 = 1$, $P_2 = 2$, and $P_{n+1} = 2P_n + P_{n-1}$ for $n \geq 2$ (see, e.g., [22, A000129]).

The next consequence of Theorem 11.3 is an unexpected connection between the (usual) central Delannoy numbers $D(n, n)$ and the enumeration of hyperbinary expansions. A *hyperbinary expansion* of an integer $n \geq 1$ is an expansion of n as a sum of powers of 2, each power being used at most twice. For instance, the hyperbinary expansions of $n = 12$ are $8+4$, $8+2+2$, $8+2+1+1$, $4+4+2+2$, $4+4+2+1+1$, five in all. The connection with Stern's diatomic sequence has long been known; in [24, Theorem 5.2] it was proved that the number of hyperbinary expansions of an integer $n \geq 2$ is given by $a(n+1)$, where $\{a(n)\}$ is Stern's sequence defined by (11.4). Note that, indeed, we have $a(12+1) = 5$.

Corollary 11.4 *For any $n \geq 1$, the central Delannoy number $D(n, n)$ is equal to the number of hyperbinary expansions of $\frac{4}{3}(2^{4n} - 1)$ that have exactly n repeated powers of 2.*

Example 11.5 Let $n = 1$. Then $\frac{4}{3}(2^{4n} - 1) = 20$, and of the $a(21) = 8$ hyperbinary expansion of 20, exactly 3 have $n = 1$ repeated power of 2, namely $16+2+2$, $16+2+1+1$, and $8+8+4$. This is consistent with $D(1, 1) = 3$.

Proof (*of Corollary* 11.4) Combining (11.29) with (11.32), we get

$$D(n, n) = D_{2,1,-1}(2n - n, n) = b_{4n+2,n}.$$

But this is the coefficient of z^n of the polynomial $a_{2,1}(\alpha_{4n+2}; z)$. This number, in turn, is the number of hyperbinary expansions of $\alpha_{4n+2} - 1$ that have exactly n repeated powers of 2, by [3] or [27]; see also [9]. Now, by (11.11),

$$\alpha_{4n+2} - 1 = \tfrac{1}{3}\left(2^{4n+2} - 1\right) - 1 = \tfrac{4}{3}\left(2^{4n} - 1\right),$$

which completes the proof.

4.3 A Variant

Next we introduce a variant of the special weighted Delannoy numbers defined before (11.28):

As before, we attach the weight 1 to any `right` step, and -1 to any `diagonal` step. To the `up` step from the origin (0, 0) to (0, 1) only, we attach the weight 1, while all other `up` steps will have weight 2, as before.

We denote the corresponding generalized Delannoy number, i.e., the sum of all the weights of the Delannoy paths from the origin to (r, s), by $\widetilde{D}_{2,1,-1}(r, s)$. The following result is analogous to Theorem 11.3; we use again the notation of (11.31).

Theorem 11.4 *For $n \geq 0$, the modified special weighted Delannoy numbers for the lattice line $r + s = n$ are given by*

$$\widetilde{D}_{2,1,-1}(n - s, s) = b_{2n+1,s}, \qquad 0 \leq s \leq n. \tag{11.35}$$

Once again, this follows from a more general result that will be proved later.

Example 11.6 We take again $n = 3$, for comparison with Example 11.4. Taking the modified weights (in bold) into account, we have
$\widetilde{D}_{2,1,-1}(3,0) = 1 \cdot 1 \cdot 1 = 1$,
$\widetilde{D}_{2,1,-1}(2,1) = 1 \cdot 1 \cdot 2 + 1 \cdot 2 \cdot 1 + \mathbf{1} \cdot 1 \cdot 1 + 1 \cdot (-1) + (-1) \cdot 1 = 3$,
$\widetilde{D}_{2,1,-1}(1,2) = \mathbf{1} \cdot 2 \cdot 1 + \mathbf{1} \cdot 1 \cdot 2 + 1 \cdot 2 \cdot 2 + \mathbf{1} \cdot (-1) + (-1) \cdot 2 = 5$,
$\widetilde{D}_{2,1,-1}(0,3) = \mathbf{1} \cdot 2 \cdot 2 = 4$.
This is consistent with $a_{2,1}(\alpha_7; z) = 1 + 3z + 5z^2 + 4z^3$.

The next corollary is actually a consequence of Corollary 11.2, rather than of Theorem 11.4; however, it can be seen as supplementing this theorem.

Corollary 11.5 *The coefficients of the polynomial $a_{2,1}(\alpha_{2n+1}; z)$ are given by*

$$b_{2n+1,s} = \sum_{j=0}^{n-s} \binom{n-s}{j} \left[\binom{s}{j} - \frac{1}{2} \binom{s-1}{j} \right] 2^{s-j}, \qquad 0 \le s \le n. \tag{11.36}$$

Proof From (11.17) we have

$$a_{2,1}(\alpha_{2n+1}; z) = a_{2,1}(\alpha_{2n+2}; z) - z\, a_{2,1}(\alpha_{2n}; z). \tag{11.37}$$

Using (11.31) and equating coefficients of z^j, we get

$$b_{2n+1,s} = b_{2n+2,s} - b_{2n,s-1}, \qquad 1 \le s \le n,$$

having written s instead of j. Finally, using (11.33) we get (11.36) after some easy manipulations.

For different explicit expansions, see Corollary 11.8 and the remark following it in Sect. 6.3. The next corollary and its proof are completely analogous to Corollary 11.3.

Corollary 11.6 *For any $n \ge 0$ we have*

$$\sum_{s=0}^{n} \widetilde{D}_{2,1,-1}(n-s,s) = F_{2n+1}. \tag{11.38}$$

5 Delannoy Paths: Polynomial Weights

5.1 The Basic Case

In this section we continue our study of Delannoy paths and numbers by attaching polynomial weights to the `up` and `diagonal` steps, while leaving all `right` steps with weight 1. The polynomial weights will depend on the starting position of each step. As consequences of the main results of this section we will obtain Theorems 11.3 and 11.4 from Sect. 4.

Given an integer $n \geq 1$, consider all Delannoy paths from the origin $(0, 0)$ to all points on the lattice line $r + s = n$, $0 \leq s \leq n$. We now attach the following weights to the individual steps, where $t \geq 1$ is an integer and z is a variable:

- `up` from (μ, ν) to $(\mu, \nu + 1)$: weight $z^{t^{2(\mu+\nu)}} + z^{t^{2(\mu+\nu)+1}}$;
- `right` from any (μ, ν) to $(\mu + 1, \nu)$: weight 1;
- `diagonal` from (μ, ν) to $(\mu + 1, \nu + 1)$: weight $-z^{t^{2(\mu+\nu)+2}}$.

As usual, we define the weight of an individual Delannoy path as the *product* of the weights of each step, and the weight of a set of paths as the *sum* of the weights of each path.

To motivate the next theorem, we consider a few examples.

Example 11.7 (a) For $n = 1$ there are only two paths:
To $(1, 0)$: weight 1; to $(0, 1)$: weight $z + z^t$.
Total weight: $1 + z + z^t$.
(b) $n = 2$. Here we have multiple steps and paths. The weights are as follows:
To $(2, 0)$: $1 \cdot 1 = 1$.
To $(1, 1)$: $(z + z^t) \cdot 1 + 1 \cdot (z^{t^2} + z^{t^3}) + (-z^{t^2}) = z + z^t + z^{t^3}$.
To $(0, 2)$: $(z + z^t) \cdot (z^{t^2} + z^{t^3}) = z^{t^2+1} + z^{t^2+t} + z^{t^3+1} + z^{t^3+t}$.
Total weight: $1 + z + z^t + z^{t^2+1} + z^{t^2+t} + z^{t^3} + z^{t^3+1} + z^{t^3+t}$.
(c) $n = 3$. The weights are as follows:
To $(3, 0)$: $1 \cdot 1 \cdot 1$.
To $(2, 1)$: $(z + z^t) \cdot 1 \cdot 1 + 1 \cdot (z^{t^2} + z^{t^3}) \cdot 1 + 1 \cdot 1 \cdot (z^{t^4} + z^{t^5}) + (-z^{t^2}) \cdot 1 + 1 \cdot (-z^{t^4})$.
To $(1, 2)$: $(z + z^t) \cdot (z^{t^2} + z^{t^3}) \cdot 1 + (z + z^t) \cdot 1 \cdot (z^{t^4} + z^{t^5}) + 1 \cdot (z^{t^2} + z^{t^3}) \cdot (z^{t^4} + z^{t^5})$

$$+(-z^{t^2}) \cdot (z^{t^4} + z^{t^5}) + (z + z^t) \cdot (-z^{t^4}).$$

To $(0, 3)$: $(z + z^t) \cdot (z^{t^2} + z^{t^3}) \cdot (z^{t^4} + z^{t^5})$.
In this case we refrain from expanding the weights, which would give a 21-term polynomial.

By comparing the total weights for $n = 1$ and $n = 2$ with the right column of Table 1, we see that they are $a_{2,t}(5; z) = a_{2,t}(\alpha_4; z)$ and $a_{2,t}(21; z) = a_{2,t}(\alpha_6; z)$, respectively. Furthermore, after expanding and adding the terms for $n = 3$, we could verify that the total weight is $a_{2,t}(85; z) = a_{2,t}(\alpha_8; z)$. This is in fact true in general, as the following result shows. Recall that $a_{2,t}(m; z)$ is the type-2 generalized Stern polynomial defined in (11.7), (11.8), and α_n is defined in (11.11). Also recall the recurrence relations (11.16), (11.17).

Theorem 11.5 *Let $n \geq 1$ and consider all Delannoy paths from $(0, 0)$ to all lattice points (r, s) on the line segment $r + s = n$, $0 \leq s \leq n$. Then with the polynomial weights as described above, the total weight is $a_{2,t}(\alpha_{2n+2}; z)$.*

For the proof we require the first part of the following lemma. To simplify notation, we set

$$f_n(z) := a_{2,t}(\alpha_n; z). \tag{11.39}$$

Lemma 11.1 *For integers $n \geq 1$ we have*

$$f_{2n+2}(z) = (z + z^t) f_{2n}(z^{t^2}) + \sum_{j=1}^{n-1} z^{t^{2j+1}} f_{2n-2j}(z^{t^{2j+2}}) + 1, \tag{11.40}$$

$$f_{2n+1}(z) = z\, f_{2n}(z^t) + \sum_{j=1}^{n-1} z^{t^{2j}} f_{2n-2j}(z^{t^{2j+1}}) + 1. \tag{11.41}$$

Proof Replacing n by $n - j$ and z by $z^{t^{2j+1}}$ $(j = 0, 1, \dots, n-1)$ in (11.16), we get with (11.39),

$$f_{2n-2j+1}(z^{t^{2j+1}}) = z^{t^{2j+1}} f_{2n-2j}(z^{t^{2j+2}}) + f_{2n-2j-1}(z^{t^{2j+3}}). \tag{11.42}$$

Taking $j = 0$ and substituting the resulting identity into (11.17), we get

$$f_{2n+2}(z) = (z + z^t) f_{2n}(z^{t^2}) + f_{2n-1}(z^{t^3}). \tag{11.43}$$

Next we substitute (11.42) with $j = 1$ into (11.43), then (11.42) with $j = 2$ into the resulting identity, and so on, thus obtaining (11.40) with the final term $f_1(z^{t^{2n+1}}) = a_{2,t}(1, z^{t^{2n+1}}) = 1$ (see Table 1), where we have used the fact that $\alpha_1 = 1$.

The proof of (11.41) is similar: We use again (11.16), replacing n by $n - j$ and this time z by $z^{t^{2j}}$ $(j = 0, 1, \dots, n-1)$. Then we have

$$f_{2n-2j+1}(z^{t^{2j}}) = z^{t^{2j}} f_{2n-2j}(z^{t^{2j+1}}) + f_{2n-2j-1}(z^{t^{2j+2}}).$$

We start with $j = 0$ and consecutively substitute the corresponding identities for $j = 1, \dots, n-1$, thus obtaining (11.41). As before, the final term reduces to 1.

Proof (*of Theorem* 11.5) We use induction on n and recall that the cases $n = 1$ and $n = 2$ were explicitly shown in Example 11.7. To these we may add the case $n = 0$ if we interpret the empty path as having weight 1; this is then consistent with $f_2(z) = 1$.

Before we continue with the induction, we fix some notation. For lattice points A, B, C, let $\Delta[A, B, C]$ be the triangle with corners A, B, C. Then all the Delannoy paths from $(0, 0)$ to the lattice line segment $r + s = n$, $0 \leq s \leq n$, lie inside the right-angled isosceles triangle

$$\Delta := \Delta[(0, 0), (n, 0), (0, n)].$$

We also consider the following right-angled isosceles triangles contained in Δ, namely

$$\Delta_j := \Delta[(j,1),(n-1,1),(j,n-j)], \qquad j=0,1,\dots,n-1.$$

The hypotenuse of each Δ_j forms a section of the lattice line segment in question, namely those lattice points (r,s) with $r+s=n$, $1\le s\le n-j$.

To continue with the induction, we assume that the statement of the theorem holds up to some $n-1$, for $n\ge 3$. By this hypothesis, if for some j, $0\le j\le n-1$, the triangle Δ_j had its lower left corner at the origin, then the total weight of all Delannoy paths in Δ_j (from the origin) would be $f_{2n-2j}(z)$. However, Δ_j is one step up and j steps to the right from this position. By the definition of the polynomial weights this means that the shift amounts to replacing z by $z^{t^{2j+2}}$, i.e., the Delannoy paths in Δ_j, starting in its lower left corner $(j,1)$, have total weight $f_{2n-2j}(z^{t^{2j+2}})$.

For the induction step we now note that all Delannoy paths in Δ that start at $(0,0)$ can be obtained as follows:

(a) One up step, followed by any path in Δ_0;
(b) for j from 1 to $n-1$:

   (i) j right steps, followed by an up step, or
   (ii) $j-1$ right steps, followed by a diagonal step, and
   (iii) the steps in (i) and (ii) followed by any path in Δ_j;

(c) n consecutive right steps.

By induction hypothesis and the definition of the weights, these steps translate into polynomials as follows:

(a) $(z+z^t)f_{2n}(z^{t^2})$,
(b) $\left((z^{t^{2j}}+z^{t^{2j+1}})+(-z^{t^{2j}})\right)f_{2n-2j}(z^{t^{2j+2}}), \quad 1\le j\le n-1,$
(c) add 1.

We note that the sum of these terms is the right-hand side of (11.40), which equals $f_{2n+2}(z)$. But this was to be shown.

We are now ready to prove Theorem 11.3. We do so by setting $t=1$ in Theorem 11.5. Then the polynomial weights introduced at the beginning of this section reduce to the following weights, which are now independent of the starting point:

- up: weight $2z$;
- right: weight 1;
- diagonal: weight $-z$.

This means that going from one (horizontal) lattice row to the next, and only then, the power of z increases by 1. Hence the paths from $(0,0)$ to (r,s) for a fixed s, $r+s=n$ and $0\le s\le n$, are exactly those whose weights are multiples of z^s. In the notation of (11.31), the sum of these weights is then $b_{2n+2,s}z^s$, which was to be shown. This completes the proof of Theorem 11.3.

5.2 A Modification

In the second part of this section we consider a modified version of the polynomial weights introduced at the beginning of this section. Given an integer $n \geq 1$, we consider again all Delannoy paths from the origin $(0, 0)$ to all lattice points on the line $r + s = n, 0 \leq s \leq n$. We attach the following modified weights to the individual steps, where once again $t \geq 1$ is an integer:

- `up` from $(0, 0)$ to $(0, 1)$: weight z;
- `up` from (μ, ν) to $(\mu, \nu + 1)$, $(\mu, \nu) \neq (0, 0)$: weight $z^{t^{2(\mu+\nu)-1}} + z^{t^{2(\mu+\nu)}}$;
- `right` from any (μ, ν) to $(\mu + 1, \nu)$: weight 1;
- `diagonal` from (μ, ν) to $(\mu + 1, \nu + 1)$: weight $-z^{t^{2(\mu+\nu)+1}}$.

In analogy to Example 11.7 we consider the first few cases:

Example 11.8 (a) $n = 1$: To $(1, 0)$: weight 1; to $(0, 1)$: weight z.
Total weight: $1 + z$.
(b) $n = 2$. The weights are as follows:
To $(2, 0)$: $1 \cdot 1 = 1$.
To $(1, 1)$: $z \cdot 1 + 1 \cdot (z^t + z^{t^2}) + (-z^t) = z + z^{t^2}$.
To $(0, 2)$: $z \cdot (z^t + z^{t^2}) = z^{t+1} + z^{t^2+1}$.
Total weight: $1 + z + z^{t+1} + z^{t^2} + z^{t^2+1}$.
(c) $n = 3$. The weights are as follows:
To $(3, 0)$: $1 \cdot 1 \cdot 1$.
To $(2, 1)$: $z \cdot 1 \cdot 1 + 1 \cdot (z^t + z^{t^2}) \cdot 1 + 1 \cdot 1 \cdot (z^{t^3} + z^{t^4}) + (-z^t) \cdot 1 + 1 \cdot (-z^{t^3})$.
To $(1, 2)$: $z \cdot (z^t + z^{t^2}) \cdot 1 + z \cdot 1 \cdot (z^{t^3} + z^{t^4}) + 1 \cdot (z^t + z^{t^2}) \cdot (z^{t^3} + z^{t^4})$
$\qquad +(-z^t) \cdot (z^{t^3} + z^{t^4}) + z \cdot (-z^{t^3})$.
To $(0, 3)$: $z \cdot (z^t + z^{t^2}) \cdot (z^{t^3} + z^{t^4})$.

By comparing again the total weights for $n = 1$ and $n = 2$ with the right column of Table 1, we see that this time they are $a_{2,t}(3; z) = a_{2,t}(\alpha_3; z)$ and $a_{2,t}(11; z) = a_{2,t}(\alpha_5; z)$, respectively. Also, after expanding and adding the terms for $n = 3$, it is easy to verify that the total weight in this case is $a_{2,t}(43; z) = a_{2,t}(\alpha_7; z)$. These are special cases of the following result, which is analogous to Theorem 11.5.

Theorem 11.6 *Let $n \geq 1$ and consider all Delannoy paths from $(0, 0)$ to all lattice points (r, s) on the line segment $r + s = n, 0 \leq s \leq n$. Then with the modified polynomial weights described above, the total weight is $a_{2,t}(\alpha_{2n+1}; z)$.*

Proof We use the same notation as in the proof of Theorem 11.5, as well as a similar approach. However, instead of induction we use the assertion of Theorem 11.5.

The proof is again based on the fact that all Delannoy paths in Δ that start at $(0, 0)$ can be obtained exactly as stated in steps (a)–(c) in the proof of Theorem 11.5. By the definition of the modified weights, this now translates into polynomials as follows:

(a) $z\, f_{2n}(z^{t^2})$,
(b) $\left((z^{t^{2j-1}} + z^{t^{2j}}) + (-z^{t^{2j-1}})\right) f_{2n-2j}(z^{t^{2j+2}}), \quad 1 \leq j \leq n - 1,$

(c) add 1.

This time the sum of these terms is the right-hand side of (11.41). But this equals $f_{2n+1}(z)$, which was to be shown.

Finally in this section, we note that Theorem 11.4 follows from Theorem 11.6 in exactly the same way as Theorem 11.3 follows from Theorem 11.5; see the proof above. The one obvious adjustment concerns the `up` step from (0, 0) to (0, 1) only, which has weight z as opposed to $2z$ for all `up` steps in the case of Theorem 11.5.

6 Further Remarks

6.1 Other Weights

Given the similarities between the sequences $\{\alpha_n\}$ and $\{\beta_n\}$ defined in (11.11), including the relations (11.13) and (11.18), (11.19), it is natural to ask whether there are results similar to Theorems 11.3–11.6 that involve the polynomials $a_{2,t}(\beta_n; z)$. This is indeed the case, as we will now briefly describe. We restrict ourselves to analogues of numerical weights, as in Sect. 4.

First, given an integer $n \geq 1$, consider again the Delannoy paths from (0, 0) to any point on the line $r + s = n$, $0 \leq s \leq n$. This time we attach the following weights:

- `up`: weight 2 for any step $(\mu, \nu) \to (\mu, \nu + 1)$, $\mu + \nu < n - 1$,
- `up`: weight 1 for any step $(\mu, \nu) \to (\mu, \nu + 1)$, $\mu + \nu = n - 1$,
- `right`: weight 1 for all steps,
- `diagonal`: weight -1 for all steps.

In other words, the weights are the same as in the case of Theorem 11.3, with the exception that any final `up` step before reaching a target lattice point has weight 1, instead of 2. We denote the associated weighted Delannoy numbers by $\mathfrak{D}_{2,1,-1}(r, s)$.

Next, we consider the same modification as in Theorem 11.4, i.e., we assign the weight 1 (rather than 2) to the `up` step $(0, 0) \to (0, 1)$, and denote the corresponding weighted Delannoy number by $\widetilde{\mathfrak{D}}_{2,1,-1}(r, s)$.

Finally, we note that $\deg a_{2,1}(\beta_n; z) = \lfloor \frac{n}{2} \rfloor$, which follows by induction from (11.18) and (11.19). With the notation

$$a_{2,1}(\beta_n; z) = \sum_{s=0}^{\lfloor \frac{n}{2} \rfloor} c_{n,s} z^s, \tag{11.44}$$

we can now state the following result.

Theorem 11.7 *For $n \geq 0$, the weighted Delannoy numbers defined above, for the lattice line $r + s = n$ with $0 \leq s \leq n$, are given by*

$$\mathfrak{D}_{2,1,-1}(n-s,s) = c_{2n+1,s}, \qquad \widetilde{\mathfrak{D}}_{2,1,-1}(n-s,s) = c_{2n,s}. \tag{11.45}$$

We leave the proofs of these identities to the reader. They can be done by induction, using the triangles in the proof of Theorem 11.5. Alternatively, analogues of Theorems 11.5 and 11.6 could also be derived, based on an analogue of Lemma 11.1 which, in turn, would easily follow from the identities (11.18) and (11.19). For explicit formulas, see Corollary 11.8 below.

6.2 A Stern Polynomial Identity

In general we have $a_{\varepsilon,t}(\alpha_n; z) \neq a_{\varepsilon,t}(\beta_n; z)$ for $\varepsilon \in \{1, 2\}$ and $t \in \mathbb{N}$, which can be seen by considering the relevant entries in Tables 1 and 2. It is therefore somewhat surprising that for the type-2 polynomials and $t = 1$ we have the following identity.

Lemma 11.2 *For all $n \in \mathbb{N}$ we have*

$$a_{2,1}(\alpha_{2n+1}; z) = a_{2,1}(\beta_{2n+1}; z). \tag{11.46}$$

Proof We show that the two sides of (11.46) satisfy the same recurrence relation, with the same initial values. The identity (11.16) with $t = 1$ gives

$$a_{2,1}(\alpha_{2n+1}; z) = z\, a_{2,1}(\alpha_{2n}; z) + a_{2,1}(\alpha_{2n-1}; z). \tag{11.47}$$

We replace n by $n - 1$ in (11.47), multiply both sides by z, and subtract it from (11.47), obtaining

$$\begin{aligned} a_{2,1}(\alpha_{2n+1}; z) = z\, a_{2,1}(\alpha_{2n}; z) &+ (1 + z)a_{2,1}(\alpha_{2n-1}; z) \\ &- z^2 a_{2,1}(\alpha_{2n-2}; z) - z\, a_{2,1}(\alpha_{2n-3}; z). \end{aligned} \tag{11.48}$$

Now (11.17), with $t = 1$ and both side multiplied by z, gives

$$z\, a_{2,1}(\alpha_{2n}; z) - z^2 a_{2,1}(\alpha_{2n-2}; z) = z\, a_{2,1}(\alpha_{2n-1}; z).$$

Subtracting this from (11.48), we get for $n \geq 2$,

$$a_{2,1}(\alpha_{2n+1}; z) = (1 + 2z)a_{2,1}(\alpha_{2n-1}; z) - z\, a_{2,1}(\alpha_{2n-3}; z). \tag{11.49}$$

Using (11.18) and (11.19) with similar manipulations, we can see that the recurrence relation (11.49) also holds for α replaced by β, for $n \geq 3$. Finally we note that $\alpha_3 = \beta_3 = 3$, and that $a_{2,1}(\alpha_5; z) = a_{2,1}(11; z) = 1 + 2z + 2z^2$, $a_{2,1}(\beta_5; z) = a_{2,1}(13; z) = 1 + 2z + 2z^2$ (see Table 1). This completes the proof.

Lemma 11.2 has an interesting application: With (11.31) and (11.35) on the one hand, and (11.44) and (11.45) on the other hand, we obtain the following consequence of (11.46).

Theorem 11.8 *For $n \in \mathbb{N}$ the modified weighted Delannoy numbers $\widetilde{D}_{2,1,-1}(r,s)$ and $\mathfrak{D}_{2,1,-1}(r,s)$, defined in Sects. 4.3 and 6.1, respectively, satisfy*

$$\widetilde{D}_{2,1,-1}(n-s,s) = \mathfrak{D}_{2,1,-1}(n-s,s), \qquad 0 \le s \le n.$$

We illustrate this with the following example, which should be compared with Examples 11.4 and 11.6.

Example 11.9 Once again we take $n = 3$. The modified weights are again in bold, and by the previous subsection we have
$\mathfrak{D}_{2,1,-1}(3,0) = 1 \cdot 1 \cdot 1 = 1,$
$\mathfrak{D}_{2,1,-1}(2,1) = 1 \cdot 1 \cdot \mathbf{1} + 1 \cdot 2 \cdot 1 + 2 \cdot 1 \cdot 1 + 1 \cdot (-1) + (-1) \cdot 1 = 3,$
$\mathfrak{D}_{2,1,-1}(1,2) = 2 \cdot 2 \cdot 1 + 2 \cdot 1 \cdot \mathbf{1} + 2 \cdot 1 \cdot \mathbf{1} + 2 \cdot (-1) + (-1) \cdot \mathbf{1} = 5,$
$\mathfrak{D}_{2,1,-1}(0,3) = 2 \cdot 2 \cdot \mathbf{1} = 4.$
This is consistent with Example 11.6.

6.3 Connections with Jacobi Polynomials

The special weighted Delannoy numbers that were the subject of Theorem 11.3 have occurred before in the literature as *asymmetric Delannoy numbers*; see [17]. Various properties were derived in [17], including (11.29) in a different notation. That paper also contains an alternative explicit expansion similar to (11.28), and the following connection with Jacobi polynomials, again in a different notation:

$$D_{2,1,-1}(n-s,s) = P_s^{(0,n-2s)}(3). \tag{11.50}$$

The Jacobi polynomials, a general class of classical orthogonal polynomials, belong to the most important special functions in mathematics; see, e.g., [23, Ch. 18]. Here we only use the following explicit expressions (see, e.g., [23, Eq. 18.5.7–8]):

$$P_k^{(a,b)}(x) = \sum_{j=0}^{k} \binom{k+a+b+j}{j}\binom{k+a}{k-j}\left(\tfrac{x-1}{2}\right)^j, \tag{11.51}$$

and the more symmetric identity

$$P_k^{(a,b)}(x) = \sum_{j=0}^{k} \binom{k+a}{j}\binom{k+b}{k-j}\left(\tfrac{x+1}{2}\right)^j\left(\tfrac{x-1}{2}\right)^{k-j}, \tag{11.52}$$

as well as the recurrence relation

$$P_k^{(a,b)}(x) - P_{k-1}^{(a,b+1)}(x) = P_k^{(a-1,b+1)}(x); \tag{11.53}$$

see, e.g., [23, Eq. 18.9.3]. We are now ready to state and prove the following results.

Theorem 11.9 *For any integers $n \geq 0$ we have*

$$a_{2,1}(\alpha_{2n+2}; z) = \sum_{k=0}^{n} P_k^{(0,n-2k)}(3) \cdot z^k, \tag{11.54}$$

$$a_{2,1}(\alpha_{2n+1}; z) = \sum_{k=0}^{n} P_k^{(-1,n+1-2k)}(3) \cdot z^k, \tag{11.55}$$

$$a_{2,1}(\beta_{2n}; z) = \sum_{k=0}^{n} P_k^{(-2,n+2-2k)}(3) \cdot z^k. \tag{11.56}$$

Recall that by (11.46) we have $a_{2,1}(\beta_{2n+1}; z) = a_{2,1}(\alpha_{2n+1}; z)$, which can be seen as supplementing (11.54)–(11.56).

Proof (*of Theorem* 11.9) The identity (11.54) is an immediate consequence of (11.50), together with (11.31) and (11.32). To obtain (11.55), we use (11.37) and (11.54) to write

$$\begin{aligned} a_{2,1}(\alpha_{2n+1}; z) &= a_{2,1}(\alpha_{2n+2}; z) - z a_{2,1}(\alpha_{2n}; z) \\ &= \sum_{k=0}^{n} P_k^{(0,n-2k)}(3) \cdot z^k - \sum_{k=0}^{n-1} P_k^{(0,n-1-2k)}(3) \cdot z^{k+1} \\ &= P_0^{(0,n)}(3) + \sum_{k=1}^{n} \left(P_k^{(0,n-2k)}(3) - P_{k-1}^{(0,n+1-2k)}(3) \right) z^k. \end{aligned}$$

The desired identity (11.55) now follows from (11.53) and the fact that $P_0^{(a,b)}(x) = 1$, independent of the parameters a, b.

To obtain (11.56), we combine the identity (11.18) for $t = 1$ with (11.46), obtaining

$$a_{2,1}(\beta_{2n}; z) = a_{2,1}(\alpha_{2n+1}; z) - z a_{2,1}(\alpha_{2n-1}; z);$$

we then get (11.56) in the same way as in the previous paragraph, this time using (11.55).

Finally we note that some intermediate steps, such as the identity (11.18), are valid only for $n \geq 2$. However, the cases $n = 0$ and $n = 1$ are easy to verify by direct computation, using Tables 1 and 2.

As an immediate consequence we get the following identities, where we include (11.50) for completeness and use the notations of (11.32), (11.35) and (11.45).

Corollary 11.7 *For all integers $n \geq 0$ and $0 \leq s \leq n$ we have*

$$D_{2,1,-1}(n-s,s) = P_s^{(0,n-2s)}(3),$$
$$\widetilde{D}_{2,1,-1}(n-s,s) = P_s^{(-1,n+1-2s)}(3),$$
$$\widetilde{\mathfrak{D}}_{2,1,-1}(n-s,s) = P_s^{(-2,n+2-2s)}(3).$$

To conclude this section, we state some explicit expansions, which follow immediately from (11.51) and Corollary 11.7.

Corollary 11.8 *For all integers $n \geq 0$ and $0 \leq s \leq n$ we have*

$$D_{2,1,-1}(n-s,s) = \sum_{j=0}^{s} \binom{n-s+j}{j}\binom{s}{s-j},$$
$$\widetilde{D}_{2,1,-1}(n-s,s) = \sum_{j=0}^{s} \binom{n-s+j}{j}\binom{s-1}{s-j},$$
$$\widetilde{\mathfrak{D}}_{2,1,-1}(n-s,s) = \sum_{j=0}^{s} \binom{n-s+j}{j}\binom{s-2}{s-j}.$$

Analogous identities can be obtained from (11.52); we leave this to the reader. For instance, the identity (11.33) can be obtained in this way.

References

1. Amdeberhan, T., Chen, X., Moll, V.H., Sagan, B.E.: Generalized Fibonacci polynomials and Fibonomial coefficients. Ann. Comb. **18**(4), 541–562 (2014)
2. Banderier, C., Schwer, S.: Why Delannoy numbers? J. Stat. Plan. Inference **135**(1), 40–54 (2005)
3. Bates, B., Mansour, T.: The q-Calkin-Wilf tree. J. Combin. Theory Ser. A **118**, 1143–1151 (2011)
4. Benjamin, A.T, Quinn, J.J.: Proofs that Really Count. The Art of Combinatorial Proof. The Dolciani Mathematical Expositions, vol. 27. Mathematical Association of America, Washington (2003)
5. Calkin, C., Wilf, H.S.: Recounting the rationals. Am. Math. Mon. **107**(4), 360–363 (2000)
6. Carlitz, L.: A problem in partitions related to the Stirling numbers. Bull. Am. Math. Soc. **70**, 275–278 (1964)
7. Cigler, J.: Some remarks and conjectures related to lattice paths in strips along the x-axis (2015). http://arxiv.org/pdf/1501.04750.pdf
8. Dilcher, K., Ericksen, L.: Reducibility and irreducibility of Stern (0, 1)-polynomials. Commun. Math. **22**, 77–102 (2014)
9. Dilcher, K., Ericksen, L.: Hyperbinary expansions and Stern polynomials. Electron. J. Combin. **22**(2), 18 (2015). Article 2.24
10. Dilcher, K., Ericksen, L.: Factors and irreducibility of generalized Stern polynomials. Integers **15**, 17 (2015). Article A50
11. Dilcher, K., Ericksen, L.: Generalized Stern polynomials and hyperbinary representations. Bull. Pol. Acad. Sci. Math. **65**, 11–28 (2017)

12. Dilcher, K., Stolarsky, K.B.: A polynomial analogue to the Stern sequence. Int. J. Number Theory **3**, 85–103 (2007)
13. Dilcher, K., Stolarsky, K.B.: Stern polynomials and double-limit continued fractions. Acta Arith. **140**, 119–134 (2009)
14. Dilcher, K., Kidwai, M., Tomkins, H.: Zeros and irreducibility of Stern polynomials. Publ. Math. Debrecen **90**(3–4), 407–433 (2017)
15. Dziemiańczuk, M.: Generalizing Delannoy numbers via counting weighted lattice paths. Integers **13**, 3 (2013). Article A54
16. Fray, R.D., Roselle, D.P.: Weighted lattice paths. Pac. J. Math. **37**, 85–96 (1971)
17. Hetyei, G.: Central Delannoy numbers and balanced Cohen-Macaulay complexes. Ann. Comb. **10**(4), 443–462 (2006)
18. Klavžar, S., Milutinović, U., Petr, C.: Stern polynomials. Adv. Appl. Math. **39**, 86–95 (2007)
19. Lehmer, D.H.: On Stern's diatomic series. Am. Math. Mon. **36**, 59–67 (1929)
20. Lucas, É.: Théorie des fonctions numériques simplement périodiques. Am. J. Math. **1**, 184–196; 197–240; 289–321 (1878). English translation http://www.fq.math.ca/Books/Complete/simply-periodic.pdf
21. Noble, R.: Asymptotics of the weighted Delannoy numbers. Int. J. Number Theory **8**(1), 175–188 (2012)
22. OEIS Foundation Inc. (2011): The On-Line Encyclopedia of Integer Sequences. http://oeis.org
23. Olver, F.W.J., et al. (eds.): NIST Handbook of Mathematical Functions. Cambridge University Press, New York (2010)
24. Reznick, B.: Some binary partition functions. In: Berndt, B.C., et al. (eds.) Analytic Number Theory, Proceedings of a conference in honor of Paul T. Bateman, pp. 451–477. Birkhäuser, Boston (1990)
25. Schinzel, A.: On the factors of Stern polynomials (remarks on the preceding paper of M. Ulas). Publ. Math. Debrecen **79**, 83–88 (2011)
26. Schinzel, A.: Stern polynomials as numerators of continued fractions. Bull. Pol. Acad. Sci. Math. **62**(1), 23–27 (2014)
27. Stanley, R.P., Wilf, H.S.: Refining the Stern diatomic sequence. Preprint (2010). Electronically available at http://www-math.mit.edu/~rstan/papers/stern.pdf
28. Ulas, M.: On certain arithmetic properties of Stern polynomials. Publ. Math. Debrecen **79**(1–2), 55–81 (2011)
29. Ulas, M.: Arithmetic properties of the sequence of degrees of Stern polynomials and related results. Int. J. Number Theory **8**, 669–687 (2012)

p-Rook Numbers and Cycle Counting in $C_p \wr S_n$

James Haglund, Jeffrey B. Remmel and Meesue Yoo

Abstract Cycle-counting rook numbers were introduced by Chung and Graham [*J. Combin. Theory Ser. B* **65** (1995), 273–290]. Cycle-counting q-rook numbers were introduced by Ehrenborg, Haglund, and Readdy [unpublished] and cycle-counting q-hit numbers were introduced by Haglund [*Adv. Appl. Math.* **17** (1996), 408–459]. Briggs and Remmel [*J. Combin. Theory Ser. A* **113** (2006), 1138–1171] introduced the theory of p-rook and p-hit numbers which is a rook theory model where the rook numbers correspond to partial permutations in $C_p \wr S_n$, the wreath product of the cyclic group C_p and the symmetric group S_n, and the hit numbers correspond to permutations in $C_p \wr S_n$. In this paper, we extend the cycle-counting q-rook numbers and cycle-counting q-hit numbers to the Briggs–Remmel model. In such a setting, we define a multivariable version of the cycle-counting q-rook numbers and cycle-counting q-hit numbers where we keep track of cycles of permutations and partial permutations of $C_p \wr S_n$ according to the signs of the cycles.

Keywords Rook numbers · Hit numbers · Cycle-counting rook numbers · Cycle-counting hit numbers · Wreath product

2010 Mathematics Subject Classification Primary 05A15 · Secondary 05E05

J. Haglund
Department of Mathematics, University of Pennsylvania, Philadelphia, PA 19104, USA
e-mail: jhaglund@math.upenn.edu

J. B. Remmel
Department of Mathematics, University of California, San Diego, La Jolla, CA 92093-0112, USA
e-mail: jremmel@ucsd.edu

M. Yoo (✉)
Applied Algebra and Optimization Research Center, Sungkyunkwan University, Suwon 16419, Republic of Korea
e-mail: meesue.yoo@skku.edu

G. E. Andrews et al. (eds.), *Lattice Path Combinatorics and Applications*, Developments in Mathematics 58, https://doi.org/10.1007/978-3-030-11102-1_12

1 Introduction

We let $[n] = \{1, \ldots, n\}$. We let $\mathbb{N} = \{0, 1, 2, \ldots\}$ denote the natural numbers and $\mathbb{P} = \{1, 2, \ldots\}$ denote the positive integers. A *board* is a subset of $\mathbb{P} \times \mathbb{P}$. We label the rows of $\mathbb{P} \times \mathbb{P}$ from bottom to top with $1, 2, 3, \ldots$, and the columns of $\mathbb{P} \times \mathbb{P}$ from left to right with $1, 2, 3, \ldots$, and (i, j) denote the square in the ith column and jth row. Given $b_1, \ldots, b_n \in \mathbb{N}$, we let $F(b_1, \ldots, b_n)$ denote the board consisting of all the cells $\{(i, j) : 1 \leq i \leq n \text{ and } 1 \leq j \leq b_i\}$. If a board B is of the form $B = F(b_1, \ldots, b_n)$, then we say that B is *skyline board* and if, in addition, $b_1 \leq b_2 \leq \cdots \leq b_n$, then we say that B is a *Ferrers board*.

Given a board $B \subseteq [n] \times [n]$, we let $\mathcal{N}_k(B)$ denote the set of all placements of k rooks in B such that no two rooks lie in the same row or column. Elements of $\mathcal{N}_k(B)$ will be called *rook placements*. For $k = 1, \ldots, n$, we let $r_k(B) = |\mathcal{N}_k(B)|$. By convention, we set $r_0(B) = 1$. We refer to $r_k(B)$ as the ***k*th rook number of** B.

Let S_n denote the symmetric group of n elements, i.e. the group of all permutations of $1, \ldots, n$ under composition. Given a permutation $\sigma = \sigma_1 \cdots \sigma_n \in S_n$, we identify each $\sigma \in S_n$ with the rook placement $\{(i, \sigma_i) : i = 1, \ldots, n\}$ on $[n] \times [n]$. We let

$$H_{k,n}(B) = |\{\sigma \in S_n : |\sigma \cap B| = k\}|.$$

We shall refer to $H_{k,n}(B)$ as the **k-th hit number of B relative to** $[n] \times [n]$.

Kaplansky and Riordan [13] proved the following fundamental relationship between the rook numbers and the hit numbers of a board $B \subseteq [n] \times [n]$.

Theorem 12.1 *For any board $B \subseteq [n] \times [n]$,*

$$\sum_{k=0}^{n} H_{k,n}(B)x^k = \sum_{k=0}^{n} r_k(B)(n-k)!(x-1)^k. \tag{12.1}$$

With each rook placement $P \in \mathcal{N}_k(B)$, we can associate a directed graph $G_P = ([n], E_P)$, where E_P is the set of (i, j) such that P has a rook in cell (i, j). We let $\text{cyc}(P)$ denote the number of cycles in the graph of P. For example, in Fig. 1, we picture a rook placement $P \in \mathcal{N}_5(B)$, where B is the 6×6 board such that $\text{cyc}(P) = 2$.

For any board $B \subseteq [n] \times [n]$, we let

$$r_k(B, y) = \sum_{P \in \mathcal{N}_k(B)} y^{\text{cyc}(P)} \text{ and}$$

$$H_{k,n}(B, y) = \sum_{\sigma \in S_n,\ |\sigma \cap B| = k} y^{\text{cyc}(P)}.$$

For $k \geq 1$, we let $(y)\uparrow_k = y(y+1)\cdots(y+k-1)$ and $(y)\downarrow_k = y(y-1)\cdots(y-k+1)$. We let $(y)\uparrow_0 = (y)\downarrow_0 = 1$. We then have the following analogue of Theorem 12.1.

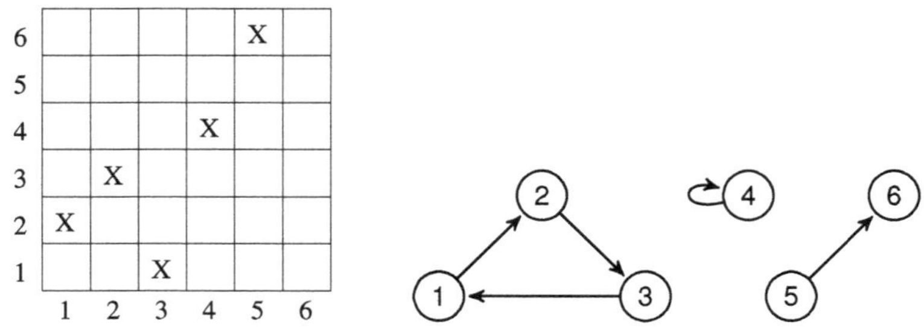

Fig. 1 Graph associated with a rook placement

Theorem 12.2 *For any board $B \subseteq [n] \times [n]$,*

$$\sum_{k=0}^{n} H_{k,n}(B, y)x^k = \sum_{k=0}^{n} r_k(B, y)(y) \uparrow_{n-k} (x-1)^k. \tag{12.2}$$

Proof First replace x by $x + 1$ in Eq. (12.2). Then, we must prove

$$\sum_{k=0}^{n} H_{k,n}(B, y)(x+1)^k = \sum_{k=0}^{n} r_k(B, y)(y) \uparrow_{n-k} x^k. \tag{12.3}$$

For (12.3), we consider configurations C which consist of a rook placement corresponding to a permutation $\sigma \in S_n$, where we circle some of the rooks that fall in $B \cap \sigma$. We let $\text{cyc}(C)$ denote the number of cycles in the graph of the underlying permutation of C and $\text{circle}(C)$ denote the number of circled rooks in C. It is then easy to see that the left-hand side of (12.3) can be interpreted as counting $y^{\text{cyc}(C)}x^{\text{circle}(C)}$ over all such configurations. The right-hand side of (12.3) can be interpreted as follows. First pick the circled rooks which correspond to a placement $Q \in \mathcal{N}_k(B)$ for some k. Then, we need to compute

$$A(Q, y) = \sum_{C} y^{\text{cyc}(C)}, \tag{12.4}$$

where the sum runs over all configurations whose set of circled rooks equals Q. This sum is easy to compute. That is, let i be the first column that does not contain a rook in Q. Then, there are $n - k$ rows to place a rook in column i that do not contain rooks in Q. We claim that there is exactly one row r where placing a rook in cell (i, r) completes a cycle in the graph of Q. That is, if there is no rook in Q which is in row i, then i is an isolated vertex in the graph of Q, so adding a rook in the cell (i, i) will give a loop on vertex i and hence increase the number of cycles by 1. Clearly, in such a situation, placing a rook in cell (i, j) for $j \neq i$ cannot complete a cycle. If there is a rook of Q in row i, then there must be a maximal length path p in the graph of Q which ends in vertex i since there are no edges coming out of the vertex i in the graph of Q. If this path starts in vertex j, then there is no rook in row

j in Q. Hence, if we add a rook to the cell (i, j), then the edge corresponding to the added rook will complete a cycle. Clearly, adding a rook to any other row in column i will not complete a cycle in this case. Thus, the placement of a rook in column i will contribute a factor $(y+n-k-1)$ to $A(Q, y)$. But then we can repeat the argument for every placement Q' which arises from Q by adding a rook in the next empty column, say column i_1. That is, for each such Q', the addition of a rook in column i_1 will contribute a factor $(y+n-k-2)$ to $A(Q, y)$. Continuing on in this way, we see that

$$A(Q, y) = (y+n-k-1)(y+n-k-2)\cdots(y) = (y)\uparrow_{n-k}.$$

Thus, another way to sum $y^{\mathrm{cyc}(C)}x^{\mathrm{circle}(C)}$ over all rook configurations is

$$\begin{aligned}
&\sum_{k=0}^{n} x^k \sum_{Q\in\mathcal{N}_k(B)} y^{\mathrm{cyc}(Q)} A(Q, y) \\
&= \sum_{k=0}^{n} x^k \sum_{Q\in\mathcal{N}_k(B)} y^{\mathrm{cyc}(Q)} (y)\uparrow_{n-k} \\
&= \sum_{k=0}^{n} x^k (y)\uparrow_{n-k} \sum_{Q\in\mathcal{N}_k} y^{\mathrm{cyc}(Q)} \\
&= \sum_{k=0}^{n} r_k(B, y)(y)\uparrow_{n-k} x^k.
\end{aligned}$$

Chung and Graham [7] proved that for any Ferrers boards $F(b_1, \ldots, b_n) \subseteq [n] \times [n]$, we have the following factorization theorem.

Theorem 12.3 *Let $B = F(b_1, \ldots, b_n) \subseteq [n] \times [n]$ be a Ferrers board. Then*

$$\prod_{i:b_i<i}(x+b_i-i+1)\prod_{i:b_i\geq i}(x+b_i-i+y) = \sum_{k=0}^{n} r_{n-k}(B, y)(x)\downarrow_k. \tag{12.5}$$

We let

$$\begin{aligned}
[n]_q &= \frac{q^n-1}{q-1} = 1+\cdots+q^{n-1}, \\
[n]_q! &= [1]_q[2]_q\cdots[n]_q, \text{ and} \\
\begin{bmatrix} n \\ k \end{bmatrix}_q &= \frac{[n]_q!}{[k]_q![n-k]_q!}
\end{aligned}$$

be the usual q-analogues of n, $n!$, and $\binom{n}{k}$. In general, we let $[x]_q = \frac{q^x-1}{q-1}$. Then for $k \geq 1$, we let $[x]_q\uparrow_k = [x]_q[x+1]_q\cdots[x+k-1]_q$ and $[x]_q\downarrow_k = [x]_q[x-1]_q\cdots[x-(k-1)]_q$. We let $[x]_q\uparrow_0 = [x]_q\downarrow_0 = 1$.

In an unpublished paper, Ehrenborg, Haglund, and Readdy [8] defined a q-analogue of the cycle-counting rook numbers $r_k(B, y, q)$ for Ferrers boards which generalized the q-analogue of the rook numbers for Ferrers boards introduced by Garsia and Remmel [9]. They proved the following generalization of Chung and Graham's theorem.

Theorem 12.4 *Let $B = F(b_1, \ldots, b_n) \subseteq [n] \times [n]$ be a Ferrers board. Then*

$$\prod_{i:b_i<i} [x + b_i - i + 1]_q \prod_{i:b_i\geq i} [x + b_i - i + y]_q = \sum_{k=0}^{n} r_{n-k}(B, y, q)[x]_q \downarrow_k . \quad (12.6)$$

Haglund [10] also extended the definition of the q-hit numbers of Garsia and Remmel [9] for Ferrers boards by defining q, x, y-hit numbers algebraically by the equation

$$\sum_{k=0}^{n} H_{k,n}(B, x, y, q)z^k = \sum_{k=0}^{n} r_{n-k}(B, y, q)[x]_q \uparrow_k z^k \prod_{i=k+1}^{n} (1 - zq^{x+i-1}). \quad (12.7)$$

Haglund [10] developed several connections between formulas for the q, x, y-hit numbers and hypergeometric series. Later, Butler [5] gave a combinatorial interpretation of $H_{k,n}(B, x, y, q)$ for Ferrers boards.

The main goal of this paper is to define analogues of cycle-counting rook numbers, cycle-counting hit numbers, and their q-analogues relative to the group $C_p \wr S_n$ which is the wreath product of the cyclic group C_p of order p with the symmetric group S_n. In particular, we extend the combinatorics of cycle-counting rook numbers and cycle-counting hit numbers to the rook theory model of Briggs and Remmel [2–4] where the rook placements correspond to partial permutations in $C_p \wr S_n$ and hit numbers correspond to permutations in $C_p \wr S_n$.

Let $\omega = e^{\frac{2\pi i}{p}}$. One can think of the group $C_p \wr S_n$ as the group of matrices under matrix multiplication where the underlying set is the set of matrices that one can form by starting with an $n \times n$ permutation matrix M and replacing 1's by powers of ω. Thus, we can think of $C_p \wr S_n$ as the group of $p^n n!$ signed permutations where there are p signs, $\omega^0 = 1, \omega, \omega^2, \ldots, \omega^{p-1}$. We will usually write the signed permutations in either one-line notation or in disjoint cycle form. For example, if $\sigma \in C_3 \wr S_8$ is the map sending $1 \to \omega 5$, $2 \to 8$, $3 \to \omega^2 3$, $4 \to \omega^2 1$, $5 \to 4$, $6 \to \omega^2 7$, $7 \to \omega 2$, and $8 \to \omega 6$, then in one-line notation,

$$\sigma = \omega 5 \ 8 \ \omega^2 3 \ \omega^2 1 \ 4 \ \omega^2 7 \ \omega 2 \ \omega 6,$$

whereas in disjoint cycle form,

$$\sigma = (\omega^2 1 \ \omega 5 \ 4)(\omega 2 \ 8 \ \omega 6 \ \omega^2 7)(\omega^2 3).$$

Fig. 2 The board B_6^3

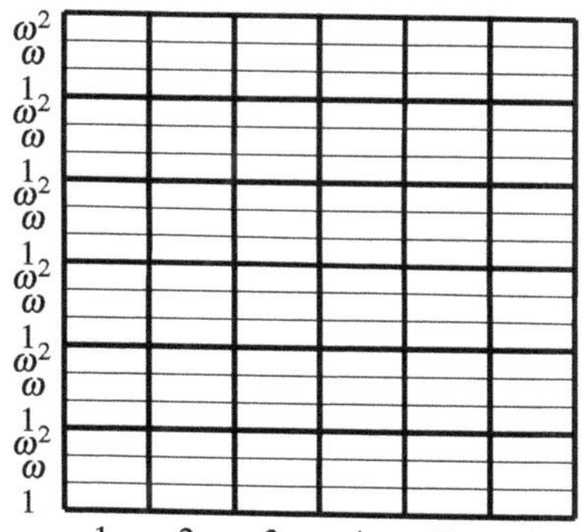

In other words, in disjoint cycle form, to determine where i is being mapped, we ignore the sign on i and only consider the sign on the element to which it is mapped. Whenever we have an r-cycle $C = (\omega^{a_0}c_0, \ldots, \omega^{a_{r-1}}c_{r-1})$ in a signed permutation in $C_p \wr S_n$, we define $sgn(C) = \prod_{i=0}^{r-1} \omega^{a_i}$. Thus, in our example,

$$sgn((\omega^2 1 \ \ \omega 5 \ \ 4)) = 1,$$
$$sgn((\omega 2 \ \ 8 \ \ \omega 6 \ \ \omega^2 7)) = \omega, \text{ and}$$
$$sgn((\omega^2 3)) = \omega^2.$$

Given $\sigma \in C_p \wr S_n$, we will write $\sigma(i)$ as $\varepsilon_i \sigma_i$, where $\sigma_i \in [n] = \{1, \ldots, n\}$, and where $\varepsilon_i = sgn(\sigma_i) \in \{1, \omega, \omega^2, \ldots, \omega^{p-1}\}$ is called the *sign* of σ_i. For each $1 \leq i \leq n$, we define $|\varepsilon_i \sigma_i| = \sigma_i$ and call this the *absolute value* of $\sigma(i)$.

Next we shall describe the rook model due to Briggs and Remmel [4] where the rook numbers correspond to partial permutations in $C_p \wr S_n$ and the hit numbers correspond to permutations in $C_p \wr S_n$.

The idea of Briggs and Remmel was to start with the $[n] \times [n]$ board and subdivide each row into p subrows. We will denote the resulting board by B_n^p. For example, if $n = 6$ and $p = 3$, then B_6^3 is pictured in Fig. 2. We shall refer to the rows of the original $[n] \times [n]$ board as levels and label the levels with $1, \ldots, n$ from bottom to top. We label the columns with $1, \ldots, n$ from left to right. Finally, within each level, we label the sublevels from bottom to top with $1, \omega, \omega^2, \ldots, \omega^{p-1}$. We let (i, j, k) denote the square in the ith column, in the jth level, and in the sublevel labelled with ω^k.

In the Briggs–Remmel model, a *board* is a subset of B_n^p. Given $b_1, \ldots, b_n \in [pn]$, we let $F(b_1, \ldots, b_n)$ denote the board consisting of all the cells $\{(i, j, k) : 1 \leq i \leq n \text{ and } 1 \leq pj + k \leq b_i\}$. If a board B is of the form $B = F(b_1, \ldots, b_n)$, then we say that B is a *skyline board* and if, in addition, $b_1 \leq b_2 \leq \cdots \leq b_n$, then we say that B is a *Ferrers board*. If $B = F(b_1, \ldots, b_n)$ is a Ferrers board and $b_{i+1} \geq rp$ whenever $(r-1)p + 1 \leq b_i \leq rp$, then we say that B is a *singleton Ferrers board*. Here, the last condition for a singleton Ferrers board in B_n^p says that whenever there are cells in level r in column i, column $i+1$ must contain all the cells in the level

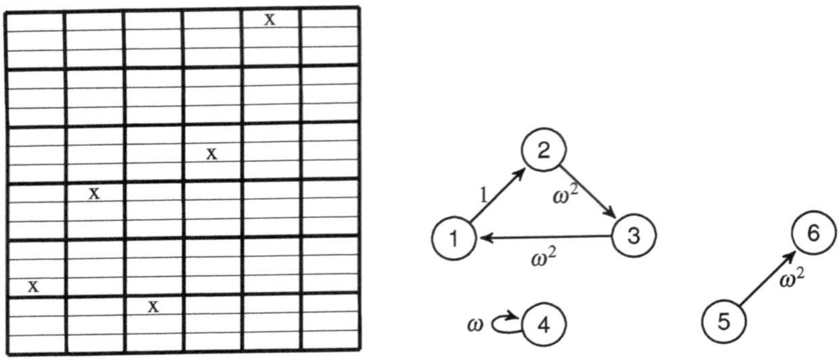

Fig. 3 Graph associated with a 3-rook placement in $\mathcal{N}_5^3(B_6^3)$

r. Finally, we shall say that a board B is a *full board* whenever, if B contains a cell (i, j, k), then it must contain the cells (i, j, r) for $r = 0, \ldots, p-1$. In other words, a Ferrers board $F(b_1, \ldots, b_n)$ is a full board if and only if b_i is a multiple of p for all $i = 1, \ldots, n$. We say that a full Ferrers board $B = F(b_1, \ldots, b_n) \subseteq B_p^n$ is *regular* if $b_i = p \cdot c_i$, where $c_i \geq i$ for $1 \leq i \leq n$.

Given a board $B \subseteq B_n^p$, we let $\mathcal{N}_k^p(B)$ denote the set of all placements of k rooks in B such that no two rooks lie in the same level or column. Elements of $\mathcal{N}_k^p(B)$ will be called *p-rook placements*. For $k = 1, \ldots, n$, we let $r_k^p(B) = |\mathcal{N}_k^p(B)|$. By convention, we set $r_0^p(B) = 1$. We refer to $r_k^p(B)$ as the ***k*th *p*-rook number of *B***. An alternative model for $r_k^p(B)$ was proposed by Wachs and Remmel [12]. In the case $p = 2$, Haglund and Remmel [11] gave yet another rook model for $r_k^p(B)$.

Given a signed permutation $\sigma = \omega^{a_1}\sigma_1 \cdots \omega^{a_n}\sigma_n \in C_p \wr S_n$, we identify σ with the p-rook placement $\{(i, \sigma_i, a_i) : i = 1, \ldots, n\}$ on B_n^p. We let

$$H_{k,n}^p(B) = |\{\sigma \in C_p \wr S_n : |\sigma \cap B| = k\}|.$$

We shall refer to $H_{k,n}^p(B)$ as the ***k*-th *p*-hit number of *B* relative to B_n^p**.

With each p-rook placement $P \in \mathcal{N}_k^p(B)$, we can associate a directed graph $G_P = ([n], E_P)$ with labelled edges, where E_P is the set of (i, j) such that P has a rook in cell (i, j, k) and we label the edge (i, j) with ω^k. For example, see Fig. 3 for the graph associated with a 3-rook placement on B_6^3. For any p-rook placement, we let $\mathrm{cyc}_i(P)$ denote the number of cycles in the graph of P such that product of labels on the cycle is ω^i.

For any board $B \subseteq B_n^p$, we let

$$r_k^p(B, y_0, \ldots, y_{p-1}) = \sum_{P \in \mathcal{N}_k^p(B)} \prod_{i=0}^{p-1} y_i^{\mathrm{cyc}_i(P)} \quad \text{and}$$

$$H_{k,n}(B, y_0, \ldots, y_{p-1}) = \sum_{\substack{\sigma \in C_p \wr S_n, \\ |\sigma \cap B| = k}} \prod_{i=0}^{p-1} y_i^{\mathrm{cyc}_i(\sigma)}.$$

The outline of the paper is as follows. In Sect. 2, we shall prove the analogues of Theorem 12.2 and Theorem 12.3 as well as give an example of cycle-counting p-Lah numbers. In Sect. 3, we shall define a q-analogue of the cycle-counting p-rook numbers and prove an analogue of the Ehrenborg, Haglund, and Readdy factorization theorem [8]. In Sect. 4, we shall define a q-analogue of the cycle-counting p-hit numbers $H_{k,n}^p[B, q, y_0, \ldots, y_{p-1}]$ for a full regular Ferrers board B. We will prove analogues of some results of Haglund [10] and Butler [6] on the q-cycle-counting rook numbers and q, x, y-hit numbers for full regular Ferrers boards which will allow us to prove that $H_{k,n}^p[B, q, y_0, \ldots, y_{p-1}]$ is always a polynomial in q with non-negative coefficients when $y_0, \ldots, y_{p-1}$ are non-negative integers. We will end Sect. 4 by giving a conjectured combinatorial interpretation of the $H_{k,n}^p[B, q, y_0, \ldots, y_{p-1}]$'s.

2 Cycle-Counting p-Rook Numbers and p-Hit Numbers.

We start this section by proving analogues of Theorem 12.2 and Theorem 12.3 for the cycle-counting p-rook and p-hit numbers.

Suppose that $p \geq 2$. Then, for $k \geq 1$, we let $(y)\uparrow_{k,p} = y(y+p)\cdots(y+p(k-1))$ and $(y)\downarrow_{k,p} = y(y-p)\cdots(y-p(k-1))$. We also let $(y)\uparrow_{0,p} = (y)\downarrow_{0,p} = 1$. We then have the following analogue of Theorem 12.2.

Theorem 12.5 *For any $p \geq 2$ and any board $B \subseteq B_n^p$,*

$$\sum_{k=0}^{n} H_{k,n}^p(B, y_0, \ldots, y_{p-1})x^k \tag{12.8}$$
$$= \sum_{k=0}^{n} r_k^p(B, y_0, \ldots, y_{p-1})(y_0 + \cdots + y_{p-1})\uparrow_{n-k,p} (x-1)^k.$$

Proof Fix $p \geq 2$. First replace x by $x+1$ in Eq. (12.8). Thus, we must prove

$$\sum_{k=0}^{n} H_{k,n}^p(B, y_0, \ldots, y_{p-1})(x+1)^k$$
$$= \sum_{k=0}^{n} r_k^p(B, y_0, \ldots, y_{p-1})(y_0 + \cdots + y_{p-1})\uparrow_{n-k,p} x^k. \tag{12.9}$$

For (12.9), we consider configurations C which consist of a rook placement corresponding to a permutation $\sigma \in C_k \wr S_n$, where we circle some of the rooks that fall in $B \cap \sigma$. We then let $\mathrm{cyc}_i(C)$ denote the number of cycles of sign ω^i in the graph of the underlying permutation of C and $\mathrm{circle}(C)$ denote the number of circled rooks in C. It is then easy to see that the left-hand side of (12.9) can be interpreted as counting $x^{\text{circle}(C)} \prod_{i=0}^{p-1} y_i^{\text{cyc}_i(C)}$ over all such configurations. The right-hand side of (12.9) can be interpreted as follows. First pick the circled rooks which correspond to a placement $Q \in \mathcal{N}_k^p(B)$ for some k. Then, we need to compute

$$A(Q, y_0, \dots, y_{p-1}) = \sum_C \prod_{i=0}^{p-1} y_i^{\text{cyc}_i(C)}, \tag{12.10}$$

where the sum runs over all configurations whose set of circled rooks equals Q. Again this sum is easy to compute. Let i be the first column that does not contain a rook in Q. Then, there are $n-k$ levels in which to place a rook in column i that do not contain rooks in Q. We claim that there is exactly one level r where placing a rook in the cell (i, r, k) for any k, $0 \le k \le p-1$ completes a cycle in the graph of Q. That is, if there is no rook in Q which is in level i, then i is an isolated vertex in the graph of Q, so adding a rook in cell (i, i, k) will give a loop on vertex i with label ω^k and hence increase the number of cycles with sign ω^k by 1. Clearly, in such a situation, placing a rook in cell (i, j, k) for $j \ne i$ and $0 \le k \le p-1$ cannot complete a cycle. If there is a rook of Q in level i, then there must be a path p of the maximal length in the graph of Q which ends in vertex i since there are no edges coming out of the vertex i in the graph of Q. If this path starts in vertex j, then there is no rook in level j in Q. Hence, if we add a rook to cell (i, j, k) for any $0 \le k \le p-1$, then this will complete a cycle. No matter what the labels are on the edges of the path from j to i in the graph corresponding to Q, there will be exactly one choice of k which results in the completed cycle having sign ω^i for any given $i \in \{0, \dots, p-1\}$. Clearly, adding a rook to any other level in column i will not complete a cycle in this case. Thus, the placement of a rook in column i will contribute a factor $(y_0 + \cdots + y_{p-1} + p(n-k-1))$ to $A(Q, y_0, \dots, y_{p-1})$. But then we can repeat the argument for every placement Q' which arises from Q by adding a rook in the next empty column, say column i_1. That is, for each such Q', the addition of a rook in column i_1 will contribute a factor $(y_0 + \cdots + y_{p-1} + p(n-k-2))$ to $A(Q, y_0, \dots, y_{p-1})$. Continuing on in this way, we see that $A(Q, y_0, \dots, y_{p-1})$ equals

$$\begin{aligned}(y_0 + \cdots + &y_{p-1} + p(n-k-1))(y_0 + \cdots + y_{p-1} + p(n-k-2)) \cdots (y_0 + \cdots + y_{p-1}) \\ &= (y_0 + \cdots + y_{p-1}) \uparrow_{n-k,p} .\end{aligned}$$

Thus, another way to sum $x^{\text{circle}(C)} \prod_{i=0}^{p-1} y_i^{\text{cyc}_i(C)}$ over all configurations is

$$\begin{aligned}&\sum_{k=0}^{n} x^k \sum_{Q \in \mathcal{N}_k^p(B)} \prod_{i=0}^{p-1} y_i^{\text{cyc}_i(Q)} A(Q, y_0, \dots, y_{p-1}) \\ &= \sum_{k=0}^{n} x^k \sum_{Q \in \mathcal{N}_k^p(B)} \prod_{i=0}^{p-1} y_i^{\text{cyc}_i(Q)} (y_0 + \cdots + y_{p-1}) \uparrow_{n-k,p}\end{aligned}$$

$$= \sum_{k=0}^{n} x^k (y_0 + \cdots + y_{p-1}) \uparrow_{n-k,p} \sum_{Q \in \mathcal{N}_k^p(B)} \prod_{i=0}^{p-1} y_i^{\mathrm{cyc}_i(Q)}$$

$$= \sum_{k=0}^{n} r_k^p(B, y_0, \ldots y_{p-1})(y_0 + \cdots + y_{p-1}) \uparrow_{n-k,p} x^k.$$

Next we shall prove a factorization theorem for cycle-counting p-rook numbers for full Ferrers boards $B \subseteq B_n^p$.

Theorem 12.6 *Let $p \geq 2$ and $B = F(b_1, \ldots, b_n)$ be a full Ferrers board contained in B_n^p. Then, we have*

$$\prod_{i:b_i<pi} (x + b_i - p(i-1)) \prod_{i:b_i \geq pi} (x + b_i - pi + y_0 + \cdots + y_{p-1})$$
$$= \sum_{k=0}^{n} r_{n-k}^p(B, y_0, \ldots, y_{p-1})(x) \downarrow_{k,p}. \quad (12.11)$$

Proof The assumption that B is a full board implies that b_i is divisible by p for all i. Since both sides of (12.11) are polynomials in x of degree n, it is enough to prove that (12.11) holds for infinitely many integers.

First we shall show that (12.11) holds for infinitely many integers px, where $x \in \mathbb{P}$. Given $x \in \mathbb{P}$, we let B_x denote the board which results by adding x-levels of length n below B. For example, if $p = 3$, $B = (3, 6, 6, 6, 9, 9)$, and $x = 6$, then the board B_x is pictured in Fig. 4. We call the boundary between B and the x-levels that we added below B the *bar*.

We let $\mathcal{N}_k^p(B_x)$ denote the set of all placements of k rooks in B_x such that there is at most one rook in each level and each column. Given a placement $P \in \mathcal{N}_n^p(B_x)$,

Fig. 4 The board B_x

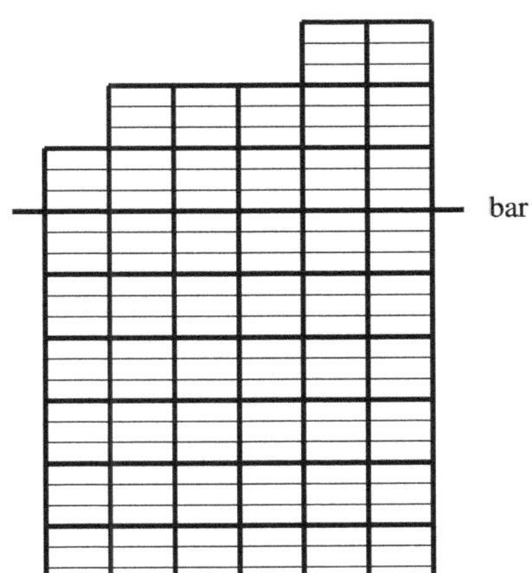

we let $wt(P) = \prod_{i=0}^{p-1} y_i^{\mathrm{cyc}_i(P \cap B)}$. Then, we claim that (12.11) where x is replaced by px arises from two different ways of computing

$$S(B, y_0, \ldots, y_{p-1}) = \sum_{P \in \mathcal{N}_n^p(B_x)} wt(P).$$

Next we prove a key lemma.

Lemma 12.1 *Suppose that $Q \in \mathcal{N}_t^p(B_x)$ is a p-rook placement of t rooks in the first $i-1$ columns of B_x. Let $D_i(Q)$ denote the set of all p-rook placements P that result from Q by adding a rook in column i. Then*

$$\sum_{P \in D_i(Q)} \prod_{l=0}^{p-1} y_l^{cyc_l(P)} =$$
$$\begin{cases} (b_i + px - p(t+1) + y_0 + \cdots + y_{p-1}) \prod_{l=0}^{p-1} y_l^{cyc_l(Q \cap B)} & \text{if } b_i \geq pi, \\ (b_i + px - pt) \prod_{l=0}^{p-1} y_l^{cyc_l(Q \cap B)} & \text{if } b_i < pi. \end{cases}$$

Proof First we claim that there is exactly one level j above the bar such that placing a rook in a cell (i, j, k) will complete a cycle in the graph of $Q \cap B$ if $b_i \geq pi$ and there is no level j above the bar such that placing a rook in a cell (i, j, k) will complete a cycle in the graph of $Q \cap B$ if $b_i < pi$. That is, suppose that $b_i \geq pi$. If there is no rook in $Q \cap B$ which is in level i, then i is an isolated vertex in the graph of $Q \cap B$, so adding a rook in cell (i, i, k) will give a loop on vertex i with label ω^k and hence increase the number of cycles with sign ω^k by 1. Clearly, in such a situation, placing a rook in cell (i, j, k) for $j \neq i$ and $0 \leq k \leq p-1$ cannot complete a cycle. If there is a rook in $Q \cap B$ in row i, then there must be a maximal length path p in the graph of $Q \cap B$ which ends in vertex i since there are no edges coming out of i in the graph of $Q \cap B$. If this path starts in vertex j, then $j \leq i \leq b_i/p$ and there is no rook in level j in $Q \cap B$ above the bar. Hence, if we add a rook to cell (i, j, k) for any $0 \leq k \leq p-1$, then it will complete a cycle. No matter what the labels are on the edges of the path from j to i in the graph corresponding to Q, there will be exactly one choice for k which results in the completed cycle having sign ω^i for any given $i \in \{0, \ldots, p-1\}$. In such a situation, we will call the level j such that adding a rook in a cell (i, j, k) completes a cycle the *special level relative to Q*. It easily follows that in this case

$$\sum_{P \in D_i(Q)} \prod_{l=0}^{p-1} y_l^{\mathrm{cyc}_l(P)} = (b_i + px - p(t+1) + y_0 + \cdots + y_{p-1}) \prod_{l=0}^{p-1} y_l^{\mathrm{cyc}_l(Q \cap B)}.$$

Alternatively, if $b_i < pi$, then we must have that $b_1 \leq \cdots \leq b_{i-1} \leq p(i-1)$ since we are assuming that B is a full Ferrers board. This implies that there can be no edge which ends in the vertex i in the graph of $Q \cap B$. Hence, i is an isolated vertex in the graph $Q \cap B$. Thus, placing a rook in the cell (i, j, k) where $j < i$ cannot create a new cycle. It easily follows that in this case

$$\sum_{P\in D_i(Q)} \prod_{l=0}^{p-1} y_l^{\mathrm{cyc}_l(P)} = (b_i + px - pt) \prod_{l=0}^{p-1} y_l^{\mathrm{cyc}_l(Q\cap B)}.$$

Now think of adding rooks column by column starting from the left to form an element $P \in \mathcal{N}_n^p(B_x)$. In the first column, we have $b_1 + px$ choices. If $b_1 \geq p$, then if we add a rook in cell $(1, 1, k)$, then we create a cycle of sign ω^k and we do not create a cycle otherwise. Thus, the first column will contribute a factor $(px + b_1 - p + y_0 + \cdots + y_{p-1})$ if $b_1 \geq p$ or a factor $(px + b_1)$ otherwise. Next if we start with a placement $Q \in \mathcal{N}_{i-1}^p(B_x)$ of $i - 1$ rooks in the first $i - 1$ columns of B_x, then we will have $px + b_i - p(i - 1)$ cells to add a rook in column i. By Lemma 12.1, our choices for placing a rook in these $px + b_i - p(i - 1)$ cells will contribute a factor $(px + b_i - pi + y_0 + \cdots + y_{p-1})$ if $b_i \geq pi$ and will contribute a factor $(px + b_i - p(i - 1))$ otherwise. Thus, it follows that

$$S(B, y_0, \ldots, y_{p-1}) = \prod_{i:b_i<pi} (px + b_i - p(i-1)) \prod_{i:b_i\geq pi} (px + b_i - pi + y_0 + \cdots + y_{p-1}).$$

On the other hand, suppose that we fix a p-rook placement $Q \in \mathcal{N}_{n-k}^p(B)$ of $n - k$ rooks above the bar. Then, we want to compute

$$B_Q = \sum_{P\in\mathcal{N}_n^p(B_x):P\cap B=Q} wt(P). \tag{12.12}$$

In this case, there will be k columns below the bar which do not contain rooks in Q. If those columns are $1 \leq i_1 < \cdots < i_k \leq n$, then we have px choices to place a rook below the bar in column i_1. Once we have placed a rook in column i_1 below the bar, we will have $px - p$ choices to add a rook below the bar in column i_2. Continuing on in this way, it is easy to see that we have $(px)(px - p)\cdots(px - p(k - 1)) = (px)\downarrow_{k,p}$ ways to extend Q to a placement in $\mathcal{N}_n^p(B_x)$. By definition, the weight of any such placement P is $\prod_{i=0}^{p-1} y_i^{\mathrm{cyc}_i(Q)}$. Thus

$$\begin{aligned}
S(B, y_0, \ldots, y_{p-1}) &= \sum_{k=0}^{n} \sum_{Q\in\mathcal{N}_{n-k}^p(B)} \prod_{i=0}^{p-1} y_i^{\mathrm{cyc}_i(Q)} (px)\downarrow_{k,p} \\
&= \sum_{k=0}^{n} (px)\downarrow_{k,p} \sum_{Q\in\mathcal{N}_{n-k}^p(B)} \prod_{i=0}^{p-1} y_i^{\mathrm{cyc}_i(Q)} \\
&= \sum_{k=0}^{n} r_{n-k}^p(B, y_0, \ldots, y_{p-1})(px)\downarrow_{k,p} .
\end{aligned}$$

A natural question here would be whether there is a similar result for singleton Ferrers boards or Ferrers boards. In the case where we set $y_i = 1$ for $i = 0, \ldots, p-1$, Briggs and Remmel [4] proved a factorization theorem for the p-rook numbers for singleton Ferrers boards, and Barrese, Loehr, Remmel and Sagan [1] proved a factorization theorem for p-rook numbers for all Ferrers boards.

As an example of an application of Theorem 12.6, we give the cycle-counting p-rook analogue of the Lah numbers. The Lah numbers $L_{n,k}$ are defined by the equation

$$(x)\uparrow_n = \sum_{k=1}^{n} L_{n,k}(x)\downarrow_k .$$

They can also be defined by the following recursion

$$L_{n+1,k} = L_{n,k-1} + (n+k)L_{n,k}, \tag{12.13}$$

with initial conditions $L_{0,0} = 1$ and $L_{n,k} = 0$ if $k < 0$ or $k > n$. The $L_{n,k}$'s have a nice rook theory interpretation, that is, $L_{n,k} = r_{n-k}(\mathcal{L}_n)$, where $\mathcal{L}_n$ is the Ferrers board consisting of n columns of height $n-1$, see [9]. From this interpretation, it is easy to see that

$$L_{n,k} = \frac{(n-1)!}{(k-1)!}\binom{n}{k}. \tag{12.14}$$

That is, to create a rook placement of $n-k$ rooks in $\mathcal{L}_n$, we first pick the $n-k$ columns that will contain the rooks. We can do this in $\binom{n}{n-k} = \binom{n}{k}$ ways. Then, we have to place the rooks in these columns starting from the left. We clearly have $n-1$ choices where to put a rook in the leftmost column, then $n-1-1$ ways to place a rook in the next column, etc. Thus, we will have $(n-1)\downarrow_{n-k} = \frac{(n-1)!}{(k-1)!}$ ways to place the rooks in the $n-k$ columns that we chose.

For the obvious cycle-counting analogue of the $L_{n,k}$'s for $C_p \wr S_n$, consider the Ferrers board $\mathcal{L}_n^p$ which consists of n columns of height $p(n-1)$. We let

$$L_{n,k}^p(y_0, \ldots y_{p-1}) = r_{n-k}^p(\mathcal{L}_n^p, y_0, \ldots, y_{p-1}). \tag{12.15}$$

In this case, (12.11) becomes

$$x(x + y_0 + \cdots + y_{p-1})\uparrow_{n-1,p} = \sum_{k=1}^{n} L_{n,k}^p(y_0, \ldots y_{p-1})(x)\downarrow_{k,p} . \tag{12.16}$$

Note that

$$
\begin{aligned}
&\sum_{k=1}^{n+1} L^p_{n+1,k}(y_0,\dots y_{p-1})(x)\downarrow_{k,p}\\
&\qquad = x(x+y_0+\dots+y_{p-1})\uparrow_{n,p}\\
&\qquad = (x+y_0+\dots+y_{p-1}+p(n-1))x(x+y_0+\dots+y_{p-1})\uparrow_{n-1,p}\\
&\qquad = (x+y_0+\dots+y_{p-1}+p(n-1))\sum_{k=1}^{n} L^p_{n,k}(y_0,\dots y_{p-1})(x)\downarrow_{k,p}\\
&\qquad = \sum_{k=1}^{n} L^p_{n,k}(y_0,\dots y_{p-1})(x)\downarrow_{k,p}(x-kp+y_0+\dots+y_{p-1}+p(n+k-1))\\
&\qquad = \sum_{k=1}^{n} L^p_{n,k}(y_0,\dots y_{p-1})(x)\downarrow_{k+1,p}\\
&\qquad\quad + \sum_{k=1}^{n} L^p_{n,k}(y_0,\dots y_{p-1})(x)\downarrow_{k,p}(y_0+\dots+y_{p-1}+p(n+k-1)).
\end{aligned}
$$

It thus follows that

$$
\begin{aligned}
&L^p_{n+1,k}(y_0,\dots y_{p-1})\\
&= L^p_{n,k-1}(y_0,\dots y_{p-1}) + (y_0+\dots+y_{p-1}+p(n+k-1))L^p_{n,k}(y_0,\dots y_{p-1}).
\end{aligned}
\tag{12.17}
$$

We also have an analogue of (12.14) in this case. That is, we want to compute

$$\sum_{P\in\mathcal{N}^p_{n-k}(\mathcal{L}^p_n)} \prod_{i=0}^{p-1} y_i^{\mathrm{cyc}_i(P)}.$$

We divide the p-rook placements in $\mathcal{N}^p_{n-k}(\mathcal{L}^p_n)$ into two sets: N_1 consisting of those p-rook placements with no rook in the last column and N_2 consisting of those p-rook placements that have a rook in the last column. For N_1, there are $\binom{n-1}{n-k} = \binom{n-1}{k-1}$ ways to choose the $n-k$ columns in which we are going to place the rooks. If $i \leq n-1$, then the height of the ith column is greater than or equal to pi. Hence, we can use Lemma 12.1 to argue that as we place the rooks in the columns from left to right, the sum of $\prod_{i=0}^{p-1} y_i^{\mathrm{cyc}_i(P)}$ over the possible placements in the $n-k$ columns that we choose is

$$
\begin{aligned}
&(p(n-2)+y_0+\dots+y_{p-1})(p(n-3)+y_0+\dots+y_{p-1})\\
&\qquad\qquad \cdots(p(k-1)+y_0+\dots+y_{p-1}).
\end{aligned}
$$

Thus

$$\sum_{P\in N_1} \prod_{i=0}^{p-1} y_i^{\mathrm{cyc}_i(P)} = \binom{n-1}{k-1}(p(k-1)+y_0+\dots+y_{p-1})\uparrow_{n-k,p}.$$

For N_2, there are $\binom{n-1}{n-k-1} = \binom{n-1}{k}$ ways to choose the columns in which we are going to place the rooks in the first $n-1$ columns. As above, the sum of $\prod_{i=0}^{p-1} y_i^{\mathrm{cyc}_i(P)}$ over the possible placements in the $n-k$ columns that we choose is

$$(p(n-2)+y_0+\cdots+y_{p-1})(p(n-3)+y_0+\cdots+y_{p-1})\cdots(pk+y_0+\cdots+y_{p-1}).$$

Once we place these rooks, we still have to place a rook in the last column. However, the height of the last column in $\mathcal{L}_n^p$ is $(n-1)p < np$. Thus, by Lemma 12.1, the factor contributed by placing the rook in the last column in the $n-1-(n-k-1) = k$ levels which are possible is pk. Thus

$$\sum_{P\in N_2}\prod_{i=0}^{p-1} y_i^{\mathrm{cyc}_i(P)} = \binom{n-1}{k}(pk)(pk+y_0+\cdots+y_{p-1})\uparrow_{n-k-1,p}.$$

Hence,

$$\begin{aligned} L_{n,k}^p &= \binom{n-1}{k-1}(p(k-1)+y_0+\cdots+y_{p-1})\uparrow_{n-k,p} \\ &\quad+\binom{n-1}{k}(pk)(pk+y_0+\cdots+y_{p-1})\uparrow_{n-k-1,p} \\ &= \binom{n-1}{k-1}(pk+y_0+\cdots+y_{p-1})\uparrow_{n-k,p}. \end{aligned}$$

3 Q-Analogues of Cycle-Counting p-Rook Numbers

In this section, we shall define q-analogues of cycle-counting p-rook numbers and prove a factorization theorem for such q-analogues for full Ferrers boards.

First, we shall recall the definitions of the q-analogues of the p-rook and p-hit numbers as defined by Briggs and Remmel [4]. Let $B = F(b_1,\ldots,b_n)$ be a Ferrers board contained in B_n^p. A rook in cell (i,j,k) is said to *rook cancel* all cells in level j that lie strictly its right and all cells that lie directly below it. Then, for any given $P \in \mathcal{N}_k^p(B)$, we let $inv_B(P)$ be the number of uncancelled cells in $B-P$. For example, in Fig. 5 we have pictured a placement in $B = F(6,9,12,12,15,15) \subseteq B_6^3$ and we have put dots in cells which are rook cancelled by rooks in P. Thus, $inv_B(P) = 30$ as there is a total of 30 squares which are not rook cancelled by rooks in P.

Suppose that $p \geq 2$. Then, for $k \geq 1$, we let

$$\begin{aligned} [y]_q\uparrow_{k,p} &= [y]_q[y+p]_q\cdots[y+p(k-1)]_q \text{ and} \\ [y]_q\downarrow_{k,p} &= [y]_q[y-p]_q\cdots[y-p(k-1)]_q. \end{aligned}$$

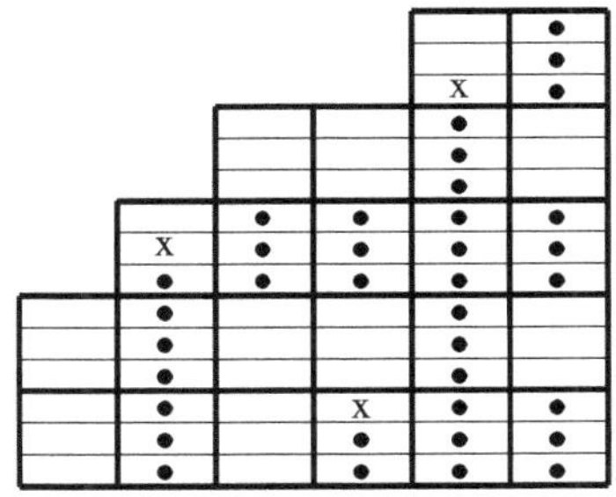

Fig. 5 An example of rook cancellation

We let $[y]_q \uparrow_{0,p} = [y]_q \downarrow_{0,p} = 1$. Then, for any Ferrers board $B = F(b_1, \ldots, b_n) \subseteq B_n^p$, Briggs and Remmel defined $r_k^p(B, q)$ by

$$r_k^p(B, q) = \sum_{P \in \mathcal{N}_k^p} q^{inv(P)} \tag{12.18}$$

and $H_{k,n}^p(B, q)$ algebraically by

$$\sum_{k=0}^{n} H_{k,n}^p(B, q)x^k = \sum_{k=0}^{n} r_k^p(B, q)[p(n-k)]_q \downarrow_{n-k,p} \prod_{\ell=n-k+1}^{n} (x - q^{p\ell}). \tag{12.19}$$

Briggs and Remmel [4] then proved the following two theorems.

Theorem 12.7 *Let $B = F(b_1, \ldots, b_n) \subseteq B_n^p$ be a Ferrers board. Then*

$$\prod_{i=1}^{n}[x + b_i - p(i-1)] = \sum_{k=0}^{n} r_{n-k}^p(B, q)[px] \downarrow_{k,p} . \tag{12.20}$$

Theorem 12.8 *Let $B = F(b_1, \ldots, b_n) \subseteq B_n^p$ be a Ferrers board. Then $H_{n,k}(B, q)$ is a polynomial in q with non-negative integer coefficients for all $k = 0, \ldots, n$.*

In fact, Briggs and Remmel proved p, q-analogues of Theorems 12.7 and 12.8 but we shall not concern ourselves with p, q-analogues in this paper.

We define the q-analogue of the cycle-counting p-rook number by

$$r_k^p(B, q, y_0, \ldots, y_{p-1}) = \sum_{P \in \mathcal{N}_k^p(B)} \left(\prod_{j=0}^{p-1} [y_j]_q^{\mathrm{cyc}_j(P)} \right) q^{\mathrm{inv}(P) + \sum_{j=0}^{p-1}(y_j - 1)E_j(P)}, \tag{12.21}$$

where

inv(P) is the number of uncancelled cells (considering one sublevel as one cell) when a rook cancels all the cells below it and all the cells to the right in the same level with the rook, and

$E_j(P)$ is the number of i's such that $b_i \geq pi$ and there is no rook from P in column i on or above $s_i^j(P)$, where $s_i^j(P)$ is the unique sublevel which, considering only rooks from P in column 1 through $i-1$ of B, completes a ω_j cycle.

Then, we have the following q-analogue of the factorization theorem.

Theorem 12.9 *Let $B = F(b_1, \ldots, b_n)$ be a full Ferrers board contained in B_n^p.*

$$\prod_{i:b_i<pi} [px + b_i - p(i-1)]_q \prod_{i:b_i \geq pi} [px + b_i - pi + y_0 + \cdots + y_{p-1}]_q$$
$$= \sum_{k=0}^{n} r_{n-k}^p(B, q, y_0, \ldots, y_{p-1})[px]_q \downarrow_{k,p} . \quad (12.22)$$

Proof It is not difficult to show that it is enough to prove (12.22) holds whenever $x, y_0, \ldots, y_{p-1}$ are positive integers. The proof is similar to the proof of Theorem 12.6. Given $x \in \mathbb{P}$, we consider the extended board B_x by adding x-levels of length n below B. Then, suppose that $y_0, \ldots, y_{p-1}$ are fixed elements of $\mathbb{P}$. For a given $P \in \mathcal{N}_n^p(B_x)$, we let

$$wt(P) = \left(\prod_{j=0}^{p-1} [y_j]_q^{\text{cyc}_j(P\cap B)}\right) q^{\text{inv}(P)+\sum_{j=0}^{p-1}(y_j-1)E_j(P\cap B)}.$$

Then, we claim that (12.22) arises by calculating

$$S(B, q, y_0, \ldots, y_{p-1}) = \sum_{P \in \mathcal{N}_n^p(B_x)} wt(P)$$

in two different ways.

First, we fix a p-rook placement $Q \in \mathcal{N}_{n-k}^p(B)$ of $n-k$ rooks in B. Then, we want to compute

$$A_Q = \sum_{P \in \mathcal{N}_n(B_x),\ P\cap B = Q} wt(P).$$

In this case, there are k columns below the bar which do not contain rooks in Q. First consider the contribution that comes from placing a rook below the bar in the first available column, reading from left to right. If we place a rook in the top cell of the first available column, then it would contribute q^0 to the weight of the rook placement. If we place that rook one cell below, then it would give q^1 and so on. Thus, our choices for placing a rook in this column contribute the weight sum

$$q^0 + q^1 + \cdots + q^{px-1} = [px]_q$$

to A_Q. Once we place a rook in the first available column, then we can use the same argument to show that our choices of placing a rook below the bar in the next available column contribute a factor $[px - p]_q$ to A_Q. By continuing in this way, we get

$$A_Q = \left(\prod_{j=0}^{p-1} [y_j]_q^{\mathrm{cyc}_j(Q)}\right) q^{\mathrm{inv}(Q)+\sum_{j=0}^{p-1}(y_j-1)E_j(Q)}[px]_q[px-p]_q\cdots[px-p(k-1)]_q.$$

Thus

$$\begin{aligned}
S(B, q, y_0, \ldots, y_{p-1}) &= \sum_{k=0}^{n} \sum_{Q \in \mathcal{N}_{n-k}^p(B)} A_Q \\
&= \sum_{k=0}^{n} [px]_q \downarrow_{k,p} \sum_{Q \in \mathcal{N}_{n-k}^p(B)} \left(\prod_{j=0}^{p-1} [y_j]_q^{\mathrm{cyc}_j(Q)}\right) q^{\mathrm{inv}(Q)+\sum_{j=0}^{p-1}(y_j-1)E_j(Q)} \\
&= \sum_{k=0}^{n} r_{n-k}^p(B, q, y_0, \ldots, y_{p-1})[px]_q \downarrow_{k,p}
\end{aligned}$$

which is the left-hand side of (12.22).

On the other hand, we can calculate $S(B, q, y_0, \ldots, y_{p-1})$ by adding rooks column by column, starting from left to right. To do this, we need an analogue of Lemma 12.1, which we state and prove separately subsequent to the current proof; see Lemma 12.2.

If we start with a placement $Q \in \mathcal{N}_{i-1}^p(B_x)$ of $i-1$ rooks in the first $i-1$ columns of B_x, then the i th column will contribute the factor $[px + b_i - pi + y_0 + \cdots + y_{p-1}]_q$ for placing a rook in the column i if $b_i \geq pi$ and will contribute a factor $[px + b_i - p(i-1)]_q$ if $b_i < pi$. Thus,

$$\begin{aligned}
&S(B, q, y_0, \ldots, y_{p-1}) \\
&\quad = \prod_{i:b_i<pi} [px + b_i - p(i-1)]_q \prod_{i:b_i \geq pi} [px + b_i - pi + y_0 + \cdots + y_{p-1}]_q,
\end{aligned}$$

which is the right-hand side of (12.22).

Lemma 12.2 *Suppose that $Q \in \mathcal{N}_t^p(B_x)$ is a p-rook placement of t rooks in the first $i-1$ columns of B_x. Let $D_i(Q)$ denote the set of all p-rook placements P that result from Q by adding a rook in column i. Then*

$$\sum_{P \in D_i(Q)} wt(P) = \begin{cases} [b_i + px - p(t+1) + y_0 + \cdots + y_{p-1}]_q wt(Q), & \text{if} b_i \geq pi, \\ [b_i + px - pt]_q wt(Q), & \text{if} b_i < pi. \end{cases} \tag{12.23}$$

Proof The proof is similar to the proof of Lemma 12.1. That is, if $b_i < pi$, then any placement of a rook in column i will not contribute to $E_j(P \cap B)$ for any j. Now, there are $px + b_i - pt$ uncancelled squares in the ith column. If we place a rook r_i in the jth uncancelled cell from the top in column i, then r_i will contribute a factor q^{j-1} to $wt(P)$ as the contribution to $inv(P)$ from r_i will be $j - 1$. Thus, in this case, the placement of r_i will contribute a factor

$$wt(Q) \sum_{j=1}^{px+b_i-pt} q^{j-1} = wt(Q)[px + b_i - pt]_q$$

to $\sum_{P \in D_i(Q)} wt(P)$.

If $b_i \geq pi$, then there is a level $l_i \leq i$ such that placing a rook r_i in level ℓ_i in column i will complete a cycle relative to the rooks in Q. Assume that if we place a rook in cell (i, l_i, s), then we complete a cycle of sign ω^{u_s}. Thus, $\omega^{u_0}, \ldots, \omega^{u_{p-1}}$ must be a rearrangement of $1, \omega, \ldots, \omega^{p-1}$. In addition, assume that there are pt_i uncancelled cells above level ℓ_i in column i. Then, as before, placing a rook in jth uncancelled cell from the top, where $j \leq pt_i$, will give a factor q^{j-1} to $\sum_{P \in D_i(Q)} wt(P)$. Thus, the placements of a rook in the top pt_i cells will give a factor

$$wt(Q)(1 + q + \cdots + q^{pt_i - 1}) = wt(Q)[pt_i]_q$$

to $\sum_{P \in D_i(Q)} wt(P)$.

Now consider the effect of placing a rook r_i in the cell $(i, l_i, p - 1)$. Then, r_i would contribute a factor

$$[y_{u_{p-1}}]_q q^{pt_i} = q^{pt_i} + \cdots + q^{pt_i + y_{u_{p-1}} - 1}$$

to $wt(P)$. Here, $[y_{u_{p-1}}]_q$ comes from the fact that we completed a cycle of sign $\omega^{u_{p-1}}$ and q^{pt_i} comes from the contribution of r_i to $inv(P)$. Note that r_i makes no contribution to $E_j(P)$ for any j in this case. Next consider the effect of placing a rook r_i in the cell $(i, l_i, p - 2)$. Then, r_i would contribute

$$[y_{u_{p-2}}]_q q^{pt_i+1} q^{y_{u_{p-1}}-1} = q^{pt_i + y_{u_{p-1}}} + \cdots + q^{pt_i + y_{u_{p-1}} + y_{u_{p-2}} - 1}$$

to $wt(P)$. Here, $[y_{u_{p-2}}]_q$ comes from the fact that we completed a cycle of sign $\omega^{u_{p-2}}$, q^{pt_i+1} comes from the contribution of r_i to $inv(P)$, and $q^{y_{u_{p-1}}-1}$ comes from the fact that the placement of r_i contributes 1 to $E_{u_{p-1}}(P)$. Next consider the effect of placing a rook r_i in the cell $(i, l_i, p - 3)$. Then, r_i would contribute

$$[y_{u_{p-3}}]_q q^{pt_i+2} q^{y_{u_{p-1}}-1+y_{u_{p-2}}-1} = q^{pt_i + y_{u_{p-1}} + y_{u_{p-2}}} + \cdots + q^{pt_i + y_{u_{p-1}} + y_{u_{p-2}} + y_{u_{p-3}} - 1}$$

to $wt(P)$. Here, $[y_{u_{p-3}}]_q$ comes from the fact that we completed a cycle of sign $\omega^{u_{p-3}}$, q^{pt_i+2} comes from the contribution of r_i to $inv(P)$, and $q^{y_{u_{p-1}}-1+y_{u_{p-2}}-1}$ comes from the fact that the placement of r_i contributes 1 to both $E_{u_{p-2}}(P)$ and $E_{u_{p-1}}(P)$. Continuing on in this way, one can show that the contribution of all the possible placements of r_i in level ℓ_i in column i contributes a factor $wt(Q)q^{pt_i}[y_0 + \cdots y_{p-1}]_q$ to $\sum_{P \in D_i(Q)} wt(P)$.

We have $px + b_i - pt - pt_i - p$ uncancelled cells below level ℓ_i in column i. If we place a rook r_i in the sth such cell reading from the top, then r_i contributes $q^{pt_i+p+s-1}q^{\sum_{j=0}^{p-1} y_i - 1} = q^{pt_i+y_0+\cdots+y_{p-1}+s-1}$ to $wt(P)$. Here, $q^{pt_i+p+s-1}$ comes from r_i contribution to $inv(P)$ and $q^{\sum_{j=0}^{p-1} y_i - 1}$ comes from the fact that r_i would contribute 1 to $E_j(P)$ for $j = 0, \ldots, p-1$. It follows that contribution to $\sum_{P \in D_i(Q)} wt(P)$ over all possible placements of rooks in the remaining $px + b_i - pt - pt_i - p$ uncancelled cells is

$$wt(Q)q^{pt_i+y_0+\cdots+y_{p-1}}[px + b_i - pt - pt_i - p]_q.$$

Hence, the total contribution to $\sum_{P \in D_i(Q)} wt(P)$ of the placements of rooks in the ith column in the case where $b_i \geq pi$ is

$$\begin{aligned} wt(Q)([pt_i]_q + q^{pt_i}[\sum_{i=0}^{p-1} y_i]_q + q^{pt_i+\sum_{i=0}^{p-1} y_i}[px + b_i - pt - pt_i - p]_q) \\ = wt(Q)[px + b_i - p(t+1) + y_0 + \cdots + y_{p-1}]_q, \end{aligned}$$

as desired.

Example 12.1 (q-cycle-counting Lah numbers) We consider the q-analogue of cycle-counting Lah numbers $L_{n,k}^p(y_0, \ldots, y_{p-1})$ for $C_p \wr S_n$. We let

$$L_{n,k}^p(q, y_0, \ldots, y_{p-1}) = r_{n-k}^p(\mathcal{L}_n^p, q, y_0, \ldots, y_{p-1}), \tag{12.24}$$

where $\mathcal{L}_n^p$ is the Ferrers board which consists of n columns of height $p(n-1)$. Then, by Theorem 12.9, we have

$$\begin{aligned} &[px]_q[px+y_0 + \cdots + y_{p-1}]_q \uparrow_{n-1,p} \\ &\quad = \sum_{k=1}^{n} [px]_q[p(x-1)]_q \cdots [p(x-k+1)]_q L_{n,k}^p(q, y_0, \ldots, y_{p-1}). \end{aligned} \tag{12.25}$$

Note that

$$\begin{aligned} &\sum_{k=1}^{n+1} L_{n+1,k}^p(q, y_0, \ldots, y_{p-1})[px]_q \downarrow_{k,p} \\ &\quad = [px]_q[px + y_0 + \cdots + y_{p-1}]_q \uparrow_{n,p} \\ &\quad = [px]_q[px + y_0 + \cdots + y_{p-1}]_q \uparrow_{n-1,p} [px + p(n-1) + y_0 + \cdots + y_{p-1}]_q \\ &\quad = \sum_{k=1}^{n} L_{n,k}^p(q, y_0, \ldots, y_{p-1})[px]_q \downarrow_{k,p} [px + p(n-1) + y_0 + \cdots + y_{p-1}]_q \end{aligned}$$

$$
\begin{aligned}
&= \sum_{k=1}^{n} L_{n,k}^{p}(q, y_0, \ldots, y_{p-1}) \\
&\quad \times [px]_q \downarrow_{k,p} [p(x-k) + p(n+k-1) + y_0 + \cdots + y_{p-1}]_q \\
&= \sum_{k=1}^{n} q^{p(n+k-1)+y_0+\cdots+y_{p-1}} L_{n,k}^{p}(q, y_0, \ldots, y_{p-1})[px]_q \downarrow_{k+1,p} \\
&\quad + \sum_{k=1}^{n} L_{n,k}^{p}(q, y_0, \ldots, y_{p-1})[px]_q \downarrow_{k,p} [p(n+k-1) + y_0 + \cdots + y_{p-1}]_q .
\end{aligned}
$$

Thus, we get the recurrence relation

$$
\begin{aligned}
L_{n+1,k}^{p}(q, y_0, \ldots, y_{p-1}) &= q^{p(n+k-1)+y_0+\cdots+y_{p-1}} L_{n,k-1}^{p}(q, y_0, \ldots, y_{p-1}) \\
&\quad + L_{n,k}^{p}(q, y_0, \ldots, y_{p-1})[p(n+k-1) + y_0 + \cdots + y_{p-1}]_q .
\end{aligned}
\tag{12.26}
$$

Using this recursion, we can also prove the following closed-form expression

$$
\begin{aligned}
&L_{n,k}^{p}(q, y_0, \ldots, y_{p-1}) \\
&\quad = q^{k(k-1)p+(k-1)(y_0+\cdots+y_{p-1})} \begin{bmatrix} n-1 \\ k-1 \end{bmatrix}_{q^p} [pk + y_0 + \cdots y_{p-1}]_q \uparrow_{n-k,p} .
\end{aligned}
\tag{12.27}
$$

4 Q-Analogues of Cycle-Counting p-Hit Numbers

Recall that a full Ferrers board $B = F(b_1, \ldots, b_n) \subseteq B_p^n$ is *regular* if $b_i = p \cdot c_i$, where $c_i \geq i$ for $1 \leq i \leq n$. The goal of this section is to define a q-analogue of cycle-counting p-hit numbers for full regular Ferrers boards and to give a conjectured combinatorial interpretation for them. Before we start, we introduce an alternate notation for a Ferrers board. Given a Ferrers board $B = F(b_1, b_2, \ldots, b_n) \subseteq B_n^p$, we will also use the notation $B = B(h_1^p, d_1; \ldots; h_t^p, d_t)$ which uses the step heights and depths as pictured in Fig. 6.

Now if $B = F(pc_1, \ldots, pc_n)$ is a regular full Ferrers board contained in B_n^p, then, in the notation $B = B(h_1^p, d_1; \ldots; h_t^p, d_t)$, $h_j^p = p \cdot h_j$ where h_j's are the number of levels of the corresponding step. Then, by Theorem 12.9,

$$
\sum_{k=0}^{n} r_{n-k}^{p}(B, q, y_0, \ldots, y_{p-1})[px]_q \downarrow_{k,p} = \prod_{i=1}^{n} [px + p(c_i - i) + y_0 + \cdots + y_{p-1}]_q .
\tag{12.28}
$$

We let the right-hand side of (12.28) be

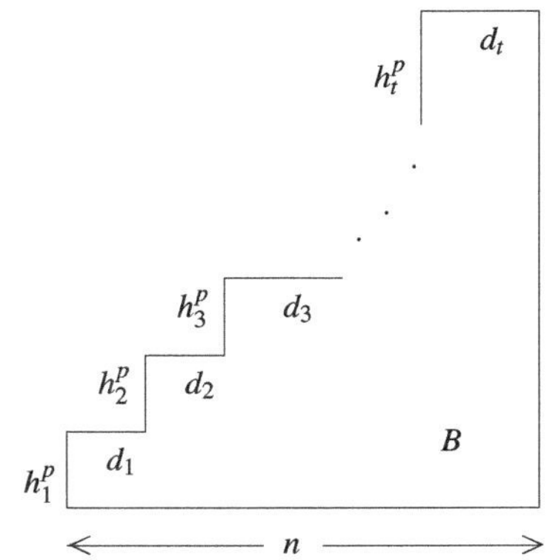

Fig. 6 Ferrers board $B = B(h_1^p, d_1; \ldots; h_t^p, d_t)$

$$\mathrm{PR}[B, x, y_0, \ldots, y_{p-1}] := \prod_{i=1}^{n} [px + p(c_i - i) + y_0 + \cdots + y_{p-1}]_q.$$

We define our q-analogue $H_{k,n}^p(B, q, y_0, \ldots, y_{p-1})$ of the cycle-counting p-hit numbers by

$$\sum_{k=0}^{n} r_{n-k}^p(B, q, y_0, \ldots, y_{p-1})[y_0 + \cdots + y_{p-1}]_q \uparrow_{k,p} z^k \times \prod_{i=k+1}^{n} (1 - zq^{y_0+\cdots+y_{p-1}+p(i-1)}) = \sum_{k=0}^{n} H_{k,n}^p(B, q, y_0, \ldots, y_{p-1})z^k. \tag{12.29}$$

Note that when $q = 1$, by changing z to z^{-1} and multiplying z^n on both sides, we can transform (12.29) to

$$\sum_{k=0}^{n} H_{n-k,n}^p(B, 1, y_0, \ldots, y_{p-1})z^k = \sum_{k=0}^{n} r_k^p(B, 1, y_0, \ldots, y_{p-1})(y_0 + \cdots + y_{p-1}) \uparrow_{n-k,p} (z-1)^k.$$

By comparing it to the result of Theorem 12.5, we can see that

$$H_{k,n}^p(B, 1, y_0, \ldots, y_{p-1}) = H_{n-k,n}^p(B, y_0, \ldots, y_{p-1}).$$

Our first goal is to give a recursion for the $H_{k,n}^p(B, q, y_0, \ldots, y_{p-1})$'s which will show that $H_{k,n}^p(B, q, y_0, \ldots, y_{p-1})$ is a polynomial in q with non-negative coefficients when $y_0, \ldots, y_{p-1}$ are non-negative integers. To derive our desired recursion of $H^p_{k,n}(B, q, y_0, \ldots, y_{p-1})$, we define a more general version of it. That is, we define $H^p_{k,n}(B, x, q, y_0, \ldots, y_{p-1})$ by

$$\sum_{k=0}^{n} H^p_{k,n}(B, x, q, y_0, \ldots, y_{p-1}) z^k$$
$$= \sum_{k=0}^{n} r^p_{n-k}(B, q, y_0, \ldots, y_{p-1})[px]_q \uparrow_{k,p} z^k \prod_{i=k+1}^{n} (1 - zq^{px+p(i-1)}).$$

Remark 12.1 We note that

$$H^p_{k,n}(B, q, y_0, \ldots, y_{p-1}) = H^p_{k,n}(B, x, q, y_0, \ldots, y_{p-1})\big|_{x=\frac{y_0+\cdots+y_{p-1}}{p}},$$

and $H^p_{k,n}(B, x, q, y_0, \ldots, y_{p-1})$ is a generalization of $H_k(x, y, B)$ as defined by Haglund in [10] and used by Butler in [5].

The following two propositions are the generalizations of the result of Haglund in [10, Lemma 5.1, Lemma 5.7].

Proposition 12.1 *Suppose $B = F(pc_1, \ldots, pc_n)$ is a regular full Ferrers board contained in B^p_n. Then, we have*

$$H^p_{k,n}(B, x, q, y_0, \ldots, y_{p-1})$$
$$= \sum_{j=0}^{k} \begin{bmatrix} n+x \\ k-j \end{bmatrix}_{q^p} \begin{bmatrix} x+j-1 \\ j \end{bmatrix}_{q^p} (-1)^{k-j} q^{p\binom{k-j}{2}} PR[B, j, y_0, \ldots, y_{p-1}], \tag{12.30}$$

where $PR[B, j, y_0, \ldots, y_{p-1}] = \prod_{i=1}^{n}[pj + p(c_i - i) + y_0 + \cdots + y_{p-1}]_q$.

Proof In the proof, we use the following short-hand notation

$$([x]_{q^p})_j = [x]_{q^p}[x+1]_{q^p} \cdots [x+j-1]_{q^p}.$$

The right-hand side of (12.30) is

$$\sum_{j=0}^{k} \begin{bmatrix} n+x \\ k-j \end{bmatrix}_{q^p} (-1)^{k-j} q^{p\binom{k-j}{2}} \begin{bmatrix} x+j-1 \\ j \end{bmatrix}_{q^p}$$
$$\times \sum_{s=0}^{n} [p]^s_q [j]_{q^p}[j-1]_{q^p} \cdots [j-s+1]_{q^p} r^p_{n-s}(B, q, y_0, \ldots, y_{p-1})$$
$$= \sum_{s=0}^{n} [p]^s_q r^p_{n-s}(B, q, y_0, \ldots, y_{p-1})$$

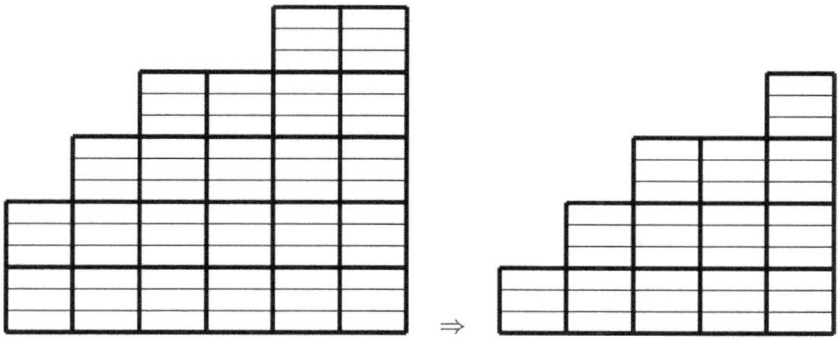

Fig. 7 $B = B(6, 1; 3, 1; 3, 2; 3, 2)$ and $B - h_1 - d_4 = B(3, 1; 3, 1; 3, 2; 3, 1)$, for $p = 3$

$$
\begin{aligned}
&\times \sum_{j \geq s} \begin{bmatrix} n+x \\ k-j \end{bmatrix}_{q^p} (-1)^{k-j} q^{p\binom{k-j}{2}} \frac{([x]_{q^p})_j}{([1]_{q^p})_j} [j]_{q^p} \cdots [j-s+1]_{q^p} \\
&= \sum_{s=0}^{n} [p]_q^s r_{n-s}^p (B, q, y_0, \ldots, y_{p-1}) \\
&\quad \times \sum_{\substack{u \geq 0 \\ j=u+s}} \begin{bmatrix} n+x \\ k-u-s \end{bmatrix}_{q^p} (-1)^{k-u-s} q^{p\binom{k-u-s}{2}} \frac{([x]_{q^p})_{u+s}}{([1]_{q^p})_{u+s}} ([u+1]_{q^p})_s \\
&= \sum_{s=0}^{n} [p]_q^s r_{n-s}^p (B, q, y_0, \ldots, y_{p-1}) \sum_{u \geq 0} \begin{bmatrix} n+x \\ k-s \end{bmatrix}_{q^p} (-1)^{k-s} q^{p\binom{k-s}{2}} \\
&\quad \times \frac{([s-k]_{q^p})_u}{([n+x-k+s+1]_{q^p})_u} \frac{([x]_{q^p})_s ([x+s]_{q^p})_u}{([s+1]_{q^p})_u} \begin{bmatrix} u+s \\ u \end{bmatrix}_{q^p} \\
&= \sum_{s=0}^{n} [p]_q^s r_{n-s}^p (B, q, y_0, \ldots, y_{p-1}) \sum_{u \geq 0} \begin{bmatrix} n+x \\ k-s \end{bmatrix}_{q^p} (-1)^{k-s} q^{p\binom{k-s}{2}} \\
&\quad \times ([x]_{q^p})_s \sum_{u \geq 0} \frac{([-k+s]_{q^p})_s ([x+s]_{q^p})_u}{([1]_{q^p})_u ([n+x-k+s+1]_{q^p})_u} \\
&= \sum_{s=0}^{n} [p]_q^s r_{n-s}^p (B, q, y_0, \ldots, y_{p-1}) \\
&\quad \times \begin{bmatrix} n+x \\ k-s \end{bmatrix}_{q^p} (-1)^{k-s} q^{p\binom{k-s}{2}} ([x]_{q^p})_s \frac{([n-k+1]_{q^p})_{k-s}}{([n+x-k+s+1]_{q^p})_{k-s}} \\
&= \sum_{s=0}^{n} [p]_q^s ([x]_{q^p})_s r_{n-s}^p (B, q, y_0, \ldots, y_{p-1}) \begin{bmatrix} n-s \\ k-s \end{bmatrix}_{q^p} (-1)^{k-s} q^{p\binom{k-s}{2}} \\
&= H_{k,n}^p (B, x, y_0, \ldots, y_{p-1}).
\end{aligned}
$$

Proposition 12.2 *Suppose* $B = B(h_1^p, d_1; h_2^p, d_2; \ldots; h_t^p, d_t)$*, where* $h_i^p = ph_i$ *for non-negative integer* h_i*,* $i = 1, \ldots, t$*, is regular full Ferrers board contained in* B_n^p*. Let* $H_i := h_1 + \cdots + h_i$*,* $D_i := d_1 + \cdots + d_i$*, and the notation* $B - h_i - d_j$ *refer to the board obtained from B by decreasing* h_i *and* d_j *by one each and leaving the other parameters fixed. (For an example of the board* $B - h_i - d_j$*, refer to Fig. 7). Then, we have the following recursion for* $H_{k,n}^p(B, x, q, y_0, \ldots, y_{p-1})$*.*

$$\begin{aligned}
&H_{k,n}^p(B, x, q, y_0, \ldots, y_{p-1}) \\
&\quad = [p]_q \left[k + H_l - D_l + d_l - 1 + \frac{y_0 + \cdots + y_{p-1}}{p}\right]_{q^p} \\
&\qquad \times H_{k,n-1}^p(B - h_l - d_l, x, y_0, \ldots, y_{p-1}) \\
&\quad + [p]_q \Bigg(-q^{p(n+x-1)}\left[k + H_l - D_l + d_l - 1 + \frac{y_0 + \cdots + y_{p-1}}{p}\right]_{q^p} \\
&\qquad + q^{p(k+H_l-D_l+d_l-2+\frac{y_0+\cdots+y_{p-1}}{p})}[n + x]_{q^p}\Bigg) \\
&\qquad \times H_{k-1,n-1}^p(B - h_l - d_l, x, q, y_0, \ldots, y_{p-1}).
\end{aligned}$$

Proof We have

$$\begin{aligned}
&H_{k,n}^p(B, x, q, y_0, \ldots, y_{p-1}) \\
&= \sum_{s=0}^{k} \begin{bmatrix} n+x \\ k-s \end{bmatrix}_{q^p} \begin{bmatrix} x+s-1 \\ s \end{bmatrix}_{q^p} (-1)^{k-s} q^{p\binom{k-s}{2}} \\
&\qquad \times \prod_{i=1}^{n} [ps + p(b_i - i) + y_0 + \cdots + y_{p-1}]_q \\
&= \sum_{s=0}^{k} \begin{bmatrix} n+x \\ k-s \end{bmatrix}_{q^p} \begin{bmatrix} x+s-1 \\ s \end{bmatrix}_{q^p} (-1)^{k-s} q^{p\binom{k-s}{2}} \\
&\quad \times [ps + p(H_l - D_l + d_l - 1) + y_0 + \cdots + y_{p-1}]_q \mathrm{PR}[B - h_l - d_l, s, y_0, \ldots, y_{p-1}] \\
&= \sum_{s=0}^{k} \begin{bmatrix} n+x \\ k-s \end{bmatrix}_{q^p} \begin{bmatrix} x+s-1 \\ s \end{bmatrix}_{q^p} (-1)^{k-s} q^{p\binom{k-s}{2}} \\
&\qquad \times \mathrm{PR}[B - h_l - d_l, s, y_0, \ldots, y_{p-1}] \\
&\qquad \times \Bigg\{[p]\left[k + H_l - D_l + d_l - 1 + \frac{y_0 + \cdots + y_{p-1}}{p}\right]_{q^p} \\
&\qquad\qquad -q^{p(s+H_l-D_l+d_l-1+\frac{y_0+\cdots+y_{p-1}}{p})}[p][k-s]_{q^p}\Bigg\} \\
&= [p]_q \left[k + H_l - D_l + d_l - 1 + \frac{y_0 + \cdots + y_{p-1}}{p}\right]_{q^p}
\end{aligned}$$

$$\times \sum_{s=0}^{k} \begin{bmatrix} n+x \\ k-s \end{bmatrix}_{q^p} \begin{bmatrix} x+s-1 \\ s \end{bmatrix}_{q^p} (-1)^{k-s} q^{p\binom{k-s}{2}} \mathrm{PR}[B-h_l-d_l, s, y_0, \ldots, y_{p-1}]$$

$$- [p]_q q^{p(s+H_l-D_l+d_l-1+\frac{y_0+\cdots+y_{p-1}}{p})} \sum_{s=0}^{k} [n+x]_{q^p} \begin{bmatrix} n+x-1 \\ k-s-1 \end{bmatrix}_{q^p}$$

$$\times \begin{bmatrix} x+s-1 \\ s \end{bmatrix}_{q^p} (-1)^{k-s} q^{p\left(\binom{k-s}{2}+s\right)} \mathrm{PR}[B-h_l-d_l, s, y_0, \ldots, y_{p-1}]$$

$$= [p]_q \left[k+H_l-D_l+d_l-1+\frac{y_0+\cdots+y_{p-1}}{p}\right]_{q^p} \sum_{s=0}^{k} \begin{bmatrix} x+s-1 \\ s \end{bmatrix}_{q^p} (-1)^{k-s}$$

$$\times \left\{ \begin{bmatrix} n+x-1 \\ k-s \end{bmatrix}_{q^p} q^{p\binom{k-s}{2}} + \begin{bmatrix} n+x-1 \\ k-s-1 \end{bmatrix}_{q^p} q^{p\left(\binom{k-s-1}{2}+n+x-1\right)} \right\}$$

$$\times \mathrm{PR}[B-h_l-d_l, s, y_0, \ldots, y_{p-1}]$$

$$- [p]_q q^{p(s+H_l-D_l+d_l-1+\frac{y_0+\cdots+y_{p-1}}{p})} \sum_{s=0}^{k-1} [n+x]_{q^p} \begin{bmatrix} n+x-1 \\ k-s-1 \end{bmatrix}_{q^p}$$

$$\times \begin{bmatrix} x+s-1 \\ s \end{bmatrix}_{q^p} (-1)^{k-s} q^{p\left(\binom{k-s-1}{2}+k-1\right)} \mathrm{PR}[B-h_l-d_l, s, y_0, \ldots, y_{p-1}]$$

$$= [p]_q \left[k+H_l-D_l+d_l-1+\frac{y_0+\cdots+y_{p-1}}{p}\right]_{q^p}$$

$$\times H^p_{k,n-1}(B-h_l-d_l, x, q, y_0, \ldots, y_{p-1})$$

$$+ [p]_q H^p_{k-1,n-1}(B-h_l-d_l, x, q, y_0, \ldots, y_{p-1})$$

$$\times \Bigg\{ q^{p(k+H_l-D_l+d_l-2+\frac{y_0+\cdots+y_{p-1}}{p})} [n+x]_{q^p}$$

$$- q^{p(n+x-1)} \left[k+H_l-D_l+d_l-1+\frac{y_0+\cdots+y_{p-1}}{p}\right]_{q^p} \Bigg\}$$

$$= [p]_q \left[k+H_l-D_l+d_l-1+\frac{y_0+\cdots+y_{p-1}}{p}\right]_{q^p}$$

$$\times H^p_{k,n-1}(B-h_l-d_l, x, q, y_0, \ldots, y_{p-1})$$

$$+ [p]_q \left[n+x-k-H_l+D_l-d_l+1-\frac{y_0+\cdots+y_{p-1}}{p}\right]_{q^p}$$

$$\times q^{p(k+H_l-D_l+d_l-2+\frac{y_0+\cdots+y_{p-1}}{p})} H^p_{k-1,n-1}(B-h_l-d_l, x, q, y_0, \ldots, y_{p-1}).$$

Proposition 12.3 *If B_j is the board*

$$B(h^p_1, d_1; \cdots; h^p_{l-1}, d_{l-1}; h^p_l - pj, d_l - j; h^p_{l+1}, d_{l+1}; \ldots; h^p_t, d_t)$$

obtained from a regular Ferrers board B by decreasing h^p_l by pj and d_l by j (here we assume that $j \le h_l, d_l$, where $h^p_l = ph_l$), then

$$H_{k,n}^p(B,x,q,y_0,\ldots,y_{p-1}) = [p]_q^j[j]_{q^p}! \sum_{s=k-j}^{k} H_{s,n}^p(B_j,x,q,y_0,\ldots,y_{p-1})$$
$$\times \begin{bmatrix} T_l-1+s \\ j-k+s \end{bmatrix}_{q^p} \begin{bmatrix} n-T_l+x-s \\ k-s \end{bmatrix}_{q^p} q^{p(k-s)(T_l+k-j-1)}, \quad (12.31)$$

where $T_l = H_l - D_{l-1} + \frac{y_0+\cdots+y_{p-1}}{p}$.

Proof The proof can be done by induction on j and by using the recursion in Proposition 12.2. The proof is similar to the proof of [10, Theorem 4.1, Theorem 5.8], and hence, we omit the details.

By using Proposition 12.2, we can derive the recursion for $H_{k,n}^p(B,q,y_0,\ldots,y_{p-1})$.

Theorem 12.10 *Suppose* $B = B(h_1^p, d_1; h_2^p, d_2; \ldots; h_t^p, d_t)$, *where* $h_i^p = ph_i$, *is regular full Ferrers board contained in* B_n^p. *Let* $H_i := h_1 + \cdots + h_i$, $D_i := d_1 + \cdots + d_i$, *and the notation* $B - h_i - d_j$ *refers to the board obtained from* B *by decreasing* h_i *and* d_j *by one each and leaving the other parameters fixed. Then, we have the following recursion for* $H_{k,n}^p(B,q,y_0,\ldots,y_{p-1})$:

$$\begin{aligned} &H_{k,n}^p(B,q,y_0,\ldots,y_{p-1}) \\ &= [p]_q \left[\frac{y_0+\cdots+y_{p-1}}{p} + k + d_t - 1\right]_{q^p} H_{k,n-1}^p(B-h_t-d_t,q,y_0,\ldots,y_{p-1}) \\ &+ [p]_q q^{p\left(\frac{y_0+\cdots+y_{p-1}}{p}+k+d_t-2\right)}[n-k-d_t+1]_{q^p} \\ &\quad \times H_{k-1,n-1}^p(B-h_t-d_t,q,y_0,\ldots,y_{p-1}), \end{aligned} \quad (12.32)$$

where h_t *and* d_t *are the height (number of levels) and the depth of the last step of* B.

We note that it follows from Theorem 12.10 that if $B = F(pc_1,\ldots,pc_n)$ is a regular full Ferrers board in B_n^p and $y_0,\ldots,y_{p-1}$ are non-negative integers, then $H_{k,n}^p(B,q,y_0,\ldots,y_{p-1})$ is a polynomial with non-negative coefficients in q. Here are some small examples.

Example 12.2 When B_1 has only one square (level) with p sublevels, i.e. $B_1 = F(p)$, then

$$\begin{aligned} &H_{0,1}^p(B_1,q,y_0,\ldots,y_{p-1}) \\ &= r_1^p = \sum_{P\in\mathcal{N}_1^p(B_1)} \left[\prod_{j=0}^{p-1}[y_j]_q^{\mathrm{cyc}_j(P)} q^{\mathrm{inv}(P)+\sum_{j=0}^{p-1}(y_j-1)E_j(P)}\right] \\ &= q^0[y_{p-1}]_q + q^{1+(y_{p-1}-1)}[y_{p-2}]_q + \cdots + q^{p-1+\sum_{j=1}^{p-1}(y_j-1)}[y_0]_q \\ &= q^0[y_{p-1}]_q + q^{y_{p-1}}[y_{p-2}]_q + q^{y_{p-1}+y_{p-2}}[y_{p-3}]_q + \cdots + q^{\sum_{j=1}^{p-1}y_j}[y_0]_q \\ &= [y_0+\cdots+y_{p-1}]_q, \end{aligned}$$

$H^p_{k,1}(B_1, q, y_0, \ldots, y_{p-1}) = 0,$ for $k > 0$.

We continue computing small examples for $n = 2$:

$H^p_{0,2}($  $, q, y_0, \ldots, y_{p-1}) = [y_0 + \cdots + y_{p-1}]_q [y_0 + \cdots + y_{p-1} + p]_q,$

$H^p_{k,2}($  $, q, y_0, \ldots, y_{p-1}) = 0,$ for $k > 0$.

Furthermore,

$H^p_{0,2}($ 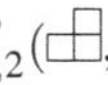 $, q, y_0, \ldots, y_{p-1}) = [y_0 + \cdots + y_{p-1}]^2_q,$

$H^p_{1,2}($ 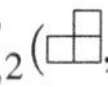 $, q, y_0, \ldots, y_{p-1}) = q^{(y_0+\cdots+y_{p-1})}[p]_q [y_0 + \cdots + y_{p-1}]_q,$

$H^p_{2,2}($ 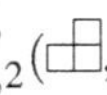 $, q, y_0, \ldots, y_{p-1}) = 0.$

Based on the q-statistics for the cycle-counting hit numbers defined by Butler in [5], we conjecture a similar q-statistic for the cycle-counting p-hit numbers. Before we make a precise statement, we need some definitions.

For a full regular Ferrers board $B \subseteq B^p_n$, let $\mathcal{N}^p(B) = \cup_{k=1}^n \mathcal{N}^p_k(B)$. For $p \in \mathcal{N}^p(B)$, note the Butler's statistic $s_{B,b}(P)$ [5] defined as the number of squares on B^p_n which neither contain a rook from P nor are cancelled, after applying the following cancellation scheme:

1. Each rook cancels all squares to the right in its row.
2. Each rook on B cancels all squares above it in its column (squares both on B and strictly above B).
3. Each rook on B which also completes a cycle cancels all squares below it in its column as well.
4. Each rook off B cancels all squares below it but above B.

Define $\text{cyc}_{\geq j}(P)$ by

$$\text{cyc}_{\geq j}(P) := \sum_{i=j}^{p-1} \text{cyc}_i(P).$$

Since $b_i \geq pi$, there exists a unique level, say u, in column i such that considering only rooks from P in column 1 through column $i-1$ of B completes a cycle. At the ith column, define $\tilde{E}_i(P)$ by

$$\tilde{E}_i(P) = \begin{cases} p, & \text{if there is no rook from } P \text{ in column } i \text{ on or above the level } u, \\ 0, & \text{if there is a rook from } P \text{ in column } i \text{ above the level } u, \\ p-1-j, & \text{if there is a rook on the level } u \text{ completing a cycle of sign } \omega_j. \end{cases}$$

Then, we conjecture the following combinatorial formula for $H^p_{k,n}(B, q, y_0, \ldots, y_{p-1})$.

Conjecture 12.1 Let $\mathcal{H}_{k,n}(B)$ be the set of all placements corresponding $\sigma \in C_p \wr S_n$ such that $|\sigma \cap B| = k$. Then, for a full regular Ferrers board $B \subseteq B_n^p$,

$$H_{k,n}^p(B, q, y_0, \ldots, y_{p-1}) = \sum_{P \in \mathcal{H}_{n-k,n}(B)} \left(\prod_{j=0}^{p-1} [y_j]_q^{\mathrm{cyc}_j(P)} \right) q^{s_{B,b}(P) + \sum_{i=1}^{n} \tilde{E}_i(P) + \sum_{j=0}^{p-1}((y_j - 1)(n - \mathrm{cyc}_{\geq j}(P)))}. \quad (12.33)$$

An obvious approach to prove Conjecture 12.1 is to give a combinatorial proof that the recursion of $H_{k,n}^p(B, q, y_0, \ldots, y_{p-1})$ in (12.32) holds. We were not able to find a natural way to partition the rook placements in $\mathcal{N}_k(B)$ to account for the two terms on the right-hand side of (12.32). Our next example will show that while we can verify the recursion holds for $B = F(p, 2p, 3p, 4p) \subset [4] \times [4p]$, the way that we can divide the partition the rook placements in B to account for the two terms on the right-hand side of (12.32) is quite complicated. Thus, we do not see how the recursion can be derived naturally by extending the rook placement corresponding to the permutations of $n - 1$ numbers.

Example 12.3 We consider a staircase board $B = F(p, 2p, 3p, 4p) \subset [4] \times [4p]$. Then $B - h_4 - d_4 = F(p, 2p, 3p)$ and the recursion (12.32) when $k = 1$ is

$$\begin{aligned} H_{1,4}^p(B, q, y_0, \ldots, y_{p-1}) &= [y_0 + \cdots + y_{p-1} + p]_q H_{1,3}^p(B - h_4 - d_4, q, y_0, \ldots, y_{p-1}) \\ &\quad + q^{y_0 + \cdots + y_{p-1}} [p]_q [3]_{q^p} H_{0,3}^p(B - h_4 - d_4, q, y_0, \ldots, y_{p-1}). \end{aligned} \quad (12.34)$$

For a rook placement $P \in \mathcal{H}_{n-k,n}(B)$, let

$$wt(P) = \left(\prod_{j=0}^{p-1} [y_j]_q^{\mathrm{cyc}_j(P)} \right) q^{s_{B,b}(P) + \sum_{i=1}^{n} \tilde{E}_i(P) + \sum_{j=0}^{p-1}[(y_j - 1)(n - \mathrm{cyc}_{\geq j}(P))]}.$$

Then for $\sigma = (1)(2)(3) \in S_3$,

•	•	X
•	X	•
X	•	•

$$H_{0,3}^p(B - h_4 - d_4, q, y_0, \ldots, y_{p-1}) = \sum_{P \in C_p \wr \sigma} wt(P) = [y_0 + \cdots + y_{p-1}]_q^3.$$

This can be extended to a placement in $\mathcal{H}_{3,4}(B)$ as follows.

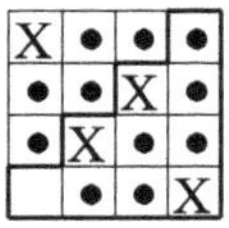

$$\sigma_1 = (14)(2)(3), \sum_{P \in C_p \wr \sigma_1} wt(P) = [y_0 + \cdots + y_{p-1}]_q^3 q^{y_0+\cdots+y_{p-1}}[p]_q, \tag{12.35}$$

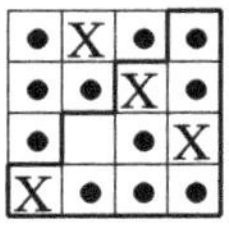

$$\sigma_2 = (1)(24)(3), \sum_{P \in C_p \wr \sigma_2} wt(P) = [y_0 + \cdots + y_{p-1}]_q^3 q^{y_0+\cdots+y_{p-1}}[p]_q, \tag{12.36}$$

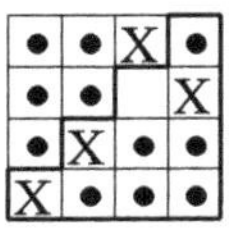

$$\sigma_3 = (1)(2)(34), \sum_{P \in C_p \wr \sigma_3} wt(P) = [y_0 + \cdots + y_{p-1}]_q^3 q^{y_0+\cdots+y_{p-1}}[p]_q. \tag{12.37}$$

There are four permutations in S_3 which can be considered for $H_{1,3}^p(B - h_4 - d_4, q, y_0, \ldots, y_{p-1})$ and they can be extended to a placement in $\mathcal{H}_{3,4}$ as follows.

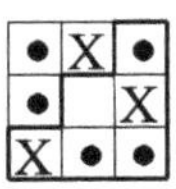

$$\alpha = (1)(23), \sum_{C_p \wr \alpha} wt(P) = [y_0 + \cdots + y_{p-1}]_q^2 q^{y_0+\cdots+y_{p-1}}[p]_q$$

$\Rightarrow$

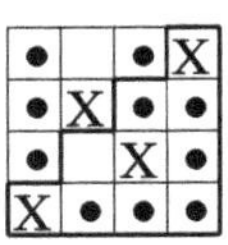

$$\alpha_1 = (1)(23)(4), \sum_{C_p \wr \alpha_1} wt(P) = [y_0 + \cdots + y_{p-1}]_q^3 q^{y_0+\cdots+y_{p-1}+p}[p]_q, \tag{12.38}$$

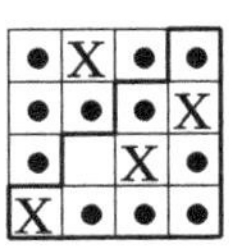

$$\alpha_2 = (1)(243), \sum_{C_p \wr \alpha_2} wt(P) = [y_0 + \cdots + y_{p-1}]_q^2 q^{2(y_0+\cdots+y_{p-1})}[p]_q^2, \tag{12.39}$$

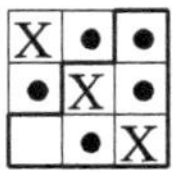

$$\beta = (13)(2), \sum_{C_p \wr \beta} wt(P) = [y_0 + \cdots + y_{p-1}]_q^2 q^{y_0+\cdots+y_{p-1}} [p]_q$$

$\Rightarrow$

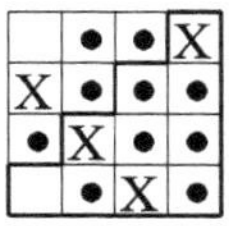

$$\beta_1 = (13)(2)(4), \sum_{C_p \wr \beta_1} wt(P) = [y_0 + \cdots + y_{p-1}]_q^3 q^{y_0+\cdots+y_{p-1}+p} [p]_q, \tag{12.40}$$

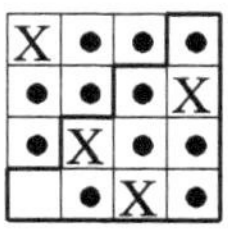

$$\beta_2 = (143)(2), \sum_{C_p \wr \beta_2} wt(P) = [y_0 + \cdots + y_{p-1}]_q^2 q^{2(y_0+\cdots+y_{p-1})} [p]_q^2, \tag{12.41}$$

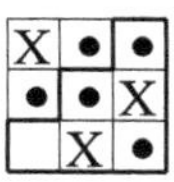

$$\gamma = (132), \sum_{C_p \wr \gamma} wt(P) = [y_0 + \cdots + y_{p-1}]_q^2 q^{2(y_0+\cdots+y_{p-1})} [p]_q^2$$

$\Rightarrow$

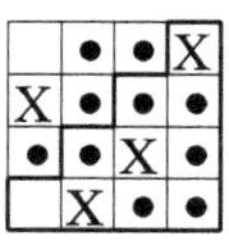

$$\gamma_1 = (132)(4), \sum_{C_p \wr \gamma_1} wt(P) = [y_0 + \cdots + y_{p-1}]_q^2 q^{2(y_0+\cdots+y_{p-1})+p} [p]_q^2, \tag{12.42}$$

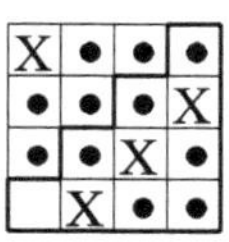

$$\gamma_2 = (1432), \sum_{C_p \wr \gamma_2} wt(P) = [y_0 + \cdots + y_{p-1}]_q q^{3(y_0+\cdots+y_{p-1})} [p]_q^3, \tag{12.43}$$

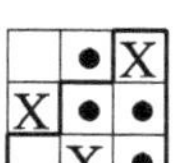

$$\delta = (12)(3), \sum_{C_p \wr \delta} wt(P) = [y_0 + \cdots + y_{p-1}]_q^2 q^{y_0+\cdots+y_{p-1}+p} [p]_q$$

$\Rightarrow$

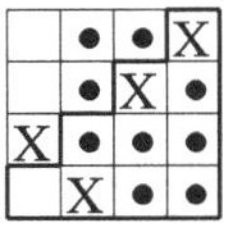

$$\delta_1 = (12)(3)(4), \sum_{C_p \wr \delta_1} wt(P) = [y_0 + \cdots + y_{p-1}]_q^3 q^{y_0+\cdots+y_{p-1}+2p}[p]_q, \tag{12.44}$$

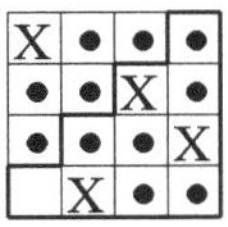

$$\delta_2 = (142)(3), \sum_{C_p \wr \delta_2} wt(P) = [y_0 + \cdots + y_{p-1}]_q^2 q^{2(y_0+\cdots+y_{p-1})}[p]_q^2. \tag{12.45}$$

(12.35) + (12.40) + (12.44) has a common factor $q^{y_0+\cdots+y_{p-1}}[p]_q[3]_{q^p}$ which is the coefficient of $H_{0,3}^p(B - h_4 - d_4, q, y_0, \ldots, y_{p-1})$ in (12.34) and the rest makes $H_{0,3}^p(B - h_4 - d_4, q, y_0, \ldots, y_{p-1})$.

$$\begin{aligned}(12.35) + (12.40) + (12.44) &= [y_0 + \cdots + y_{p-1}]_q^3 q^{y_0+\cdots+y_{p-1}}[p]_q(1 + q^p + q^{2p})\\ &= q^{y_0+\cdots+y_{p-1}}[y_0 + \cdots + y_{p-1}]_q^3 [p]_q [3]_{q^p}\\ &= q^{y_0+\cdots+y_{p-1}}[p]_q[3]_{q^p} H_{0,3}^p(B - h_4 - d_4, q, y_0, \ldots, y_{p-1}),\end{aligned}$$

Similarly, ((12.36) + (12.45)), ((12.37) + (12.41)), ((12.39) + (12.43)) and((12.38) + (12.42)) have a common factor $[y_0 + \cdots + y_{p-1} + p]_q$ which is the coefficient of $H_{1,3}^p(B - h_4 - d_4, q, y_0, \ldots, y_{p-1})$ in (12.34).

$$\begin{aligned}&((12.36) + (12.45)) + ((12.37) + (12.41))\\ &\qquad\qquad + ((12.39) + (12.43)) + ((12.38) + (12.42))\\ &= \left([y_0 + \cdots + y_{p-1}]_q^2 q^{y_0+\cdots+y_{p-1}}[p]_q([y_0 + \cdots + y_{p-1}]_q + q^{y_0+\cdots+y_{p-1}}[p]_q)\right)\\ &\quad + \left([y_0 + \cdots + y_{p-1}]_q^2 q^{y_0+\cdots+y_{p-1}}[p]_q([y_0 + \cdots + y_{p-1}]_q + q^{y_0+\cdots+y_{p-1}}[p]_q)\right)\\ &\quad + \left([y_0 + \cdots + y_{p-1}]_q q^{2(y_0+\cdots+y_{p-1})}[p]_q^2([y_0 + \cdots + y_{p-1}]_q + q^{y_0+\cdots+y_{p-1}}[p]_q)\right)\\ &\quad + \left([y_0 + \cdots + y_{p-1}]_q^2 q^{y_0+\cdots+y_{p-1}+p}[p]_q([y_0 + \cdots + y_{p-1}]_q + q^{y_0+\cdots+y_{p-1}}[p]_q)\right)\\ &= ([y_0 + \cdots + y_{p-1}]_q + q^{y_0+\cdots+y_{p-1}}[p]_q)\\ &\qquad \times \Big\{[y_0 + \cdots + y_{p-1}]_q q^{y_0+\cdots+y_{p-1}}[p]([y_0 + \cdots + y_{p-1}]_q + q^{y_0+\cdots+y_{p-1}}[p])\\ &\qquad\quad + [y_0 + \cdots + y_{p-1}]_q^2 q^{y_0+\cdots+y_{p-1}}[p](1 + q^p)\Big\}\\ &= [y_0 + \cdots + y_{p-1} + p]_q q^{y_0+\cdots+y_{p-1}}[p]_q\\ &\quad \times [y_0 + \cdots + y_{p-1}]_q \left([y_0 + \cdots + y_{p-1} + p]_q + [2]_q[y_0 + \cdots + y_{p-1}]_q\right)\\ &= [y_0 + \cdots + y_{p-1} + p]_q H_{1,3}^p(B - h_4 - d_4, q, y_0, \ldots, y_{p-1}).\end{aligned}$$

References

1. Barrese, K., Loehr, N., Remmel, J., Sagan, B.E.: m-Level rook placements. J. Combin. Theory Ser. A **124**, 130–165 (2014)
2. Briggs, K.S.: A combinatorial interpretation for p, q-hit numbers. Discret. Math. **308**, 6078–6103 (2008)
3. Briggs, K.S., Remmel, J.B.: A p, q-analogue of a formula of Frobenius. Electron. J. Combin. **10**, 26 (2003). Article #R9
4. Briggs, K.S., Remmel, J.B.: m-Rook numbers and a generalization of a formula of Frobenius to $\mathcal{C}_m \S \mathcal{S}_n$. J. Combin. Theory Ser. A **113**, 1138–1171 (2006)
5. Butler, F.: Rook theory and cycle-counting permutation statistics. Adv. Appl. Math. **33**, 655–675 (2004)
6. Butler, F.: Symmetry and unimodality f the q, x, y-hit numbers. Sém. Lothar. Combin. **54A**, 11 (2006). Article B54Ae
7. Chung, F., Graham, R.: On the cover polynomial of a digraph. J. Combin. Theory Ser. B **65**, 273–290 (1995)
8. Ehrenborg, R., Haglund, J., Readdy, M.: Colored juggling patterns and weighted rook placements. unpublished manuscript (1996)
9. Garsia, A.M., Remmel, J.B.: Q-Counting rook configurations and a formula of Frobenius. J. Combin. Theory Ser. A **41**, 246–275 (1986)
10. Haglund, J.: Rook theory and hypergeometric series. Adv. Appl. Math. **17**, 408–459 (1996)
11. Haglund, J., Remmel, J.B.: Rook theory for perfect matchings. Adv. App. Math. **27**, 438–481 (2001)
12. Remmel, J.B., Wachs, M.: Rook theory, generalized Stirling numbers and (p, q)-analogues. Electron. J. Combin. **11**, 48 (2004). Article #R84
13. Kaplansky, I., Riordan, J.: The problem of rooks and its applications. Duke Math. J. **13**, 259–268 (1946)

Asymptotic Behaviour of Certain q-Poisson, q-Binomial and Negative q-Binomial Distributions

Andreas Kyriakoussis and Malvina Vamvakari

Abstract We study the asymptotic behaviour of a class of discrete q-distributions. Specifically, the pointwise convergence of the Heine distribution to a deformed continuous Stieltjes–Wigert distribution and that of the Euler distribution to a deformed Gaussian distribution are established. Note that the Heine distribution is the limiting behaviour of both the q-binomial distribution of the first kind (q-Binomial I) and the negative q-binomial distribution of the first kind (negative q-Binomial I). Also, the Euler distribution is the limiting behaviour of both the q-binomial distribution of the second kind (q-Binomial II) and the negative q-binomial distribution of the second kind (negative q-Binomial II). In this paper, we also establish, by pointwise convergence, the deformed Gaussian approximation of the q-Binomial II and negative q-Binomial II distributions. The limiting behaviour of the q-Binomial I and the negative q-Binomial I distributions have been already studied by the authors.

Keywords Stirling's formula · q-factorials · Pointwise convergence · Continuous Stieltjes–Wigert distribution · q-distributions · q-Binomial I and II · Negative q-Binomial I and II · Heine and Euler distributions

2010 Mathematics Subject Classification Primary: 60C05 · Secondary: 05A30 · 33D90.

1 Introduction and Preliminaries

Kemp [3–5] introduced two q-Poisson distributions, called Heine and Euler distributions, with probability functions (p.f.) given by

---

A. Kyriakoussis · M. Vamvakari (✉)
Department of Informatics and Telematics, Harokopion University, Tavros, Greece
e-mail: mvamv@hua.gr

A. Kyriakoussis
e-mail: akyriak@hua.gr

G. E. Andrews et al. (eds.), *Lattice Path Combinatorics and Applications*, Developments in Mathematics 58, https://doi.org/10.1007/978-3-030-11102-1_13

$$f_X^H(x) = e_q(-\lambda)\frac{q^{\binom{x}{2}}\lambda^x}{[x]_q!}, \quad x = 0, 1, 2, \ldots, \ 0 < q < 1, \ 0 < \lambda < \infty \tag{13.1}$$

and

$$f_X^E(x) = E_q(-\lambda)\frac{\lambda^x}{[x]_q!}, \quad x = 0, 1, 2, \ldots, \ 0 < q < 1, \ 0 < \lambda(1-q) < 1, \tag{13.2}$$

respectively, where

$$e_q(z) := \sum_{n=0}^{\infty} \frac{(1-q)^n z^n}{(q;q)_n} = \sum_{n=0}^{\infty} \frac{z^n}{[n]_q!} = \frac{1}{((1-q)z;q)_\infty}, \quad |z| < 1, \tag{13.3}$$

$$E_q(z) := \sum_{n=0}^{\infty} \frac{(1-q)^n q^{\binom{n}{2}} z^n}{(q;q)_n} = \sum_{n=0}^{\infty} \frac{q^{\binom{n}{2}} z^n}{[n]_q!} = ((1-q)z;q)_\infty, \quad |z| < 1, \tag{13.4}$$

and

$$[n]_q! = [1]_q[2]_q \cdots [n]_q = \prod_{k=1}^{n} \frac{1-q^k}{(1-q)^n} = \frac{(q;q)_n}{(1-q)^n}, \quad 0 < q < 1, \quad [t]_q = \frac{1-q^t}{1-q}, \tag{13.5}$$

and he gave many applications. Both q-Poisson distributions are unimodal and log-concave with the Euler distribution being infinitely divisible but not the Heine distribution. Moreover, the Heine distribution is underdispersed but the Euler distribution is overdispersed.

Charalambides [1] reproduced the Heine distribution as direct approximation, as $n \to \infty$, of the q-Binomial I and the negative q-Binomial I distributions, with probability functions given by

$$f_X^B(x) = \binom{n}{x}_q q^{\binom{x}{2}}\theta^x \prod_{j=1}^{n}(1+\theta q^{j-1})^{-1}, \quad x = 0, 1, \ldots, n, \tag{13.6}$$

and

$$f_X^{NB}(x) = \binom{n+x-1}{x}_q q^{\binom{x}{2}}\theta^x \prod_{j=1}^{n+x}(1+\theta q^{j-1})^{-1}, \quad x = 0, 1, \ldots, \tag{13.7}$$

respectively, where $\theta > 0$ and $0 < q < 1$.

Moreover, Charalambides [1] reproduced the Euler distribution as direct approximation, as $n \to \infty$, of the q-Binomial II and the negative q-Binomial II distributions, with probability functions given by

$$f_X^{BS}(x) = \binom{n}{x}_q \theta^x \prod_{j=1}^{n-x}(1-\theta q^{j-1}), \quad x = 0, 1, \ldots, n, \tag{13.8}$$

and

$$f_X^{NBS}(x) = \binom{n+x-1}{x}_q \theta^x \prod_{j=1}^{n}(1-\theta q^{j-1}), \quad x = 0, 1, \ldots, \tag{13.9}$$

respectively, where $0 < \theta < 1$ and $0 < q < 1$ or $1 < q < \infty$ with $\theta q^{n-1} < 1$.

Kyriakoussis and Vamvakari [6, 7] proved limit theorems for the q-binomial distribution of the first kind (13.6) and negative q-Binomial distribution of the first kind (13.7) for constant q, by using pointwise convergence in a "q-analogous sense" of the classical de Moivre–Laplace limit theorem. Specifically in [6], for the needs of their study they established a q-Stirling formula for $n \to \infty$ of the q-factorial of order n, defined by by relation (13.5). Analytically, for constant q with $0 < q < 1$, we have

$$[n]_q! = \frac{q^{-1/8}(2\pi(1-q))^{1/2}}{(q \log q^{-1})^{1/2}} \frac{q^{\binom{n}{2}} q^{-n/2}[n]_{1/q}^{n+1/2}}{\prod_{j=1}^{\infty}(1+(q^{-n}-1)q^{j-1})} \left(1 + O(n^{-1})\right). \tag{13.10}$$

Then, the pointwise convergence of the q-binomial distribution of the first kind to a deformed continuous Stieltjes–Wigert distribution was established. In detail, transferred from the random variable X of the q-binomial distribution (13.6) to the equal-distributed deformed random variable $Y = [X]_{1/q}$ and for $n \to \infty$, the q-binomial distribution of the first kind was approximated by a deformed standardized continuous Stieltjes–Wigert distribution as follows:

$$\begin{aligned} f_X^B(x) \cong & \frac{q^{1/8}(\log q^{-1})^{1/2}}{(2\pi)^{1/2}} \left(q^{-3/2}(1-q)^{1/2}\frac{[x]_{1/q}-\mu_q}{\sigma_q} + q^{-1}\right)^{1/2} \\ & \cdot \exp\left(\frac{1}{2\log q} \log^2\left(q^{-3/2}(1-q)^{1/2}\frac{[x]_{1/q}-\mu_q}{\sigma_q} + q^{-1}\right)\right), \\ & x \geq 0, \end{aligned} \tag{13.11}$$

where $\theta = \theta_n$, for $n = 0, 1, 2, \ldots$, such that $\theta_n = q^{-\alpha n}$ with $0 < a < 1$ constant and μ_q and σ_q^2 the mean value and variance of the random variable Y, respectively. A similar asymptotic result has been provided in [7] for the negative q-binomial distribution of the first kind, as an application in a more general context concerning pointwise convergence of a family of confluent q-Chu–Vandermonde distributions.

Note that, with $q := q(n)$ a sequence of n with $q(n) \to 1$ as $n \to \infty$, Vamvakari [9] studied the effect of this assumption on the $q(n)$-analogue of the Stirling type and on the asymptotic behaviour of the $q(n)$-Binomial I distribution. Specifically, the following $q(n)$-analogue of the Stirling type has been provided, leading to the proof of a deformed Gaussian limiting behaviour for the $q(n)$-Binomial distribution of the first kind,

$$[n]_q! = \frac{(2\pi(1-q))^{1/2}}{(q\log q^{-1})^{1/2}} \frac{q^{\binom{n}{2}} q^{-n/2} [n]_{1/q}^{n+1/2}}{\prod_{j=1}^{\infty}(1+(q^{-n}-1)q^{j-1})} \left(1 + O\left(q^n(1-q)\right)\right), \tag{13.12}$$

where $q = q(n)$ with $q(n) \to 1$ as $n \to \infty$ and $q(n)^n = \Omega(1)$.

In this paper, we study the pointwise convergence of both the Heine and Euler distributions, when $\lambda \to \infty$. Specifically, we prove that the Heine distribution converges to a deformed continuous Stieltjes–Wigert distribution and the Euler distribution to a deformed Gaussian distribution. Moreover, the pointwise convergence of the q-binomial distribution of the second kind and the negative q-binomial distribution of the second kind to this deformed Gaussian distribution are proved.

2 Main Results

2.1 Continuous Limiting Behaviour of the Heine Distribution

In this section, we transfer from the random variable X of the Heine distribution (13.1) to the equal-distributed deformed random variable $Y = [X]_{1/q}$, and, using the q-analogue of Stirling's formula in (13.10), we establish the convergence of the Heine distribution to a deformed standardized continuous Stieltjes–Wigert distribution. Initially, we need to compute the mean value and the variance of the random variable Y, say μ_q^H and $(\sigma_q^H)^2$, respectively.

Proposition 13.1 *The mean and the variance of the random variable $Y = [X]_{1/q}$ are given by*

$$\mu_q^H = E\left([X]_{1/q}\right) = \lambda \quad \textit{and} \quad (\sigma_q^H)^2 = V\left([X]_{1/q}\right) = \lambda^2 q^{-1}(1-q) + \lambda, \tag{13.13}$$

respectively.

Proof The q-mean of the Heine distribution is equal to

$$\begin{aligned} \mu_q^H = E(Y) = E([X]_{1/q}) &= \sum_{x=0}^{\infty} [x]_{1/q} f_X(x) \\ &= e_q(-\lambda) \sum_{x=0}^{\infty} [x]_{1/q} \frac{q^{\binom{x}{2}} \lambda^x}{[x]_q!}. \end{aligned} \tag{13.14}$$

Since

$$[x]_{1/q} = q^{-x+1}[x]_q, \quad q^{-x+1} q^{\binom{x}{2}} = q^{\binom{x-1}{2}}, \quad [x]_q/[x]_q! = [x-1]_q!,$$

this can be rewritten as

$$\mu_q^H = e_q(-\lambda)\lambda \sum_{x=1}^{\infty} \frac{q^{\binom{x-1}{2}}\lambda^{x-1}}{[x-1]_q!}. \tag{13.15}$$

Using (13.3), we obtain the formula of the q-mean in (13.13).

For the evaluation of the q-variance, we need to find the second order moment of the random variable $Y = [X]_{1/q}$, which is given by

$$\begin{aligned} E[Y^2] = E[[X]_{1/q}^2] &= \sum_{x=0}^{\infty} [x]_{1/q}^2 f_X(x) \\ &= e_q(-\lambda) \sum_{x=0}^{\infty} [x]_{1/q}^2 \frac{q^{\binom{x}{2}}\lambda^x}{[x]_q!}. \end{aligned} \tag{13.16}$$

Since

$$[x]_q = [x-1]_q + q^{x-1}, \quad q^{-2x+2}q^{\binom{x}{2}} = q^{-1}q^{\binom{x-2}{2}},$$

Equation (13.16) becomes

$$\begin{aligned} E[Y^2] &= e_q(-\lambda) \sum_{x=0}^{\infty} [x]_{1/q}^2 \frac{q^{\binom{x}{2}}\lambda^x}{[x]_q!} \\ &= e_q(-\lambda) \sum_{x=0}^{\infty} [x]_q [x-1]_q q^{-2x+2} \frac{q^{\binom{x}{2}}\lambda^x}{[x]_q!} + e_q(-\lambda) \sum_{x=0}^{\infty} [x]_q q^{-x+1} \frac{q^{\binom{x}{2}}\lambda^x}{[x]_q!} \\ &= e_q(-\lambda) q^{-1}\lambda^2 \sum_{x=2}^{\infty} \frac{q^{\binom{x-2}{2}}\lambda^{x-2}}{[x-2]_q!} + e_q(-\lambda)\lambda \sum_{x=1}^{\infty} \frac{q^{\binom{x-1}{2}}\lambda^{x-1}}{[x-1]_q!} \\ &= q^{-1}\lambda^2 + \lambda. \end{aligned} \tag{13.17}$$

So,

$$(\sigma_q^H)^2 = V(Y) = V([X]_{1/q}) = q^{-1}\lambda^2 + \lambda - \lambda^2. \tag{13.18}$$

Next we prove that the Heine distribution (13.1) converges to a deformed standardized continuous Stieltjes–Wigert distribution. The continuous Stieltjes–Wigert distribution has probability density function

$$v_W^{SW}(w) = \frac{q^{1/8}}{\sqrt{2\pi \log q^{-1}}\, w} e^{\frac{(\log w)^2}{2\log q}}, \quad w > 0, \tag{13.19}$$

with mean value $\mu^{SW} = q^{-1}$ and standard deviation $\sigma^{SW} = q^{-3/2}(1-q)^{1/2}$.

The following lemma can be proved in an elementary way.

Lemma 13.1 *Let the random variable W be distributed according to the Stieltjes–Wigert probability distribution function (p.d.f) in* (13.19). *Then, the random variable* $X = \frac{1}{\log q^{-1}} \log\left((q^{-1}-1)(\sigma_q^H Z + \mu_q^H) + 1\right)$, *where* $Z = \frac{W-\mu^{SW}}{\sigma^{SW}}$, *has p.d.f.*

$$u_q^{DSW}(x) = \frac{q^{-7/8}}{\sigma_q^H (2\pi)^{1/2}} \left(\frac{\log q^{-1}}{q^{-1}-1}\right)^{1/2} \cdot \left(q^{-3/2}(1-q)^{1/2}\frac{[x]_{1/q}-\mu_q^H}{\sigma_q^H} + q^{-1}\right)^{-1/2} q^{-x}$$
$$\cdot \exp\left(\frac{1}{2\log q}\log^2\left(q^{-3/2}(1-q)^{1/2}\frac{[x]_{1/q}-\mu_q^H}{\sigma_q^H} + q^{-1}\right)\right), \quad x \geq 0. \quad (13.20)$$

Note that the random variable X is a deformation and standardization of (13.19).

Theorem 13.1 *For $\lambda \to \infty$, the Heine distribution given by the p.f. in* (13.1) *is approximated by the deformed standardized continuous Stieltjes–Wigert distribution* (13.20). *That is,*

$$f_X^H(x) \cong \frac{q^{-7/8}}{\sigma_q^H (2\pi)^{1/2}} \left(\frac{\log q^{-1}}{q^{-1}-1}\right)^{1/2} \left(q^{-3/2}(1-q)^{1/2}\frac{[x]_{1/q}-\mu_q^H}{\sigma_q^H} + q^{-1}\right)^{-1/2} q^{-x}$$
$$\cdot \exp\left(\frac{1}{2\log q}\log^2\left(q^{-3/2}(1-q)^{1/2}\frac{[x]_{1/q}-\mu_q^H}{\sigma_q^H} + q^{-1}\right)\right), \quad x \geq 0, \quad (13.21)$$

where μ_q^H and σ_q^H are given by (13.13).

Proof Using the q-Stirling formula (13.10), the p.f. of the q-Poisson distribution is approximated by

$$f_X^H(x) = e_q(-\lambda)\frac{q^{\binom{x}{2}}\lambda^x}{[x]_q!}$$
$$\cong e_q(-\lambda)\frac{q^{-1/8}(q\log q^{-1})^{1/2}}{(2\pi(1-q))^{1/2}}\lambda^x q^{x/2}[x]_{1/q}^{-(x+1/2)}\prod_{j=1}^{\infty}(1+(q^{-x}-1)q^{j-1}). \quad (13.22)$$

From the random variable

$$Z = \frac{[X]_{1/q}-\mu_q^H}{\sigma_q^H}$$

or

$$X = \frac{1}{\log q^{-1}}\log\left((q^{-1}-1)(\sigma_q^H Z + \mu_q^H) + 1\right),$$

with μ_q^H and σ_q^H given in (13.13), we get

$$\begin{aligned}[x]_{1/q} &= \sigma_q^H z + \mu_q^H \\ &= \left[\lambda^2 q^{-1}(1-q) + \lambda\right]^{1/2} z + \lambda \\ &= \lambda \left[\left(q^{-1}(1-q) + \frac{1}{\lambda} \right)^{1/2} z + 1 \right], \end{aligned} \tag{13.23}$$

and, for large λ ($\lambda \to \infty$), we have

$$[x]_{1/q} \cong \lambda q (q^{-3/2}(1-q)^{1/2} z + q^{-1}). \tag{13.24}$$

Furthermore, by the previous two equations, we get

$$q^{-x} = \frac{(1-q)\lambda}{q} \left[\left(q^{-1}(1-q) + \frac{1}{\lambda} \right)^{1/2} z + 1 \right] + 1 \tag{13.25}$$

and

$$q^{-x} \cong (1-q)\lambda \left(q^{-3/2}(1-q)^{1/2} z + q^{-1} \right). \tag{13.26}$$

Moreover, by (13.25), we find

$$x = \frac{1}{\log q^{-1}} \log \left(\frac{(1-q)\lambda}{q} \left[\left(q^{-1}(1-q) + \frac{1}{\lambda} \right)^{1/2} z + 1 \right] + 1 \right) \tag{13.27}$$

and

$$x \cong \frac{1}{\log q^{-1}} \log \left((1-q)\lambda \left(q^{-3/2}(1-q)^{1/2} z + q^{-1} \right) \right). \tag{13.28}$$

Finally, by (13.23), we get

$$\begin{aligned}[x]_{1/q}^x &= \lambda^x \left[\left(q^{-1}(1-q) + \frac{1}{\lambda} \right)^{1/2} z + 1 \right]^x \\ &= \lambda^x \exp \left(x \log \left[\left(q^{-1}(1-q) + \frac{1}{\lambda} \right)^{1/2} z + 1 \right] \right) \end{aligned} \tag{13.29}$$

and

$$\begin{aligned}[x]_{1/q}^x \cong \lambda^x \exp \Bigg(&\frac{1}{\log q^{-1}} \log \left((1-q)\lambda \left(q^{-3/2}(1-q)^{1/2} z + q^{-1} \right) \right) \\ &\log \left(q \left(q^{-3/2}(1-q)^{1/2} z + q^{-1} \right) \right) \Bigg). \end{aligned} \tag{13.30}$$

or

$$[x]_{1/q}^{x} \cong \lambda^x \exp\left(-\log\lambda(1-q)\left(q^{-3/2}(1-q)^{1/2}z+q^{-1}\right)\right.$$
$$\left.+\frac{1}{\log q^{-1}}\log\lambda(1-q)\log\left(q^{-3/2}(1-q)^{1/2}z+q^{-1}\right)\right)$$
$$\cdot\exp\left(\frac{1}{\log q^{-1}}\log^2\left(q^{-3/2}(1-q)^{1/2}z+q^{-1}\right)\right). \quad (13.31)$$

We also need to estimate the products

$$\prod_{j=1}^{\infty}(1+(q^{-x}-1)q^{j-1}) \quad \text{and} \quad \prod_{j=1}^{\infty}(1+\lambda(1-q)q^{j-1}).$$

Since the first product is estimated as

$$\prod_{j=1}^{\infty}(1+(q^{-x}-1)q^{j-1}) \cong \prod_{j=1}^{\infty}\left(1+\lambda(1-q)\left(q^{-3/2}(1-q)^{1/2}z+q^{-1}\right)q^{j-1}\right), \quad (13.32)$$

and

$$\prod_{j=1}^{\infty}\left(1+\lambda(1-q)\left(q^{-3/2}(1-q)^{1/2}z+q^{-1}\right)q^{j-1}\right)$$
$$=\exp\left(\sum_{j=1}^{\infty}\log\left(1+\lambda(1-q)\left(q^{-3/2}(1-q)^{1/2}z+q^{-1}\right)q^{j-1}\right)\right), \quad (13.33)$$

where the function

$$h(x)=\log\left(1+\lambda(1-q)\left(q^{-3/2}(1-q)^{1/2}z+q^{-1}\right)q^{x-1}\right)$$

has continuous derivatives in $[1,\infty)$ to all orders, we can apply the Euler–Maclaurin summation formula (see [8, p. 1090]) to the sum in (13.33). So,

$$\sum_{j=1}^{\infty}\log\left(1+\lambda(1-q)\left(q^{-3/2}(1-q)^{1/2}z+q^{-1}\right)q^{j-1}\right)$$
$$=\int_{1}^{\infty}\log\left(1+\lambda(1-q)\left(q^{-3/2}(1-q)^{1/2}z+q^{-1}\right)q^{u-1}\right)du$$
$$+\frac{1}{2}\log\left(1+\lambda(1-q)\left(q^{-3/2}(1-q)^{1/2}z+q^{-1}\right)\right)$$

$$+\sum_{k=1}^{m}\frac{\beta_{2k}}{(2k)!}h^{(2k-1)}\left(1+\lambda(1-q)\left(q^{-3/2}(1-q)^{1/2}z+q^{-1}\right)\right)+R_k, \tag{13.34}$$

where

$$|R_k|\le\frac{|\beta_{2k}|}{(2k)!}\int_1^\infty|g^{(2k)}\left(1+\lambda(1-q)\left(q^{-3/2}(1-q)^{1/2}z+q^{-1}\right)q^{u-1}\right)|\,du, \tag{13.35}$$

with β_k the Bernoulli numbers.

Now, expressing the integral appearing in (13.34) through the dilogarithm function, we get

$$\int_1^\infty\log\left(1+\lambda(1-q)\left(q^{-3/2}(1-q)^{1/2}z+q^{-1}\right)q^{u-1}\right)du$$
$$=\frac{1}{\log q}\mathrm{Li}_2\left(-\lambda(1-q)\left(q^{-3/2}(1-q)^{1/2}z+q^{-1}\right)\right), \tag{13.36}$$

where

$$\mathrm{Li}_2(y)=\sum_{k\ge1}\frac{y^k}{k^2}$$

is the dilogarithm function.

The dilogarithm satisfies Landen's identity

$$\mathrm{Li}_2(-y)=-\mathrm{Li}_2\left(\frac{y}{y+1}\right)-\frac{1}{2}\log^2(1+y). \tag{13.37}$$

Applying Landen's identity to (13.36), we obtain

$$\int_1^\infty\log\left(1+\lambda(1-q)\left(q^{-3/2}(1-q)^{1/2}z+q^{-1}\right)q^{u-1}\right)du$$
$$=\frac{1}{2\log q^{-1}}\log^2\left(\lambda(1-q)\left(q^{-3/2}(1-q)^{1/2}z+q^{-1}\right)\right)$$
$$+\frac{1}{\log q^{-1}}\mathrm{Li}_2\left(\frac{\lambda(1-q)\left(q^{-3/2}(1-q)^{1/2}z+q^{-1}\right)}{\lambda(1-q)\left(q^{-3/2}(1-q)^{1/2}z+q^{-1}\right)+1}\right)$$
$$=\frac{1}{2\log q^{-1}}\log^2\lambda(1-q)$$
$$+\frac{1}{2\log q^{-1}}\log^2\left(q^{-3/2}(1-q)^{1/2}z+q^{-1}\right)$$

$$
+ \frac{1}{\log q^{-1}} \log \lambda(1-q) \log \left(q^{-3/2}(1-q)^{1/2}z + q^{-1}\right)
$$
$$
+ \frac{1}{\log q^{-1}} \mathrm{Li}_2 \left(\frac{\lambda(1-q)\left(q^{-3/2}(1-q)^{1/2}z + q^{-1}\right)}{\lambda(1-q)\left(q^{-3/2}(1-q)^{1/2}z + q^{-1}\right) + 1} \right). \tag{13.38}
$$

Next, we estimate the sum and the quantity R_k appearing in (13.34):

$$
\begin{aligned}
\sum_{k=1}^{m} & \frac{\beta_{2k}}{(2k)!} h^{(2k-1)} \left(1 + \lambda(1-q)\left(q^{-3/2}(1-q)^{1/2}z + q^{-1}\right)\right) + R_k \\
&= \frac{\beta_2}{2} h' \left(1 + \lambda\left(q^{-3/2}(1-q)^{1/2}z + q^{-1}\right)\right) + R_1 + O(\lambda^{-2}) \\
&= \frac{\beta_2 \log q}{2} \frac{\lambda(1-q)\left(q^{-3/2}(1-q)^{1/2}z + q^{-1}\right)}{1 + \lambda(1-q)\left(q^{-3/2}(1-q)^{1/2}z + q^{-1}\right)} + O(\lambda^{-1}).
\end{aligned} \tag{13.39}
$$

So, by applying (13.38) and (13.39) to (13.34), we obtain

$$
\begin{aligned}
\sum_{j=1}^{\infty} & \log\left(1 + \lambda(1-q)\left(q^{-3/2}(1-q)^{1/2}z + q^{-1}\right) q^{j-1}\right) \\
&= \frac{1}{2\log q^{-1}} \log^2 \lambda(1-q) \\
&\quad + \frac{1}{2\log q^{-1}} \log^2 \left(q^{-3/2}(1-q)^{1/2}z + q^{-1}\right) \\
&\quad + \frac{1}{\log q^{-1}} \log \lambda(1-q) \log\left(q^{-3/2}(1-q)^{1/2}z + q^{-1}\right) \\
&\quad + \frac{1}{\log q^{-1}} \mathrm{Li}_2 \left(\frac{\lambda(1-q)\left(q^{-3/2}(1-q)^{1/2}z + q^{-1}\right)}{\lambda(1-q)\left(q^{-3/2}(1-q)^{1/2}z + q^{-1}\right) + 1} \right) \\
&\quad + \frac{1}{2} \log\left(1 + \lambda(1-q)\left(q^{-3/2}(1-q)^{1/2}z + q^{-1}\right)\right) \\
&\quad + \frac{\beta_2 \log q}{2} \frac{\lambda(1-q)\left(q^{-3/2}(1-q)^{1/2}z + q^{-1}\right)}{1 + \lambda(1-q)\left(q^{-3/2}(1-q)^{1/2}z + q^{-1}\right)} + O(\lambda^{-1}).
\end{aligned} \tag{13.40}
$$

Working similarly with the sum appearing in the product

$$
\prod_{j=1}^{\infty} \left(1 + \lambda(1-q)q^{j-1}\right) = \exp\left(\sum_{j=1}^{\infty} \log\left(1 + \lambda(1-q)q^{j-1}\right) \right), \tag{13.41}
$$

we obtain

$$\sum_{j=1}^{\infty} \log\left(1+\lambda(1-q)q^{j-1}\right)$$
$$= \frac{1}{2\log q^{-1}} \log^2(\lambda(1-q)) + \frac{1}{\log q^{-1}} \mathrm{Li}_2\left(\frac{\lambda(1-q)}{\lambda(1-q)+1}\right) + \frac{1}{2}\log(1+\lambda(1-q)) + \frac{\beta_2 \log q}{2} \frac{\lambda(1-q)}{1+\lambda(1-q)} + O(\lambda^{-1}). \tag{13.42}$$

Substituting the previous approximations (13.24), (13.26), (13.28), (13.30), (13.40), (13.42) in the p.f. $f_X^H(x)$ in (13.22), we derive the approximation

$$\begin{aligned}
f_X^H(x) \cong{}& \frac{q^{-1/8}(q\log q^{-1})^{1/2}}{(2\pi(1-q))^{1/2}}(1-q)^{-1/2}\lambda^{-1/2}\left(q^{-3/2}(1-q)^{1/2}z+q^{-1}\right)^{-1/2} \\
&\cdot \lambda^{-1/2}q^{-1/2}(q^{-3/2}(1-q)^{1/2}z+q^{-1})^{-1/2} \\
&\cdot \exp\left(\log\lambda(1-q)\left(q^{-3/2}(1-q)^{1/2}z+q^{-1}\right)\right. \\
&\qquad \left. - \frac{1}{\log q^{-1}}\log\lambda(1-q)\log\left(q^{-3/2}(1-q)^{1/2}z+q^{-1}\right)\right) \\
&\cdot \exp\left(-\frac{1}{\log q^{-1}}\log^2\left(q^{-3/2}(1-q)^{1/2}z+q^{-1}\right)\right) \\
&\cdot \exp\left(\frac{1}{2\log q^{-1}}\log^2\lambda(1-q) + \frac{1}{2\log q^{-1}}\log^2\left(q^{-3/2}(1-q)^{1/2}z+q^{-1}\right)\right) \\
&\cdot \exp\left(\frac{1}{\log q^{-1}}\log\lambda(1-q)\log\left(q^{-3/2}(1-q)^{1/2}z+q^{-1}\right)\right. \\
&\qquad \left. + \frac{1}{\log q^{-1}}\mathrm{Li}_2\left(\frac{\lambda(1-q)\left(q^{-3/2}(1-q)^{1/2}z+q^{-1}\right)}{\lambda(1-q)\left(q^{-3/2}(1-q)^{1/2}z+q^{-1}\right)+1}\right)\right) \\
&\cdot \exp\left(\frac{1}{2}\log\left(1+\lambda(1-q)\left(q^{-3/2}(1-q)^{1/2}z+q^{-1}\right)\right)\right. \\
&\qquad \left. + \frac{\beta_2\log q}{2}\frac{\lambda(1-q)\left(q^{-3/2}(1-q)^{1/2}z+q^{-1}\right)}{1+\lambda(1-q)\left(q^{-3/2}(1-q)^{1/2}z+q^{-1}\right)} + O(\lambda^{-1})\right) \\
&\cdot \exp\left(-\frac{1}{2\log q^{-1}}\log^2(\lambda(1-q)) - \frac{1}{\log q^{-1}}\mathrm{Li}_2\left(\frac{\lambda(1-q)}{\lambda(1-q)+1}\right)\right. \\
&\qquad \left. -\frac{1}{2}\log(1+\lambda(1-q)) - \frac{\beta_2\log q}{2}\frac{\lambda(1-q)}{1+\lambda(1-q)} + O(\lambda^{-1})\right)
\end{aligned}$$

$$z = \frac{[x]_{1/q} - \mu_q^H}{\sigma_q^H}, \quad x \geq 0. \tag{13.43}$$

Carrying out all the manipulations and using the approximation

$$\exp\left(\frac{1}{\log q^{-1}}\mathrm{Li}_2\left(\frac{\lambda(1-q)\left(q^{-3/2}(1-q)^{1/2}z+q^{-1}\right)}{\lambda(1-q)\left(q^{-3/2}(1-q)^{1/2}z+q^{-1}\right)+1}\right)\right)$$
$$\cdot\exp\left(\frac{\beta_2\log q}{2}\frac{\lambda(1-q)\left(q^{-3/2}(1-q)^{1/2}z+q^{-1}\right)}{1+\lambda(1-q)\left(q^{-3/2}(1-q)^{1/2}z+q^{-1}\right)}+O(\lambda^{-1})\right)$$
$$\cdot\exp\left(-\frac{1}{\log q^{-1}}\mathrm{Li}_2\left(\frac{\lambda(1-q)}{\lambda(1-q)+1}\right)-\frac{\beta_2\log q}{2}\frac{\lambda(1-q)}{1+\lambda(1-q)}+O(\lambda^{-1})\right)\to 1,$$
$$\text{as }\ \lambda\to\infty,$$

the previous expression, reduces to

$$f_X^H(x)\cong\frac{q^{-1/8}(q\log q^{-1})^{1/2}}{(2\pi(1-q))^{1/2}}(1-q)^{-1/2}\lambda^{-1/2}\left(q^{-3/2}(1-q)^{1/2}z+q^{-1}\right)^{-1/2}$$
$$\cdot\lambda^{-1/2}q^{-1/2}(q^{-3/2}(1-q)^{1/2}z+q^{-1})^{-1/2}$$
$$\cdot\lambda^{3/2}(1-q)^{3/2}\left(q^{-3/2}(1-q)^{1/2}z+q^{-1}\right)^{3/2}$$
$$\cdot\exp\left(-\frac{1}{2\log q^{-1}}\log^2\left(q^{-3/2}(1-q)^{1/2}z+q^{-1}\right)\right)$$
$$\cdot\lambda^{-1/2}(1-q)^{-1/2},$$
$$z=\frac{[x]_{1/q}-\mu_q^H}{\sigma_q^H},\quad x\geq 0. \tag{13.44}$$

Next, rearranging the terms of the last asymptotic formula, we derive the approximation

$$f_X^H(x)\cong\frac{q^{-1/8}(q\log q^{-1})^{1/2}}{(2\pi(1-q))^{1/2}}(1-q)^{-1/2}\lambda^{-1}q^{-1/2}\left(q^{-3/2}(1-q)^{1/2}z+q^{-1}\right)^{-1/2}$$
$$\cdot\lambda(1-q)\left(q^{-3/2}(1-q)^{1/2}z+q^{-1}\right)$$
$$\cdot\exp\left(-\frac{1}{2\log q^{-1}}\log^2\left(q^{-3/2}(1-q)^{1/2}z+q^{-1}\right)\right),$$
$$z=\frac{[x]_{1/q}-\mu_q^H}{\sigma_q^H},\quad x\geq 0. \tag{13.45}$$

From (13.13) and (13.26), we have

$$\sigma_q^H\cong\lambda q^{1/2}(1-q)^{1/2}\quad\text{and}\quad q^{-x}\cong\lambda(1-q)\left(q^{-3/2}(1-q)^{1/2}z+q^{-1}\right),$$

respectively. So,

$$f_X^H(x) \cong \frac{Cq^{-1/8}}{\sigma_q^H (2\pi)^{1/2}} \left(\frac{\log q^{-1}}{q^{-1}-1}\right)^{1/2} \cdot \left(q^{-3/2}(1-q)^{1/2}\frac{[x]_{1/q}-\mu_q^H}{\sigma_q^H}+q^{-1}\right)^{-1/2} q^{-x} \cdot \exp\left(\frac{1}{2\log q}\log^2\left(q^{-3/2}(1-q)^{1/2}\frac{[x]_{1/q}-\mu_q^H}{\sigma_q^H}+q^{-1}\right)\right), \quad x \geq 0,$$

where C is a normalizing constant. By Lemma 13.1, $C = q^{-3/4}$, and the proof is completed.

2.2 Continuous Limiting Behaviour of the Euler Distribution

In this section, we transfer from the random variable X of the Euler distribution (13.2) to the equal-distributed deformed random variable $Y = [X]_q$, and using the q-analogue of Stirling's formula in (13.12), we establish the convergence of the Euler distribution to a deformed standardized continuous Gaussian distribution. Initially, we need to compute the mean value and the variance of the random variable Y, say μ_q^E and $(\sigma_q^E)^2$, respectively. Using Charalambides' q-factorial moments of the Euler distribution [1], we easily derive that the mean and the variance of Y are given by

$$\mu_q^E = E\left([X]_q\right) = \lambda \quad \text{and} \quad (\sigma_q^E)^2 = V\left([X]_q\right) = \lambda\left(1-\lambda(1-q)\right), \tag{13.46}$$

respectively.

The following lemma can be proved in an elementary way.

Lemma 13.2 *Let the random variable Z be distributed according to the standardized Gaussian distribution. Then, the random variable*

$$X = \frac{1}{\log q}\log\left(1-(1-q)(\sigma_q^E Z + \mu_q^E)\right)$$

has p.d.f.

$$v_q^{DG}(x) = \frac{(\log q^{-1})^{1/2}}{\sigma_q^E (2\pi(1-q))^{1/2}} q^x \exp\left(-\frac{1}{2}\left(\left(\frac{1-q}{\log q^{-1}}\right)^{1/2} \cdot \frac{[x]_q - \mu_q^E}{\sigma_q^E}\right)^2\right). \tag{13.47}$$

Note that the random variable has a deformed standardized Gaussian distribution.

Theorem 13.2 *Let $q := q(\lambda)$ with $1 - 1/\lambda < q < 1$. Then, for $\lambda \to \infty$ with $\lambda(1-q) \to 0$, the Euler distribution given by the p.f. in* (13.2) *is approximated by a deformed standardized Gaussian distribution as follows:*

$$f_X^E(x) \cong \frac{(\log q^{-1})^{1/2}}{\sigma_q^E (2\pi(1-q))^{1/2}} q^x \exp\left(-\frac{1}{2}\left(\left(\frac{1-q}{\log q^{-1}}\right)^{1/2} \cdot \frac{[x]_q - \mu_q^E}{\sigma_q^E}\right)^2\right), \quad x \geq 0, \tag{13.48}$$

where μ_q^E and $(\sigma_q^E)^2$ are given by (13.46).

Proof Using the q-Stirling formula (13.12), for $q = q(\lambda)$ with $q(\lambda) \to 1$, as $\lambda \to \infty$ and $q(\lambda)^\lambda = \Omega(1)$, the Euler distribution (13.2) is approximated by

$$\begin{aligned} f_X^E(x) &= E_q(-\lambda)\frac{\lambda^x}{[x]_q!} \\ &\cong E_q(-\lambda)\frac{(q\log q^{-1})^{1/2}}{(2\pi(1-q))^{1/2}}\lambda^x q^{-\binom{x}{2}} q^{x/2}[x]_{1/q}^{-(x+1/2)} \prod_{j=1}^{\infty}(1+(q^{-x}-1)q^{j-1}) \end{aligned}$$

or

$$f_X^E(x) \cong E_q(-\lambda)\frac{(\log q^{-1})^{1/2}}{(2\pi(1-q))^{1/2}}\lambda^x q^{x^2/2} q^{-x/2}[x]_q^{-(x+1/2)} \prod_{j=1}^{\infty}(1+q(q^{-x}-1)q^{j-1}). \tag{13.49}$$

Consider the random variable $[X]_q = \frac{1-q^X}{1-q}$ and the q-standardized random variable $Z = \frac{[X]_q - \mu_q^E}{\sigma_q^E}$ with μ_q^E and σ_q^E given by (13.46). Then, the following relations are easily derived:

$$\begin{aligned} [x]_q &= \mu_q^E\left(\frac{\sigma_q^E}{\mu_q^E}z + 1\right) \\ &= \lambda\left(1 + (1-\lambda(1-q))^{1/2}\lambda^{-1/2}z\right), && (13.50) \\ q^x &= 1-(1-q)\mu_q^E\left(\frac{\sigma_q^E}{\mu_q^E}z+1\right) \\ &= 1-(1-q)\lambda\left(1+(1-\lambda(1-q))^{1/2}\lambda^{-1/2}z\right), && (13.51) \\ x &= \frac{1}{\log q}\log\left(1-(1-q)\mu_q^E\left(\frac{\sigma_q^E}{\mu_q^E}z+1\right)\right) \\ &= \frac{1}{\log q}\log\left(1-(1-q)\lambda\left(1+(1-\lambda(1-q))^{1/2}\lambda^{-1/2}z\right)\right), && (13.52) \\ q^{x^2/2} &= \exp\left(\frac{1}{2\log q}\log^2\left(1-(1-q)\mu_q^E\left(\frac{\sigma_q^E}{\mu_q^E}z+1\right)\right)\right) \end{aligned}$$

$$= \exp\left(\frac{1}{2\log q}\log^2\left(1-(1-q)\lambda\left(1+(1-\lambda(1-q))^{1/2}\lambda^{-1/2}z\right)\right)\right), \tag{13.53}$$

and

$$\begin{aligned}
[x]_q^x &= (\mu_q^E)^x \exp\left(\frac{1}{\log q}\log\left(1-(1-q)\mu_q^E\left(\frac{\sigma_q^E}{\mu_q^E}z+1\right)\right)\log\left(\frac{\sigma_q^E}{\mu_q^E}z+1\right)\right)\\
&= \lambda^x \exp\left(\frac{1}{\log q}\log\left(1-(1-q)\lambda\left(1+(1-\lambda(1-q))^{1/2}\lambda^{-1/2}z\right)\right)\right.\\
&\qquad\left.\cdot\log\left(1+(1-\lambda(1-q))^{1/2}\lambda^{-1/2}z\right)\right). \qquad (13.54)
\end{aligned}$$

Expanding the logarithms appearing in (13.54) in a Taylor series and carrying out all the manipulations, we get

$$\begin{aligned}
&\exp\left(\frac{1}{\log q}\log\left(1-(1-q)\lambda\left(1+(1-\lambda(1-q))^{1/2}\lambda^{-1/2}z\right)\right)\right.\\
&\qquad\left.\cdot\log\left(1+(1-\lambda(1-q))^{1/2}\lambda^{-1/2}z\right)\right)\\
&\quad= \exp\left(\frac{1-q}{\log q^{-1}}\frac{z^2}{2}(1+\lambda(1-q))^{1/2}+O(\lambda(1-q))\right). \qquad (13.55)
\end{aligned}$$

Moreover, we have

$$\begin{aligned}
\prod_{j=1}^{\infty}(1+q(q^{-x}-1)q^{j-1}) &= \prod_{j=1}^{\infty}(1+q(1-q)q^{-x}[x]_q q^{j-1})\\
&= \prod_{j=1}^{\infty}\left(1+q(1-q)\mu_q^E\left(1-(1-q)\mu_q^E\left(\frac{\sigma_q^E}{\mu_q^E}z+1\right)\right)^{-1}\left(\frac{\sigma_q^E}{\mu_q^E}z+1\right)q^{j-1}\right)\\
&= \exp\left(\sum_{j=1}^{\infty}\log\left(1+q(1-q)\mu_q^E\left(1-(1-q)\mu_q^E\left(\frac{\sigma_q^E}{\mu_q^E}z+1\right)\right)^{-1}\right.\right.\\
&\qquad\left.\left.\cdot\left(\frac{\sigma_q^E}{\mu_q^E}z+1\right)q^{j-1}\right)\right) \qquad (13.56)
\end{aligned}$$

with

$$\begin{aligned}
&(1-q)\mu_q^E\left(1-(1-q)\mu_q^E\left(\frac{\sigma_q^E}{\mu_q^E}z+1\right)\right)^{-1}\left(\frac{\sigma_q^E}{\mu_q^E}z+1\right)\\
&\quad= (1-q)\lambda\left(1-(1-q)\lambda\left(1+(1-\lambda(1-q))^{1/2}\lambda^{-1/2}z\right)\right)^{-1}\\
&\qquad\cdot\left(1+(1-\lambda(1-q))^{1/2}\lambda^{-1/2}z\right), \qquad (13.57)
\end{aligned}$$

which is derived by applying the Euler–Maclaurin summation formula (see Odlyzko [8, p 1090]) to the sum in (13.56) as follows:

$$\sum_{j=1}^{\infty} \log\left(1+q(1-q)\mu_q^E\left(1-(1-q)\mu_q^E\left(\frac{\sigma_q^E}{\mu_q^E}z+1\right)\right)^{-1}\left(\frac{\sigma_q^E}{\mu_q^E}z+1\right)q^{j-1}\right)$$
$$= \frac{1}{2\log q^{-1}}\log^2\left(1+q(1-q)\mu_q^E\left(1-(1-q)\mu_q^E\left(\frac{\sigma_q^E}{\mu_q^E}z+1\right)\right)^{-1}\left(\frac{\sigma_q^E}{\mu_q^E}z+1\right)\right)$$
$$+\frac{1}{\log q^{-1}}\mathrm{Li}_2\left(\frac{q(1-q)\mu_q^E\left(1-(1-q)\mu_q^E\left(\frac{\sigma_q^E}{\mu_q^E}z+1\right)\right)^{-1}\left(\frac{\sigma_q^E}{\mu_q^E}z+1\right)}{q(1-q)\mu_q^E\left(1-(1-q)\mu_q^E\left(\frac{\sigma_q^E}{\mu_q^E}z+1\right)\right)^{-1}\left(\frac{\sigma_q^E}{\mu_q^E}z+1\right)+1}\right)$$
$$+\frac{1}{2}\log\left(1+q(1-q)\mu_q^E\left(1-(1-q)\mu_q^E\left(\frac{\sigma_q^E}{\mu_q^E}z+1\right)\right)^{-1}\left(\frac{\sigma_q^E}{\mu_q^E}z+1\right)\right)$$
$$+\frac{\beta_2\log q}{2}\frac{q(1-q)\mu_q^E\left(1-(1-q)\mu_q^E\left(\frac{\sigma_q^E}{\mu_q^E}z+1\right)\right)^{-1}\left(\frac{\sigma_q^E}{\mu_q^E}z+1\right)}{1+q(1-q)\mu_q^E\left(1-(1-q)\mu_q^E\left(\frac{\sigma_q^E}{\mu_q^E}z+1\right)\right)^{-1}\left(\frac{\sigma_q^E}{\mu_q^E}z+1\right)}$$
$$+O\left(\lambda(1-q)\right) \tag{13.58}$$

with

$$(1-q)\mu_q^E\left(1-(1-q)\mu_q^E\left(\frac{\sigma_q^E}{\mu^E u_q}z+1\right)\right)^{-1}\left(\frac{\sigma_q^E}{\mu_q^E}z+1\right)$$
$$=(1-q)\lambda\left(1-(1-q)\lambda\left(1+(1-\lambda(1-q))^{1/2}\lambda^{-1/2}z\right)\right)^{-1}$$
$$\cdot\left(1+(1-\lambda(1-q))^{1/2}\lambda^{-1/2}z\right),$$

where Li_2 is the dilogarithm function and β_2 the second Bernoulli number.

Furthermore, working similarly with the sum appearing in the product

$$\prod_{j=1}^{\infty}\left(1-\lambda(1-q)q^{j-1}\right)=\exp\left(\sum_{j=1}^{\infty}\log\left(1-\lambda(1-q)q^{j-1}\right)\right), \tag{13.59}$$

we obtain

$$\sum_{j=1}^{\infty}\log\left(1-\lambda(1-q)q^{j-1}\right)$$
$$=-\frac{1}{2\log q^{-1}}\log^2\left(1-\lambda(1-q)\right)-\frac{1}{\log q^{-1}}\mathrm{Li}_2\left(\frac{-\lambda(1-q)}{1-\lambda(1-q)}\right)$$
$$+\frac{1}{2}\log\left(1-\lambda(1-q)\right)+\frac{\beta_2\log q^{-1}}{2}\frac{\lambda(1-q)}{1-\lambda(1-q)}+O(\lambda(1-q)). \tag{13.60}$$

Applying the previous estimations (13.53)–(13.60) to the p.f. $f^E(x)$ (13.49), we get

$$
\begin{aligned}
f_X^E(x) \cong{}& \frac{(\log q^{-1})^{1/2}}{(2\pi(1-q))^{1/2}} \\
&\cdot \exp\left(\frac{1}{2\log q}\log^2\left(1-(1-q)\lambda\left(1+(1-\lambda(1-q))^{1/2}\lambda^{-1/2}z\right)\right)\right) \\
&\cdot \left(1-(1-q)\lambda\left(1+(1-\lambda(1-q))^{1/2}\lambda^{-1/2}z\right)\right)^{-1/2} \\
&\cdot \lambda^{-1/2}\left(1+(1-\lambda(1-q))^{1/2}\lambda^{-1/2}z\right)^{-1/2} \\
&\cdot \exp\left(-\frac{1-q}{\log q^{-1}}\frac{z^2}{2}(1+\lambda(1-q))^{1/2}+O(\lambda(1-q))\right) \\
&\cdot \exp\left(\frac{1}{2\log q^{-1}}\log^2\left(1+\frac{q(1-q)\lambda\left(1+(1-\lambda(1-q))^{1/2}\lambda^{-1/2}z\right)}{\left(1-(1-q)\lambda\left(1+(1-\lambda(1-q))^{1/2}\lambda^{-1/2}z\right)\right)}\right)\right) \\
&\cdot \exp\left(\frac{1}{\log q^{-1}}\mathrm{Li}_2\left(\frac{\frac{q(1-q)\lambda\left(1+(1-\lambda(1-q))^{1/2}\lambda^{-1/2}z\right)}{\left(1-(1-q)\lambda\left(1+(1-\lambda(1-q))^{1/2}\lambda^{-1/2}z\right)\right)}}{\frac{q(1-q)\lambda\left(1+(1-\lambda(1-q))^{1/2}\lambda^{-1/2}z\right)}{\left(1-(1-q)\lambda\left(1+(1-\lambda(1-q))^{1/2}\lambda^{-1/2}z\right)\right)}+1}\right)\right) \\
&\cdot \exp\left(\frac{1}{2}\log\left(1+\frac{q(1-q)\lambda\left(1+(1-\lambda(1-q))^{1/2}\lambda^{-1/2}z\right)}{\left(1-(1-q)\lambda\left(1+(1-\lambda(1-q))^{1/2}\lambda^{-1/2}z\right)\right)}\right)\right) \\
&\cdot \exp\left(\frac{\beta_2\log q}{2}\frac{\frac{q(1-q)\lambda\left(1+(1-\lambda(1-q))^{1/2}\lambda^{-1/2}z\right)}{\left(1-(1-q)\lambda\left(1+(1-\lambda(1-q))^{1/2}\lambda^{-1/2}z\right)\right)}}{1+\frac{q(1-q)\lambda\left(1+(1-\lambda(1-q))^{1/2}\lambda^{-1/2}z\right)}{\left(1-(1-q)\lambda\left(1+(1-\lambda(1-q))^{1/2}\lambda^{-1/2}z\right)\right)}}+O\left(\lambda(1-q)\right)\right) \\
&\cdot \exp\left(\frac{1}{2\log q^{-1}}\log^2(1-\lambda(1-q))+\frac{1}{\log q^{-1}}\mathrm{Li}_2\left(\frac{-\lambda(1-q)}{1-\lambda(1-q)}\right)\right) \\
&\cdot \exp\left(-\frac{1}{2}\log(1-\lambda(1-q))\right) \\
&\cdot \exp\left(-\frac{\beta_2\log q^{-1}}{2}\frac{\lambda(1-q)}{1-\lambda(1-q)}+O(\lambda(1-q))\right), \quad z=\frac{[x]_q-\mu_q^E}{\sigma_q^E}, \; x\geq 0.
\end{aligned}
\tag{13.61}
$$

Note that the following approximation holds:

$$
\begin{aligned}
&\exp\left(\frac{1}{2\log q}\log^2\left(1-(1-q)\lambda\left(1+(1-\lambda(1-q))^{1/2}\lambda^{-1/2}z\right)\right)\right) \\
&\cdot \left(1-(1-q)\lambda\left(1+(1-\lambda(1-q))^{1/2}\lambda^{-1/2}z\right)\right)^{-1/2} \\
&\cdot \left(1+(1-\lambda(1-q))^{1/2}\lambda^{-1/2}z\right)^{-1/2} \\
&\cdot \exp\left(\frac{1}{2\log q^{-1}}\log^2\left(1+\frac{q(1-q)\lambda\left(1+(1-\lambda(1-q))^{1/2}\lambda^{-1/2}z\right)}{\left(1-(1-q)\lambda\left(1+(1-\lambda(1-q))^{1/2}\lambda^{-1/2}z\right)\right)}\right)\right) \\
&\cdot \exp\left(\frac{1}{\log q^{-1}}\mathrm{Li}_2\left(\frac{\frac{q(1-q)\lambda\left(1+(1-\lambda(1-q))^{1/2}\lambda^{-1/2}z\right)}{\left(1-(1-q)\lambda\left(1+(1-\lambda(1-q))^{1/2}\lambda^{-1/2}z\right)\right)}}{\frac{q(1-q)\lambda\left(1+(1-\lambda(1-q))^{1/2}\lambda^{-1/2}z\right)}{\left(1-(1-q)\lambda\left(1+(1-\lambda(1-q))^{1/2}\lambda^{-1/2}z\right)\right)}+1}\right)\right)
\end{aligned}
$$

$$
\cdot \exp\left(\frac{1}{2}\log\left(1+\frac{q(1-q)\lambda\left(1+(1-\lambda(1-q))^{1/2}\lambda^{-1/2}z\right)}{\left(1-(1-q)\lambda\left(1+(1-\lambda(1-q))^{1/2}\lambda^{-1/2}z\right)\right)}\right)\right)
$$

$$
\cdot \exp\left(\frac{\beta_2\log q}{2}\frac{\frac{q(1-q)\lambda\left(1+(1-\lambda(1-q))^{1/2}\lambda^{-1/2}z\right)}{\left(1-(1-q)\lambda\left(1+(1-\lambda(1-q))^{1/2}\lambda^{-1/2}z\right)\right)}}{1+\frac{q(1-q)\lambda\left(1+(1-\lambda(1-q))^{1/2}\lambda^{-1/2}z\right)}{\left(1-(1-q)\lambda\left(1+(1-\lambda(1-q))^{1/2}\lambda^{-1/2}z\right)\right)}}+O\left(\lambda(1-q)\right)\right)
$$

$$
\cdot \exp\left(\frac{1}{2\log q^{-1}}\log^2\left(1-\lambda(1-q)\right)+\frac{1}{\log q^{-1}}\mathrm{Li}_2\left(\frac{-\lambda(1-q)}{1-\lambda(1-q)}\right)\right)
$$

$$
\cdot \exp\left(-\frac{\beta_2\log q^{-1}}{2}\frac{\lambda(1-q)}{1-\lambda(1-q)}+O(\lambda(1-q))\right)\to 1,
$$

$$
\text{as } \lambda\to\infty \quad \text{with} \quad \lambda(1-q)\to 0.
$$

From (13.46) and (13.51), we get $\sigma_q^E=\lambda^{1/2}(1-\lambda(1-q))^{1/2}$ and $q^x\cong 1$. So,

$$
f_X^E(x)\cong\frac{(\log q^{-1})^{1/2}}{\sigma_q^E(2\pi(1-q))^{1/2}}q^x\exp\left(-\frac{1}{2}\left(\left(\frac{1-q}{\log q^{-1}}\right)^{1/2}\cdot\frac{[x]_q-\mu_q^E}{\sigma_q^E}\right)^2\right),\quad x\geq 0,
$$

and by Lemma 13.2 the proof is completed.

Remark 13.1 Possible realizations of the sequence $q:=q(\lambda)$ considered in Theorem 13.2 above are among others

$$
q(\lambda)=1-\frac{\alpha}{\lambda^{\beta}},\quad \alpha>1,\ \beta>1
$$

or

$$
q(\lambda)=1-\exp(-\lambda).
$$

2.2.1 Continuous Limiting Behaviour of the q-Binomial and Negative q-Binomial Distributions of the Second Kind

In this section, we transfer from the random variable X of the q-Binomial II distribution (13.8) to the equal-distributed deformed random variable $Y=[X]_q$, and using the q-analogue of Stirling's formula in (13.12), we establish the convergence of the q-Binomial II distribution to a deformed standardized continuous Gaussian distribution. Initially, we need to compute the mean value and the variance of the random variable Y, say μ_q^{BS} and $(\sigma_q^{BS})^2$, respectively. Using Charalambides' q-factorial moments of the q-Binomial II distribution [1], we easily derive that the mean and variance of Y, are given by

$$
\mu_q^{BS}=E\left([X]_q\right)=[n]_q\theta \quad\text{and}\quad (\sigma_q^{BS})^2=V\left([X]_q\right)=[n]_q\theta\left(1-\theta\left(1-q^{n-1}\right)\right), \tag{13.62}
$$

respectively.

Theorem 13.3 *Let $\theta = \theta_n = (1-q)^a$, $0 < a < 1$, $q = q(n)$ with $q(n) \to 1$, as $n \to \infty$, and $q(n)^n = \Omega(1)$. Then, for $n \to \infty$, the q-binomial distribution of the second kind given by the p.f. in* (13.8) *is approximated by a deformed standardized Gaussian distribution as follows:*

$$f_X^{BS}(x) \cong \frac{(\log q^{-1})^{1/2}}{\sigma_q^{BS}(2\pi(1-q))^{1/2}} q^x \exp\left(-\frac{1}{2}\left(\left(\frac{1-q}{\log q^{-1}}\right)^{1/2} \cdot \frac{[x]_q - \mu_q^{BS}}{\sigma_q^{BS}}\right)^2\right),$$
$$x \geq 0, \tag{13.63}$$

where μ_q^{BS} and $\sigma_q^{BS^2}$ are the mean value and the variance of the random variable $[X]_q$, respectively, given in (13.62).

Proof Using the q-Stirling formula in (13.12), for $q = q(n)$, with $q(n) \to 1$ as $n \to \infty$, and $q(n)^n = \Omega(1)$, the q-binomial distribution of the second kind (13.8) is approximated by

$$\begin{aligned} f_X^{BS}(x) &= \binom{n}{x}_q \theta^x \prod_{j=1}^{n-x}(1-\theta q^{j-1}) \\ &\cong \prod_{j=1}^{\infty}(1-\theta q^{j-1}) \frac{(q \log q^{-1})^{1/2}}{(2\pi(1-q))^{1/2}} \left(\frac{\theta}{1-q}\right)^x q^{-\binom{x}{2}} q^{-x/2} [x]_{1/q}^{-(x+1/2)} \\ &\qquad \cdot \prod_{j=1}^{\infty}(1+q(q^{-x}-1)q^{j-1}) \end{aligned}$$

or

$$\begin{aligned} f_X^{BS}(x) &\cong \prod_{j=1}^{\infty}(1-\theta q^{j-1}) \frac{(\log q^{-1})^{1/2}}{(2\pi(1-q))^{1/2}} \left(\frac{\theta}{1-q}\right)^x q^{x^2/2} q^{-x/2} [x]_q^{-(x+1/2)} \\ &\qquad \cdot \prod_{j=1}^{\infty}(1+q(q^{-x}-1)q^{j-1}). \end{aligned} \tag{13.64}$$

Consider the random variable $[X]_q = \frac{1-q^X}{1-q}$ and the q-standardized random variable $Z = \frac{[X]_q - \mu_q^{BS}}{\sigma_q^{BS}}$, with μ_q^{BS} and σ_q^{BS} given by (13.62). Then the following relations are easily derived:

$$\begin{aligned}
[x]_q &= \mu_q^{BS}\left(\frac{\sigma_q^{BS}}{\mu_q^{BS}}z+1\right)\\
&= [n]_q\theta\left(1+\frac{\left(1-\theta\left(1-q^{n-1}\right)\right)^{1/2}}{([n]_q\theta)^{1/2}}z\right),
\end{aligned} \tag{13.65}$$

$$\begin{aligned}
q^x &= 1-(1-q)\mu_q^{BS}\left(\frac{\sigma_q^{BS}}{\mu_q^{BS}}z+1\right)\\
&= 1-(1-q)[n]_q\theta\left(1+\frac{\left(1-\theta\left(1-q^{n-1}\right)\right)^{1/2}}{([n]_q\theta)^{1/2}}z\right),
\end{aligned} \tag{13.66}$$

$$\begin{aligned}
x &= \frac{1}{\log q}\log\left(1-(1-q)\mu_q^{BS}\left(\frac{\sigma_q^{BS}}{\mu_q^{BS}}z+1\right)\right)\\
&= \frac{1}{\log q}\log\left(1-(1-q)[n]_q\theta\left(1+\frac{\left(1-\theta\left(1-q^{n-1}\right)\right)^{1/2}}{([n]_q\theta)^{1/2}}z\right)\right),
\end{aligned} \tag{13.67}$$

$$\begin{aligned}
q^{x^2/2} &= \exp\left(\frac{1}{2\log q}\log^2\left(1-(1-q)\mu_q^{BS}\left(\frac{\sigma_q^{BS}}{\mu_q^{BS}}z+1\right)\right)\right)\\
&= \exp\left(\frac{1}{2\log q}\log^2\left(1-(1-q)[n]_q\theta\left(1+\frac{\left(1-\theta\left(1-q^{n-1}\right)\right)^{1/2}}{([n]_q\theta)^{1/2}}z\right)\right)\right),
\end{aligned} \tag{13.68}$$

and

$$\begin{aligned}
[x]_q^x &= (\mu_q^{BS})^x\exp\left(\frac{1}{\log q}\log\left(1-(1-q)\mu_q^{BS}\left(\frac{\sigma_q^{BS}}{\mu_q^{BS}}z+1\right)\right)\log\left(\frac{\sigma_q^{BS}}{\mu_q^{BS}}z+1\right)\right)\\
&= ([n]_q\theta)^x\exp\left(\frac{1}{\log q}\log\left(1-(1-q)[n]_q\theta\left(1+\frac{\left(1-\theta\left(1-q^{n-1}\right)\right)^{1/2}}{([n]_q\theta)^{1/2}}z\right)\right)\right.\\
&\qquad\left.\cdot\log\left(1+\frac{\left(1-\theta\left(1-q^{n-1}\right)\right)^{1/2}}{([n]_q\theta)^{1/2}}z\right)\right).
\end{aligned} \tag{13.69}$$

Expanding the logarithms appearing in (13.69) in Taylor series and carrying out all the manipulations, we get

$$\exp\left(\frac{1}{\log q}\log\left(1-(1-q)[n]_q\theta\left(1+\frac{\left(1-\theta\left(1-q^{n-1}\right)\right)^{1/2}}{([n]_q\theta)^{1/2}}z\right)\right)\right.$$
$$\left.\cdot\log\left(1+\frac{\left(1-\theta\left(1-q^{n-1}\right)\right)^{1/2}}{([n]_q\theta)^{1/2}}z\right)\right)$$
$$=\exp\left(\frac{1-q}{\log q^{-1}}\frac{z^2}{2}\left(1+(1-q)[n]_q\theta)\right)^{1/2}+O((1-q)[n]_q\theta))\right). \tag{13.70}$$

Moreover, we have

$$\prod_{j=1}^{\infty}(1+q(q^{-x}-1)q^{j-1})=\prod_{j=1}^{\infty}(1+q(1-q)q^{-x}[x]_q q^{j-1})$$
$$=\prod_{j=1}^{\infty}\left(1+q(1-q)\mu_q^{BS}\left(1-(1-q)\mu_q^{BS}\left(\frac{\sigma_q^{BS}}{\mu_q^{BS}}z+1\right)\right)^{-1}\left(\frac{\sigma_q^{BS}}{\mu_q^{BS}}z+1\right)q^{j-1}\right)$$
$$=\exp\left(\sum_{j=1}^{\infty}\log\left(1+q(1-q)\mu_q^{BS}\left(1-(1-q)\mu_q^{BS}\left(\frac{\sigma_q^{BS}}{\mu_q^{BS}}z+1\right)\right)^{-1}\right.\right.$$
$$\left.\left.\cdot\left(\frac{\sigma_q^{BS}}{\mu_q^{BS}}z+1\right)q^{j-1}\right)\right), \tag{13.71}$$

with

$$(1-q)\mu_q^{BS}\left(1-(1-q)\mu_q^{BS}\left(\frac{\sigma_q^{BS}}{\mu_q^{BS}}z+1\right)\right)^{-1}\left(\frac{\sigma_q^{BS}}{\mu_q^{BS}}z+1\right)$$
$$=(1-q)[n]_q\theta\left(1-(1-q)[n]_q\theta\left(1+\frac{\left(1-\theta\left(1-q^{n-1}\right)\right)^{1/2}}{([n]_q\theta)^{1/2}}z\right)\right)^{-1}$$
$$\cdot\left(1+\frac{\left(1-\theta\left(1-q^{n-1}\right)\right)^{1/2}}{([n]_q\theta)^{1/2}}z\right), \tag{13.72}$$

which is derived by applying the Euler–Maclaurin summation formula (see Odlyzko [8, p. 1090]) to the sum in (13.56) as follows:

$$\sum_{j=1}^{\infty}\log\left(1+q(1-q)\mu_q^{BS}\left(1-(1-q)\mu_q^{BS}\left(\frac{\sigma_q^{BS}}{\mu_q^{BS}}z+1\right)\right)^{-1}\left(\frac{\sigma_q^{BS}}{\mu_q^{BS}}z+1\right)q^{j-1}\right)$$
$$=\frac{1}{2\log q^{-1}}\log^2\left(1+q(1-q)\mu_q^{BS}\left(1-(1-q)\mu_q^{BS}\left(\frac{\sigma_q^{BS}}{\mu_q^{BS}}z+1\right)\right)^{-1}\right.$$

$$\cdot\left(\frac{\sigma_q^{BS}}{\mu_q^{BS}}z+1\right)\Bigg)$$

$$+\,\mathrm{Li}_2\left(\frac{q(1-q)\mu_q^{BS}\left(1-(1-q)\mu_q^{BS}\left(\frac{\sigma_q^{BS}}{\mu_q^{BS}}z+1\right)\right)^{-1}\left(\frac{\sigma_q^{BS}}{\mu_q^{BS}}z+1\right)}{q(1-q)\mu_q^{BS}\left(1-(1-q)\mu_q^{BS}\left(\frac{\sigma_q^{BS}}{\mu_q^{BS}}z+1\right)\right)^{-1}\left(\frac{\sigma_q}{\mu_q}z+1\right)+1}\right)$$

$$+\,\frac{1}{2}\log\left(1+q(1-q)\mu_q^{BS}\left(1-(1-q)\mu_q^{BS}\left(\frac{\sigma_q^{BS}}{\mu_q^{BS}}z+1\right)\right)^{-1}\left(\frac{\sigma_q^{BS}}{\mu_q^{BS}}z+1\right)\right)$$

$$+\,\frac{\beta_2\log q}{2}\frac{q(1-q)\mu_q^{BS}\left(1-(1-q)\mu_q^{BS}\left(\frac{\sigma_q^{BS}}{\mu_q^{BS}}z+1\right)\right)^{-1}\left(\frac{\sigma_q^{BS}}{\mu_q^{BS}}z+1\right)}{1+q(1-q)\mu_q^{BS}\left(1-(1-q)\mu_q^{BS}\left(\frac{\sigma_q^{BS}}{\mu_q^{BS}}z+1\right)\right)^{-1}\left(\frac{\sigma_q^{BS}}{\mu_q^{BS}}z+1\right)}$$

$$+\,O\left(\log q\right), \tag{13.73}$$

with

$$(1-q)\mu_q\left(1-(1-q)\mu_q^{BS}\left(\frac{\sigma_q^{BS}}{\mu_q^{BS}}z+1\right)\right)^{-1}\left(\frac{\sigma_q^{BS}}{\mu_q^{BS}}z+1\right)$$
$$=(1-q)[n]_q\theta\left(1-(1-q)[n]_q\theta\left(1+\frac{\left(1-\theta\left(1-q^{n-1}\right)\right)^{1/2}}{([n]_q\theta)^{1/2}}z\right)\right)^{-1}$$
$$\cdot\left(1+\frac{\left(1-\theta\left(1-q^{n-1}\right)\right)^{1/2}}{([n]_q\theta)^{1/2}}z\right),$$

where Li_2 is the dilogarithm function and β_2 the second Bernoulli number.

Furthermore, working similarly with the sum appearing in the product

$$\prod_{j=1}^{\infty}\left(1-\theta q^{j-1}\right)=\exp\left(\sum_{j=1}^{\infty}\log\left(1-\theta q^{j-1}\right)\right), \tag{13.74}$$

we obtain

$$\sum_{j=1}^{\infty}\log\left(1-\theta q^{j-1}\right)=-\frac{1}{2\log q^{-1}}\log^2(1-\theta)-\frac{1}{\log q^{-1}}\mathrm{Li}_2\left(\frac{-\theta}{1-\theta}\right)$$
$$+\frac{1}{2}\log(1-\theta)+\frac{\beta_2\log q^{-1}}{2}\frac{\theta}{1-\theta}+O(\theta). \tag{13.75}$$

Applying the previous expressions (13.68)–(13.75) to the p.f. $f_X^{BS}(x)$ in (13.64), and carrying out all the necessary manipulations, we derive our desired asymptotic result (13.63). By Lemma 13.2, the proof is completed.

Remark 13.2 Possible realizations of the sequences $\theta := \theta_n$ and $q := q(n)$ considered in Theorem 13.3 above are

$$\theta_n = n^{-1/2} \text{ and } q(n) = 1 - \frac{\alpha}{n}, \quad \alpha > 0,$$

or

$$\theta_n = (\log n)^{-1/2} \text{ and } q(n) = 1 - 1/\log n.$$

Next we transfer from the random variable X of the distribution (13.9) to the equal-distributed deformed random variable $Y = [X]_q$. Using Charalambides' q-factorial moments of the negative q-Binomial II distribution [1], we easily derive that the mean and the variance of the random variable Y are given by

$$\mu_q^{NBS} = E\left([X]_q\right) = \frac{[n]_q\,\theta}{1 - \theta q^n} \tag{13.76}$$

and

$$\begin{aligned}(\sigma_q^{NBS})^2 &= V\left([X]_q\right) \\ &= \frac{[n]_q[n+1]_q\theta^2}{(1-\theta q^n)(1-\theta q^{n+1})} + \frac{[n]_q\,\theta(1-\theta)}{(1-\theta q^n)(1-\theta q^{n+1})} - \frac{[n]_q^2\,\theta^2}{(1-\theta q^n)^2},\end{aligned} \tag{13.77}$$

respectively. Considering the random variable $[X]_q = \frac{1-q^X}{1-q}$ and the q-standardized random variable $Z = \frac{[X]_q - \mu_q^{NBS}}{\sigma_q^{NBS}}$, with μ_q^{NBS} and σ_q^{NBS} given by (13.76) and (13.77), respectively, and working analogously as in the proofs of Theorems 13.2 and 13.3, we obtain the following theorem concerning the asymptotic behaviour of the negative q-binomial distribution of the second kind.

Theorem 13.4 *Let $\theta = \theta_n = (1-q)^a$, $0 < a < 1$, $q = q(n)$, with $q(n) \to 1$ as $n \to \infty$, and $q(n)^n = \Omega(1)$. Then, for $n \to \infty$, the negative q-binomial distribution of the second kind given by the p.f.* (13.9) *is approximated by a deformed standardized Gaussian distribution as follows:*

$$f_W^{NBS}(x) \cong \frac{(\log q^{-1})^{1/2}}{\sigma_q^{NBS}(2\pi(1-q))^{1/2}}\, q^x \exp\left(-\frac{1}{2}\left(\left(\frac{1-q}{\log q^{-1}}\right)^{1/2} \cdot \frac{[x]_q - \mu_q^{NBS}}{\sigma_q^{NBS}}\right)^2\right), \quad x \geq 0, \tag{13.78}$$

where μ_q^{NBS} and $(\sigma_q^{NBS})^2$ are given by (13.76) *and* (13.77), *respectively.*

3 Concluding Remarks

In this paper, we studied the pointwise convergence of the Heine and Euler distributions when $\lambda \to \infty$. Specifically, we proved that the Heine distribution converges to a deformed standardized continuous Stieltjes–Wigert distribution, and that the Euler distribution converges to a deformed standardized Gaussian distribution. Moreover, the pointwise convergence of the q-binomial distribution of the second kind and the negative q-binomial distribution of the second kind to this deformed standardized Gaussian distribution were proved.

Acknowledgements The authors would like to thank the anonymous referee and Professor Christian Krattenthaler for their helpful and insightful comments and suggestions.

References

1. Charalambides, C.A.: Discrete q-distributions on Bernoulli trials with geometrically varying success probability. J. Stat. Plan. Inference **140**, 2355–2383 (2010)
2. Ismail, M.E.H.: Classical and Quantum Orthogonal Polynomials in One Variable. Cambridge University Press, Cambridge (2005)
3. Kemp, A.W., Newton, J.: Certain state-dependent processes for dichotomized parasite populations. J. Appl. Probab. **27**, 251–258 (1990)
4. Kemp, A.W.: Heine-Euler Extensions of the Poisson distribution. Commun. Stat. Theory Meth. **21**, 571–588 (1992)
5. Kemp, A.W.: Steady-state Markov chain models for the Heine and Euler distributions. J. Appl. Probab. **29**, 869–876 (1992)
6. Kyriakoussis, A., Vamvakari, M.G.: On a q-analogue of the Stirling formula and a continuous limiting behaviour of the q-Binomial distribution-Numerical calculations. Method. Comput. Appl. Probab. **15**, 187–213 (2013)
7. Kyriakoussis, A., Vamvakari, M.G.: Continuous Stieltjes-Wigert limiting behaviour of a family of confluent q-Chu-Vandermonde distributions. Axioms **3**, 140–152 (2014)
8. Odlyzko, A.M.: Asymptotic enumeration methods. In: Graham, R.L., Grötschel, M., Lovász, L. (eds.) Handbook of Combinatorics, pp. 1063–1229. Elsevier Science Publishers, Amsterdam (1995)
9. Vamvakari, M.G.: On continuous limiting behaviour for the $q(n)$-Binomial distribution with $q(n) \to 1$ as $n \to \infty$. Appl. Math. **3**, 2101–2108 (2012)

Asymptotic Estimates for Queueing Systems with Time-Varying Periodic Transition Rates

Barbara Margolius

Abstract We consider the $M_t/M_t/1$ queue, the multi-server queue ($M_t/M_t/c_t$), and queues with jumps of size one and two. Results are extensible to more general ergodic quasi-birth-death processes (QBDs) with time-varying periodic transition rates of period one. The estimates are asymptotic in the level of the process (the length of the queue). These asymptotic estimates highlight the connections between the asymptotic periodic distribution of a stable queue with time-varying rates and the same type of queue with constant rates. The estimates can also be used to approximate other performance measures such as the waiting time distribution. We illustrate the method with several examples.

Keywords $M_t/M_t/1$ queues · Multi-server queues · Queues with jumps · Waiting time distribution · Generating function

2010 Mathematics Subject Classification Primary: 60K25 · Secondary: 60J10 60J80

1 Introduction

Systems with time-varying periodic rates are pervasive. They include telephone call centers, hospital emergency rooms, airports, any system which exhibits seasonal behavior whether natural or man-made, ambulances, police and fire service and many, many others. The recent paper by Schwarz, Selinka and Stolletz [7] provides both a useful survey of applications and a survey of methods for analyzing queueing systems with time-varying parameters.

In this paper, we consider the $M_t/M_t/1$ queue, the multi-server queue ($M_t/M_t/c_t$), and queues with jumps of size one and two. Results are extensible to more general ergodic quasi-birth-death processes (QBDs) with time-varying periodic transition rates of period one. The estimates are asymptotic in the level of the process (the

B. Margolius (✉)
College of Sciences and Health Professions, Cleveland State University, Cleveland, OH, USA
e-mail: B.Margolius@csuohio.edu

G. E. Andrews et al. (eds.), *Lattice Path Combinatorics and Applications*, Developments in Mathematics 58, https://doi.org/10.1007/978-3-030-11102-1_14

length of the queue). These asymptotic estimates highlight the connections between the asymptotic periodic distribution of a stable queue with time-varying rates and the same type of queue with constant rates. The estimates can also be used to approximate other performance measures such as the waiting time distribution. We illustrate the method with several examples.

The basic approach is to solve for the generating function of the queueing system using the assumption that if the process is in its asymptotic periodic distribution at some time t, then the generating function at time t will equal the generating function at time $t+1$. That is, we assume the system is in the periodic analog of steady state. For more detail on when the asymptotic periodic distribution exists, see [5]. This will yield a function for the generating function in terms of an integral equation. We find the poles of this function to create our asymptotic estimates. The poles of the generating function depend only on the evolution of the system over the course of a single period.

For the single-server queue, these estimates take a particularly simple form. Let $\bar{\lambda}$ be the average arrival rate in a time period and $\bar{\mu}$ be the average departure rate. Then an asymptotic estimate for the probability that there are n in the queue at time t is given by $\pi_n(t) \approx f(t)\left(\frac{\bar{\lambda}}{\bar{\mu}}\right)^n$. For constant rates, the formula is exact and $f(t) = \pi_0 = 1 - \frac{\bar{\lambda}}{\bar{\mu}}$. In general, $f(t)$ depends on $\pi_0(t)$. Given $\pi_0(t)$, $f(t)$ can be easily computed for any stable periodic $M_t/M_t/1$ queue. Similar formulas can be developed using this approach for more complex quasi-birth-and-death processes.

In Sect. 2, we find the transient solution of the $M_t/M_t/1$ queue up to an integral equation. In Sect. 2.1, we find the generating function for the single-server queue with periodic rates and use that to find the asymptotic periodic distribution of the number in the system in terms of an integral equation. In Sect. 2.2, we obtain asymptotic estimates for the distribution of the number in the queue at time t within the period. Section 2.3 provides numerical examples for the single-server queue with time-varying transition rates. In Sect. 3 we find similar quantities for the multi-server queue with time-varying transition rates. In Sect. 4, we illustrate the method for another type of queue.

2 $M_t/M_t/1$ Queue Example

Consider a single-server queue with time-varying arrival and departure rates. Let X_t represent the number in the queue at time t, and let $X_s = i$ give the length of the queue at some given initial time s. Define $p_{i,n}(t) = P\{X_t = n | X_s = i\}$.

We have the Chapman–Kolmogorov equations:

$$\dot{p}_{i,0}(t) = -\lambda(t) p_{i,0}(t) + \mu(t) p_{i,1}(t)$$
$$\dot{p}_{i,n}(t) = \lambda(t) p_{i,n-1}(t) - (\lambda(t) + \mu(t)) p_{i,n}(t) + \mu(t) p_{i,n+1}(t),$$

with $p_{i,n}(s) = \delta_{i=n}$ for some initial time s. We define the generating function $P(z, s, t) = \sum_{n=0}^{\infty} z^n p_{i,n}(t)$, then $P(z, s, t)$ satisfies the ordinary differential equation:

$$\dot{P}(z, s, t) = (\lambda(t)(z-1) + \mu(t)(z^{-1}-1))P(z, s, t) + (\mu(t) - z^{-1}\mu(t))p_{i,0}(t).$$

This has solution

$$P(z, s, t) = \int_s^t \mu(u)(1 - z^{-1})p_{i,0}(u)\Phi(z, u, t)du + P(z, s, s)\Phi_Y(z, s, t),$$

where $\Phi(z, u, t)$ is defined below. Note that since $p_{i,n}(s) = \delta_{i=n}$, $P(z, s, s) = z^i$.

We define the randomized random walk, Y_t with jumps to the left occurring at rate $\mu(t)$ and to the right at rate $\lambda(t)$. Let

$$M(s, t) = \int_s^t \mu(u)du$$

and

$$\Lambda(s, t) = \int_s^t \lambda(u)du,$$

then

$$P\{Y_t = n + k | Y_s = k\} = e^{-(M(s,t)+\Lambda(s,t))}\left(\frac{\Lambda(s, t)}{M(s, t)}\right)^{n/2} I_n(2\sqrt{\Lambda(s, t)M(s, t)}),$$

where $I_n(\cdot)$ is the modified Bessel function of the first kind. We denote the generating function for Y_t as $\Phi_Y(z, s, t)$ with

$$\Phi_Y(z, s, t) = \sum_{n=-\infty}^{\infty} P\{Y_t = n + i | Y_s = i\}z^n = e^{\Lambda(s,t)(z-1)+M(s,t)(z^{-1}-1)}.$$

Furthermore, we define

$$\phi_n(s, t) = P\{Y_t = n + i | Y_s = i\}.$$

Note that $\phi_n(s, t)$ does not depend on i, the location of the random walk at time s.

$\Phi_Y(z, s, t)$ is the solution of the evolution equation:

$$\frac{\partial}{\partial t}\Phi_Y(z, s, t) = \Phi_Y(z, s, t)\left(\lambda(t)(z-1) + \mu(t)(z^{-1}-1)\right)$$

and

$$\Phi(z, s, s) = \mathbf{I}.$$

We will use the notation $[z^n](A(z))$ to represent the coefficient a_n of z^n in the series $A(z) = \sum_n a_n z^n$ (see Flajolet and Sedgewick [1]). The transient solution for the single-server queue with time-varying transition rates is then given by

$$p_{i,n}(t) = [z^n]P(z,s,t) = \int_s^t p_{i,0}(u)\mu(u)(\phi_n(u,t) - \phi_{n+1}(u,t))du + \phi_{n-i}(s,t).$$

To find $p_{i,0}(u)$, we solve the Volterra equation

$$p_{i,0}(t) = [z^0]P(z,s,t) = \int_s^t p_{i,0}(u)\mu(u)(\phi_0(u,t) - \phi_1(u,t))du + \phi_{-i}(s,t).$$

2.1 Periodic

Now suppose transition rates are periodic with period one. In this case, we use $\pi_n(t)$ to designate the asymptotic periodic probability of n in the queue at time t rather than $p_{i,n}(t)$ for the transient probability. We wish to solve directly for the asymptotic periodic solution. In that case, $P(z, t-1, t) = P(z, t-1, t-1)$, so

$$P(z,t-1,t) = \int_{t-1}^t \pi_0(u)\mu(u)(1 - z^{-1})\Phi_Y(z,u,t)du\,(1 - \Phi(z,t-1,t))^{-1} \tag{14.1}$$

Now,

$$(I - \Phi(z,t-1,t))^{-1} = \sum_{k=0}^{\infty} \Phi(z,t-1,t)^k = \sum_{k=0}^{\infty} \Phi(z,t,t+k),$$

and

$$\Phi(z,u,t)\Phi(z,t,t+k) = \Phi(z,u,t+k),$$

so

$$\pi_n(t) = [z^n]P(z,t-1,t) = \int_{t-1}^t \pi_0(u)\mu(u)\sum_{k=0}^{\infty}\big(\phi_n(u,t+k) - \phi_{n+1}(u,t+k)\big)\,du.$$

Although we have an explicit formula for $\phi_n(u, t+k) = P\{Y(u, t+k) = n\}$, computation of these quantities is very cumbersome and in general convergence is slow.

In the next section, we describe asymptotic methods for approximating these infinite sums.

2.2 Asymptotic Methods

Define $\bar{\lambda} = \Lambda(0, 1) = \int_{t-1}^{t} \lambda(u)du$ and $\bar{\mu} = M(0, 1) = \int_{t-1}^{t} \mu(u)du$. These represent the expected number of steps to the right ($\bar{\lambda}$) or to the left ($\bar{\mu}$) during one period for the random walk Y_t. We have zeros in the denominator of equation (14.1), where

$$\bar{\lambda}(z-1) + \bar{\mu}(z^{-1} - 1) = 0.$$

This expression has two zeros: $z = 1$, and $z = \bar{\mu}/\bar{\lambda}$. $z = 1$ is a root of both the numerator and the denominator, so we factor it out:

$$P(z, t-1, t) = \int_{t-1}^{t} \pi_0(u) \frac{\mu(u)}{\bar{\mu} - \bar{\lambda} z} \Phi_Y(z, u, t) du \times \left(\sum_{k=1}^{\infty} \frac{(\bar{\lambda}(z-1) + \bar{\mu}(z^{-1} - 1))^{k-1}}{k!} \right)^{-1}.$$

We can rewrite this as

$$P(z, t-1, t) = \sum_{j=0}^{\infty} \left(\frac{\bar{\lambda}}{\bar{\mu}} \right)^{j} \int_{t-1}^{t} \pi_0(u) \frac{\mu(u)}{\bar{\mu}} \Phi_Y(z, u, t) du z^j \times \left(\sum_{k=1}^{\infty} \frac{(\bar{\lambda}(z-1) + \bar{\mu}(z^{-1} - 1))^{k-1}}{k!} \right)^{-1}. \tag{14.2}$$

This is suggestive of the relation between the asymptotic periodic solution for the queue with time-varying parameters and the steady-state solution for the constant rate queue which is given by

$$\pi_j = \pi_0 \left(\frac{\lambda}{\mu} \right)^j = \left(1 - \frac{\lambda}{\mu} \right) \left(\frac{\lambda}{\mu} \right)^j.$$

To obtain the asymptotic estimate, we use the following result (see [1, p. 228, IV.2]):

$$[z^n] \frac{h(z)}{(1-z)} \sim h(1).$$

So

$$[z^n] \frac{h(z)}{\left(1 - \frac{\bar{\lambda}}{\bar{\mu}} z\right)} \sim h\left(\frac{\bar{\mu}}{\bar{\lambda}} \right).$$

Hence an asymptotic estimate is given by

$$\pi_n(t) \approx \frac{1}{\bar{\mu}} \int_{t-1}^{t} \pi_0(u)\mu(u)\Phi_Y\left(\frac{\bar{\mu}}{\bar{\lambda}}, u, t\right) du \left(\frac{\bar{\lambda}}{\bar{\mu}}\right)^n$$
$$= \frac{1}{\bar{\mu}} \int_{t-1}^{t} \pi_0(u)\mu(u) \exp\left\{\left(\frac{\Lambda(u,t)}{\bar{\lambda}} - \frac{M(u,t)}{\bar{\mu}}\right)(\bar{\mu} - \bar{\lambda})\right\} du \left(\frac{\bar{\lambda}}{\bar{\mu}}\right)^n. \tag{14.3}$$

We have shown that

$$\pi_n(t) \sim \left(\frac{\bar{\lambda}}{\bar{\mu}}\right)^n f(t),$$

where

$$f(t) = \frac{1}{\bar{\mu}} \int_{t-1}^{t} \pi_0(u)\mu(u) \exp\left\{\left(\frac{\Lambda(u,t)}{\bar{\lambda}} - \frac{M(u,t)}{\bar{\mu}}\right)(\bar{\mu} - \bar{\lambda})\right\} du. \tag{14.4}$$

Note that, if $\mu(t) = \bar{\mu}$ and $\lambda(t) = \bar{\lambda}$, the expression simplifies to

$$\pi_n = \pi_0 \left(\frac{\lambda}{\mu}\right)^n = \left(1 - \frac{\lambda}{\mu}\right)\left(\frac{\lambda}{\mu}\right)^n.$$

If we let $\alpha(z) = \bar{\lambda}(z-1) + \bar{\mu}(z^{-1} - 1)$, then we obtain

$$\left(\sum_{k=1}^{\infty} \frac{(\bar{\lambda}(z-1) + \bar{\mu}(z^{-1} - 1))^{k-1}}{k!}\right)^{-1} = \frac{\alpha(z)}{1 - e^{\alpha(z)}}.$$

This has Maclaurin series in α of

$$\frac{\alpha}{1 - e^{\alpha}} = 1 - \frac{\alpha}{2} + \frac{\alpha^2}{12} - \frac{\alpha^4}{720} + \frac{\alpha^6}{30240} + \cdots = \sum_{n=0}^{\infty} \frac{B_n}{n!}\alpha^n,$$

where the B_n's are the Bernoulli numbers. See, for example, [6, p. 289]. Additional terms of an asymptotic expansion for the number in queue may be obtained using this expansion, since as $z \to \bar{\mu}/\bar{\lambda}$, $\alpha \to 0$. In what follows, we will use the asymptotic estimate given in Eq. (14.3).

2.3 Numerical Examples for the $M_t/M_t/1$ Queue

We consider several numerical examples:

These examples collectively illustrate the convergence behavior toward the asymptotic estimates for the probabilities of the number in queue for three different traffic intensities: $\rho = 0.75$ (examples 1–3), $\rho = 0.9$ (examples 4–5), and

Examples	1	2	3
Arrival rate $\lambda(t)$	$3 - 2\cos(2\pi t)$	$0.3 - 0.2\cos(2\pi t)$	$3 - 2\cos(2\pi t)$
Departure rate $\mu(t)$	$4 + 2\cos(2\pi t)$	$0.4 + 0.2\cos(2\pi t)$	$4 - 2\cos(2\pi t)$
$\bar{\lambda}$	3	0.3	3
$\bar{\mu}$	4	0.4	4
$\rho = \frac{\bar{\lambda}}{\bar{\mu}}$	0.75	0.75	0.75

Examples	4	5	6
Arrival rate $\lambda(t)$	$3.6 - 2.8\cos(2\pi t)$	$0.36 - 0.28\cos(2\pi t)$	$0.4 - 0.3\cos(2\pi t)$
Departure rate $\mu(t)$	$4 + 2\cos(2\pi t)$	$0.4 + 0.2\cos(2\pi t)$	$4 - 2\cos(2\pi t)$
$\bar{\lambda}$	3.6	0.36	0.4
$\bar{\mu}$	4	0.4	4
$\rho = \frac{\bar{\lambda}}{\bar{\mu}}$	0.9	0.9	0.1

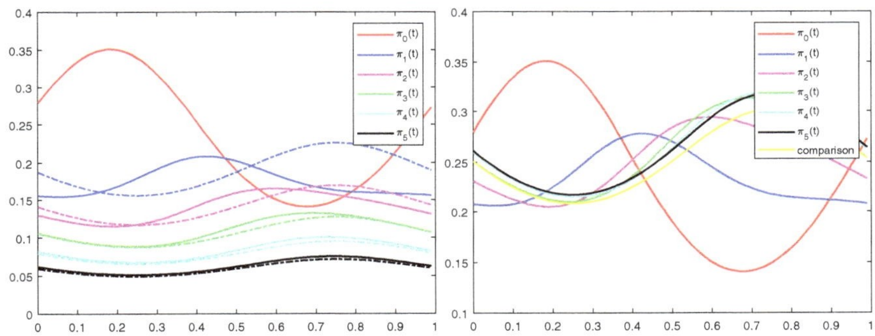

Fig. 1 Example 1. The graph on the left shows $\pi_j(t)$, $j = 0, \ldots, 5$ (solid lines) and the asymptotic estimates for each of these probabilities (dashed lines). The graph on the right shows $\pi_j(t)\left(\frac{\bar{\mu}}{\bar{\lambda}}\right)^j$, $j = 0, \ldots, 5$ and the asymptotic estimate for $f(t) = \pi_j(t)\left(\frac{\bar{\mu}}{\bar{\lambda}}\right)^j$ which does not depend on j

$\rho = 0.1$ (example 6); for different rates and for arrival and service intensities that move together or do not. For each of the examples, the convergence to the asymptotic estimate is fairly rapid. The behavior of periodic probabilities for some sets of parameters is close to the constant rate steady state probabilities, but for other examples (those with greater rates), the probabilities show greater variability within the period (Fig. 1).

For each of these six examples, we provide two graphs. One shows the asymptotic periodic probabilities of having j in the queue for $j = 0$–5 (solid lines) compared to the asymptotic estimate of the probability shown as dashed lines. The second graph shows the estimate for the function $f(t)$ used in the asymptotic estimates for each of probabilities zero to five compared to the exact $f(t)$, where $f(t)$ is given in Eq. (14.4). As the number in the queue increases, it can be seen that the quality of the asymptotic estimate improves (Fig. 2).

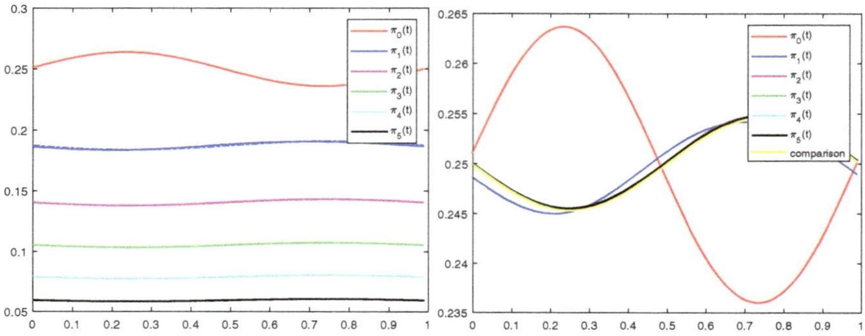

Fig. 2 Example 2. The graph on the left shows $\pi_j(t)$, $j = 0, \ldots, 5$ (solid lines) and the asymptotic estimates for each of these probabilities (dashed lines). The graph on the right shows $\pi_j(t)\left(\frac{\bar{\mu}}{\bar{\lambda}}\right)^j$, $j = 0, \ldots, 5$ and the asymptotic estimate for $f(t) = \pi_j(t)\left(\frac{\bar{\mu}}{\bar{\lambda}}\right)^j$ which does not depend on j

3 Example: Multi-server Queue

Next, we consider the multi-server queue with time-varying periodic rates and c servers. The analysis parallels the analysis for the single-server queue (Fig. 3).

We have the Chapman–Kolmogorov equations

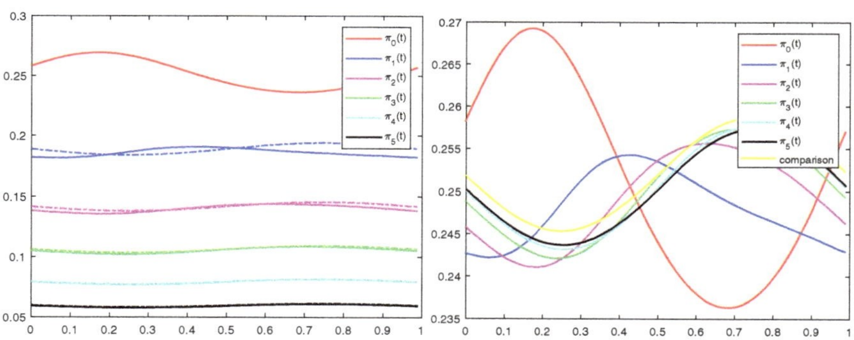

Fig. 3 Example 3. The graph on the left shows $\pi_j(t)$, $j = 0, \ldots, 5$ (solid lines) and the asymptotic estimates for each of these probabilities (dashed lines). The graph on the right shows $\pi_j(t)\left(\frac{\bar{\mu}}{\bar{\lambda}}\right)^j$, $j = 0, \ldots, 5$ and the asymptotic estimate for $f(t) = \pi_j(t)\left(\frac{\bar{\mu}}{\bar{\lambda}}\right)^j$ which does not depend on j

$$\dot{p}_{i,0}(t) = -\lambda(t)p_{i,0}(t) + \mu(t)p_{i,1}(t)$$
$$\dot{p}_{i,1}(t) = \lambda(t)p_{i,0}(t) - (\lambda(t) + \mu(t))p_{i,1}(t) + 2\mu(t)p_{i,2}(t)$$
$$\vdots$$
$$\dot{p}_{i,j}(t) = \lambda(t)p_{i,j-1}(t) - (\lambda(t) + j\mu(t))p_{i,j}(t) + (j+1)\mu(t)p_{i,j+1}(t), \quad 0 < j < c$$
$$\vdots$$
$$\dot{p}_{i,c-1}(t) = \lambda(t)p_{i,c-2}(t) - (\lambda(t) + (c-1)\mu(t))p_{i,c-1}(t) + c\mu(t)p_{i,c}(t)$$
$$\dot{p}_{i,n}(t) = \lambda(t)p_{i,n-1}(t) - (\lambda(t) + c\mu(t))p_{i,n}(t) + c\mu(t)p_{i,n+1}(t), \quad n \geq c.$$

We define the generating function $P(z,s,t) = \sum_{n=0}^{\infty} z^n p_{i,n}(t)$. The function $P(z,s,t)$ satisfies the ordinary differential equation

$$\begin{aligned}\dot{P}(z,s,t) = \lambda(t)\sum_{n=0}^{\infty} p_{i,n}(t)(z^{n+1} - z^n) \\ - \mu(t)\sum_{n=0}^{c-1} np_{i,n}(t)z^n - c\mu(t)\sum_{n=c}^{\infty} p_{i,n}(t)z^n \\ + \mu(t)\sum_{n=0}^{c-1} np_{i,n}(t)z^{n-1} + c\mu(t)\sum_{n=c}^{\infty} p_{i,n}(t)z^{n-1}\end{aligned}$$

In line 1 on the right-hand side of the preceding equation, we replace the infinite sum with the generating function for the number in queue. In lines 2 and 3, we add and subtract like quantities so that we can rewrite the expressions in terms of the generating function for the number in queue:

$$\begin{aligned}\dot{P}(z,s,t) = \lambda(t)(z-1)P(z,s,t) \\ + \mu(t)\sum_{n=0}^{c-1}(c-n)p_{i,n}(t)z^n - c\mu(t)\sum_{n=0}^{\infty} p_{i,n}(t)z^n \\ + \mu(t)\sum_{n=0}^{c-1}(n-c)p_{i,n}(t)z^{n-1} + c\mu(t)\sum_{n=0}^{\infty} p_{i,n}(t)z^{n-1}.\end{aligned}$$

We replace the series in lines 2 and 3 of the right-hand side of the previous equation with the generating function for the number in queue:

$$\begin{aligned}\dot{P}(z,s,t) = \big(\lambda(t)(z-1) + c\mu(t)(z^{-1} - 1)\big)P(z,s,t) \\ + \mu(t)\sum_{n=0}^{c-1}(c-n)z^n(1 - z^{-1})p_{i,n}(t).\end{aligned}$$

This differential equation has solution

$$P(z,s,t) = \int_s^t \mu(u) \sum_{n=0}^{c-1} (c-n) z^n (1-z^{-1}) p_{i,n}(u) \Phi_Y(z,u,t) du + P(z,s,s)\Phi_Y(z,s,t),$$

where we denote the generating function for Y_t as $\Phi_Y(z,s,t)$ and

$$\Phi_Y(z,s,t) = \sum_{n=-\infty}^{\infty} P\{Y_t = n\} z^n = e^{\Lambda(s,t)(z-1)+cM(s,t)(z^{-1}-1)}.$$

$\Phi_Y(z,s,t)$ is the solution of the evolution equation

$$\frac{\partial}{\partial t}\Phi_Y(z,s,t) = \Phi_Y(z,s,t)\left(\lambda(t)(z-1) + c\mu(t)(z^{-1}-1)\right)$$

and

$$\Phi(z,s,s) = \mathbf{I}.$$

The transient distribution for number in queue is given by

$$\begin{aligned} p_{i,j}(s,t) &= [z^j]P(z,s,t) \\ &= \int_s^t \mu(u) \sum_{n=0}^{c-1} (c-n) p_{i,n}(s,u) \left(\phi_{j-n}(u,t) - \phi_{j-n+1}(u,t)\right) du \\ &\quad + \phi_{j-i}(s,t). \end{aligned}$$

Let

$$\mathbf{p}_i(s,t) = \left[\, p_{i,0}(s,t)\ p_{i,1}(s,t) \cdots p_{i,c-1}(s,t) \,\right]$$

and

$$\phi_i(s,t) = \left[\, \phi_{-i}(s,t)\ \phi_{-i+1}(s,t) \cdots \phi_{-i+c-1}(s,t) \,\right].$$

Define the matrix function $\mathbf{K}(s,t)$

$$\mathbf{K}(s,t) = \left[\, \mathbf{k}_0(s,t)\ \mathbf{k}_1(s,t) \ldots \mathbf{k}_{c-1}(s,t) \,\right],$$

where $\mathbf{k}_j(u,t)$ is the column vector

$$\mathbf{k}_j(u,t) = \mu(s) \begin{bmatrix} c(\phi_j(s,t) - \phi_{j+1}(s,t)) \\ (c-1)(\phi_{j-1}(s,t) - \phi_j(s,t)) \\ \vdots \\ 2(\phi_{j-c+2}(s,t) - \phi_{j-c+3}(s,t)) \\ (\phi_{j-c+1}(s,t) - \phi_{j-c+2}(s,t)) \end{bmatrix}.$$

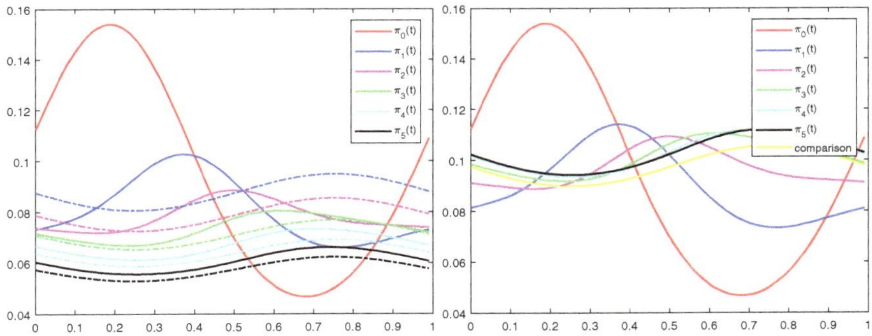

Fig. 4 Example 4. The graph on the left shows $\pi_j(t)$, $j = 0, \ldots, 5$ (solid lines) and the asymptotic estimates for each of these probabilities (dashed lines). The graph on the right shows $\pi_j(t)\left(\frac{\bar{\mu}}{\bar{\lambda}}\right)^j$, $j = 0, \ldots, 5$ and the asymptotic estimate for $f(t) = \pi_j(t)\left(\frac{\bar{\mu}}{\bar{\lambda}}\right)^j$ which does not depend on j

Then

$$\mathbf{p}_i(s,t) = \int_s^t \mathbf{p}_i(s,u)\mathbf{K}(u,t)du + \phi_i(s,t),$$

and

$$p_{i,j}(s,t) = \int_s^t \mathbf{p}_i(s,u)\mathbf{k}_j(u,t)du + \phi_{j-i}(s,t).$$

To compute these transient probabilities, we discretize each component of the matrix function $\mathbf{K}(s,t)$ so that we have a matrix of matrices. Let m be the mesh size, then we weight each component matrix by $\frac{1}{m}$.

In the case where transition rates are periodic only the first m rows of each component matrix need be computed. Subsequent blocks of m rows are equal to preceding blocks shifted m columns to the right. We also weight the diagonal and the top row of each component matrix by $\frac{1}{2}$. These weights allow us to use matrix multiplication to apply the trapezoidal rule of numerical integration to solve for the transient probabilities (Fig. 4).

3.1 Periodic

Now suppose transition rates are periodic with period one. In this case, we use $\pi_n(t)$ to designate the asymptotic periodic probability of n in the queue at time t rather than $p_{i,n}(t)$ for the transient probability. We wish to solve directly for the asymptotic periodic solution. In that case, $P(z, t-1, t) = P(z, t-1, t-1)$, so

$$P(z,t-1,t) = \int_{t-1}^{t} \mu(u) \sum_{n=0}^{c-1} (c-n) z^n (1-z^{-1}) \pi_n(u) \Phi_Y(z,u,t) du \, (1-\Phi(z,t-1,t))^{-1}. \tag{14.5}$$

The coefficient on z^j gives us the following integral formula for the periodic probability, $\pi_j(t)$, of j in the queue at time t within the period:

$$\pi_j(t) = [z^j] P(z,t-1,t) = \int_{t-1}^{t} \mu(u) \sum_{n=0}^{c-1} (c-n) \pi_n(u) \sum_{k=0}^{\infty} \left(\phi_{j-n}(u,t+k) - \phi_{j-n+1}(u,t+k)\right) du.$$

We have zeros in the denominator of equation (14.5), where

$$\bar{\lambda}(z-1) + c\bar{\mu}(z^{-1}-1) = 0.$$

This expression has two zeros: $z = 1$, and $z = c\bar{\mu}/\bar{\lambda}$. The value $z = 1$ is a root of both the numerator and the denominator, so we factor it out:

$$P(z,t-1,t) = \int_{t-1}^{t} \sum_{n=0}^{c-1} (c-n) z^n \pi_n(u) \frac{\mu(u)}{c\bar{\mu} - \bar{\lambda} z} \Phi_Y(z,u,t) du \times \left(\sum_{k=1}^{\infty} \frac{(\bar{\lambda}(z-1) + c\bar{\mu}(z^{-1}-1))^{k-1}}{k!} \right)^{-1}.$$

We can rewrite this as

$$P(z,t-1,t) = \int_{t-1}^{t} \sum_{n=0}^{c-1} (c-n) z^n \pi_n(u) \frac{\mu(u)}{c\bar{\mu}} \Phi_Y(z,u,t) du \left(\sum_{k=1}^{\infty} \frac{(\bar{\lambda}(z-1) + c\bar{\mu}(z^{-1}-1))^{k-1}}{k!} \right)^{-1} \sum_{j=0}^{\infty} \left(\frac{\bar{\lambda}}{c\bar{\mu}}\right)^j z^j. \tag{14.6}$$

An asymptotic estimate is then given by

$$\pi_j(t) \approx \int_{t-1}^{t} \frac{\mu(u)}{c\bar{\mu}} \Phi_Y\left(\frac{c\bar{\mu}}{\bar{\lambda}}, u, t\right) \sum_{n=0}^{c-1} (c-n) \pi_n(u) du \left(\frac{\bar{\lambda}}{c\bar{\mu}}\right)^{j-n}$$

$$= \int_{t-1}^{t} \frac{\mu(u)}{c\bar{\mu}} \exp\left\{\left(\frac{\Lambda(u,t)}{\bar{\lambda}} - \frac{M(u,t)}{\bar{\mu}}\right)(c\bar{\mu} - \bar{\lambda})\right\} \sum_{n=0}^{c-1} (c-n)\pi_n(u) \left(\frac{\bar{\lambda}}{c\bar{\mu}}\right)^{j-n} du. \tag{14.7}$$

We may estimate $\pi_j(t)$ as

$$\pi_j(t) \approx f(t) \left(\frac{\bar{\lambda}}{c\bar{\mu}}\right)^j,$$

where

$$f(t) = \int_{t-1}^{t} \frac{\mu(u)}{c\bar{\mu}} \exp\left\{\left(\frac{\Lambda(u,t)}{\bar{\lambda}} - \frac{M(u,t)}{\bar{\mu}}\right)(c\bar{\mu} - \bar{\lambda})\right\} \sum_{n=0}^{c-1} (c-n)\pi_n(u) \left(\frac{c\bar{\mu}}{\bar{\lambda}}\right)^{n} du.$$

The resulting expression for $f(t)$ in the multi-server case is analogous to that in the single-server case in Eq. (14.4). For a c server queue, the expression depends on the first c periodic probabilities for number in queue. The expression reduces to the single-server expression when $c = 1$. The queue length probabilities are asymptotically geometric with rate $\frac{\bar{\lambda}}{c\bar{\mu}}$.

4 Example: The Queue with Transitions of Size One and Two

For this queueing system, the classical single-server queueing system, $M/M/1$ is generalized to allow transition rates of size two in addition to the standard transition rates of size one (Fig. 5).

In terms of the queueing models, these systems each allow customers to arrive or be served instantly in pairs as well as individually (Fig. 6).

Krinik and Shun [2] have derived the steady-state distributions explicitly and determined a condition for the existence of a steady-state distribution. Assuming that a steady-state condition prevails, they determined the canonical performance measures, including expressions for the average number of customers in either system or queue. They also derived formulae for the average waiting time that a customer spends in the system or queue.

In this example, we generalize their model to allow transition rates to vary periodically with period of length one.

We have the Chapman–Kolmogorov equations with $\beta(t)$ and $\gamma(t)$ giving the rates at which transitions of size two occur:

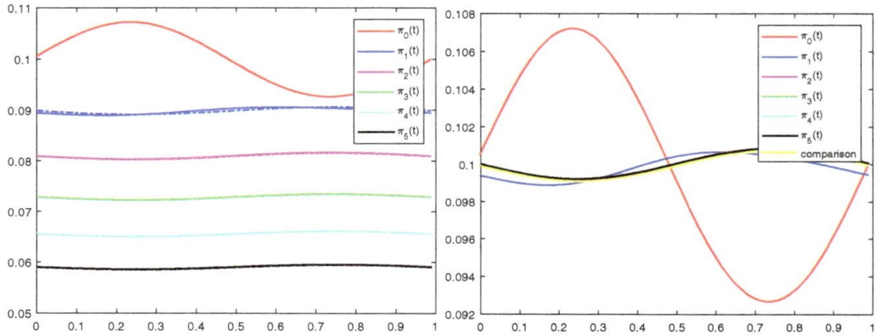

Fig. 5 Example 5. The graph on the left shows $\pi_j(t)$, $j = 0, \ldots, 5$ (solid lines) and the asymptotic estimates for each of these probabilities (dashed lines). The graph on the right shows $\pi_j(t)\left(\frac{\bar{\mu}}{\bar{\lambda}}\right)^j$, $j = 0, \ldots, 5$ and the asymptotic estimate for $f(t) = \pi_j(t)\left(\frac{\bar{\mu}}{\bar{\lambda}}\right)^j$ which does not depend on j

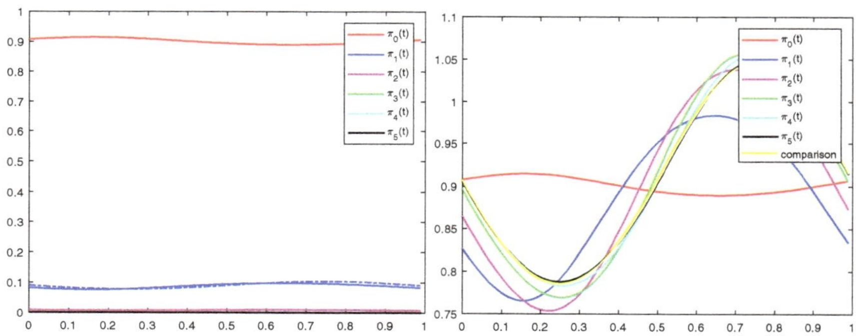

Fig. 6 Example 6. The graph on the left shows $\pi_j(t)$, $j = 0, \ldots, 5$ (solid lines) and the asymptotic estimates for each of these probabilities (dashed lines). The graph on the right shows $\pi_j(t)\left(\frac{\bar{\mu}}{\bar{\lambda}}\right)^j$, $j = 0, \ldots, 5$ and the asymptotic estimate for $f(t) = \pi_j(t)\left(\frac{\bar{\mu}}{\bar{\lambda}}\right)^j$ which does not depend on j

$$
\begin{aligned}
\dot{p}_{i,0}(t) &= -(\lambda(t) + \beta(t))p_{i,0}(t) + \mu(t)p_{i,1}(t) + \gamma(t)p_{i,2}(t) \\
\dot{p}_{i,1}(t) &= \lambda(t)p_{i,0}(t) - (\lambda(t) + \beta(t) + \mu(t))p_{i,1}(t) + \mu(t)p_{i,2}(t) + \gamma(t)p_{i,3}(t) \\
\dot{p}_{i,n}(t) &= \beta(t)p_{i,n-2}(t) + \lambda(t)p_{i,n-1}(t) \\
&\quad -(\lambda(t) + \beta(t) + \mu(t) + \gamma(t))p_{i,n}(t) + \mu(t)p_{i,n+1}(t) + \gamma(t)p_{i,n+2}(t).
\end{aligned}
$$

We define the generating function $P(z,t) = \sum_{n=0}^{\infty} z^n p_{i,n}(t)$. Then we have

$$
\begin{aligned}
\dot{P}(z,t) =& (z^2\beta(t) + z\lambda(t) - (\lambda(t) + \beta(t) + \mu(t) + \gamma(t)) + z^{-1}\mu(t) + z^{-2}\gamma(t))P(z,t) \\
& + \gamma(t)(z - z^{-1})p_{i,1}(t) + (\gamma(t) + \mu(t) - z^{-1}\mu(t) - \gamma(t)z^{-2})p_{i,0}(t).
\end{aligned}
$$

This has solution

$$P(z, s, t) = \int_0^t \big(\gamma(t)(z - z^{-1})p_{i,1}(u) + (\gamma(t)(1 - z^{-2}) + \mu(t)(1 - z^{-1}))p_{i,0}(u)\big) \times \Phi(z, u, t)du + P(z, s, s)\Phi(z, s, t), \tag{14.8}$$

where

$$\begin{aligned}
\Phi(z, s, t) &= \exp\left\{\int_s^t (z^2\beta(u) + z\lambda(u) - (\lambda(u) + \beta(u) + \mu(u) + \gamma(u)) + z^{-1}\mu(u) + z^{-2}\gamma(u))du\right\} \\
&= \exp\left\{\int_s^t (z\lambda(u) - (\lambda(u) + \mu(u)) + z^{-1}\mu(u))du\right\} \\
&\quad \times \exp\left\{\int_s^t (z^2\beta(u) - (\beta(u) + \gamma(u)) + z^{-2}\gamma(u))du\right\} \\
&= \Phi_Y(z, s, t)\Phi_X(z^2, s, t),
\end{aligned}$$

and X_t and Y_t are the randomized random walks. For the walk X_t steps to the right occur at rate $\beta(t)$ and to the left at rate $\gamma(t)$. For the walk Y_t steps to the right occur at rate $\lambda(t)$ and to the left at rate $\mu(t)$. $\Phi_X(z, s, t)$ and $\Phi_Y(z, s, t)$ are the generating functions for the randomized random walks X_t and Y_t, respectively. Expanding the generating function in terms of coefficients on z^n, we have

$$\begin{aligned}
\Phi(z, s, t) &= \sum_{n=-\infty}^{\infty} z^n \phi_n(s, t) \\
&= \sum_{n=-\infty}^{\infty} z^n \sum_{j=-\infty}^{\infty} P\{X_t = j | X_s = 0\} P\{Y_t = n - 2j | Y_s = 0\}.
\end{aligned}$$

Assume that $p_{i,j}(s) = \delta_{j=i}$. Matching coefficients on z_n, we see that

$$\begin{aligned}
p_{i,0}(t) = \int_s^t \big(&p_{i,1}(u)\gamma(u)\,(\phi_{-1}(u, t) - \phi_1(u, t)) \\
&+ p_{i,0}(u)\,((\gamma(u) + \mu(u))\phi_0(u, t) - \mu(u)\phi_1(u, t) - \gamma(u)\phi_2(u, t))\big)\,du \\
&+ \phi_{-i}(s, t) \\
p_{i,1}(t) = \int_s^t \big(&p_{i,1}(u)\gamma(u)\,(\phi_0(u, t) - \phi_2(u, t)) \\
&+ p_{i,0}(u)\,((\gamma(u) + \mu(u))\phi_1(u, t) - \mu(u)\phi_2(u, t) - \gamma(u)\phi_3(u, t))\big)\,du \\
&+ \phi_{-i+1}(s, t),
\end{aligned}$$

and more generally

$$p_{i,n}(t) = \int_s^t \left[p_{i,1}(u)\gamma(u)\left(\phi_{i,n-1}(u,t) - \phi_{i,n+1}(u,t)\right)\right.$$
$$\left. + p_{i,0}(u)\left((\gamma(u)+\mu(u))\phi_n(u,t) - \phi_{n+1}(u,t)\mu - \gamma\phi_{n+2}(u,t)\right)\right] du$$
$$+ \phi_{-i+n}(s,t).$$

Now suppose that transition rates are time-varying and periodic. Further suppose that our generating function is for the asymptotic periodic distribution of the number in the system. Then we have the generating function

$$P(z,t-1,t) = \int_{t-1}^t \left(\pi_1(u)\gamma(u)(z - z^{-1})\right.$$
$$\left. + \pi_0(u)(\gamma(u)(1-z^{-2}) + \mu(u)(1-z^{-1}))\right) \Phi(z,u,t)du$$
$$\times (1 - \Phi(z,t-1,t))^{-1} \quad (14.9)$$

and

$$\pi_0(t) = \int_0^t \left[\pi_1(u)\gamma(u)\sum_{k=0}^{\infty}\left(\phi_{-1}(u,t+k) - \phi_1(u,t+k)\right)\right.$$
$$+ \pi_0(u)\left((\gamma(u)+\mu(u))\sum_{k=0}^{\infty}\phi_0(u,t+k)\right.$$
$$\left.\left. - \sum_{k=0}^{\infty}\phi_1(u,t+k)\mu(u) - \gamma(u)\sum_{k=0}^{\infty}\phi_2(u,t+k)\right)\right] du,$$

$$\pi_1(t) = \int_0^t \left[\pi_1(u)\gamma(u)\sum_{k=0}^{\infty}(\phi_0(u,t+k) - \phi_2(u,t+k))\right.$$
$$+ \pi_0(u)\left((\gamma(u)+\mu(u))\sum_{k=0}^{\infty}\phi_1(u,t+k)\right.$$
$$\left.\left. - \mu(u)\sum_{k=0}^{\infty}\phi_2(u,t+k) - \gamma(u)\sum_{k=0}^{\infty}\phi_3(u,t+k)\right)\right] du,$$

and more generally

$$\pi_n(t) = \int_0^t \left[\pi_1(u)\gamma(u)\sum_{k=0}^{\infty}\left(\phi_{n-1}(u,t+k) - \sum_{k=0}^{\infty}\phi_{n+1}(u,t+k)\right)\right.$$
$$+ \pi_0(u)\left((\gamma(u)+\mu(u))\sum_{k=0}^{\infty}\phi_n(u,t+k)\right.$$

$$-\sum_{k=0}^{\infty}\phi_{n+1}(u,t+k)\mu(u)-\gamma(u)\sum_{k=0}^{\infty}\phi_{n+2}(u,t+k)\Bigg)\Bigg]du.$$

These expressions are difficult to evaluate numerically. So as in the $M_t/M_t/1$ example, we apply an asymptotic estimate of the transition probabilities.

The first step is to factor out $z-1$ since one is a root of both the numerator and the denominator. Second, we follow Krinik and Shun and find the roots of the denominator. Then following the approach outlined by Sedgewick and Flajolet [1], we approximate the integrand as a sum of geometric series. We will need to do a partial fractions decomposition.

Factorization of the numerator of Eq. (14.9) by $z-1$ yields

$$\int_{t-1}^{t}\left[\gamma(u)(1+z^{-1})p_1(u)+(\gamma(u)(z^{-1}+z^{-2})+\mu(u)z^{-1})p_0(u)\right]\Phi(z,u,t)du.$$

The denominator is zero when $\Phi(z,t-1,t)=1$, that is, when

$$\left(\bar{\beta}z^2+\bar{\lambda}z-(\bar{\beta}+\bar{\lambda}+\bar{\mu}+\bar{\gamma})+\bar{\mu}z^{-1}+\bar{\gamma}z^{-2}\right)=0.$$

One root occurs at $z=1$:

$$\begin{aligned}&\left(\bar{\beta}z^2+\bar{\lambda}z-(\bar{\beta}+\bar{\lambda}+\bar{\mu}+\bar{\gamma})+\bar{\mu}z^{-1}+\bar{\gamma}z^{-2}\right)\\&\qquad=(z-1)(\bar{\beta}z+\bar{\lambda}+\bar{\beta}-(\bar{\mu}+\bar{\gamma})z^{-1}-\bar{\gamma}z^{-2}).\end{aligned}$$

Factorization of the denominator of equation (14.9) by $z-1$ yields

$$\begin{aligned}&(\bar{\beta}z+\bar{\lambda}+\bar{\beta}-(\bar{\mu}+\bar{\gamma})z^{-1}-\bar{\gamma}z^{-2})\\&\qquad\times\sum_{k=1}^{\infty}\frac{\left(\bar{\beta}z^2+\bar{\lambda}z-(\bar{\beta}+\bar{\lambda}+\bar{\mu}+\bar{\gamma})+\bar{\mu}z^{-1}+\bar{\gamma}z^{-2}\right)^{k-1}}{k!}.\end{aligned}$$

There are three other real roots of the denominator. These were computed by Krinik and Shun [2, Lemma 1.1]. Their reciprocals are given by

$$\frac{1}{r_1}=-\frac{a}{3}-2\sqrt{U}\cos\left(\frac{\theta}{3}\right),\quad \frac{1}{r_2}=-\frac{a}{3}-2\sqrt{U}\cos\left(\frac{\theta}{3}+\frac{4\pi}{3}\right),$$
$$\frac{1}{r_3}=-\frac{a}{3}-2\sqrt{U}\cos\left(\frac{\theta}{3}+\frac{2\pi}{3}\right),$$

where $U=\frac{a^2-3b}{9}$, $V=\frac{2a^3-9ab+27c}{54}$, $\theta=\cos^{-1}\left(\frac{V}{\sqrt{U^3}}\right)$, $a=\frac{\bar{\mu}+\bar{\gamma}}{\bar{\gamma}}$, $b=-\frac{\bar{\beta}+\bar{\gamma}}{\bar{\gamma}}$, $c=-\frac{\bar{\beta}}{\bar{\gamma}}$ and $\frac{1}{r_1}<-1<\frac{1}{r_2}<0<\frac{1}{r_3}$.

We are considering stable queues, so r_1 is also a root of the numerator. We apply a partial fractions decomposition to

$$\frac{1}{\left(1-\frac{z}{r_2}\right)\left(1-\frac{z}{r_3}\right)} = \frac{-r_3}{(r_2-r_3)\left(1-\frac{z}{r_2}\right)} + \frac{r_2}{(r_2-r_3)\left(1-\frac{z}{r_3}\right)}.$$

Define

$$H(z,t) = -z^2 \int_{t-1}^{t} \left[\gamma(u)(1+z^{-1})p_1(u) + (\gamma(u)(z^{-1}+z^{-2}) \right.$$
$$\left. +\mu(u)z^{-1})p_0(u)\right] \Phi(z,u,t)du \left(\bar{\gamma}\left(1-\frac{z}{r_1}\right)(r_2-r_3)\right)^{-1}.$$

So we may write our asymptotic estimate as

$$\pi_n(t) \approx \frac{-r_3 H(r_2,t)}{r_2^n} + \frac{r_2 H(r_3,t)}{r_3^n}.$$

This solution is analogous to that obtained by Krinik and Shun for the steady-state distribution. They had

$$\pi_n = \frac{c_2}{r_2^n} + \frac{c_3}{r_3^n}$$

for constants c_2 and c_3 which they give explicitly in their paper [2].

5 Asymptotic Estimates for Level Independent Quasi-Birth-Death Processes

The same method can be used to obtain estimates for the level distribution for QBDs with level independent transitions. Such QBDs will have infinitesimal generator with block tri-diagonal structure:

$$\mathbf{Q}(t) = \begin{bmatrix} \mathbf{B}(t) & \mathbf{A}_1(t) & 0 & 0 & \cdots \\ \mathbf{A}_{-1}(t) & \mathbf{A}_0(t) & \mathbf{A}_1(t) & 0 & \cdots \\ 0 & \mathbf{A}_{-1}(t) & \mathbf{A}_0(t) & \mathbf{A}_1(t) & \cdots \\ 0 & 0 & \mathbf{A}_{-1}(t) & \mathbf{A}_0(t) & \cdots \\ \vdots & \vdots & \vdots & \vdots & \ddots \end{bmatrix}, \tag{14.10}$$

where $\mathbf{A}_{-1}(t), \mathbf{A}_0(t), \mathbf{A}_1(t)$ and $\mathbf{B}(t)$ are square matrices of order m, and m represents the number of phases.

We partition $\boldsymbol{\pi}(t)$ by levels into subvectors $\boldsymbol{\pi}_n(t)$, $n \geq 0$, where $\boldsymbol{\pi}_n(t)$ has m components. The QBD system satisfies the Chapman–Kolmogorov forward equations

$$\dot{\boldsymbol{\pi}}_0(t) = \boldsymbol{\pi}_0(t)\mathbf{B}(t) + \boldsymbol{\pi}_1(t)\mathbf{A}_{-1}(t)$$
$$\dot{\boldsymbol{\pi}}_n(t) = \boldsymbol{\pi}_{n-1}(t)\mathbf{A}_1(t) + \boldsymbol{\pi}_n(t)\mathbf{A}_0(t) + \boldsymbol{\pi}_{n+1}(t)\mathbf{A}_{-1}(t),$$

with the additional requirement that

$$\sum_{n=0}^{\infty} \boldsymbol{\pi}_n(t)\mathbf{1} = 1.$$

For periodic rates with period of length one, if stability conditions are met, there will be a solution of the Chapman–Kolmogorov equations such that

$$\boldsymbol{\pi}_n(t) = \boldsymbol{\pi}_n(t+k),$$

$k \in \mathbb{Z}$.

The generating function for the random walk corresponding to this QBD satisfies

$$\Phi(z, s, t) = \sum_{n=-\infty}^{\infty} \boldsymbol{\phi}_n(s, t) z^n,$$
$$\frac{\partial}{\partial t}\Phi(z, u, t) = \Phi(z, u, t)\left(z^{-1}\mathbf{A}_{-1}(t) + \mathbf{A}_0(t) + z\mathbf{A}_1(t)\right),$$

where $\phi_n(s, t)$ is an $m \times m$ matrix of transition probabilities. The (i, j) component represents the probability of traveling to phase j by time t and remaining there until at least time t and traveling to a level n units to the right of the level occupied at time s given that the random walk process was in phase i at time s. For more details on the set up and analysis of such systems with time-varying periodic transitions, see [4] or [5]. For more background on quasi-birth-death processes in general see [3].

The generating function for the levels of the QBD solves the differential equation

$$\frac{\partial}{\partial t}P(z, s, t) = P(z, s, t)\left(z^{-1}\mathbf{A}_{-1}(t) + \mathbf{A}_0(t) + z\mathbf{A}_1(t)\right) + \boldsymbol{\pi}_0(t)\left(\mathbf{B}(t) - z^{-1}\mathbf{A}_{-1}(t) - \mathbf{A}_0(t)\right),$$

so

$$P(z, s, t) = \int_s^t \boldsymbol{\pi}_0(u)\left(\mathbf{B}(u) - z^{-1}\mathbf{A}_{-1}(u) - \mathbf{A}_0(u)\right)\Phi(z, u, t)du + P(z, s, s)\Phi(z, s, t),$$

and for the periodic case with period 1,

$$P(z, t-1, t) = \int_{t-1}^{t} \pi_0(u) \left(\mathbf{B}(u) - z^{-1}\mathbf{A}_{-1}(u) - \mathbf{A}_0(u)\right) \Phi(z, u, t) du \times (\mathbf{I} - \Phi(z, t-1, t))^{-1}.$$

There will be poles in the determinant of the matrix $(\mathbf{I} - \Phi(z, t-1, t))$. These poles will reveal the geometric behavior of the level distribution.

6 Conclusion

This approach to the analysis of time-varying queues with periodic transition rates offers considerable promise for improving the understanding of the behavior of such systems. In particular, it shows that such queues are asymptotically geometric in the queue length distribution. Future work will involve extending these results and further analyzing quasi-birth-and-death problems that fit this framework. For scalar queueing models, computation of the roots is straightforward. For quasi-birth-and-death processes, computation of the roots is feasible in special cases, but is challenging for general QBDs.

References

1. Flajolet, P., Sedgewick, R.: Analytic Combinatorics. Cambridge University Press, Cambridge (2009)
2. Krinik, A.C., Shun, K.: Markov processes with constant transition rates of size one and two. J. Stat. Theory Pract. **5**(3), 475–495 (2011)
3. Latouche, G., Ramaswami, V.: Introduction to Matrix Analytic Methods in Stochastic Modeling. ASA-SIAM Series on Statistics and Applied Probability, vol. 5. Society for Industrial and Applied Mathematics (1999)
4. Margolius, B.H.: Transient and periodic solution to the time-inhomogeneous quasi-birth process. Queueing Syst. **56**(3–4), 183–194 (2007)
5. Margolius, B.H.: The matrices *R* and *G* of matrix analytic methods and the time-inhomogeneous periodic quasi-birth-and-death process. Queueing Syst. **60**(1–2), 131–151 (2008)
6. Paulsen, W.: Asymptotic Analysis and Perturbation Theory. A Chapman and Hall book, Taylor and Francis (2014)
7. Schwarz, J.A., Selinka, G., Stolletz, R.: Performance analysis of time-dependent queueing systems: Survey and classification. Omega **63**, 170–189 (2016)

A Combinatorial Analysis of the $M/M^{[m]}/1$ Queue

Guven Mercankosk and Gopalan M. Nair

Abstract Neuts' Matrix Geometric Method makes use of the left-skip free characteristic of M/G/1-type Markov chains and determines the first passage distribution matrix G by solving a non-linear matrix equation. In this paper, we focus on the k-step first passage problem. In particular, we identify three associated matrices, namely the matrix G_k, the conditional first passage probability matrix P_k, and the first passage count matrix T_k. The reformulation allows for combinatorial techniques. Specifically, we refer to an extension of Takács' ballot theorem. We note that the matrix P_k exhibits some ballot properties. In the case of the $M/M^{[m]}/1$ queue, we establish the special structure of the count matrix T_k using lattice path arguments. Furthermore, we obtain a closed-form expression for the G matrix, where the first passage probabilities are expressed in terms of generalized hypergeometric functions.

Keywords $M/M^{[m]}/1$ queue · Ballot theorems · Combinatorial techniques · Generalized hypergeometric functions · Lattice paths

2010 Mathematics Subject Classification Primary: 60K25 · Secondary: 05A05 · 05A19

1 Introduction

Neuts [6] makes use of the left-skip free characteristic of M/G/1 type Markov chains to define a fundamental period and determines the associated *first passage distribution matrix G* by solving a non-linear matrix functional equation. In this paper, we limit our focus to M/G/1 type Markov chains embedded at the service termination epochs of a queueing system where service is carried out in batches of size m. In

---

An initial draft of this paper was completed in 2006; Dr. Guven Mercankosk, who was an academic staff at the University of Western Australia, passed away on January 4, 2007.

G. Mercankosk · G. M. Nair (✉)
The University of Western Australia, Crawley, Perth, WA 6009, Australia
e-mail: gopalan.nair@uwa.edu.au

G. E. Andrews et al. (eds.), *Lattice Path Combinatorics and Applications*,
Developments in Mathematics 58, https://doi.org/10.1007/978-3-030-11102-1_15

doing so, we reformulate the first passage problem in a way to allow for combinatorial techniques.

The state space of a Markov chain of M/G/1 type consists of semi-infinite strips of the form $\{(i, j) : i \geq 0,\ 0 \leq j < m\}$. The set of states $\{(i, j) : 0 \leq j < m\}$ is referred to as *level* i, and the state (i, j) is referred to as *phase* j in *level* i. Neuts [7] defines G_{jl}, $0 \leq jl < m$, the entries of the first passage distribution matrix $G = [G_{jl}]$, as the probability that level $i - 1$, $i \geq 1$, is eventually reached for the first time, by a visit to $(i - 1, l)$, given that the Markov chain started in state (i, j). Neuts [7] also defines the $m \times m$ matrix $G_{k,jl}$, $0 \leq j, l < m$, as the probability that level $i - 1$, $i \geq 1$, is reached for the first time in exactly k transitions, by a visit to $(i - 1, l)$, given that the Markov chain started in state (i, j).

Our formulation starts by conditioning $G_{k,jl}$ by the exact number of arrivals needed for the first passage from (i, j) to $(i - 1, l)$ for the first time in k steps. We observe that the associated conditional first passage probabilities $P_{k,jl}$ are of ballot type, and an extension of Takács' ballot theorem shows that the conditional first passage probability matrix P_k exhibits some ballot properties, which can then be used in conjunction with other combinatorial techniques. More specifically, we identify the distinct arrival patterns that are problem specific. Then the problem is equivalent to counting the number of arrival patterns that satisfy the first passage conditions, and we thereby define the conditional first passage count matrix T_k.

For the case of an $M/M^{[m]}/1$ queue, the associated count matrix T_k has a special structure of identical rows. We develop a lattice path formulation for proving this special structure. Using the diagonal properties of the matrix P_k and the special structure of the matrix T_k, we obtain a closed-form solution for the matrix G_k. Summing G_k over k yields closed-form expressions in terms of hypergeometric functions for the entries of the G matrix.

This paper is organized as follows: In Sect. 2, we provide an illustration of the first passage problem and outline the problem reformulation that allows for combinatorial techniques. We then formally define the matrices, G_k, P_k, and T_k. In Sect. 3, for the sake of completeness, we restate the extended ballot theorem and elaborate on the diagonal properties of the P_k matrix. In Sect. 4, we further limit our scope to the $M/M^{[m]}/1$ queue. We first introduce a lattice path formulation and prove the special structure of identical rows for the count matrix T_k. We then apply the extended ballot theorem to obtain a closed-form solution for the first passage matrix G. Section 5 concludes the paper.

2 Reformulation of the First Passage Problem

In this section, we first provide a summary of the essential elements of the matrix-analytic method pioneered by Neuts [7]. After stating the common structure of transition probability matrices of Markov chains of M/G/1 type, we review the concept of fundamental period of a Markov chain and present an associated first passage problem as our starting point. We then outline our approach and elaborate on its combinatorial nature.

2.1 The Canonical Form

The canonical form refers to the common structure of transition probability matrices of Markov chains of M/G/1 type. Let Q denote a transition probability matrix in canonical form of the Markov chain with state space $\{(i, j) : i \geq 0;\ 0 \leq j < m\}$, then it is given by

$$Q = \begin{bmatrix} B_0 & B_1 & B_2 & B_3 & \cdots \\ A_0 & A_1 & A_2 & A_3 & \cdots \\ 0 & A_0 & A_1 & A_2 & \cdots \\ 0 & 0 & A_0 & A_1 & \cdots \\ \vdots & \vdots & \vdots & \vdots & \end{bmatrix},$$

where the matrices A_γ, B_γ, $\gamma \geq 0$, are square matrices of order m. The matrix Q is also referred to as a stochastic matrix of M/G/1 type as its rows add up to 1. The structure of the matrix Q implies that the chain is left-skip free for levels. That is, any path leading from a state in a higher level to a state in a lower level must visit every intermediate level at least once.

2.2 The Fundamental Period

In order to have a recurrent chain Q, it is necessary that the level $i - 1$ is eventually reached from any state in level i with probability one. Accordingly, for a Markov chain of M/G/1 type, the fundamental period is defined as the first passage time taken to visit level $i - 1$ for the first time, having started in level i for $i \geq 1$. For the purpose of our discussion, we restrict our attention to the number of transitions during a fundamental period.

From the definition of G and G_k, $k \geq 1$, it follows that

$$G = \sum_{k=1}^{\infty} G_k.$$

The matrix sequence $\{G_k\}_{k\geq 1}$ is referred to as the matrix density of the number of transitions in the fundamental period of visiting $(i - 1, l)$ from (i, j). It has been shown in [7] that $\{G_k\}_{k\geq 1}$ satisfies the matrix equations

$$G_1 = A_0, \quad G_k = \sum_{\nu=1}^{k-1} A_\nu G_{k-1}^{(\nu)} \quad \text{for } k \geq 2, \tag{15.1}$$

where

$$G_r^{(\nu)} = \sum_{i=1}^{r-\nu+1} G_i^{(1)} G_{r-i}^{(\nu-1)}. \tag{15.2}$$

Note that $G_{r,jl}^{(\nu)}$ is the conditional probability that level i, $i \geq 0$, is reached for the first time by a visit to (i, l) in exactly r transitions, given that the Markov chain started in $(i + \nu, j)$. The matrix G has been proven by Neuts [5] to be the minimal non-negative solution to the non-linear matrix equation

$$G = A(G) = \sum_{k=0}^{\infty} A_k G^k. \tag{15.3}$$

One can obtain the non-linear matrix equation (15.3) by summing the matrix sequence $\{G_k\}_{k\geq 1}$ as defined by (15.1).

2.3 Service in Batches of Size *m*

Returning our attention to M/G/1 type Markov chains embedded at the service termination epochs of a queueing system where service is carried out in batches of size m, let $V(i, j; i-1, l)$ denote the number of batch service completions in a fundamental period of visiting $(i-1, l)$ from (i, j). Note that

$$G_{k,jl} = \Pr\{V(i, j; i-1, l) = k\}.$$

Also, let $A(t, t+\tau)$ denote the number of arrivals during an interval covering τ successive service intervals during a fundamental period and let R_t represent the number in the system just after a service completion at time t. Again, note that we necessarily and exactly have $(k-1)m + l - j$ arrivals during a fundamental period of k service completions, that is

$$A(t, t+k) = (k-1)m + l - j.$$

Conditioning $G_{k,jl}$ on the number of arrivals during the fundamental period, we have

$$\begin{aligned} G_{k,jl} = \Pr\{V(i, j; i-1, l) = k \mid A(t, t+k) = (k-1)m + l - j\} \\ \times \Pr\{A(t, t+k) = (k-1)m + l - j\}. \end{aligned} \tag{15.4}$$

Let $P_{k,jl}$ denote the conditional probability on the right-hand side of (15.4). Then, we have

$$G_{k,jl} = P_{k,jl} \times \Pr\{A(t, t+k) = (k-1)m + l - j\}. \tag{15.5}$$

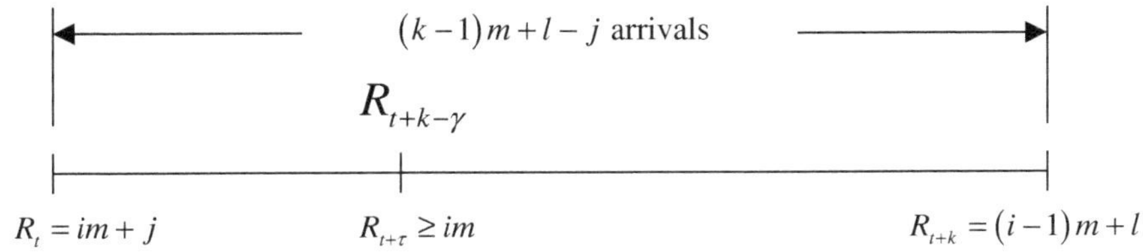

Fig. 1 First passage from level i to level $i-1$ in k transitions

Given $(k-1)m + l - j$ arrivals during a fundamental period of k service completions we also require that the associated arrival pattern, as illustrated in Fig. 1, to satisfy $R_{t+\tau} \geq im$, for $\tau = 0, 1, \ldots, k-1$.

That is,

$$\{V(i, j; i-1, l) = k \mid A(t, t+k) = (k-1)m + l - j\}$$
$$= \{R_{t+\tau} \geq im \text{ for } \tau = 0, 1, \ldots, k-1 \mid A(t, t+k) = (k-1)m + l - j\}. \tag{15.6}$$

Therefore we have

$$P_{k,jl} = \Pr\{R_{t+\tau} \geq im \text{ for } \tau = 0, 1, \ldots, k-1 \mid A(t, t+k) = (k-1)m + l - j\}. \tag{15.7}$$

We further note that the number in the system just after a batch service completion, $R_{t+\tau}$, during a fundamental period is related to the number in the system at the start of the fundamental period, R_t, as

$$R_{t+\tau} = R_t + A(t, t+\tau) - \tau m, \quad \text{for } \tau = 0, 1, \ldots, k-1. \tag{15.8}$$

2.4 *The Conditional First Passage Probabilities* $P_{k,jl}$

The change of variables $\gamma \leftarrow k - \tau$, in (15.7), and use of (15.8) lead to

$$P_{k,jl} = \Pr\{\max_{1\leq\gamma\leq k} [A(t+k-\gamma, t+k) - \gamma m] < l - (m-1)$$
$$|A(t, t+k) = k(m-1) + l - j\}, \tag{15.9}$$

where $0 \leq j, l < m$ and $A(t, t+k) < km$. Note that each entry of the matrix P_k as given by (15.9) is of *ballot type* and, for $m = 1$, the matrix P_k reduces to a scalar as

$$P_k = \Pr\{\max_{1\leq\gamma\leq k} [A(t+k-\gamma, t+k) - \gamma] < 0 \mid A(t, t+k) = k - 1\}. \tag{15.10}$$

The expression given in (15.10) was used by Takács [8] and is known to be equal to $1/k$ under any cyclically interchangeable arrival patterns. Consequently, the scalar first passage probabilities in k transitions for Poisson arrivals are given by

$$G_k = \int_0^{\infty} e^{-\lambda x} \frac{(\lambda x)^{k-1}}{(k-1)!} dB_k(x), \tag{15.11}$$

where $B_x(x)$ denotes the kth iterated convolution of the service time distribution $B(x)$ with itself.

2.5 The First Passage Counts $T_{k,jl}$

In determining the conditional probabilities given in (15.9), we generally encounter an associated combinatorial problem. In fact, it is an arrangement problem of $(k-1)m + l - j$ arrival epochs and k service termination epochs in a way that the condition given in (15.6) for k transition first passage is satisfied. So, we define $T_{k,jl}$ to denote the number of arrival patterns that satisfy the condition (15.6).

The associated arrangement problem is generally specific to the queueing system in hand. In the case of an M/D$^{[m]}$/1 queue, the arrivals occur according to the Poisson distribution. So, each of the $(k-1)m + l - j$ arrivals is uniformly distributed over k service periods. With deterministic service times, we have $k^{(k-1)m+l-j}$ equally likely arrival patterns. Noting that $T_{k,jl}$ denotes the number of arrival patterns that satisfy the condition (15.6), the problem is reduced to counting such arrival patterns.

It can be shown (see [2]) that the numbers $T_{k,jl}$ for an M/D$^{[m]}$/1 queue satisfy the recurrence

$$\begin{aligned} T_{1,jl} &= 0, \quad \text{if } l < j, \\ T_{1,jl} &= 1, \quad \text{if } l \geq j, \\ T_{k,j0} &= \sum_{r=1}^{k-1} \binom{(k-1)m - j - 1}{rm - j - 1} T_{r,j(m-1)r} T_{k-r,00}, \\ T_{k,jl} &= k T_{k,j(l-1)} + \sum_{r=1}^{k-1} \binom{(k-1)m - j - 1}{rm - j - 1} T_{r,j(m-1)} T_{k-r,0l}, \end{aligned} \tag{15.12}$$

and consequently we may determine the ballot probabilities defined in (15.9) as

$$P_{k,jl} = \frac{T_{k,jl}}{k^{(k-1)m+l-j}}. \tag{15.13}$$

In tabulating the matrix-sequence P_k for a given $m > 1$, we note that the entries $P_{k,jl}$ are independent of not only the arrival rate λ but also of the length of the deterministic service time. This in turn allows the matrix-sequence P_k associated with any M/D$^{[m]}$/1 queue for a specified $m > 1$ to be pre-computed and stored.

Finally, the first passage probabilities $G_{k,jl}$ from level i to level $i-1$ in k transitions for an M/D$^{[m]}$/1 queue is given by

$$G_{k,jl} = P_{k,jl} \frac{(\lambda k)^{(k-1)m+l-j}}{[(k-1)m+l-j]!} e^{-\lambda x}, \tag{15.14}$$

where the deterministic service time has been taken as unity.

2.6 A Numerical Example

We next provide a numerical example to illustrate the computation of the matrix triple $\{T_k, P_k, G_k\}$, when $k = 3$, for the M/D$^{[3]}$/1 queue, and highlight an important property.

The matrix entries $T_{3,jl}$, for $j, l = 0, 1, 2$, can be computed using (15.12) as

$$T_3 = \begin{bmatrix} 42 & 393 & 2187 \\ 26 & 213 & 1065 \\ 15 & 106 & 474 \end{bmatrix}$$

Using this and Eq. (15.13), the entries of P_3 (independent of arrival rate λ) can be obtained:

$$P_3 = \begin{bmatrix} 14/243 & 131/729 & 1/3 \\ 26/243 & 71/243 & 355/729 \\ 5/27 & 106/243 & 158/243 \end{bmatrix}.$$

Hence, using (15.14), we have

$$G_3 = \begin{bmatrix} \frac{7}{120}\lambda^6 & \frac{131}{1680}\lambda^7 & \frac{243}{4480}\lambda^8 \\ \frac{13}{60}\lambda^5 & \frac{71}{240}\lambda^6 & \frac{71}{336}\lambda^7 \\ \frac{5}{8}\lambda^4 & \frac{53}{60}\lambda^5 & \frac{79}{120}\lambda^6 \end{bmatrix} e^{-3\lambda}.$$

Alternatively, without paying attention to numerical efficiency, we may use (15.1) for calculating the matrix G_k and then P_k and T_k using (15.14) and (15.13) in order.

Note an important property: $P_{3,02} = 1/3$, $P_{3,01} + P_{3,12} = 2/3$, and finally on the main diagonal we have $P_{3,00} + P_{3,11} + P_{3,22} = 3/3$. This is an interesting combinatorial structure, which is explained by a generalization of the ballot theorem, and our objective is to exploit any such combinatorial structures available in order to achieve numerical efficiency. As a matter of fact, this is the approach we take for the rest of the paper in obtaining a closed-form solution for the M/M$^{[m]}$/1 queue.

3 Takács' Ballot Theorem

In this section, we first state two results extending Takács' ballot theorem. These results first appeared in [3], but are repeated here without proof for the sake of completeness. We then elaborate on the upper diagonal properties of the conditional first passage matrix P_k.

3.1 Extended Results

Theorem 15.1 *Let $n_1, n_2, \ldots, n_k$ be non-negative integers with sum $n_1 + \cdots + n_k = n < km$. Consider the k cyclic permutations of $(n_1, n_2, \ldots, n_k)$. For $d = 0, 1, \ldots, m-1$, let C_d denote the number of cyclic permutations for which the sum of the first r elements is less than $rm - d$ for all $r = 1, 2, \ldots, k$.*

- *If a cyclic permutation of $(n_1, n_2, \ldots, n_k)$ contributes to C_d, then it also contributes to $C_0, C_1, \ldots, C_{d-1}$.*
- *Furthermore, we have $C_0 + C_1 + \cdots + C_{m-1} = km - n$ for $d = 0, 1, \ldots, m-1$.*

Theorem 15.2 *Let $\nu_1, \nu_2, \ldots, \nu_k$ be cyclically interchangeable random variables taking on non-negative integer values. Set $N_s = \nu_1 + \cdots + \nu_s$ for $1 \le s \le k$ with $N_k < km$. Then we have*

$$\sum_{d=0}^{m-1} \Pr\{\max_{1\le s\le k}[N_s - sm] < -d \mid N_k\} = m - \frac{N_k}{k}. \tag{15.15}$$

Corollary 15.1 *More specifically, for $N_k = km - x$, where $1 \le x \le m$, we have*

$$\sum_{d=0}^{x-1} \Pr\{\max_{1\le s\le k}[N_s - sm] < -d \mid N_k\} = \frac{x}{k}. \tag{15.16}$$

According to Theorem 15.1, the sum on the left-hand side of (15.15) simplifies to the left-hand side of (15.16) as the second bullet point leads to

$$C_0 + C_1 + \cdots + C_{m-1} = x < m,$$

and by the first bullet point, we necessarily have

$$C_x = C_{x+1} = \cdots = C_{m-1} = 0.$$

Then the right-hand side of (15.16) trivially follows from (15.15).

3.2 Upper Diagonal Properties of the Matrix P_k

We next return our attention to (15.9). Using Corollary 15.1, we can derive an important property for the upper diagonal entries in the conditional first passage matrix P_k. Let E_d represent the event

$$\{\max_{1\leq\gamma\leq k}[A(t+k-\gamma,t+k)-\gamma m]<-d\},$$

and $A_{k,x}=\{A(t,t+k)=(k-1)m+x\}$. Then P_k takes the form

$$P_k=\begin{bmatrix} \Pr\{E_{m-1}|A_{k,0}\} & \Pr\{E_{m-2}|A_{k,1}\} & \cdots & \Pr\{E_1|A_{k,m-2}\} & \Pr\{E_0|A_{k,m-1}\} \\ \Pr\{E_{m-1}|A_{k,-1}\} & \Pr\{E_{m-2}|A_{k,0}\} & \cdots & \Pr\{E_1|A_{k,m-3}\} & \Pr\{E_0|A_{k,m-2}\} \\ \Pr\{E_{m-1}|A_{k,-2}\} & \Pr\{E_{m-2}|A_{k,-1}\} & \cdots & \Pr\{E_1|A_{k,m-4}\} & \Pr\{E_0|A_{k,m-3}\} \\ \vdots & \vdots & & \vdots & \vdots \\ \Pr\{E_{m-1}|A_{k,-m+1}\} & \Pr\{E_{m-2}|A_{k,-m+2}\} & \cdots & \Pr\{E_1|A_{k,-1}\} & \Pr\{E_0|A_{k,0}\} \end{bmatrix} \quad (15.17)$$

By taking $x=1$ in (15.16), we observe that the entry $P_{k,0(m-1)}$, at the top-right corner of the matrix P_k is always $1/k$, resembling Takács' result. As a matter of fact, it holds, by Theorem 15.2, for any cyclically interchangeable arrival patterns, which is a weaker condition than Poisson arrivals. Furthermore, by taking $x=m$ in (15.16), the main diagonal entries of P_k always add up to m/k.

In general, numbering off diagonals away from the top-right corner and including the main diagonal, for the xth off diagonal, $1\leq x\leq m$, we have

$$\sum_{y=0}^{x-1}P_{k,y(m-x+y)}=\frac{x}{k}, \quad (15.18)$$

which again holds for any cyclically interchangeable arrival patterns. Note also that the equality given in (15.18) is independent of the service time distribution. For more details on this result see [3, Theorem 6]. These upper diagonal properties enable us to evaluate P_k explicitly for the M/M$^{[m]}$/1 queue.

4 Application of the Extended Ballot Theorem

So far, all is applicable for the M/M$^{[m]}$/1 queue. We can even replace Poisson arrivals by any cyclically interchangeable arrival process. Here, we look at the case where the service time distribution is negative exponential with parameter μ. The inter-arrival distribution parameter is taken as λ.

Note that we can easily obtain closed-form expressions for matrices T_1, P_1, and G_1. Indeed, the fundamental period, from (i, j) to $(i-1, l)$ in one step, is only possible when $l-j\geq 0$. Hence, $T_{1,jl}=P_{1,jl}=G_{1,jl}=0$ for $l<j$. Furthermore, when $l-j\geq 0$, there is only one possible arrival pattern which also leads to a

fundamental period in one step. Therefore, we have $T_{1,jl} = P_{1,jl} = 1$ for $l \geq j$. Using the relation in (15.5), we obtain

$$G_{1,jl} = P_{1,jl} \times \Pr\{A(t, t+1) = l - j\} = \frac{\lambda^{l-j}\mu}{(\lambda+\mu)^{l-j+1}}. \tag{15.19}$$

For $k > 1$, we observe that the first passage counting matrix T_k has a special structure. We use a lattice path formulation (see [4]) to prove the generality of this special structure. Then, using the extended ballot results, we obtain closed-form expressions for the matrices T_k, P_k, and G_k.

4.1 A Lattice Path Formulation

Suppose there are u events occurring from one Poisson process, say $P1$, and v events occurring from another independent Poisson process, say $P2$, over an interval of specified length. The number of different ways these events can be arranged in the interval depends neither on the process' parameters nor on the length of the interval. In fact, the number of possible arrangements is

$$C(u, u+v) = \frac{(u+v)!}{u!\,v!}.$$

In terms of lattice paths, each arrangement corresponds to a path from (0, 0) to (u, v) where events from process $P1$ and $P2$ are represented by horizontal and vertical units, respectively.

For the case of an M/M$^{[m]}$/1 queue, we are interested in a fundamental period of k transitions. Let, as before, $T_{k,jl}$ denote number of arrival patterns that result in a fundamental period of k transitions starting in phase j and ending in phase l. Consider an arrival pattern contributing to $T_{k,jl}$ over the interval. We accordingly have $(k-1)m + l - j$ arrival events and $k-1$ service completion events. Therefore, by taking $P1$ as the arrival process and $P2$ as the service completion process, for each arrival pattern the associated lattice path starts at (0, 0) and ends at $((k-1)m + l - j, k-1)$. We note that the very last event by definition has to be a service completion event and therefore it has been left out without loss of generality.

Note that the conditional first passage probabilities $P_{k,jl}$ for an M/M$^{[m]}$/1 queue are then related to first passage counts $T_{k,jl}$ by

$$P_{k,jl} = \frac{T_{k,jl}}{\binom{(k-1)m+l-j+k-1}{k-1}}. \tag{15.20}$$

In the lattice path context, we represent the fundamental period requirement by a straight line that the path representing an arrival pattern cannot cross upwards. Recall that $A(t, t+\tau)$ denotes the number of arrivals during an interval covering τ

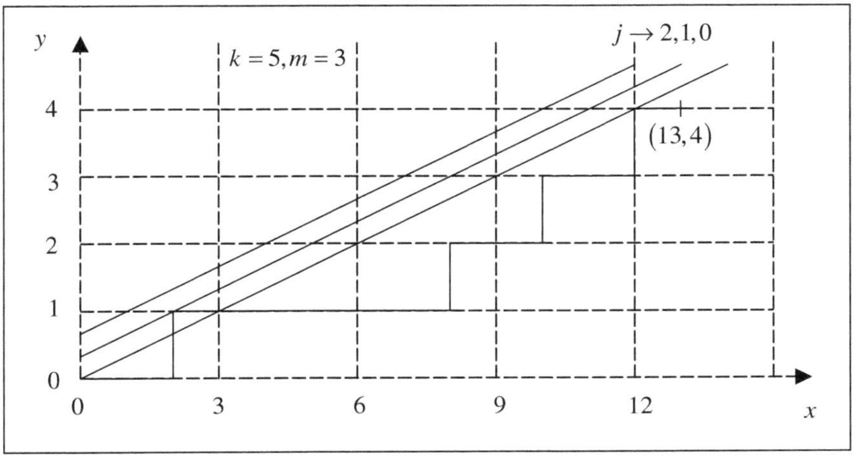

Fig. 2 Lattice path representation of an arrival pattern

successive transitions during a fundamental period and $R_{t+\tau}$ represents the number in the system (queue length) just after the transition at $t+\tau$. Formally, we need $R_{t+\tau}$ to stay at or above level i for $\tau = 0, 1, \ldots, k-1$. In other words, we need $R_t + A(t, t+\tau) - \tau m \geq im$. This means that $im + j + A(t, t+\tau) - \tau m \geq im$. Hence we must have $A(t, t+\tau) \geq \tau m - j$. Replacing $A(t, t+\tau)$ by x and τ by y, we see that an arrival pattern satisfies the requirements of a fundamental period if it does not cross the line

$$y = \frac{1}{m}x + \frac{j}{m}.$$

Since $0 \leq j < m$, we have m such barriers to consider for the matrix T_k as illustrated in Fig. 2.

4.2 The M/M$^{[m]}$/1 Queue

Lemma 15.1 *For an M/M$^{[m]}$/1 queue, with $k > 1$, $T_{k,0l} = T_{k,jl}$ for $j = 1, 2, \ldots, m-1$.*

Proof It suffices to show that for every path contributing to $T_{k,0l}$ one can construct a path contributing to $T_{k,jl}$ and vice versa. Note that a path contributing to $T_{k,0l}$ consists of $(k-1)m + l$ horizontal steps, and the first vertical step can occur only after m horizontal steps, as otherwise it would cross the associated line $y = xm$. In fact, any vertical step, say the yth, $2 \leq y < k$, can occur only after ym horizontal steps. Similarly, if a path contributes to $T_{k,jl}$, it consists of $(k-1)m + l - j$ horizontal steps, its first vertical step may occur only after $m - j$ horizontal steps, and any other vertical step, say the yth, $2 \leq y < k$, can occur only after $ym - j$ horizontal steps.

For a path contributing to $T_{k,0l}$, by removing its first j horizontal steps and shifting the remaining path horizontally to the origin, we obtain a path with $(k-1)m + l - j$ horizontal steps and with none of its vertical steps crossing the line $y = (x + j)m$. Hence, the path obtained as such contributes to $T_{k,jl}$. Conversely, let us consider a path contributing to $T_{k,jl}$. Inserting j horizontal steps at the beginning results in a path that does not cross the line $y = xm$ and increases the number of horizontal steps to $(k-1)m + l$ in total. Hence, the new path contributes to $T_{k,0l}$. This completes the proof of the lemma.

Theorem 15.3 *For an M/M*$^{[m]}$*/1 queue, with* $k > 1$, *we have*

$$T_{k,jl} = \frac{l+1}{(k-1)m+l+1}\binom{(k-1)m+l+k-1}{k-1}. \tag{15.21}$$

Proof Since the expression given in (15.21) does not depend on j, we prove the theorem, without loss of generality, for $j = 0$.

By the diagonal property of the matrix P_k as expressed in (15.18), we have

$$\sum_{y=0}^{x-1} P_{k,y(m-x+y)} = \sum_{y=0}^{x-1} \frac{T_{k,y(m-x+y)}}{\binom{(k-1)m+m-x+y-y+k-1}{k-1}} = \frac{x}{k}, \tag{15.22}$$

where we also make use (15.20). Since $T_{k,y(m-x+y)} = T_{k,0(m-x+y)}$ by Lemma 15.1, and using the change of variables $l \leftarrow m - x$, one can rewrite (15.22) as

$$\sum_{y=0}^{m-l-1} T_{k,0(l+y)} = T_{k,0l} + T_{k,0(l+1)} + \cdots + T_{k,0(m-1)} = \frac{m-l}{k}\binom{(k-1)m+l+k-1}{k-1}.$$

Hence

$$\begin{aligned}
T_{k,0l} &= \sum_{y=0}^{m-l-1} T_{k,0(l+y)} - \sum_{y=0}^{m-(l+1)-1} T_{k,0(l+1+y)} \\
&= \frac{m-l}{k}\binom{(k-1)m+l+k-1}{k-1} - \frac{m-(l+1)}{k}\binom{(k-1)m+(l+1)+k-1}{k-1} \\
&= \frac{1}{k}\left\{(m-l) - \left[\frac{(k-1)m+l+k}{(k-1)m+l+1}\right](m-l-1)\right\}\binom{(k-1)m+l+k-1}{k-1} \\
&= \frac{l+1}{(k-1)m+l+1}\binom{(k-1)m+l+k-1}{k-1}.
\end{aligned}$$

This completes the proof of the theorem.

In what follows, we obtain an explicit expression for the entries of the first passage distribution matrix G in terms of hypergeometric functions. The definition and an identity for hypergeometric functions are given in the Appendix. Note that these functions can be evaluated in mathematical packages such as *MATHEMATICA* and *MAPLE*.

Theorem 15.4 *For an M/M$^{[m]}$/1 queue, the first passage distribution matrix* $G = [G_{jl}]$ *is given by*

$$G_{jl} = \begin{cases} p^{l-j} q \; {}_{m+1}F_m \left(a_1^l, a_2^l, \ldots, a_{m+1}^l; b_1^l, b_2^l, \ldots, b_m^l; z\right), & \text{if } j \le l, \\ p^{l-j} q \left[{}_{m+1}F_m \left(a_1^l, a_2^l, \ldots, a_{m+1}^l; b_1^l, b_2^l, \ldots, b_m^l; z\right) - 1 \right], & \text{if } j > l, \end{cases} \tag{15.23}$$

where

$$\begin{aligned} a_i^l &= \frac{l+i}{m+1}, \quad \text{for } i = 1, 2, \ldots, m+1, \\ b_i^l &= \frac{l+1+i}{m}, \quad \text{for } i = 1, 2, \ldots, m, \\ z &= \frac{(m+1)^{m+1}}{n^m} p^m q, \\ (p, q) &= \left(\frac{\lambda}{\lambda+\mu}, \frac{\mu}{\lambda+\mu} \right). \end{aligned}$$

Proof From the discussion preceding (15.19), for an M/M$^{[m]}$/1 queue, we have

$$G_{1,jl} = \begin{cases} p^{l-j} q, & \text{if } j \le l, \\ 0, & \text{if } j > l. \end{cases}$$

Further noting that the probability of having $(k-1)m + l - j$ arrivals from a Poisson process with parameter λ over k exponentially distributed intervals each with parameter μ is given by

$$\Pr\{A(t, t+k) = (k-1)m + l - j\} = \binom{(k-1)m + l - j + k - 1}{k-1} p^{(k-1)m+l-j} q^k, \tag{15.24}$$

by substituting (15.24) in (15.5) and making use of (15.20) and (15.21), for $k > 1$ we arrive at

$$G_{k,jl} = \frac{l+1}{(k-1)m + l + 1} \binom{(k-1)m + l + k - 1}{k-1} p^{(k-1)m+l-j} q^k. \tag{15.25}$$

Hence, for $j \leq l$, the fundamental period probability G_{jl} is given by

$$\begin{aligned}
G_{jl} &= \sum_{k\geq 1} G_{k,jl} \\
&= \sum_{k\geq 1} \frac{l+1}{(k-1)m+l+1}\binom{(k-1)m+l+k-1}{k-1} p^{(k-1)m+l-j} q^k \\
&= \sum_{k\geq 0} \frac{l+1}{km+l+1}\binom{km+l+k}{k} p^{km+l-j} q^{k+1} \\
&= p^{l-j} q \sum_{k\geq 0} \frac{l+1}{km+l+1}\binom{km+l+k}{k} \left[p^m q\right]^k .
\end{aligned}$$

By taking $x = p^m q$ in (15.27), the result follows for $j \leq l$.

On the other hand, for $j > l$, the probability G_{jl} is given by

$$\begin{aligned}
G_{jl} &= \sum_{k\geq 2} G_{k,jl} \\
&= \sum_{k\geq 2} \frac{l+1}{(k-1)m+l+1}\binom{(k-1)m+l+k-1}{k-1} p^{(k-1)m+l-j} q^k \\
&= \sum_{k\geq 1} \frac{l+1}{km+l+1}\binom{km+l+k}{k} p^{km+l-j} q^{k+1} \\
&= p^{l-j} q \left[-1+1+\sum_{k\geq 1} \frac{l+1}{km+l+1}\binom{km+l+k}{k} p^{km} q^k\right] \\
&= p^{l-j} q \left[{}_{m+1}F_m\left(a_1^l, a_2^l, \ldots, a_{m+1}^l; b_1^l, b_2^l, \ldots, b_m^l; z\right) - 1\right]
\end{aligned}$$

where $\{a_1^l, \ldots, a_{m+1}^l\}$, $\{b_1^l, \ldots, b_m^l\}$, and z are given in the theorem. This completes the proof of the theorem.

5 Conclusion

This paper presents a combinatorial approach to Neuts' first passage problem for the left-skip free Markov chains. The key to the presented approach is an extension of Takács' ballot theorem. The particular application of the extension to the M/M$^{[m]}$/1 queue yields closed-form expressions for the entries of the first passage distribution matrix G. Preliminary results suggest that Takács' elegant combinatorial methods can be carried beyond the M/G/1 queue.

Acknowledgements We would like to thank Sri Gopal Mohanty for fruitful discussions and valuable suggestions.

Appendix: A Result on Generalized Hypergeometric Functions

First define $(a)_r$, called *Pochhammer symbol* by $(a)_0 = 1$ and

$$(a)_r = a(a+1)\cdots(a+r-1).$$

Then the *generalized hypergeometric function* ${}_pF_q(a_1,\ldots,a_p;b_1,\ldots,b_q;x)$ is defined by

$${}_pF_q(a_1,\ldots,a_p;b_1,\ldots,b_q;x) = \sum_{r=0}^{\infty} \frac{(a_1)_r\,(a_2)_r\cdots(a_p)_r}{(b_1)_r\,(b_2)_r\cdots(b_q)_r}\frac{x^r}{r!}. \tag{15.26}$$

See [1] for detailed properties of these functions. We need the following identities in order to establish a closed-form for the first passage distribution matrix G.

Lemma 15.2 *If m and l are positive integers such that $m > 1$ and $0 \le l < m$, then*

$$\sum_{k\ge 0} \frac{l+1}{km+l+1}\binom{km+l+k}{k}x^k = {}_{m+1}F_m(a_1,\ldots,a_{m+1};b_1,\ldots,b_m;u), \tag{15.27}$$

where

$$a_i = \frac{l+i}{m+1}, \quad \text{for } i = 1,2,\ldots,m+1,$$
$$b_i = \frac{l+i+1}{m}, \quad \text{for } i = 1,2,\ldots,m,$$
$$u = \frac{(m+1)^{m+1}}{m^m}x.$$

Proof This amounts to a routine verification which is left to the reader.

References

1. Bell, W.W.: Special Functions for Scientists and Engineers. D. Van Nostrand Company Ltd., London (1968)
2. Blake, I.F., Konheim, A.G.: Big buckets are (are not) better. J. Assoc. Comput. Mach. **24**, 590–606 (1977)
3. Mercankosk, G., Nair, G.M., Soet, W.J.: Combinatorial techniques for M/G/1-type queue. J. Appl. Probab. **38**, 722–736 (2001)
4. Mohanty, S.G.: Lattice Path Counting and Applications. Academic, New York (1979)
5. Neuts, M.F.: The Markov renewal branching process. In: Clarke, A.B. (ed.) Conference on Mathematical Methods in the Theory of Queues, Kalamazoo MI, pp. 1–21. Springer, New York (1974)

6. Neuts, M.F.: Queues solvable without Rouché's theorem. Oper. Res. **27**, 767–781 (1978)
7. Neuts, M.F.: Structured Stochastic Matrices of M/G/1 Type and Applications. Marcel Dekker, New York (1989)
8. Takács, L.: Combinatorial Methods in the Theory of Stochastic Processes. Wiley, New York (1967)

Laws Relating Runs, Long Runs, and Steps in Gambler's Ruin, with Persistence in Two Strata

Gregory J. Morrow

Abstract Define a certain gambler's ruin process $\mathbf{X}_j$, $j \geq 0$, such that the increments $\varepsilon_j := \mathbf{X}_j - \mathbf{X}_{j-1}$ take values ± 1 and satisfy $P(\varepsilon_{j+1} = 1 | \varepsilon_j = 1, |\mathbf{X}_j| = k) = P(\varepsilon_{j+1} = -1 | \varepsilon_j = -1, |\mathbf{X}_j| = k) = a_k$, all $j \geq 1$, where $a_k = a$ if $0 \leq k \leq f - 1$, and $a_k = b$ if $f \leq k < N$. Here, $0 < a, b < 1$ denote persistence parameters and $f, N \in \mathbb{N}$ with $f < N$. The process starts at $\mathbf{X}_0 = m \in (-N, N)$ and terminates when $|\mathbf{X}_j| = N$. Denote by $\mathscr{R}'_N$, $\mathscr{U}'_N$, and $\mathscr{L}'_N$, respectively, the numbers of runs, long runs, and steps in the meander portion of the gambler's ruin process. Define $X_N := \left(\mathscr{L}'_N - \frac{1-a-b}{(1-a)(1-b)} \mathscr{R}'_N - \frac{1}{(1-a)(1-b)} \mathscr{U}'_N \right) / N$ and let $f \sim \eta N$ for some $0 < \eta < 1$. We show $\lim_{N \to \infty} E\{e^{itX_N}\} = \hat{\varphi}(t)$ exists in an explicit form. We obtain a companion theorem for the last visit portion of the gambler's ruin.

Keywords Runs · Generating function · Excursion · Gambler's ruin · Last visit · Meander · Persistent random walk · Generalized Fibonacci polynomial

2010 Mathematics Subject Classification Primary: 60F05 · Secondary: 05A15

1 Introduction

Define a gambler's ruin process $\{\mathbf{X}_j, \ j \geq 0\}$, with values in $\mathbb{Z} \cap [-N, N]$, such that the increments $\varepsilon_j := \mathbf{X}_j - \mathbf{X}_{j-1}$ take values ± 1 and satisfy $P(\varepsilon_{j+1} = 1 | \varepsilon_j = 1, |\mathbf{X}_j| = k) = P(\varepsilon_{j+1} = -1 | \varepsilon_j = -1, |\mathbf{X}_j| = k) = a_k$, all $j \geq 1$, where $a_k = a$ if $0 \leq k \leq f - 1$, and $a_k = b$ if $f \leq k < N$. Here, $0 < a, b < 1$ denote persistence parameters and $f, N \in \mathbb{N}$ with $f < N$. The process starts at some fixed level $m \in (-N, N)$ and terminates at an epoch j when $|\mathbf{X}_j| = N$. For initial probabilities, take $\pi_+ = P(\varepsilon_j = 1) = \pi_- = P(\varepsilon_j = -1) = \frac{1}{2}$. We call the two ranges of values $|k| \leq f - 1$ and $f \leq |k| < N$ as strata for the two persistence parameter values a and b, respectively. In gambling, $\mathbf{X}_j$ denotes a fortune after j games on which the

G. J. Morrow (✉)
University of Colorado, Colorado Springs, CO 80918, USA
e-mail: gmorrow@uccs.edu

G. E. Andrews et al. (eds.), *Lattice Path Combinatorics and Applications*,
Developments in Mathematics 58, https://doi.org/10.1007/978-3-030-11102-1_16

gambler makes unit bets. If $a, b > \frac{1}{2}$, then any run of fortune tends to keep going in the same direction. Thus, for example, a win [loss] resulting in fortune k for some $|k| \leq f - 1$ is followed by another win [loss] with probability a, whereas a change in fortune occurs with probability $1 - a$. Henceforth, we shall simply refer to $\{\mathbf{X}_j = \mathbf{X}_j^N\}$ as the gambler's ruin process, with or without mention of the parameters a, b, f, and N. Note that $\{\mathbf{X}_j\}$ is the classical fair gambler's ruin process in case $a = b = \frac{1}{2}$, with symmetric boundaries N and $-N$. For the *homogeneous* case $a = b$, the increments, $\{\varepsilon_j, j \geq 0\}$, form a strictly stationary process with zero means, where the correlation between ε_j and ε_{j+1} is $2a - 1$. If $a = b$ and also $N = \infty$, then $\{\mathbf{X}_j^\infty\}$ becomes a symmetric persistent (or correlated) random walk on $\mathbb{Z}$ that is recurrent by [19, Thm. 8.1].

Physical models of persistence often consider the velocity of a particle either staying the same or being changed according to a random collision process [1, 17, 21]; in our model, the velocity only takes values ± 1. Our introduction of strata corresponds to a change in medium over which the persistence parameter, or likelihood of the velocity staying the same, would deterministically change. In [21], the authors obtain a Wiener limit for the normalized sum of velocities under a random environment, that includes our deterministic model. Our aim is different since we want results for discrete statistics that have no analogue in the Wiener process. In this context, our stratified model seems to be new.

We define a nearest neighbor path of length n in $\mathbb{Z}$ to be a sequence $\Gamma = \Gamma_0, \Gamma_1, \ldots, \Gamma_n$, where $\Gamma_j \in \mathbb{Z}$ and $\delta_j := \Gamma_j - \Gamma_{j-1}$ satisfies $|\delta_j| = 1$ for all $j = 1, \ldots, n$. We also call n the number of steps of Γ. We connect successive lattice points $(j - 1, \Gamma_{j-1})$ and (j, Γ_j) in the plane by straight line segments and term this connected union of straight line segments the *lattice path*. See Figs. 1 and 2. We define the number of runs along Γ as the number of inclines, either straight line ascents or descents, of maximal extent along the lattice path; the length of a run is the number of steps in such a maximal ascent or descent. A *long run* is itself a run that consists of at least two steps; in gambling terminology, a long run means that the run of fortune does not immediately change direction. A *short run* is on the other hand a run of length exactly one, so every run is either a long run or a short run. In Fig. 2, the lattice path shown has 15 runs, with 7 short runs and 8 long runs. An *excursion* is a nearest neighbor path that starts and ends at $m = 0$, $\Gamma_0 = \Gamma_n = 0$, but for which $\Gamma_j \neq 0$ for $1 \leq j \leq n - 1$. A positive excursion is an excursion whose graph lies above the x-axis save for its endpoints. For a positive excursion path, the number of runs is just twice the number of peaks, where a peak at lattice point (j, Γ_j) corresponds to $\delta_j = 1$ and $\delta_{j+1} = -1$.

The *last visit* is defined as

$$\mathscr{L}_N := \max\{j \geq 0 : j = 0, \text{ or } \mathbf{X}_j = 0 \text{ for some } j \geq 1\}. \tag{16.1}$$

The *meander* is the portion of the process that extends from the epoch of the last visit $\mathscr{L}_N$ until the gambler's ruin process terminates. So the meander process never returns to the level $m = 0$. See Fig. 1. It is shown by [14] that, for $a = b = \frac{1}{2}$, if $\mathscr{R}_N$ denotes the total number of runs over all excursions of the absolute value process

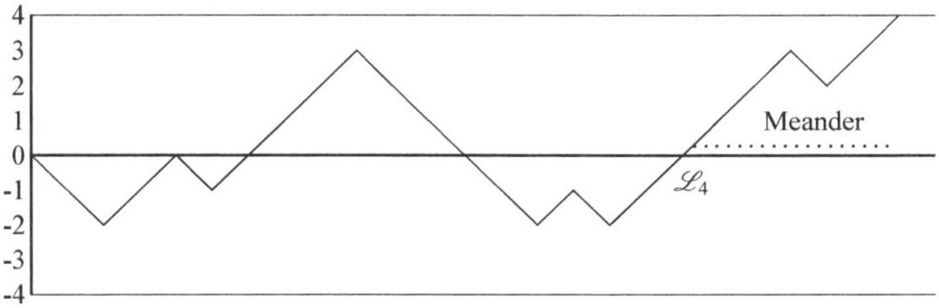

Fig. 1 Last visit and meander; $N = 4$

$\{|\mathbf{X}_j|\}$ until the last visit, then, with the order N scaling, it holds that $(\mathscr{L}_N - 2\mathscr{R}_N)/N$ converges in law. Also, if $\mathscr{R}'_N$ and $\mathscr{L}'_N$ denote respectively the number of runs and steps over the meander portion of the process, then $(\mathscr{L}'_N - 2\mathscr{R}'_N)/N$ converges in law to a density $\varphi(x) = (\pi/4)\text{sech}^2(\pi x/2)$, $-\infty < x < \infty$, with characteristic function $\int_{-\infty}^{\infty} \varphi(x)e^{ixt}dx = t/\sinh(t)$. We first generalize this result for the meander case. Let $\mathscr{R}'_N, \mathscr{V}'_N, \mathscr{L}'_N$ denote, respectively, the numbers of runs, short runs, and steps, in the meander portion of the gambler's ruin; for the lattice path of Fig. 1, we have $\mathscr{R}'_4 = 3$, $\mathscr{V}'_4 = 1$, $\mathscr{L}'_4 = 6$. Define the following scaled random variable over the meander:

$$X_N := \frac{1}{N}\left(\mathscr{L}'_N - \frac{2-a-b}{(1-a)(1-b)}\mathscr{R}'_N + \frac{1}{(1-a)(1-b)}\mathscr{V}'_N\right). \tag{16.2}$$

Theorem 16.1 *Let $f = \eta N$ for some fixed $0 < \eta < 1$. Denote $\kappa_1 := \frac{\eta\sigma_1}{1-b}$ and $\kappa_2 := \frac{(1-\eta)\sigma_2}{1-a}$, with $\sigma_1 = \sqrt{a + b^2 - 2ab}$ and $\sigma_2 = \sqrt{b + a^2 - 2ab}$. Let X_N be defined by* (16.2). *Then, $\lim_{N\to\infty} E\{e^{itX_N}\} = \hat{\varphi}(t)$, where*

$$\begin{aligned}(b\kappa_1\sigma_2 + a\kappa_2\sigma_1)t/\hat{\varphi}(t) &:= a\sigma_1\cosh(\kappa_1 t)\sinh(\kappa_2 t)\\ &+ b\sigma_2\sinh(\kappa_1 t)\cosh(\kappa_2 t) + i(b-a)^2\sinh(\kappa_1 t)\sinh(\kappa_2 t).\end{aligned} \tag{16.3}$$

In Theorem 16.1, we obtain that $\varphi(x) := \frac{1}{2\pi}\int_{-\infty}^{\infty} e^{-itx}\hat{\varphi}(t)dt$ is real since the complex conjugate of $\varphi(x)$ is equal to itself; observe this by making a change of variables $t \to -t$ after conjugation of the integral.

We have a bivariate result for the homogeneous case as follows. Define

$$Y_{1,N} := \frac{1}{N}\left(\mathscr{R}'_N - \frac{1}{(1-a)}\mathscr{V}'_N\right);\quad Y_{2,N} := \frac{1}{N}\left(\mathscr{L}'_N - \frac{1}{(1-a)}\mathscr{R}'_N\right) - Y_{1,N}. \tag{16.4}$$

Corollary 16.1 *Suppose $a = b$. Then the limiting joint characteristic function of the random variables $Y_{1,N}$ and $Y_{2,N}$ is:*

$$\lim_{N\to\infty} E\{e^{isY_{1,N}+itY_{2,N}}\} = \frac{\sqrt{(1-a)s^2 + at^2}}{\sinh(\sqrt{(1-a)s^2 + at^2})}. \tag{16.5}$$

Remark 16.1 Let $a = b$, and define $X_{\zeta,N} := \frac{1}{N}\left(\mathscr{L}'_N - \frac{1+\zeta}{(1-a)}\mathscr{R}'_N + \frac{\zeta}{(1-a)^2}\mathscr{V}'_N\right)$. Then by setting $s = (1 - a - \zeta)t/(1 - a)$ in (16.5) we obtain, for all $\zeta \in \mathbb{R}$,

$$\lim_{N\to\infty} E\{e^{itX_{\zeta,N}}\} = \frac{A_\zeta t}{\sinh(A_\zeta t)}; \quad A_\zeta := \sqrt{[(2\zeta - 1)a + (1 - \zeta)^2]/(1 - a)}.$$

Example 16.1 As a special case of Theorem 16.1, consider $b = 1 - a$ and $\eta = a$. Then $\kappa_i = \sigma_j = \sigma := \sqrt{1 - 3a + 3a^2}$, for all $i, j = 1, 2$. In this case, we have

$$\hat{\varphi}(t) = \sigma^2 t / \left\{\sinh(\sigma t)[\sigma \cosh(\sigma t) + i(1 - 2a)^2 \sinh(\sigma t)]\right\}. \tag{16.6}$$

The complex factor of the denominator of $\hat{\varphi}(t)$ in (16.6) is equal to zero if and only if $e^{2\sigma t} = \frac{-\sigma+(1-2a)^2 i}{\sigma+(1-2a)^2 i}$. The smallest root is $t = \frac{i}{2\sigma}(\pi - \arctan \frac{2\sigma(1-2a)^2}{\sigma^2-(1-2a)^4})$, with $\sigma^2 - (1 - 2a)^4 = a(1 - a)(5 - 16a + 16a^2) > 0$. Thus, we can analytically continue $\hat{\varphi}(t)$ to a suitably chosen ball of positive radius ϵ_0 about the origin such that $\sup_{|\xi|\le\epsilon_0} \|\hat{\varphi}(\cdot + i\xi)\|_2 < \infty$. It follows by [18, Thm. IX.13] that the inverse Fourier transform $\varphi(x)$ of $\hat{\varphi}(t)$ has exponential decay, meaning $e^{\epsilon|x|}\varphi(x)$ is square integrable for any $\epsilon < \epsilon_0$. However, the probability density, $\varphi(x)$, is not symmetric in x under (16.6) with $a \neq \frac{1}{2}$; see Fig. 4 at the end of the paper; see also [15] for computational details.

We now introduce the definitions of the excursion statistics to further describe our results. For the definitions in this paragraph, we assume $\mathbf{X}_0 = 0$ and $N = \infty$. Define the index j, or step, of first return of $\{\mathbf{X}_j\}$ to the origin by $\mathbf{L} := \inf\{j \ge 1 : \mathbf{X}_j = 0\}$. Define the excursion sequence from the origin by $\boldsymbol{\Gamma} := \{\mathbf{X}_j, j = 0, \ldots, \mathbf{L}\}$; again $\mathbf{L}$ is the number of steps of $\boldsymbol{\Gamma}$. Define the *height* $\mathbf{H}$ of the excursion $\boldsymbol{\Gamma}$ as the maximum absolute value of the path over this excursion:

$$\mathbf{H} := \max\{|\mathbf{X}_j| : j = 1, \ldots, \mathbf{L}\}. \tag{16.7}$$

Also define $\mathbf{R}$ as the number of runs along $\boldsymbol{\Gamma}$, and further define $\mathbf{V}$ as the number of short runs along $\boldsymbol{\Gamma}$. Thus, officially $\mathbf{U} := \mathbf{R} - \mathbf{V}$ is the number of long runs along $\boldsymbol{\Gamma}$. In Fig. 1, there are 4 excursions until the last visit to the origin, with respective heights: 2, 1, 3, 2. The numbers of runs in the excursions of the absolute value process $\{|\mathbf{X}_j|\}$ until the last visit of Fig. 1, wherein negative excursions are reflected into positive excursions, are: 2, 2, 2, 4. The corresponding numbers of short runs in this last visit portion of the absolute value process are: 0, 2, 0, 2.

The first motivation of the present paper is to show how the method of [14] extends to the three statistics, runs, short runs, and steps, in the homogeneous setting ($a = b$). As a particular result, we find the following Corollary 16.2, which connects the present work with a certain combinatorial domain in the study of Dyck paths. Note that the generating function method which drives the present study depends heavily on a *return to the level* 1 type recurrence approach that has been applied extensively in the field of lattice path combinatorics; see [2–4, 7, 9, 10, 12]. Let P_a

denote the probability for the homogeneous model with persistence parameter a. We obtain the following symmetry for the joint distribution of the excursion statistics.

Corollary 16.2 *Let $a = b$ and assume $\mathbf{X}_0 = 0$ and $N = \infty$. Then for all $n \geq 2$ there holds:*

$$(1-a)P_a(\mathbf{L} = 2n, \mathbf{R} = 2k, \mathbf{U} = \ell) = aP_{1-a}(\mathbf{L} = 2n, \mathbf{L} - \mathbf{R} = 2k, \mathbf{U} = \ell). \tag{16.8}$$

In particular, if $a = \frac{1}{2}$, then $E\{e^{ir\mathbf{R}}e^{is\mathbf{U}}e^{it\mathbf{L}}\} - E\{e^{ir(\mathbf{L}-\mathbf{R})}e^{is\mathbf{U}}e^{it\mathbf{L}}\} = \frac{1}{2}e^{2it}$ $(e^{2ir} - 1)$.

Corollary 16.2 extends the known result for the simple symmetric random walk that $P(\mathbf{L} = 2n, \mathbf{R} = 2k) = P(\mathbf{L} = 2n, \mathbf{L} - \mathbf{R} = 2k), n \geq 2$. Our proof depends on algebraic manipulation of the generating function; see Sect. 3.5.1.

The second motivation is to extend the persistence model to the case of two distinct strata $a \neq b$. This *full model*, together with its solution, has interesting features, which include:

1. its intrinsic value as physical model; cf. [1, 21],
2. completely explicit formulae throughout for key polynomials, identities, and generating functions;
3. new limiting distributions for a scaling of order N in both the meander and the last visit portions of the gambler's ruin.

We finally state a companion result to Theorem 16.1, again for the full model, that gives a scaling limit of order N over the last visit portion of the gambler's ruin. Let $\mathscr{R}_N$ and $\mathscr{V}_N$ denote the total number of runs and short runs of the absolute value process $\{|\mathbf{X}_j|\}$ until the epoch of the last visit, $\mathscr{L}_N$, defined by (16.1). Define $\mathscr{M}_N$ as the number of consecutive excursions of height at most $N-1$ of the absolute value process $\{|\mathbf{X}_j|\}$ until $\mathscr{L}_N$. In Fig. 1, we have $\mathscr{M}_4 = 4, \mathscr{R}_4 = 10, \mathscr{V}_4 = 4, \mathscr{L}_4 = 18$.

Define:

$$\mathscr{X}_N := \frac{1}{N}\left(\mathscr{L}_N - \frac{2-a-b}{(1-a)(1-b)}\mathscr{R}_N + \frac{1}{(1-a)(1-b)}\mathscr{V}_N - \frac{a(b-a)}{(1-a)(1-b)}\mathscr{M}_N\right). \tag{16.9}$$

Theorem 16.2 *Let $f \sim \eta N$, as $N \to \infty$, for some fixed $0 < \eta < 1$. Let $\mathscr{X}_N$ be defined by* (16.9). *Let also κ_j, $j = 1, 2$, and σ_j, $j = 1, 2$, be as defined in Theorem 16.1. Let $\hat{\varphi}$ be defined by* (16.3). *Then, $\lim_{N\to\infty} E\{e^{it\mathscr{X}_N}\} = \hat{\psi}(t)/\hat{\varphi}(t)$, where*

$$(ab\sigma_1\sigma_2)/\hat{\psi}(t) := ab\sigma_1\sigma_2\cosh(\kappa_1 t)\cosh(\kappa_2 t) + a^2\sigma_1^2\sinh(\kappa_1 t)\sinh(\kappa_2 t) + ia\sigma_1(b-a)^2\cosh(\kappa_1 t)\sinh(\kappa_2 t).$$

Theorem 16.1, Corollary 16.1, and Theorem 16.2 are proved in Sect. 4, naturally following Sect. 3 on building blocks for the proofs.

2 Elements of the Proof

Recall the definitions of the excursion statistics in (16.7) and following. We define the conditional joint probability generating function of the excursion statistics for runs, short runs, and steps given the height is at most N by

$$K_N(a,b) := E\{r^{\mathbf{R}} y^{\mathbf{V}} z^{\mathbf{L}} | \mathbf{X}_0 = 0,\ \mathbf{H} \le N\}. \tag{16.10}$$

To calculate (16.10), our proofs feature bivariate Fibonacci polynomials $\{q_n(x,\beta)\}$ and $\{w_n(x,\beta)\}$, defined as follows.

Definition 16.1 Define sequences $q_n(x,\beta)$ and $w_n(x,\beta)$ generated by the following recurrence relations, valid for $n \ge 1$.

$$q_{n+1} = \beta q_n - x q_{n-1},\ q_0 = 0, q_1 = 1;\ w_{n+1} = \beta w_n - x w_{n-1},\ w_0 = 1, w_1 = 1. \tag{16.11}$$

Here, $\beta, x \in \mathbb{C}$. The polynomials $q_n(x,\beta)$ generalize the univariate Fibonacci polynomials $F_n(x) = q_n(x,1)$, see [10, p. 327]; also $w_n(x,1) = F_{n+1}(x)$. In the case of steps alone in the classical fair gamblers ruin problem ($a = b = \frac{1}{2}$; $\beta = 1$ and $x = \frac{1}{4}z^2$ in (16.11)), the $\{q_n = F_n(x)\}$ are classically *numerator* polynomials, and the $\{w_n = F_{n+1}(x)\}$ are the *denominator* polynomials for the excursion generating function of height less than n, namely $K_{n-1}(\frac{1}{2},\frac{1}{2})$ with $r = y = 1$ in (16.10), [6], [10, Sect. V.4.3]. Here, numerator and denominator refer to the convergent of a continued fraction representation of K_∞. See [4] for an interesting direction on excursions with different step sets besides the classical steps ± 1.

We write an *interlacing* property of any two-term recurrence $v_{n+1} = \beta v_n - x v_{n-1}$, $n \ge 1$, with coefficients β and x independent of n:

$$v_{n+1}v_{n-1} - v_n^2 = \beta^{-1}x^{n-1}(v_3 v_0 - v_2 v_1),\ \beta \ne 0; \tag{16.12}$$

see [14, Eqs. (2.7)–(2.8)]. Note that when $v_0 = 0,\ v_1 = 1$, the polynomials $v_n = v_n(\beta, -x)$ are called the generalized Fibonacci polynomials in β and $-x$, and by standard generating function techniques, the fundamental sequences (16.11) have closed formulae given as follows: cf. [20, Eqs. (2.1) and (2.3)]; or [14, Eqs. (2.11)–(2.12)]. Define $\alpha := \sqrt{\beta^2 - 4x}$. Then, for all $n \ge 1$, and with $q_0(x,\beta) = 0$,

$$q_n(x,\beta) = \frac{2^{-n}}{\alpha}\left((\beta+\alpha)^n - (\beta-\alpha)^n\right);\ w_n(x,\beta) = q_n(x,\beta) - x q_{n-1}(x,\beta). \tag{16.13}$$

The formula for w_n follows from that of q_n, for $n \ge 1$, since $q_1 - x q_0 = 1 = w_1$, and $q_2 - x q_1 = \beta - x = w_2$.

We need some additional notation to describe our method as follows. For any pair of integers $m, n \in (-N, N)$ with $m \ne n$, we define the following *first passage* length for the process $\{\mathbf{X}_j\}$ that starts at $\mathbf{X}_0 = m$:

$$\mathbf{L}_{m,n} := \inf\{j \geq 1 : \mathbf{X}_j = n \text{ or } |\mathbf{X}_j| = N\}. \tag{16.14}$$

For any starting level, $\mathbf{X}_0 = m$, let $\boldsymbol{\Gamma}_{m,n} := \{\mathbf{X}_j, j = 0, \ldots, \mathbf{L}_{m,n}\}$ denote the ordinary first passage path from level m to either level n or to the boundary of the gambler's ruin process. For our key definition (16.15), additional conditions are placed on the first passage path to make it *one-sided.*

Denote by $\mathbf{R}_{m,n}$ the number of runs and by $\mathbf{V}_{m,n}$ the number of short runs, respectively, along $\boldsymbol{\Gamma}_{m,n}$, where $\mathbf{L}_{m,n}$ denotes the number of steps along this path. For $n > m$, define $g_{m,n} = g_{m,n}(a, b)$ as the following *upward* conditional joint probability generating function for these counting statistics given two conditions on the path: (1) The path is a one-sided first passage path that starts at m and stays at or above level m until it reaches level n, and (2) the first two steps of this path are both in the positive direction. If still $n > m$, then we also define the analogous *downward* conditional joint generating function $g_{n,m}$:

$$\begin{aligned} g_{m,n} &:= E(r^{\mathbf{R}_{m,n}} y^{\mathbf{V}_{m,n}} z^{\mathbf{L}_{m,n}} | \varepsilon_1 = \varepsilon_2, \mathbf{X}_0 = m, \mathbf{X}_j \geq m,\ j = 0, \ldots, \mathbf{L}_{m,n}). \\ g_{n,m} &:= E(r^{\mathbf{R}_{n,m}} y^{\mathbf{V}_{n,m}} z^{\mathbf{L}_{n,m}} | \varepsilon_1 = \varepsilon_2, \mathbf{X}_0 = n, \mathbf{X}_j \leq n,\ j = 0, \ldots, \mathbf{L}_{n,m}). \end{aligned} \tag{16.15}$$

The condition that the first two steps be in the same direction in the definition (16.15) arises due to the inclusion of the statistic $\mathbf{V}_{m,n}$ in the analysis. The path in Fig. 2 is a downward, first passage path from level 5 to level 0.

Let $n > m$. In the formulation of the recurrence for $g_{m,n}$, we must take account of the unconditional probability that a first passage from level m to level n remains at or above the starting level; we must also define the corresponding probability $\rho_{n,m}$, as follows.

$$\begin{aligned} \rho_{m,n} &:= P(\mathbf{X}_j \geq m, j = 0, \ldots, \mathbf{L}_{m,n} | \mathbf{X}_0 = m); \\ \rho_{n,m} &:= P(\mathbf{X}_j \leq n, j = 0, \ldots, \mathbf{L}_{n,m} | \mathbf{X}_0 = n). \end{aligned} \tag{16.16}$$

For $a = b = \frac{1}{2}$, the probability $\rho_{n,0} = \rho_{0,n}$ is determined by the classical solution of the probability of ruin started from fortune n on the interval $[0, n+1]$. For $a = b$, $\rho_{m,n}$ depends only on $k = n - m$ and is determined by $\rho_{m,m+\ell} = \frac{1}{2}(\ell - (\ell - 1)a)^{-1}$, see [13, Eq. (2.4)].

There are many calculations used to establish various formulae by the help of certain key definitions. We reserve the phrase *direct calculation* to mean that computer algebra (*Mathematica* [22]) is used to help verify the results. The companion document [15] to the present paper provides details of the verifications. In our approach, the complication of a second stratum is solved by finding the right formulae and then rendering a proof; we often utilize induction based on the proposed formulae. Our proofs may be termed elementary, since we use path decompositions to establish explicit formulae for the conditional generating functions $g_{m,n}$.

Our method for the full model is to show that the appropriate denominators $\{\overline{w}_{m,n}\}$ of the conditional generating functions $g_{m,n}$, together with certain singly indexed numerators $\{\overline{q}_n\}$, give rise also to a nice representation of (16.10); see Theorem 16.3, in which our approach involves conditioning on the height $\mathbf{H} = n$ of an excursion. For the homogeneous case of Proposition 16.5, the formula for (16.10) follows in a standard way of dealing with a finite continued fraction. In the homogeneous case, an alternative approach based on the format of [10], Proposition V.3, could probably be devised. Yet we need a closed formula for the one-sided first passage generating function $g_{0,N}$ to handle the meander in the full model, and this leads us to take an approach via recurrences proper, not only for $g_{m,n}$ but for $\overline{w}_{m,n}$. Accordingly, by Propositions 16.3 and 16.4, we obtain our main results with the help of trigonometric substitutions and direct calculations.

3 Proofs of the Building Blocks

3.1 Recurrence for $g_{m,n}$

We first establish the general recurrence relations governing the upward and downward generating functions of (16.15). The condition *initial two steps the same* on the trajectory of the lattice path yields immediately that

$$g_{m,m+2}(a,b) := rz^2, \; g_{m+2,m}(a,b) := rz^2, \; m \geq 0. \tag{16.17}$$

The path decomposition of [14] handles runs and steps; here, we extend that approach for short runs as well. It is convenient to focus on $g_{n,0}$ with some $n \geq 3$; see the definition (16.15). Fig. 2 is an illustration of one lattice path counted by $g_{5,0}$.

Let U or D stand for one step up or down, respectively, in a lattice path, and let $(UD)^\ell$ be shorthand for $UDUD\cdots$ with ℓ repetitions of the pattern UD for some $\ell \geq 0$. Since any downward lattice path from n to 0 must first reach the level $m = 1$, we have an initial factor $g_{n,1}$ in a product formula for $g_{n,0}$.

Any section of a lattice path for $g_{n,1}$, which must end in DD, is followed by a sequence of steps of the form $(UD)^\ell UU$ or by a *terminal* sequence $(UD)^\ell D$. To handle transitions that do not start UU or DD, we introduce:

$$\begin{aligned} &\omega(a,b) := 1-(1-a)(1-b)r^2y^2z^2, \; k(a,b) := (a+b-ab)/\omega(a,b), \\ &\tau(a,b) := 1+(1-a)(1-b)r^2z^2y(1-y); \; h(a,b) := \tau(a,b)/\omega(a,b). \end{aligned} \tag{16.18}$$

Denote $\mathbf{1} = (1,1,1)$ and evaluation of any function $u(a,b)$ at (r,y,z) by $u(a,b)[r,y,z]$. For brevity, we may write u_a in place of $u(a,a)$. By (16.18), $k(a,b)[\mathbf{1}] = \tau(a,b)[\mathbf{1}] = 1$. Thus, $k(a,b)$ is a probability generating function; the term $h(a,b)/h(a,b)[\mathbf{1}]$ is as well. In our discussion of $g_{n,0}$, if $f \geq 3$, then $k_a = k(a,a)$ accounts for a generating factor for an *upward preamble* $(UD)^\ell$ from level $m = 1$, succeeding DD and preceding UU; in this case, $k_a = c\sum_{\ell=0}^{\infty}((1-a)^2r^2y^2z^2)^{\ell} = c/\omega_a$, where $c = a(2-a)$. If instead $f = 2$ is the change of stratum parameter, then we obtain $k(a,b)$ in place of k_a due to the fact that now a change in direction at level $m = 2$ occurs with probability $(1-b)$ while a change in direction at level $m = 1$ occurs with probability $(1-a)$. To handle the dependence on f, we define

$$[a,b]_m^+ := \begin{cases} (a,a), & \text{if } m \le f-2 \\ (a,b), & \text{if } m = f-1, \\ (b,b), & \text{if } m \ge f; \end{cases} \qquad [a,b]_n^- := \begin{cases} (a,a), & \text{if } n \le f-1, \\ (a,b), & \text{if } n = f, \\ (b,b), & \text{if } n \ge f+1. \end{cases} \tag{16.19}$$

Let us suppose that the continuation of the path after the first downward passage to level $m = 1$ is not yet passing into a terminal sequence, so takes the form $(UD)^kUU\dots$. Starting thus from UU, the path makes an upward first passage to level n again (or not), and the pattern "up to level n and down to level 1" repeats for an indefinite number of times, $\ell \ge 0$. To handle the probability associated with the turning of the path downward from a level it will no longer exceed in the future of the path, or in turning from the bottom level $m = 1$ to upwards (in the return to level 1), we define the *turning probability at altitude m* by

$$\gamma_m := \begin{cases} 1-a, & \text{if } m \le f-1, \\ 1-b, & \text{if } m \ge f. \end{cases} \tag{16.20}$$

By definition (16.16), it now follows that $g_{n,0} = cg_{n,1}\lambda_{1,n}\lambda_{1,n-1}\cdots\lambda_{1,3}zh[a,b]_1^+$, with $\lambda_{1,n} := \sum_{\ell=0}^{\infty}(4\gamma_1\gamma_n\rho_{1,n}\rho_{n,1}k[a,b]_1^+k[a,b]_n^-g_{1,n}g_{n,1})^{\ell}$, or

$$\lambda_{1,n} = \frac{1}{1-4\gamma_1\gamma_n\rho_{1,n}\rho_{n,1}k[a,b]_1^+k[a,b]_n^-g_{1,n}g_{n,1}}.$$

Here, the factor of 4 arises due to the fact that the stationary probabilities for first step up and down, namely $\pi_+ = \frac{1}{2}$ and $\pi_- = \frac{1}{2}$, get replaced by γ_1 and γ_n, respectively, in $\rho_{1,n}$ and $\rho_{n,1}$. The factor $k[a,b]_n^-$ takes account of a *downward preamble* succeeding UU and preceding DD from the maximum possible level $M_2 \ge 3$ in the remainder of the downward lattice path. Here, the successive maximum levels $n = M_1 \ge M_2 \ge \cdots \ge M_r$ over the whole future of the path, determined in turn from the points of each of its returns to level $m = 1$ from the previous such maximum, are the *future maxima* (cf. [14]) of a downward path from level $n \ge 3$ to level $m = 0$. See Fig. 2, in which we have $M_1 = 5$, and $M_2 = 4$, $M_3 = 3$; there is no second future maximum of level 4, for example, because there is no return to level 1 between the two peaks at level 4, but instead, we see a downward preamble $(DU)^1$ at M_2. By definition, we have $M_r \ge 3$, and the downward path goes into a *terminal* sequence after a return to level 1 from M_r; the terminal sequence is of form $(UD)^kD$; see Fig. 2. Eventually, in the beginning, the path will never rise to level n again, but to lower future maxima at levels $3 \le m \le n-1$; thus, the product $\lambda_{1,n}\lambda_{1,n-1}\dots\lambda_{1,3}$. The factor $zh[a,b]_1^+$ corresponds to the terminal sequence. Now replace $m=1$ by $m \ge 1$ for a final destination level $m-1$, to obtain the following downward recurrence relation for any $m < n-1$:

$$g_{n,m-1} = czh[a,b]_m^+ g_{n,m} \prod_{j=m+2}^{n} \lambda_{m,j}. \tag{16.21}$$

for a normalization constant c such that $g_{n,m-1}[\mathbf{1}] = 1$. Here, we officially define $\lambda_{m,j} = \lambda_{m,j}(a,b)$:

$$\lambda_{m,j} := \frac{1}{1 - 4\gamma_m\gamma_j\rho_{m,j}\rho_{j,m}k[a,b]_j^- k[a,b]_m^+ g_{m,j}g_{j,m}}, \quad m+2 \le j. \tag{16.22}$$

By symmetric arguments, we also obtain the upward recurrence relation for any $m < n-1$:

$$g_{m,n+1} = czh[a,b]_n^- g_{m,n} \prod_{j=m}^{n-2} \lambda_{j,n}, \tag{16.23}$$

where c denotes a generic normalization constant. Each factor $\lambda_{m,n}/\lambda_{m,n}[\mathbf{1}]$ defined by (16.22) is a probability generating function for a class of paths starting and ending at the same level n (say), with probability of first step given by the turning probability (at n). Each path besides the empty path makes a positive number of consecutive down–up transitions of type "a first passage downward transition to m followed immediately by a first passage upward transition to n". See Fig. 3, where a path starts at level $n=3$ at the first marker γ_n, makes exactly 2 down–up transitions between $n=3$ and $m=0$, and ends at the third marker γ_n. We obtain the same generating function if the paths instead start and end at m, with up–down transitions.

By (16.21)–(16.23), we retrieve a closed recurrence for $g_{m,n}$. Indeed, by (16.23), for $m < n-2$, we simply have $g_{m,n+1}/g_{m+1,n+1} = c_1 g_{m,n}\lambda_{m,n}/g_{m+1,n}$, for a normalization constant c_1. Hence,

$$g_{m,n+1} = c_1 g_{m,n} g_{m+1,n+1} (g_{m+1,n})^{-1} \lambda_{m,n}, \quad n-m \ge 3. \tag{16.24}$$

Similarly, by applying (16.21), $g_{n,m-1}/g_{n-1,m-1} = c_2 g_{n,m}\lambda_{m,n}/g_{n-1,m}$, for a normalization constant c_2 and $m < n-2$. Hence,

$$g_{n,m-1} = c_2 g_{n,m} g_{n-1,m-1} (g_{n-1,m})^{-1} \lambda_{m,n}, \quad n-m \ge 3. \tag{16.25}$$

Observe that the factor $\lambda_{m,n}$ of (16.22), with $m+2 < n$, appears exactly the same in both (16.21) and (16.23), and again in (16.24)–(16.25).

We introduce some notation for the basic method to calculate (16.10), which consists of conditioning on $\{\mathbf{H} = n\}$. In the remainder of this section, we assume $f \ge 3$. Let G_n denote the conditional joint probability generating function of the

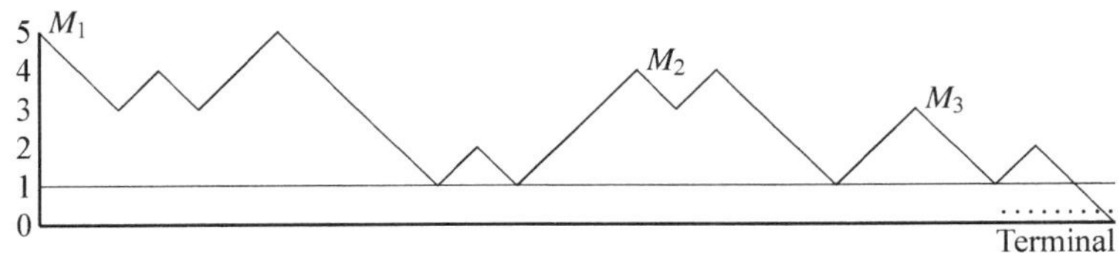

Fig. 2 Downward transition with future maxima $M_1 = 5$, $M_2 = 4$, $M_3 = 3$

number of runs, short runs, and steps in an excursion given that the height is $\mathbf{H} = n$ for some $1 \leq n < N$:

$$G_n := E(r^{\mathbf{R}} y^{\mathbf{V}} z^{\mathbf{L}} | \mathbf{H} = n, \mathbf{X}_0 = 0), \quad n \geq 1. \tag{16.26}$$

In definition (16.26), the condition is that after the first step from $m = 0$, the path does not return to the x-axis until it terminates, but that also, for a positive excursion, the path reaches the specified height, n, as a maximum. Now we work with positive excursions. We consider an initial sequence $U(UD)^\ell UU$ that brings a lattice path for the first time to level $m = 3$ while never returning to level $m = 0$. The joint generating function for the numbers of runs, short runs, and steps, for only the part $U(UD)^\ell$ of this initial sequence is simply $J_a := a(2-a)zh_a$, with h_a defined by (16.18). Now, to make a positive excursion that starts at level $m = 0$ and reaches a level $n \geq 3$ for a first time, we also consider any upward path $\Gamma_{1,n}^+$ for $g_{1,n}$ that starts at level $m = 1$ with UU. We link the initial sequence $U(UD)^\ell UU$ and $\Gamma_{1,n}^+$ together by making them overlap on the end UU of the initial sequence and beginning of $\Gamma_{1,n}^+$. Thus, the factor of G_n corresponding to a lattice path first reaching level $n \geq 3$ is given by $J_a g_{1,n}$. The remaining factor corresponds to a downward preamble from level n followed by a downward path from level n to level 0. Hence,

$$G_n = a(2-a)zh_a g_{1,n} k[a,b]_n^- g_{n,0}, \; n \geq 3, \; f \geq 3. \tag{16.27}$$

Moreover, by symmetry, we have $g_{0,-n} = g_{0,n}$ for all $n \geq 2$. Hence, the joint generating function of the meander statistics is:

$$E\{r^{\mathscr{R}'_N} y^{\mathscr{V}'_N} z^{\mathscr{L}'_N}\} = a(2-a)zh_a g_{1,N}, \; f \geq 3. \tag{16.28}$$

3.2 Formula for $\rho_{m,n}$

In this section, we establish a formula for $\rho_{m,n}$ as defined by (16.16). Note that $1 - \rho_{1,N}$ is the probability of ruin for the gambler's ruin persistence model with two strata on $[0, N]$ in case $\mathbf{X}_0 = 1$. The novelty of our approach, based on induction, is unnecessary if $a = b$, since by [13] a difference equation will solve the probability of ruin in this case.

The method we use to establish a formula is based first on the future maxima construction of Sect. 3.1; only in (16.16), there is no condition on upward or downward paths starting UU or DD. In place of $\lambda_{m,j}$ of (16.22), here define:

$$u_{m,j} := \frac{1}{1 - 4\gamma_m \gamma_j \rho_{m,j} \rho_{j,m}}, \quad m+1 \le j. \tag{16.29}$$

Let $m < n$. By the way, we developed the formulae (16.21)–(16.23), we have

$$\text{(i)} \quad \rho_{m,n+1} = (1-\gamma_n)\rho_{m,n} \prod_{j=m}^{n-1} u_{j,n};$$
$$\text{(ii)} \quad \rho_{n,m-1} = (1-\gamma_m)\rho_{n,m} \prod_{j=m+1}^{n} u_{m,j}. \tag{16.30}$$

The factor $(1-\gamma_n)$ in (16.30)(i) gives the probability (a or b) of the last step in any one-sided first passage path from level m to level n; a similar comment applies to (16.30)(ii). See Fig. 3. By the same method as shown to obtain (16.24)–(16.25), we have by (16.29)–(16.30) that

$$\text{(i)} \quad \rho_{m,n+1} = \frac{\rho_{m,n}\rho_{m+1,n+1}}{\rho_{m+1,n}} u_{m,n};$$
$$\text{(ii)} \quad \rho_{n,m-1} = \frac{\rho_{n,m}\rho_{n-1,m-1}}{\rho_{n-1,m}} u_{m,n}. \tag{16.31}$$

With the help of (16.31), we will now develop a closed recurrence for $\rho_{m,n}$. We first make a definition to establish a convenient form of $\rho_{m,n}$.

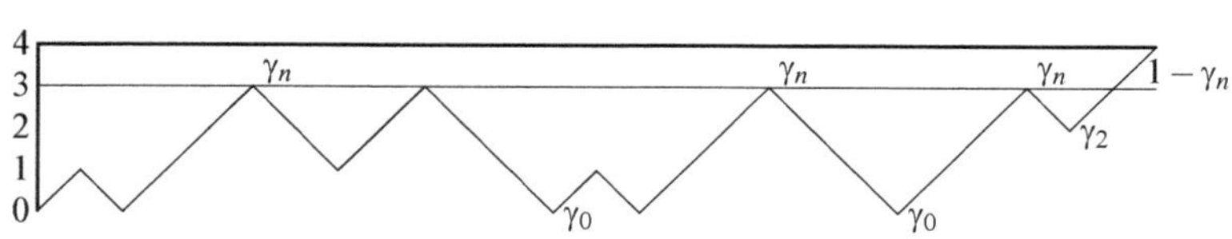

Fig. 3 Illustration of $\rho_{0,n+1} = (1-\gamma_n)\rho_{0,n} \prod_{j=0}^{n-1} u_{j,n}$ for $n = 3$. For the path shown, $u_{0,n}$ has 2 down–up transitions and $u_{1,n}$ has none, while $u_{2,n}$ has 1 down–up transition

Definition 16.2 Let $\rho_{m,n}$ be defined by (16.16). We define a denominator term $\Pi_{m,n}$ for $\rho_{m,n}$ as follows, with $m < n$ in all cases:

(I) (1) $\rho_{m,n} = \frac{1}{2}\frac{b/a}{\Pi_{m,n}}$, $m \le f-1$; (2) $\rho_{m,n} = \frac{1}{2}\frac{1}{\Pi_{m,n}}$, $f \le m$.
(II) (1) $\rho_{n,m} = \frac{1}{2}\frac{b/a}{\Pi_{n,m}}$, $n \le f-1$; (2) $\rho_{n,m} = \frac{1}{2}\frac{1}{\Pi_{n,m}}$, $f \le n$.

Proposition 16.1 *The terms $\Pi_{m,n}$ determined by Definition 16.2 satisfy:*

(I) *Between strata formulae:*

(1) $\Pi_{f-\ell,f+j} = j + \ell(\frac{b}{a}) - (\ell + j - 1)b,\ \ell \geq 1,\ j \geq 0;$
(2) $\Pi_{f+j,f-\ell} = (j+1) + (\ell - 1)(\frac{b}{a}) - (\ell + j - 1)b,\ \ell \geq 1,\ j \geq 0.$

(II) *Within-stratum formulae:*

(1) $\Pi_{m,m+\ell} = \Pi_{m+\ell,m} = \frac{b}{a}\{\ell - (\ell - 1)a\},\ m < m + \ell \leq f - 1;$
(2) $\Pi_{m,m+j} = \Pi_{m+j,m} = j - (j-1)b,\ f \leq m < m + j.$

Remark 16.2 If $a = b$, we have $\Pi_{m,m+\ell} = \Pi_{m+\ell,m} = \ell - (\ell - 1)a$ in all cases of Proposition 16.1, consistent with Definition 16.2 and [13, Eq. (2.4)].

Proof (*of Proposition* 16.1) By Remark 16.2, and Definition 16.2, the within-stratum formulae (II)(1), (2) hold in general. The proof of the between strata cases proceeds by induction on $n - m$, where we assume $n > m$ throughout. Recall that the first step ε_1 of the gambler's ruin process is determined by $\pi_+ = P(\varepsilon_j = 1) = \frac{1}{2}$. We have $\rho_{m,m+1} = \rho_{m,m-1} = \frac{1}{2}$, so the case $n - m = 1$ is easily checked. We next verify the cases $n - m = 2$ for $\Pi_{m,n}$ and $\Pi_{n,m}$ in (I)(1), (2). We apply (16.29)–(16.30) with $u_{m,m+1} = (1 - \gamma_m\gamma_{m+1})^{-1}$. Thus, for all m, $\rho_{m,m+2} = \frac{1}{2}$ $(1 - \gamma_{m+1})/(1 - \gamma_m\gamma_{m+1})$. In particular, by (16.20), $\rho_{f-1,f+1} = \frac{1}{2}b/(1 - (1-a)$ $(1-b)) = \frac{1}{2}(b/a)/\left(1 + \frac{b}{a} - b\right)$. This gives the correct form for the denominator in (I)(1) by Definition 16.2(I)(1). We apply direct calculation to check the other between strata cases. Thus, all the cases $n - m = 2$ have been verified.

Assume by induction that all statements of the proposition hold for $2 \leq n - m \leq k$ for some $k \geq 2$. We wish to show the following induction step:

$$\text{Both } (i): \Pi_{m,n+1}, \text{ and } (ii): \Pi_{n,m-1}, \text{ conform to statements (I)(1) and (I)(2), respectively, for all } m \leq f-1 \text{ and } n \geq f, \text{ with } n - m = k+1. \tag{16.32}$$

There are two boundary cases, $\Pi_{f-k-1,f}$ and $\Pi_{f+k,f-1}$, that aren't covered formally by this scheme. However, both of these cases actually fall under the within-stratum regime. For example, in the calculation of $\rho_{f-k-1,f}$, the one-sided first passage path from $f - k - 1$ to f never oscillates between levels $f - 1$ and f, so the probability $\rho_{f-k-1,f}$ is governed by a single stratum design. Hence, $\rho_{f-k-1,f} = \frac{1}{2}/(k + 1 - ka) = \frac{1}{2}(b/a)/\left\{(k+1)\left(\frac{b}{a}\right) - kb\right\}$, consistent with (I)(1). For the other boundary case, by similar reasoning, $\rho_{f+k,f-1} = \frac{1}{2}/(k + 1 - kb)$, consistent with (I)(2). So the boundary cases have been resolved for all k.

We proceed with our argument for establishing (16.32). By our ranges for m and n, we have $\gamma_m = 1 - a$ and $\gamma_n = 1 - b$ throughout. By Definition 16.2(I), we compute $u_{m,n}$ by (16.29) under (16.32) as follows.

$$u_{m,n} = \left\{1 - \gamma_m\gamma_n \frac{b/a}{\Pi_{m,n}\Pi_{n,m}}\right\}^{-1} = \frac{\Pi_{m,n}\Pi_{n,m}}{\Pi_{m,n}\Pi_{n,m} - \gamma_m\gamma_n(b/a)}. \tag{16.33}$$

Now write $m = f - \ell$ and $n = f + j$ for some $\ell \geq 1$ and $j \geq 0$ with $\ell + j = k \geq 2$. By the induction hypothesis, we can write $\Pi_{m,n} = j + \ell(\frac{b}{a}) - (\ell + j - 1)b$ and also $\Pi_{n,m} = (j+1) + (\ell - 1)(\frac{b}{a}) - (\ell + j - 1)b$. Now, by direct calculation, we have a simple identity for the denominator of the right-hand side of (16.33):

$$\Pi_{m,n}\Pi_{n,m} - \gamma_m\gamma_n(b/a) = \left\{j + 1 + \ell(\frac{b}{a}) - (\ell + j)b\right\} \Pi_{m+1,n}, \tag{16.34}$$

where we applied the induction hypothesis for $\Pi_{m,n}$, $\Pi_{n,m}$, and $\Pi_{m+1,n}$. Now rewrite (16.31)(i) by applying Definition 16.2 and (16.33)–(16.34), as follows. We have that $\rho_{m,n+1}$ is given by:

$$\left[\frac{\frac{1}{2}b/a}{\Pi_{m,n}}\right]\left[\frac{\frac{1}{2}b/a}{\Pi_{m+1,n+1}}\right]\left[\frac{\frac{1}{2}b/a}{\Pi_{m+1,n}}\right]^{-1} \frac{\Pi_{m,n}\Pi_{n,m}}{\left\{j + 1 + \ell(\frac{b}{a}) - (\ell + j)b\right\} \Pi_{m+1,n}}, \tag{16.35}$$

with the caveat that if $m + 1 = f$, then the factor b/a in the second and third factors on the left is replaced by 1. Finally, we use that, by the induction hypothesis and all statements of the proposition themselves, we have $\Pi_{m+1,n+1} = \Pi_{n,m}$ for all $n > m$ under (16.32). Therefore, simply by cancelation of 3 Π–factors, (16.35) yields $\rho_{m,n+1} = \frac{\frac{1}{2}b/a}{\{j+1+\ell(\frac{b}{a})-(\ell+j)b\}}$. Thus, by Definition 16.2(I)(1), the induction step (16.32) has been verified for case (i). The argument for the downward case (ii) is wholly similar to the upward case (i). In fact by direct calculation, the relevant identity in place of (16.34) is the same except with $\Pi_{n-1,m}$, in place of $\Pi_{m+1,n}$. And in (16.35), the roles of b/a and 1 are reversed. Thus, $\rho_{f+j,f-\ell-1} = \frac{1}{2}/\left\{j + 1 + \ell(\frac{b}{a}) - (\ell + j)b\right\}$, as required. Therefore, the induction step (16.32) has been verified.

3.3 *The Denominators* $\overline{w}_{m,n}$ *of* $g_{m,n}$

We first consider the homogeneous case $a = b$ and establish formulae for the denominators $w_n^*(a)$ of $g_{0,n}(a, a)$ defined by (16.15), where of course $g_{m,n}(a, a)$ depends only on $|n - m|$. We will abbreviate $g_n := g_{0,n}$ without confusion for this homogeneous case. Denote $\rho_n := \rho_{0,n} = \frac{1}{2}/(n - (n-1)a)$, by Definition 16.2 and Proposition 16.1, and $\lambda_n := \lambda_{m,m+n} = \{1 - 4(1-a)^2 k_a^2 \rho_n^2 g_n^2\}^{-1}$, defined by (16.22). By (16.24), we have

$$g_{n+1} = c_1 g_n^2 g_{n-1}^{-1} \lambda_n, \quad n \geq 3; \quad g_3 = czh_a g_2 \lambda_2. \tag{16.36}$$

We now establish that, for a certain sequence of polynomials

$$\{w_n^*(a) = w_n^*(a; r, y, z), n \geq 1\},$$

with constant coefficient 1, we have

$$g_n = C_{n,a}\omega_a r z^n \tau_a^{n-2}/w_n^*(a), n \geq 2;\ C_{n,a} := a^{n-2}(n-(n-1)a)/(2-a). \tag{16.37}$$

The proposed formula (16.37) holds for $n = 2$ with $w_2^*(a) := \omega_a$, since $C_{2,a} = 1$. We also define $w_1^*(a) := 1$. Motivated by the idea that $w_n^*(a)$ satisfies a Fibonacci recurrence, we introduce

$$x_a := a^2 z^2 \tau_a^2;\ \ \beta_a := 1 + z^2(a^2 - (1-a)^2 r^2 (y^2 + a^2(1-y)^2 z^2)), \tag{16.38}$$

and we define

$$w_{n+1}^*(a) = \beta_a w_n^*(a) - x_a w_{n-1}^*(a),\ n \geq 2. \tag{16.39}$$

The form of (16.38) used to make the definition (16.39) may be guessed by taking account of (16.12) together with the proposed form (16.37). That is, we already have defined $w_1^*(a)$ and $w_2^*(a)$, consistent with (16.37), and we can derive g_3 via (16.36). So we will have thereby guessed $w_3^*(a)$. We can likewise predict $w_4^*(a)$. But (16.12) gives that the appropriate x_a for (16.39) is

$$x_a = \left(w_3^*(a)^2 - w_4^*(a)w_2^*(a)\right) / \left(w_2^*(a)^2 - w_3^*(a)w_1^*(a)\right).$$

Once we have x_a, we find β_a via (16.39), and we also extend the definition (16.39) to $n = 0$ by solving (16.39) backward: $w_0^*(a) := (\beta_a - \omega_a)/x_a$. We define also the associated numerators $\{q_n^*(a)\}$ defined by the Fibonacci recurrence $q_{n+1}^*(a) = \beta_a q_n^*(a) - x_a q_{n-1}^*(a),\ n \geq 1$, with initial conditions

$$q_0^*(a) := -(1-y)(1+y+(1-a)^2 r^2 y^2 z^2 (1-y))/\tau_a^2,\ \ q_1^*(a) := y^2. \tag{16.40}$$

By the choice of $q_1^*(a)$, we obtain the form $K_1 = r^2 z^2 q_1^*(a)/w_1^*(a)$ for (16.10). By the choice of $q_0^*(a)$, we obtain by direct computation an interlacing form $w_{n+1}^* q_n^* - w_n^* q_{n+1}^* = a^2 z^2 x_a^{n-1}$ at $n = 0$. By direct computation to check the initial conditions for Fibonacci recurrences, we have:

$$\begin{aligned} q_n^*(a) &= c_1 q_n(x_a, \beta_a) + c_2 w_n(x_a, \beta_a),\ \ c_2 := q_0^*(a), c_1 = y^2 - c_2; \\ w_n^*(a) &= c_1' q_n(x_a, \beta_a) + c_2' w_n(x_a, \beta_a),\ \ c_2' := w_0^*(a), c_1' = 1 - c_2'. \end{aligned} \tag{16.41}$$

We first verify (16.37) for $n = 3$. By (16.18) and (16.23), we have $g_3 = czh_a g_2 \lambda_2 = cz(\tau_a/\omega_a) r z^2 (1 - a^2(1-a)^2 r^2 z^4/\omega_a^2)^{-1}$, since $\rho_2 = \frac{1}{2}/(2-a)$ and $k_a = a(2-a)/\omega_a$. This yields $g_3 = C_{3,a}\omega_a r z^3 \tau_a / w_3^*(a)$, by direct computation. Now assume by induction that (16.37) holds with m in place of n for all $2 \leq m \leq n$, for some $n \geq 3$. Then by (16.36), we have $g_{n+1} = c_{n+1}[\omega_a r z^n \tau_a^{n-2}/w_n^*]^2 [\omega_a r z^{n-1} \tau_a^{n-3}/w_{n-1}^*]^{-1} \lambda_n$, where c_{n+1} incorporates both the constant c_1 of (16.36) and the factor $C_{n,a}^2/C_{n-1,a}$. By direct substitution of the induction hypothesis, and taking care to write the term g_n^2 that appears in λ_n in terms of x_a via $\tau_a^2 = a^{-2} z^{-2} x_a$, so that

$$\lambda_n = \left(1 - a^2(1-a)^2 r^2 z^4 x_a^{n-2}/w_n^*(a)^2\right)^{-1},$$

we obtain

$$g_{n+1} = c_{n+1}\omega_a r z^{n+1}\tau_a^{n-1} w_{n-1}^*(a)/\{w_n^*(a)^2 - a^2(1-a)^2 r^2 z^4 x_a^{n-2}\}. \quad (16.42)$$

To compute the denominator in this last expression, we note the following.

Lemma 16.1 *Let* $w_n^*(a)$ *be defined as the solution to* (16.39). *Then for all* $n \geq 1$, *we have:* $w_n^*(a)^2 - w_{n+1}^*(a)w_{n-1}^*(a) = a^2(1-a)^2 r^2 z^4 x_a^{n-2}$.

Proof By the definition (16.39) and by (16.12), we have:

$$w_n^*(a)^2 - w_{n+1}^*(a)w_{n-1}^*(a) = -\beta_a^{-1} x_a^{n-1}(w_3^*(a)w_0^*(a) - w_2^*(a)w_1^*(a)). \quad (16.43)$$

By direct calculation, $w_3^*(a)w_0^*(a) - w_2^*(a)w_1^*(a) = -a^2 r^2 z^4 (1-a)^2 \beta_a / x_a$. Hence, the lemma follows by substitution of this last formula into (16.43).

Up to the form of the constant $C_{n,a}$, relation (16.37) now follows by induction from (16.42) and Lemma 16.1. To verify the constant, we need only verify the claim: $w_n^*(a)[\mathbf{1}] = a^{n-1}(n - (n-1)a)$. This is easily verified by induction, (16.39), and direct computation. Hence, we have verified (16.37).

Now turn to the full model. We recursively define an array of functions $\{\overline{w}_{m,n} = \overline{w}_{m,n}(a, b)\}$ such that $\overline{w}_{m,n}$ will turn out to be the denominator polynomial with constant term 1 for the rational expression of $g_{m,n}$. We first define initial cases:

$$\overline{w}_{m,m+2} := \omega[a,b]_m^+, \ m \geq 0; \ \ \overline{w}_{n,n-2} := \omega[a,b]_n^-, \ n-2 \geq 0;$$
$$\overline{w}_{m,m+1} = \overline{w}_{m+1,m} = 1, \ m \geq 0. \quad (16.44)$$

For example, if $m \leq f - 2$, then $[a,b]_m^+ = (a,a)$, so $\overline{w}_{m,m+2} := \omega_a$. We require a generalization of x_a, and β_a of (16.38) to make our definition of $\{\overline{w}_{m,n}\}$ for two strata, as follows. Define

$$x(a,b) := b^2 z^2 \tau^2(a,b); \ \ \beta(a,b) = \beta_b - (b-a)b^2(1-b)r^2(1-y)^2 z^4. \quad (16.45)$$

for β_b defined by (16.38). Here, we note that $\tau(a,b)$ is symmetric in a and b, so $x(b,a) = a^2 z^2 \tau^2(a,b)$ for $\tau(a,b)$ defined by (16.18).

Definition 16.3 Denote $\overline{w}_{m,n} = \overline{w}_{m,n}(a,b)$.

(I) Define the upward denominator $\overline{w}_{m,n}$ for all $n - m \geq 2$ by:

(1) $\overline{w}_{m,m+\ell} := w_\ell^*(a)$, $m < m + \ell \leq f$; $\overline{w}_{m,m+\ell} := w_\ell^*(b)$, $f \leq m < m + \ell$;
(2) $\overline{w}_{f-\ell,f+1} := \frac{1-b}{1-a} w_{\ell+1}^*(a) + \frac{b-a}{1-a} w_\ell^*(a)$, $1 \leq \ell \leq f$;
(3) $\overline{w}_{m,f+2} := \beta(a,b)\overline{w}_{m,f+1} - x(a,b)\overline{w}_{m,f}$, $m \leq f - 1$;
(4) $\overline{w}_{m,f+j+1} := \beta_b \overline{w}_{m,f+j} - x_b \overline{w}_{m,f+j-1}$, $m \leq f - 1$, $j \geq 2$.

(II) Define the downward denominator $\overline{w}_{n,m}$ for all $n - m \geq 2$ by:

(1) $\overline{w}_{m+\ell,m} := w^*_\ell(a),\ m < m+\ell \leq f-1;\ \overline{w}_{m+\ell,m} := w^*_\ell(b),\ f-1 \leq m;$
(2) $\overline{w}_{f+j,f-2} := \frac{1-a}{1-b} w^*_{j+2}(b) + \frac{a-b}{1-b} w^*_{j+1}(b),\ 0 \leq j;$
(3) $\overline{w}_{n,f-3} := \beta(b,a)\overline{w}_{n,f-2} - x(b,a)\overline{w}_{n,f-1},\ f \leq n;$
(4) $\overline{w}_{n,f-\ell-2} := \beta_a \overline{w}_{n,f-\ell-1} - x_a \overline{w}_{n,f-\ell},\ f \leq n,\ \ell \geq 2.$

Notice that in Definition 16.3(II), we are effectively reversing the roles of a and b from (I). In case $a = b$, we simply have $\overline{w}_{m,n} = w^*_{|n-m|}(a),\ |n-m| \geq 2.$

We write the first step of *crossing over the threshold of the stratum* in either upward or downward directions as a linear combination of two successive homogeneous case solutions. For the next step over the threshold, we use the *mixed* parameters for x and β, and for further steps, we use the appropriate homogeneous parameters for x and β. With no crossing over a stratum, the homogeneous solution is shown. Finally, Definition 16.3 and (16.44) are consistent. For example, in part (I)(2) of the definition, we find:

$$\overline{w}_{f-1,f+1} = \frac{1-b}{1-a} w^*_2(a) + \frac{b-a}{1-a} w^*_1(a) = \frac{1-b}{1-a}\omega(a,a) + \frac{b-a}{1-a} = \omega(a,b).$$

3.3.1 Interlacing Identity and Closed Formula for $\overline{w}_{m,n}$

To establish a formula for $g_{m,n}$, we will employ an interlacing identity for the denominators $\overline{w}_{m,n}$. Define the interlacing bracket:

$$[\overline{w}]_{m,n} := \overline{w}_{m,n}\overline{w}_{m+1,n+1} - \overline{w}_{m,n+1}\overline{w}_{m+1,n},\ m \leq n-2. \tag{16.46}$$

It actually suffices to consider only the upward direction for $[\overline{w}]_{m,n}$, since by Lemma 16.2, the natural corresponding downward definition,

$$[\overline{w}]_{n,m} := \overline{w}_{n,m}\overline{w}_{n-1,m-1} - \overline{w}_{n,m-1}\overline{w}_{n-1,m},\quad m \leq n-2,$$

satisfies $[\overline{w}]_{n,m} = [\overline{w}]_{m,n}$.

Proposition 16.2 *The following identities hold for* $[\overline{w}]_{m,n}$*:*

(1) $[\overline{w}]_{f-\ell,f+j} = a^2r^2z^4(1-a)(1-b)x_a^{\ell-2}x(a,b)x_b^{j-1},\quad \ell \geq 2,\ j \geq 1;$
(2) $[\overline{w}]_{f-\ell,f} = a^2r^2z^4(1-a)(1-b)x_a^{\ell-2},\quad \ell \geq 2;$
(3) $[\overline{w}]_{f-1,f+j} = b^2r^2z^4(1-a)(1-b)x_b^{j-1},\quad j \geq 1;$
(4) $[\overline{w}]_{m,m+\ell} = a^2r^2z^4(1-a)^2x_a^{\ell-2},\ m+\ell \leq f-1;$
(5) $[\overline{w}]_{m,m+j} = b^2r^2z^4(1-b)^2x_b^{j-2},\ f \leq m,\ j \geq 2.$

Proof By Definition 16.3(I)(1) and by Lemma 16.1, we have that statements (4)–(5) of the proposition hold. Next fix $\ell \geq 2$ and notice that the case $j = 0$ in (1) is similar to the case of statement (2), the difference being $x(a,b) \neq x_b$. We will first verify (2). Thus, we write, using the Definition 16.3 and (16.46), that $[\overline{w}]_{f-\ell,f}$ is given by:

$$w_\ell^*(a)\{\frac{1-b}{1-a}w_\ell^*(a)+\frac{b-a}{1-a}w_{\ell-1}^*(a)\}-\{\frac{1-b}{1-a}w_{\ell+1}^*(a)+\frac{b-a}{1-a}w_\ell^*(a)\}w_{\ell-1}^*(a). \tag{16.47}$$

The $w_\ell^*(a)w_{\ell-1}^*(a)$ terms cancel in (16.47). Thus, by (16.47) and Lemma 16.1, $[\overline{w}]_{f-\ell,f}=\frac{1-b}{1-a}\left(w_\ell^*(a)^2-w_{\ell+1}^*(a)w_{\ell-1}^*(a)\right)=a^2(1-a)(1-b)r^2z^4x_a^{\ell-2}$. Hence, statement (2) is proved.

We now turn to statement (1). Fix $\ell\ge 2$ and let $j\ge 0$. Denote $[a,b]_0=(a,b)$ and $[a,b]_j=(b,b)$ for $j\ge 1$. Thus, by Definition 16.3(I)(3)–(4) and (16.46),

$$\begin{aligned}[\overline{w}]_{f-\ell,f+j+1}&=\overline{w}_{f-\ell,f+j+1}\left\{\beta[a,b]_j\overline{w}_{f-\ell+1,f+j+1}-x[a,b]_j\overline{w}_{f-\ell+1,f+j}\right\}\\&\quad-\left\{\beta[a,b]_j\overline{w}_{f-\ell,f+j+1}-x[a,b]_j\overline{w}_{f-\ell,f+j}\right\}\overline{w}_{f-\ell+1,f+j+1}.\end{aligned} \tag{16.48}$$

Now the terms of (16.48) involving $\beta[a,b]_j$ cancel, and we obtain from (16.48) and (16.46) that

$$[\overline{w}]_{f-\ell,f+j+1}=x[a,b]_j[\overline{w}]_{f-\ell,f+j}. \tag{16.49}$$

Now put $j=0$ in (16.49) and conclude by (2) and (16.49) that statement (1) holds for the initial case $j=1$ for the given fixed $\ell\ge 2$. Now for the same fixed index ℓ, take statement (1) as an induction hypothesis for induction on $j\ge 1$. We have just established this induction hypothesis for $j=1$. Thus, verify by (16.49) again that the induction step holds since $x[a,b]_j=x(b,b)=x_b$ for all $j\ge 1$. Thus, statement (1) is proved.

Finally, we turn to statement (3). We note that (16.48)–(16.49) continue to hold by Definition 16.3(I)(3) with $\ell=1$ as long as $j\ge 1$. Now we compute by (16.44)–(16.45), Definition 16.3(I)(3), and the interlacing bracket definition (16.46) that, since by (16.44), $\overline{w}_{f-1,f+1}=\omega(a,b)$, while by Definition 16.3, $\overline{w}_{f,f+2}=w_2^*(b)=\omega(b,b)$,

$$\begin{aligned}[\overline{w}]_{f-1,f+1}&=\omega(a,b)\omega(b,b)-(\beta(a,b)\omega(a,b)-x(a,b)\cdot 1)\cdot 1\\&=\omega(a,b)\left(\omega(b,b)-\beta(a,b)\right)+x(a,b)=b^2(1-a)(1-b)r^2z^4,\end{aligned} \tag{16.50}$$

where at the last step, we make a direct calculation based on the definitions in (16.18) and (16.45). Now take statement (3) as an induction hypothesis for induction on $j\ge 1$. By (16.50) have established this induction hypothesis for $j=1$. Thus, verify by (16.49) with $\ell=1$ and $j\ge 1$ that the induction step holds since $x[a,b]_j=x(b,b)=x_b$ for all $j\ge 1$. So, statement (3) is proved.

We turn to the task of obtaining a closed formula for $\overline{w}_{m,n}$. By Definition 16.3(I)(4), given $m=f-\ell<f$, $\overline{w}_{m,f+1}$ and $\overline{w}_{m,f+2}$ form the initial conditions for a recurrence $\overline{w}_{m,f+j+1}:=\beta_b\overline{w}_{m,f+j}-x_b\overline{w}_{m,f+j-1}$, $j\ge 2$. Put $m=f-\ell$ for some $\ell\ge 1$. We denote the vector of these upward initial conditions across the stratum threshold

by the 2×1 vector $\mathbf{W}(\ell)$. Then we define a 2×2 matrix $Q(b)$, and for each $\ell < f$, a 2×1 vector $\mathbf{d} = \mathbf{d}(\ell)$ by

$$Q(b) := \begin{bmatrix} q_1^*(b) & w_1^*(b) \\ q_2^*(b) & w_2^*(b) \end{bmatrix}, \; \mathbf{W}(\ell) := \begin{bmatrix} \overline{w}_{f-\ell,f+1} \\ \overline{w}_{f-\ell,f+2} \end{bmatrix} = Q(b)\mathbf{d}; \; \mathbf{d} := \begin{bmatrix} d_1(\ell) \\ d_2(\ell) \end{bmatrix}. \tag{16.51}$$

By Definitions 16.3(I)(1–3), we can write each term of the right side of the recurrence of (I)(3) using (I)(1–2) in terms of $w_\ell^*(a)$ and $w_{\ell+1}^*(a)$ as follows: $\overline{w}_{f-\ell,f+1} = \frac{1-b}{1-a} w_{\ell+1}^*(a) + \frac{b-a}{1-a} w_\ell^*(a)$ and $\overline{w}_{f-\ell,f} = w_\ell^*(a)$, so

$$\overline{w}_{f-\ell,f+2} = \beta(a,b) \left(\frac{1-b}{1-a} w_{\ell+1}^*(a) + \frac{b-a}{1-a} w_\ell^*(a) \right) - x(a,b) w_\ell^*(a).$$

We combine terms with the notation $\kappa(a,b) := \left(\frac{b-a}{1-a}\right) \beta(a,b) - x(a,b)$. Thus,

$$\mathbf{W}(\ell) = B \begin{bmatrix} w_\ell^*(a) \\ w_{\ell+1}^*(a) \end{bmatrix}; \quad B = \begin{bmatrix} \frac{b-a}{1-a} & \frac{1-b}{1-a} \\ \kappa(a,b) & \frac{1-b}{1-a}\beta(a,b) \end{bmatrix}. \tag{16.52}$$

By equating the two expressions for the vector $\mathbf{W}(\ell)$ in (16.51) and (16.52), we recover

$$\mathbf{d}(\ell) = \begin{bmatrix} d_1(\ell) \\ d_2(\ell) \end{bmatrix} = M \begin{bmatrix} w_\ell^*(a) \\ w_{\ell+1}^*(a) \end{bmatrix}; \quad M := Q(b)^{-1} B. \tag{16.53}$$

Here, it is clear that the entries of the matrix $M = (\mu_{i,j})$, with $\mu_{i,j} = \mu_{i,j}(a,b)$ $1 \le i, j \le 2$, do not depend on ℓ. We note by direct calculation from (16.51) that $\det(Q(b)) = -b^2 z^2$, so we have a straightforward formula for M via (16.51) and (16.53).

Proposition 16.3 *Let $d_1(\ell)$ and $d_2(\ell)$ be defined by* (16.51)–(16.53). *Then*

$$\overline{w}_{f-\ell,f+j} = d_1(\ell) q_j^*(b) + d_2(\ell) w_j^*(b), \; \ell \ge 1, \; j \ge 1. \tag{16.54}$$

Proof Fix $\ell \ge 1$. By Definition 16.3(I)(4), for all $j \ge 2$ it holds that $\overline{w}_{f-\ell,f-j+1} = \beta_b \overline{w}_{f-\ell,f+j} - x_b \overline{w}_{f-\ell,f+j-1}$. But if we denote the right side of (16.54) by v_j, then also $v_{j+1} = \beta_b v_j - x_b v_{j-1}$, $j \ge 2$, because by construction each of $\{q_j^*(b)\}$ and $\{w_j^*(b)\}$ satisfy the same two-term recurrence, and the coefficients $d_1(\ell)$ and $d_2(\ell)$ in (16.54) are independent of j. Also by definition (16.51), for any given $\ell \ge 1$, (16.54) holds for $j = 1$ and $j = 2$, that is, $v_j = \overline{w}_{f-\ell,f+j}$, $j = 1, 2$. Hence, we have $v_j = \overline{w}_{f-\ell,f+j}$ for all $j \ge 1$. Since ℓ was arbitrary, the proof is complete.

Lemma 16.2 *For all $1 \le m < n$, there holds: $\overline{w}_{m,n} = \overline{w}_{n-1,m-1}$.*

Proof Notice that the lemma holds in the initial cases $n - m = 1, 2$ by (16.44). Also, if $f \le m < n$ or $1 \le m < n \le f$, then the statement holds by (I)(1) and (II)(1) in Definition 16.3. So consider now $\overline{w}_{f-\ell,f+j}$ for $1 \le \ell < f$ and $j \ge 1$. Our method is to prove that the statement: $(\mathrm{H})_{\ell,j}$: $\overline{w}_{f-\ell,f+j} = \overline{w}_{f+j-1,f-\ell-1}$, holds for both the initial cases $\ell = 1$ and $\ell = 2$, and all $j \geq 1$.

We first establish $(\mathrm{H})_{\ell,j}$ for $\ell = 1$ and all $j \geq 1$. On the one hand, write $\overline{w}_{f-1,f+j}$ by (16.54) with $\ell = 1$, and on the other hand, write $\overline{w}_{f+j-1,f-2}$ by Definition 16.3(II)(2), as follows.

$$\overline{w}_{f-1,f+j} = d_1(1)q_j^*(b) + d_2(1)w_j^*(b); \overline{w}_{f+j-1,f-2} = \frac{1-a}{1-b}w_{j+1}^*(b) + \frac{a-b}{1-b}w_j^*(b). \tag{16.55}$$

By (16.51), (16.53) and direct calculation, we have that $d_1(1) = \mu_{1,1}w_1^*(a) + \mu_{1,2}w_2^*(a) = -(1-a)(1-b)r^2z^2$, and $d_2(1) = \mu_{2,1}w_1^*(a) + \mu_{2,2}w_2^*(a) = 1$. Therefore, by substitution into (16.55), we find that the two expressions in (16.55) are equal if and only if

$$-(1-b)^2r^2z^2q_j^*(b) = w_{j+1}^*(b) - w_j^*(b). \tag{16.56}$$

By direct computation, we check that (16.56) is true at both $j = 1$ and $j = 2$. Thus, since $\{q_j^*(b)\}$ and $\{w_j^*(b)\}$ each satisfy the same Fibonacci recurrence, (16.56) holds for all $j \geq 1$.

Next we establish that $(\mathrm{H})_{\ell,j}$ holds with $\ell = 2$ and all $j \geq 1$. Write $\overline{w}_{f-2,f+j}$ by (16.54) with $\ell = 2$, and write $\overline{w}_{f+j-1,f-3}$ by Definition 16.3(II)(3), as follows.

$$\begin{aligned}
&\text{(i)} \quad \overline{w}_{f-2,f+j} = d_1(2)q_j^*(b) + d_2(2)w_j^*(b);\\
&\text{(ii)} \quad \overline{w}_{f+j-1,f-3} = \beta(b,a)\overline{w}_{f+j-1,f-2} - x(b,a)\overline{w}_{f+j-1,f-1},
\end{aligned} \tag{16.57}$$

with $\overline{w}_{f+j-1,f-2} = \frac{1-a}{1-b}w_{j+1}^*(b) + \frac{a-b}{1-b}w_j^*(b)$; $\overline{w}_{f+j-1,f-1} = w_j^*(b)$. By (16.51) and (16.53), we directly verify that $d_1(2) = -(1-a)(1-b)r^2z^2\beta(b,a)$; $d_2(2) = \beta(b,a) - x(b,a)$. To verify that the expressions (i) and (ii) in (16.57) are equal, we substitute $d_1(2)$ and $d_2(2)$, and obtain, after a little algebra in which $x(b,a)x_j^*(b)$ cancels on the two sides, the condition

$$-(1-b)^2r^2z^2\beta(b,a)q_j^*(b) = \beta(b,a)\left(w_{j+1}^*(b) - w_j^*(b)\right), \quad \text{for all } j \geq 1.$$

This is obviously equivalent to the condition (16.56). Hence, the two expressions in (16.57) are equal for all $j \geq 1$, so $(\mathrm{H})_{\ell,j}$ holds also at $\ell = 2$ for all $j \geq 1$.

Finally, fix any $j \geq 1$. We appeal to (16.53) and (16.54) and to Definition 16.3(II)(4), to obtain, for any $\ell \geq 3$,

$$\begin{aligned}
\overline{w}_{f-\ell,f+j} &= \left(\mu_{1,1}w_\ell^*(a) + \mu_{1,2}w_{\ell+1}^*(a)\right)q_j^*(b)\\
&\qquad + \left(\mu_{2,1}w_\ell^*(a) + \mu_{2,2}w_{\ell+1}^*(a)\right)w_j^*(b);\\
\overline{w}_{f+j-1,f-\ell-1} &= \beta_a\overline{w}_{f+j-1,f-\ell} - x_a\overline{w}_{f+j-1,f-\ell+1}.
\end{aligned} \tag{16.58}$$

For any $\ell \geq 1$, write $u_\ell := \overline{w}_{f-\ell,f+j}$ and $v_\ell := \overline{w}_{f+j-1,f-\ell-1}$ for the two lines of (16.58). Since u_ℓ is a linear combination of two successive terms of the sequence

$\{w_\ell^*(a)\}$, it follows that,$\{u_\ell\}$ itself satisfies the recursion $u_{\ell+1} = \beta_a u_\ell - x_a u_{\ell-1}$, $\ell \geq 2$. But also $\{v_\ell\}$ satisfies the same recurrence. Moreover, we proved that $(\mathrm{H})_{\ell,j}$ holds for $\ell = 1$ and $\ell = 2$, so we have $u_1 = v_1$, and $u_2 = v_2$. Therefore, we have $u_\ell = v_\ell$ for all $\ell \geq 1$. Thus, by (16.58), $(\mathrm{H})_{\ell,j}$ is proved for all $\ell \geq 1$. Since $j \geq 1$ was arbitrary, $(\mathrm{H})_{\ell,j}$ is true for all ℓ, $j \geq 1$.

Lemma 16.3 *The following identities hold.*

(1) $\overline{w}_{f-\ell,f+j}[\mathbf{1}] = a^\ell b^{j-1} \Pi_{f-\ell,f+j}$, *for all* $\ell \geq 1$, $j \geq 1$.
(2) $\overline{w}_{f+j,f-\ell}[\mathbf{1}] = a^{\ell-1} b^j \Pi_{f+j,f-\ell}$, *for all* $\ell \geq 2$, $j \geq 0$.
(3) $q_\ell^*(a)[\mathbf{1}] = \ell a^{\ell-1}$, $w_\ell^*(a)[\mathbf{1}] = a^{\ell-1}(\ell - (\ell-1)a)$; *for all* $\ell \geq 1$.

Proof At $(r, y, z) = \mathbf{1}$, we have $\beta_a = 2a$ and $x_a = a^2$. Thus, $\alpha = 0$ in (16.13). Therefore by (16.13), $q_\ell^*(a)[\mathbf{1}] = \lim_{\alpha\to 0} \frac{2^{-\ell}}{\alpha}\{(2a+\alpha)^\ell - (2a-\alpha)^\ell\} = \ell a^{\ell-1}$. Thus, by the second formula of (16.13), we obtain $w_\ell^*(a)[\mathbf{1}]$ by $x_a[\mathbf{1}] = a^2$, so (3) is proved. Now apply (16.54), also at $(r, y, z) = \mathbf{1}$. By (16.53) and direct calculation, $d_1(\ell)[\mathbf{1}] = -(1-a)(1-b)\ell a^{\ell-1}$, and $d_2(\ell)[\mathbf{1}] = a^{\ell-1}[\ell - (\ell-1)a]$. Now plug in $q_j^*(b)[\mathbf{1}]$ and $w_j^*(b)[\mathbf{1}]$ from (3), into (16.54) to obtain formula (1) from Proposition 16.3 after direct simplification. The proof of (2) follows from (1) and Lemma 16.2, in view of Definition 16.2.

3.4 Closed Formula for $g_{m,n}$

Proposition 16.4 *We have the following formulae for* $g_{m,n}$.

(I) *The formulae for upward between-strata cases,* $j \geq 1$ *and* $\ell \geq 2$:

(1) $$g_{f-\ell,f+j} = \frac{\omega(a,a)}{2-a} r z^{j+\ell} \tau(a,b)[a\tau(a,a)]^{\ell-2}[b\tau(b,b)]^{j-1} \times \left(a\Pi_{f-\ell,f+j}/\overline{w}_{f-\ell,f+j}\right),$$

(2) $$g_{f-1,f+j} = \frac{\omega(a,b)}{a+b-ab} r z^{j+1}[b\tau(b,b)]^{j-1}\left(a\Pi_{f-1,f+j}/\overline{w}_{f-1,f+j}\right);$$

(II) *The formulae for downward between-strata cases,* $j \geq 1$ *and* $\ell \geq 2$:

(1) $$g_{f+j,f-\ell} = \frac{\omega(b,b)}{2-b} r z^{j+\ell} \tau(a,b)[a\tau(a,a)]^{\ell-2}[b\tau(b,b)]^{j-1} \times \left(a\Pi_{f+j,f-\ell}/\overline{w}_{f+j,f-\ell}\right),$$

(2) $$g_{f,f-\ell} = \frac{\omega(a,b)}{a+b-ab} r z^{\ell}[a\tau(a,a)]^{\ell-2}\left(a\Pi_{f,f-\ell}/\overline{w}_{f,f-\ell}\right);$$

(III) *The formulae for within-stratum cases:*

(1) $$g_{m,m+\ell} = g_{m+\ell,m} = \frac{\omega(a,a)}{2-a} r z^\ell [a\tau(a,a)]^{\ell-2}\left(\tfrac{a}{b}\Pi_{m,m+\ell}/w_\ell^*(a)\right),$$
$m < m + \ell \leq f - 1$;
(a) $$g_{f-\ell,f} = \frac{\omega(a,a)}{2-a} r z^\ell [a\tau(a,a)]^{\ell-2}\left(\tfrac{a}{b}\Pi_{f-\ell,f}/w_\ell^*(a)\right),\ \ell \geq 1;$$

(2) $$g_{m,m+j} = g_{m+j,m} = \frac{\omega(b,b)}{2-b} r z^j [b\tau(b,b)]^{j-2}\left(\Pi_{m,m+j}/w_j^*(b)\right),$$
$f \leq m < m + j$;
(a) $$g_{f+j,f-1} = \frac{\omega(b,b)}{2-b} r z^{j+1}[b\tau(b,b)]^{j-1}\left(\Pi_{f+j,f-1}/w_{j+1}^*(b)\right),\ j \geq 1.$$

Furthermore, the following identity holds for all $n \geq m+2$, *where* $\lambda_{m,n}$ *is defined by* (16.22).

$$\lambda_{m,n} = \frac{\overline{w}_{m,n}\overline{w}_{m+1,n+1}}{\overline{w}_{m,n+1}\overline{w}_{m+1,n}}. \tag{16.59}$$

Remark 16.3 Since $\tau(a,b)[\mathbf{1}] = 1$ for all a and b, one easily checks by Definition 16.2 and Lemma 16.3 that the formulae of Proposition 16.4 all yield the evaluation $g_{m,n}[\mathbf{1}] = 1$. The factor $\Pi_{m,n}$ appears in Lemma 16.3 the same as it does in the statements (I)–(II), so these factors cancel at $\mathbf{1}$.

Remark 16.4 All formulae in (III) hold by (16.37) for the homogeneous case. For example, in the statement (III)(1), we have $\frac{a}{b}\Pi_{m,m+\ell} = \ell - (\ell-1)a$, so there is no dependence on b.

Proof (*of Proposition* 16.4) Recall by (16.17) that $g_{m,m+2} = g_{m+2,m} = rz^2$. One can easily check by Definitions 16.2 and (16.44) that in each of (I)(1) with $j=1$, and (II)(2) with $\ell = 2$, the formulae reduce to rz^2. By (16.24)–(16.25), we must calculate a term $\lambda_{m,n}$ defined by (16.22). The term $\lambda_{m,n}$ is the same in both (16.24) and (16.25), so we only consider $m < n$ in (16.22). The structure of the proof is to first establish (16.59) for $n = m+2$ and to establish the initial cases $n - m = 3$ of the statements (I)–(II) of the proposition. Following this, an induction step will be established for all cases at once, wherein an inductive step for (16.59) shall be the main stepping stone of the proof.

Thus, consider first $n := m+2$ in (16.22). We consider 4 cases: (i) $n \leq f-1$; (ii) $m = f-2,\ n = f$; (iii) $m = f-1,\ n = f+1$; (iv) $m \geq f$. We verify by (16.18)–(16.20), Definition 16.2, Propositions 16.1, 16.2, and direct calculation, that in all cases (i)–(iv), $\lambda_{m,n} = \frac{\omega[a,b]^+_m \omega[a,b]^+_{m+1}}{\omega[a,b]^+_m \omega[a,b]^+_{m+1} - [\overline{w}]_{m,n}}$. Verification of this initial identity by direct calculation suffices for (16.59), since for the numerator we have by (16.44) that $\overline{w}_{m,n} = \omega[a,b]^+_m$ and $\overline{w}_{m+1,n+1} = \omega[a,b]^+_{m+1}$, and since for the denominator, we have by definition (16.46) that $\overline{w}_{m,n}\overline{w}_{m+1,n+1} - [\overline{w}]_{m,n} = \overline{w}_{m+1,n}\overline{w}_{m,n+1}$.

We turn to the initial conditions for (I)–(II). There are again four cases to consider. We conform with the notation of (16.21) and (16.23). For the upward cases, we write the lower index m and the upper index $m+3$. For the downward cases, we write the upper index $m+2$ and the lower index $m-1$. The cases are (I.1) $m = f-2,\ m+3 = f+1$; (I.2) $m = f-1,\ m+3 = f+2$; (II.1) $m+2 = f+1,\ m-1 = f-2$; (II.2) $m+2 = f,\ m-1 = f-3$. We use direct calculation of $g_{m,m+3}$ or $g_{m+2,m-1}$ for the upward and downward cases, respectively. Besides the formulae (16.21) and (16.23), we use $\lambda_{m,m+2}$ given by (16.59), where $\overline{w}_{m+1,m+2} = 1$ by definition (16.44). Since the denominator of $\lambda_{m,m+2}$ in each case is $\overline{w}_{m,m+3} = \overline{w}_{m+2,m-1}$ by Lemma 16.2, we compute $p_{m,m+3} := (1/c)g_{m,m+3}\overline{w}_{m,m+3}$ and $p_{m+2,m-1} := (1/c)g_{m+2,m-1}\overline{w}_{m,m+3}$ in the upwards and downwards cases, respectively. Schematically, since $g[\mathbf{1}] = 1$, we have $p/p[\mathbf{1}] = g\overline{w}/\overline{w}[\mathbf{1}] = numerator/\overline{w}[\mathbf{1}]$, where $numerator$ stands for the stated formula without the denominator $\overline{w}$. By cancelation of the Π–factors as in Remark 16.3, we match $p/p[\mathbf{1}]$ with $(numerator/\Pi)/(\overline{w}[\mathbf{1}]/\Pi)$ for verification by direct calculation.

We now proceed by induction on all cases of the proposition at once, where we assume that all statements hold for $g_{m,n}$ and $g_{n,m}$ with $2 \leq n - m \leq k$, for some $k \geq 3$. By the above, we have established all the initial cases, $k = 3$, for this hypothesis; as noted earlier, the case $n - m = 2$ is trivial. We now apply the formulae (16.24) and (16.25) to establish an induction step in each of the upward (I)(1), (2) and downward (II)(1), (2) cases, respectively. We are allowed to use any of the statements of (III) by Remark 16.4. Notice that for the range of indices we must now consider, in all cases $n \geq f$ and $m \leq f - 1$, so by (16.20), $\gamma_m \gamma_n = (1 - a)(1 - b)$.

Consider first (I)(1). Let first (i) $n + 1 = m + k + 1$, for some $m \leq f - 2$ and $n \geq f + 1$; there is another subcase (ii) $n = f$ that we handle as a special case by direct calculation below. We rewrite (16.24) for easy reference:

$$g_{m,n+1} = c_1 g_{m,n} g_{m+1,n+1} (g_{m+1,n})^{-1} \lambda_{m,n}.$$

In the definition (16.22), we have by (16.18)–(16.20) that $k[a, b]_m^+ = k(a, a) = \frac{a(2-a)}{\omega(a,a)}$, $k[a, b]_n^- = k(b, b) = \frac{b(2-b)}{\omega(b,b)}$. Also, by Definition 16.2 and Proposition 16.1, $4\rho_{m,n}\rho_{n,m} = \frac{(b/a)}{\Pi_{m,n}\Pi_{n,m}} = \frac{ab}{(a\Pi_{m,n})(a\Pi_{n,m})}$. Therefore by (16.22) and the induction hypothesis (I)(1), for $g_{m,n}$, and (II)(1), for $g_{n,m}$, the expression (†) $1 - 1/\lambda_{m,n}$, is written as

$$\frac{\gamma_m \gamma_n ab}{\left(a\Pi_{m,n}\right)\left(a\Pi_{n,m}\right)} \frac{ab(2-a)(2-b)}{\omega(a,a)\omega(b,b)} g_{m,n} g_{n,m}$$
$$= \frac{\gamma_m \gamma_n a^2 r^2 z^{2j+2\ell} b^2 \tau^2(a,b)[a^2\tau_{\bar{a}}^2]^{\ell-2}[b^2\tau_{\bar{b}}^2]^{j-1}}{\overline{w}_{m,n}\overline{w}_{n,m}}. \tag{16.60}$$

Now apply (16.38) and (16.45) to write x_a, $x(a, b) = b^2 z^2 \tau^2(a, b)$, and x_b, using all but 4 powers of z. So the numerator of the right member of (16.60) is simply the interlacing bracket $[\overline{w}]_{m,n}$ of Proposition 16.2(1). Thus, after writing $\overline{w}_{n,m} = \overline{w}_{m-1,n+1}$ by Lemma 16.2, and applying the bracket definition (16.46), we have established that (†) is given by $\frac{\overline{w}_{m,n}\overline{w}_{m+1,n+1} - \overline{w}_{m,n+1}\overline{w}_{m+1,n}}{\overline{w}_{m,n}\overline{w}_{m+1,n+1}}$, so (16.59) holds. Finally, apply (16.24) and the induction hypothesis (I)(1) and (16.59). Since the lower index $m + 1$ is the same in both the numerator and denominator of the ratio $g_{m+1,n+1}/g_{m+1,n}$, we obtain, by (I)(1) for $m + 1 \leq f - 2$, or by (I)(2) for $m + 1 = f - 1$, that $g_{m+1,n+1}/g_{m+1,n} = cz[b\tau(b, b)]\overline{w}_{m+1,n}/\overline{w}_{m+1,n+1}$. Thus,

$$g_{m,n+1} = c_1 \frac{g_{m,n} g_{m+1,n+1}}{g_{m+1,n}} \frac{\overline{w}_{m,n}\overline{w}_{m+1,n+1}}{\overline{w}_{m,n+1}\overline{w}_{m+1,n}} = cz[b\tau(b,b)]\left(g_{m,n}\overline{w}_{m,n}\right)/\overline{w}_{m,n+1}.$$

Hence, by plugging in the numerator $p_{m,n} := g_{m,n}\overline{w}_{m,n}$ (ignoring the constants) from the induction hypothesis for (I)(1), the induction step for (I)(1)(i), including (16.59), is complete by Remark 16.3.

To recapitulate, in general, there are two steps, where for the upward and downward cases we conform to the recurrences (16.24) and (16.25), respectively.

1. Establish (16.59) by showing that the numerator in the analogue of the right-hand member of (16.60) gives a bracket $[\overline{w}_{m,n}] = [\overline{w}_{n,m}]$ from Proposition 16.2, for the parameters m, n of $\lambda_{m,n}$.
2. Establish that when the induction hypothesis is applied, the condition

$$\text{(u)} \quad \frac{p_{m,n}\,p_{m+1,n+1}}{p_{m,n}[\mathbf{1}]\,p_{m+1,n+1}[\mathbf{1}]} - \frac{p_{m+1,n}\,p_{m,n+1}}{p_{m+1,n}[\mathbf{1}]\,p_{m,n+1}[\mathbf{1}]} = 0,$$

is verified for (I), and condition

$$\text{(d)} \quad \frac{p_{n,m}\,p_{n-1,m-1}}{p_{n,m}[\mathbf{1}]\,p_{n-1,m-1}[\mathbf{1}]} - \frac{p_{n-1,m}\,p_{n,m-1}}{p_{n-1,m}[\mathbf{1}]\,p_{n,m-1}[\mathbf{1}]} = 0,$$

is verified for (II).

For all the remaining cases of the induction steps in (I)–(II), including the subcase (I)(1)(ii), we proceed by direct calculation to check the details of these 2 Steps.

In Step 1, it is implicit that the factors of $\omega(a, b)$ that occur variously by substitution from factors $k(a, b)$ in the formula for $\lambda_{m,n}$, and also from the numerators of $g_{m,n}$ and $g_{n,m}$, cancel one another in every case. This is borne out in the direct calculations, where the pattern of substitutions from the induction hypothesis is shown. By Remark 16.3, conditions (u)–(d) are equivalent to showing, for the ratio $\frac{p_{m+1,n+1}}{p_{m+1,n}} = cz\tau$, in the upward case, or $\frac{p_{n-1,m-1}}{p_{n-1,m}} = cz\tau$, in the downward case, that the factor of $z\tau$ completes the form of the numerator $p_{m,n+1}$ [respectively $p_{n,m-1}$] as one extra factor of the numerator form $p_{m,n}$ [respectively $p_{n,m}$]. Here, the factor τ depends on subcases; it is $\tau(a, b)$ in subcases (I)(1)(ii), and in (II)(1)(ii): $n \geq f + 2$, $m = f - 1$. We show the pattern of substitutions for (u)–(d) in the direct calculations [15].

3.5 *Generating Function of the Excursion Statistics*

We derive a closed formula for $K_N(a, b)$ of (16.10). Recall by (16.41) that $\{q_n^*(a)\}$ and $\{w_n^*(a)\}$ share a common Fibonacci recurrence: $v_{n+1} = \beta_a v_n - x_a v_{n-1}$, $n \geq 1$. We extend the $\{q_n^*(a)\}$ from the homogeneous model to the full model as $\{\overline{q}_n\}$, analogous to $\{\overline{w}_{0,n}\}$ of Definition 16.3, except with $\overline{q}_n$ we start the stratum crossing at index $n = f$ rather than $n = f + 1$.

Definition 16.4 Define $\overline{q}_n$ for all $n \geq 1$ by:

(1) $\overline{q}_n := q_n^*(a)$, $1 \leq n < f$;
(2) $\overline{q}_f := \frac{1-b}{1-a} q_f^*(a) + \frac{b-a}{1-a} q_{f-1}^*(a)$;
(3) $\overline{q}_{f+1} := \beta(a, b)\overline{q}_f - x(a, b)\overline{q}_{f-1}$;
(4) $\overline{q}_{f+j+1} := \beta_b \overline{q}_{f+j} - x_b \overline{q}_{f+j-1}$, $j \geq 1$.

Denote the single indexed bracket (cf. Casorati determinant, see [11])

$$[\overline{w}, \overline{q}]_n := \overline{w}_{n,0}\overline{q}_{n+1} - \overline{q}_n\overline{w}_{n+1,0}, \ n \geq 1; \tag{16.61}$$

We note that the homogeneous case of (16.61) is written as

$$[w^*(a), q^*(a)]_n := w_n^*(a)q_{n+1}^*(a) - q_n^*(a)w_{n+1}^*(a), \ n \geq 1.$$

Lemma 16.4 *The following identities hold:*

(1) $[w^*(a), q^*(a)]_n = a^2z^2x_a^{n-1}$, $n \geq 1$;
(2) $[\overline{w}, \overline{q}]_{f-1} = \frac{1-b}{1-a}[w^*(a), q^*(a)]_{f-1} = \frac{1-b}{1-a}a^2z^2x_a^{f-2}$;
(3) $[\overline{w}, \overline{q}]_{f+j-1} = \frac{1-b}{1-a}a^2z^2x_a^{f-2}x(a,b)x_b^{j-1}$, *for all* $j \geq 1$.

Before we can prove Lemma 16.4, we write a formula for $\overline{q}_n$ as follows.

Lemma 16.5 *Let* $M := Q(b)^{-1}B$, *for* $Q(b)$ *defined by* (16.51) *and* B *defined by* (16.52). *Then,*

$$\overline{q}_{f+j-1} = \begin{bmatrix} q_j^*(b) & w_j^*(b) \end{bmatrix} M \begin{bmatrix} q_{f-1}^*(a) \\ q_f^*(a) \end{bmatrix}, \quad \textit{for all } j \geq 1. \tag{16.62}$$

Proof The proof is almost the same as the proof of Proposition 16.3. Define $Q(b)$ as before in (16.51), but write now a revision $\mathbf{W}_q(f)$ of $\mathbf{W}(\ell)$, and write $\mathbf{W}_q(f)$ in two ways as follows.

$$\mathbf{W}_q(f) := \begin{bmatrix} \overline{q}_f \\ \overline{q}_{f+1} \end{bmatrix} = Q(b)\mathbf{d}_q(f); \quad \mathbf{W}_q(f) = B\begin{bmatrix} q_{f-1}^*(a) \\ q_f^*(a) \end{bmatrix}. \tag{16.63}$$

We have B given by (16.52), because by Definitions 16.3 and 16.4 the equations that define B in (16.63) are the same as those defining B in (16.52). By equating the two expressions for the vector $\mathbf{W}_q(f)$ in (16.63), we recover $\mathbf{d}_q(f) = M\begin{bmatrix} q_{f-1}^*(a) \\ q_f^*(a) \end{bmatrix}$. Now the formula (16.62) follows because by (16.63) and the definition of $Q(b)$ we have established the formula for $j = 1, 2$. Therefore, by the fact that either side of (16.62) satisfies the same recurrence $v_{j+1} = \beta_b v_j - x_b v_{j-1}$, $j \geq 2$, we have that both sides are equal as stated.

Proof (*of Lemma* 16.4) We first prove statement (1). By the simple fact that $\{q_n^*(a)\}$ and $\{w_n^*(a)\}$ satisfy the same Fibonacci recurrence, we have:

$$[w^*(a), q^*(a)]_n = w_n^*(\beta_a q_n^* - x_a q_{n-1}^*) - q_n^*(\beta_a w_n^* - x_a w_{n-1}^*) = x_a[w^*, q^*]_{n-1}$$

holds for all $n \geq 1$, where we suppressed the a in q_n^* and w_n^*. By direct calculation from (16.38)–(16.40) and (16.61), $[w^*(a), q^*(a)]_0 = w_0^*(a)q_1^*(a) - q_0^*(a)w_1^*(a) = a^2z^2/x_a$. Therefore, since we may iterate the one-step recursion for $[w^*(a), q^*(a)]_n$, we obtain statement (1) of the lemma.

Next, by Definitions 16.3 and 16.4,

$$\begin{aligned}[\overline{w},\overline{q}]_{f-1} &= \overline{w}_{1,f}\overline{q}_f - \overline{q}_{f-1}\overline{w}_{1,f+1} \\ &= w^*_{f-1}\left(\frac{1-b}{1-a}q^*_f + \frac{b-a}{1-a}q^*_{f-1}\right) - q^*_{f-1}\left(\frac{1-b}{1-a}w^*_f + \frac{b-a}{1-a}w^*_{f-1}\right) \\ &= \frac{1-b}{1-a}[w^*,q^*]_{f-1}.\end{aligned}$$

Therefore, statement (2) of the lemma follows by statement (1).

Finally, note that by Lemma 16.2, we have $\overline{w}_{n,0} = \overline{w}_{1,n+1}$. Further by (16.53) and Proposition 16.3 with $\ell = f-1$, we may write

$$\overline{w}_{1,f+j} = \begin{bmatrix} q^*_j(b) & w^*_j(b) \end{bmatrix} M \begin{bmatrix} w^*_{f-1}(a) \\ w^*_f(a) \end{bmatrix}. \tag{16.64}$$

Now write $n = f + j - 1$ for some $j \geq 1$. Also for simplicity abbreviate $q_0 = q^*_{f-1}(a)$, $q_1 = q^*_f(a)$, $w_0 = w^*_{f-1}(a)$, $w_1 = w^*_f(a)$, $u = q^*_j(b)$, $v = w^*_j(b)$, $U = q^*_{j+1}(b)$, $V = w^*_{j+1}(b)$. By (16.61), Lemma 16.5, and (16.64),

$$[\overline{w},\overline{q}]_n = \begin{bmatrix} u & v \end{bmatrix} M \left\{ \begin{bmatrix} w_0 \\ w_1 \end{bmatrix} \begin{bmatrix} U & V \end{bmatrix} M \begin{bmatrix} q_0 \\ q_1 \end{bmatrix} - \begin{bmatrix} q_0 \\ q_1 \end{bmatrix} \begin{bmatrix} U & V \end{bmatrix} M \begin{bmatrix} w_0 \\ w_1 \end{bmatrix} \right\}.$$

Denote $M = (\mu_{i,j})$, and calculate the expression under the curly brackets as follows: $(w_0q_1 - q_0w_1)\begin{bmatrix} \mu_{1,2} & \mu_{2,2} \\ -\mu_{1,1} & -\mu_{2,1} \end{bmatrix}\begin{bmatrix} U \\ V \end{bmatrix}$. Therefore, after multiplying through by $\begin{bmatrix} u & v \end{bmatrix} M$, we obtain:
$[\overline{w},\overline{q}]_n = (w_0q_1 - q_0w_1)\begin{bmatrix} u & v \end{bmatrix}\begin{bmatrix} 0 & \det(M) \\ -\det(M) & 0 \end{bmatrix}\begin{bmatrix} U \\ V \end{bmatrix}$. Putting back our variables, we thus have

$$[\overline{w},\overline{q}]_n = [w^*(a),q^*(a)]_{f-1}(-\det(M))[w^*(b),q^*(b)]_j, \text{ valid for all } j \geq 1.$$

Also, by direct calculation, $\det(M) = -\frac{1-b}{1-a}\tau(a,b)^2$. Thus, by (16.45) and statement (1), the statement (3) of the lemma is proved.

Theorem 16.3 *The conditional generating function* (16.10) *has the following formula.*

$$K_N(a,b) = C_{N,a,b}\frac{r^2z^2\overline{q}_N}{\overline{w}_{1,N+1}}, \quad N \geq 1,$$

where $\overline{q}_N$ and $\overline{w}_{1,N+1}$ are given by Lemma 16.5 and (16.64)*, respectively, each with $j = N - f + 1$, and where $C_{N,a,b} = (1-a)^{\frac{(N-f)a+(f-1)b-(N-1)ab}{(N-f)a+(f-1)b-Nab}}$.*

In the homogeneous case $a = b$, Theorem 16.3 takes the following form.

Proposition 16.5 *Suppose $a = b$. Then $K_N(a)$ of* (16.10) *has the following formula.* $K_N(a) = \frac{1}{N}(N - (N-1)a)\frac{r^2z^2q_N^*(a)}{w_N^*(a)}$, $N \geq 1$, *where $q_N^*(a)$ and $w_N^*(a)$ are defined by* (16.38)–(16.41).

Proof (*of Proposition* 16.5 *and Theorem* 16.3) First let $a = b$. The proof parallels the construction of convergents to a continued fraction; see [5, Ch. III]. By $G_n(a)$ of (16.27) and by formula (16.37), for all $n \geq 3$,

$$G_n(a) = a(2-a)zh_a k_a g_{n-1} g_n = c_{n,a} r^2 z^{2n} \tau_a^{2n-4} / [w_n^*(a) w_{n-1}^*(a)], \quad (16.65)$$

for $c_{n,a} := a^2(2-a)^2 C_{n,a} C_{n-1,a}$, with $C_{n,a}$ given by (16.37), where we have written $h_a k_a \omega_a^2 = a(2-a)\tau_a$ by the definitions (16.18). Further, for the full model we have:

$$\begin{aligned} P(\mathbf{H} = n) &= 4a\rho_{1,n}\gamma_n\rho_{n,0}, \ n \geq 2; \\ P(\mathbf{H} \geq n+1) &= 2a\rho_{1,n+1}, \ n \geq 2; \\ P(H = 1) &= 2\frac{1}{2}(1-a) = 1-a. \end{aligned} \quad (16.66)$$

By first principles, we find $G_1(a) = r^2y^2z^2$ and $G_2(a) = r^2z^4k_a$. Therefore, by (16.10) and (16.26), $P(\mathbf{H} \leq N)K_N(a) = \sum_{n=1}^{N} G_n(a)P(\mathbf{H} = n)$ is written by:

$$(1-a)r^2y^2z^2 + \frac{a(1-a)}{2-a}r^2z^4k_a + \sum_{n=3}^{N} c_{n,a}P(\mathbf{H} = n)\frac{r^2z^{2n}\tau_a^{2n-4}}{w_n^*(a)w_{n-1}^*(a)}. \quad (16.67)$$

By (16.65)–(16.66) and direct calculation, $c_{n,a}P(\mathbf{H} = n) = (1-a)a^{2n-2}$. Also, by (16.18) and (16.38), $a^{2n-2}r^2z^{2n}\tau_a^{2n-4} = a^2r^2z^4x_a^{n-2}$ while $k_a = \frac{a(2-a)}{w_2^*(a)}$, since $\omega_a = w_2^*(a)$. Therefore, by (16.67), $P(\mathbf{H} \leq N)K_N(a)$ is written:

$$(1-a)r^2z^2\left(y^2 + \frac{a^2z^2}{w_2^*(a)} + \sum_{n=3}^{N} \frac{a^2z^2x_a^{n-2}}{w_n^*(a)w_{n-1}^*(a)}\right). \quad (16.68)$$

By (16.38)–(16.41) and direct calculation, we have that $y^2 = q_1^*(a)/w_1^*(a)$ and $y^2 + a^2z^2/w_2^*(a) = q_2^*(a)/w_2^*(a)$. But, by Lemma 16.4(1), we may write

$$\frac{q_n^*}{w_n^*} - \frac{q_{n-1}^*}{w_{n-1}^*} = \frac{[w^*(a), q^*(a)]_{n-1}}{w_n^*w_{n-1}^*} = \frac{a^2z^2x_a^{n-2}}{w_n^*w_{n-1}^*}, \quad n \geq 1,$$

where we suppressed the dependence on a in q_n^* and w_n^*. Therefore, the sum in (16.68) telescopes. Therefore, for all $N \geq 1$ the right side of (16.68) becomes: $(1-a)r^2z^2q_N^*(a)/w_N^*(a)$. Finally, apply (16.66) and Proposition 16.1 to compute $P(\mathbf{H} \leq N) = \frac{N(1-a)}{N-(N-1)a}$, so that, by (16.68), Proposition 16.5 is proved.

We now indicate the additional steps required to prove the Theorem 16.3. First, with $N = f - 1$ in Proposition 16.5, and by Lemma 16.4, (16.68) yields:

$$\sum_{n=1}^{f-1} G_n P(\mathbf{H} = n) = (1-a)r^2z^2 \left(y^2 + \sum_{n=2}^{f-1} \frac{[w^*(a), q^*(a)]_{n-1}}{w_n^*(a)w_{n-1}^*(a)} \right), \tag{16.69}$$

where here and in the rest of the proof we abbreviate $G_n = G_n(a, b)$. Next by (16.27) and Proposition 16.4, (III)(1)(a) and (II)(2),

$$G_f = a(2-a)zh_a k(a, b) g_{1,f} g_{f,0} = c_f \frac{r^2 z^{2f} \, (a\tau_a)^{2f-4}}{w_{f-1}^*(a)\overline{w}_{f,0}}. \tag{16.70}$$

Moreover, by (16.27) and Proposition 16.4, (I)(1) and (II)(1), for all $j \geq 1$, G_{f+j} becomes:

$$a(2-a)zh_a k(b, b) g_{1,f+j} g_{f+j,0} = c_{f+j} \frac{r^2 z^{2f+2j} \, (a\tau_a)^{2f-4} \, \tau(a, b)^2 \, (b\tau_b)^{2j-2}}{\overline{w}_{1,f+j}\overline{w}_{f+j,0}}. \tag{16.71}$$

In (16.70) and (16.71), the constants c_f and c_{f+1}, respectively, can be determined from Lemma 16.3 since $G_n[\mathbf{1}] = 1$. Indeed, we find in this way, and by Definition 16.2, Proposition 16.1, (16.66), and direct calculation that $c_f P(\mathbf{H} = f) = a^2(1-b)$ and $c_{f+j} P(H = f + j) = a^2b^2(1-b)$, $j \geq 1$. Thus, by (16.38), (16.45), and (16.70)–(16.71), and since $w_{f-1}^*(a) = \overline{w}_{1,f}$, for all $j \geq 0$, $G_{f+j} P(\mathbf{H} = f + j)$ equals

$$a^2(1-b)\frac{r^2z^4x_a^{f-2}}{\overline{w}_{1,f}\overline{w}_{f,0}}, \text{ if } j = 0; \quad a^2(1-b)\frac{r^2z^4x_a^{f-2}x(a, b)x_b^{j-1}}{\overline{w}_{1,f+j}\overline{w}_{f+j,0}}, \text{ if } j \geq 1. \tag{16.72}$$

Therefore, by (16.69), (16.72) and Lemma 16.4, for all $j \geq 0$ there holds:

$$\sum_{n=1}^{f+j} G_n P(\mathbf{H} = n) = (1-a)r^2z^2 \left(y^2 + \sum_{n=2}^{f+j} \frac{[\overline{w}, \overline{q}]_{n-1}}{\overline{w}_{1,n}\overline{w}_{n,0}} \right), \tag{16.73}$$

where the fraction $\frac{1-b}{1-a}$ enters to form $[\overline{w}, \overline{q}]_{n-1}$ when $n = f + j$ for $j \geq 0$ because we have factored out $(1-a)$ from the entire sum on the right. But by the definition (16.61) and Lemma 16.2, we have that $\frac{[\overline{w},\overline{q}]_{n-1}}{\overline{w}_{1,n}\overline{w}_{n,0}} = \frac{\overline{q}_n}{\overline{w}_{1,n+1}} - \frac{\overline{q}_{n-1}}{\overline{w}_{1,n}}$. Hence, the sum in (16.73) telescopes, and thereby, we finally obtain $K_N(a, b) = P(\mathbf{H} \leq N)^{-1}(1 - a)\frac{r^2z^2\overline{q}_N}{\overline{w}_{1,N+1}}$, $N \geq f$, where $P(\mathbf{H} \leq N)^{-1} = \frac{(N+1-f)a+(f-1)b-(N-1)ab}{(N+1-f)a+(f-1)b-Nab}$ by (16.66) and direct calculation.

3.5.1 Proof of Corollary 16.2

The unconditional joint generating function of the excursion statistics is $K := E\{r^{\mathbf{R}}y^{\mathbf{V}}z^{\mathbf{L}}\}$. We develop a simple representation of K in the homogeneous case, as follows.

Corollary 16.3 *Let* $a = b$ *and define* $\alpha_a := \sqrt{\beta_a^2 - 4x_a}$ *for* x_a *and* β_a *given by* (16.38). *Then* $K = \lim_{N\to\infty} K_N(a) = \left(1 - \frac{1}{2}\beta_a - \frac{1}{2}\alpha_a\right)/(1-a)$.

Proof (*of Corollary* 16.3) Since we have explicitly seen in the proof of Proposition 16.5 that if $a = b$, then $P(\mathbf{H} \leq N) = \frac{N(1-a)}{N-(N-1)a}$, we have that the persistent random walk is recurrent: $\lim_{N\to\infty} P(\mathbf{H} \leq N) = 1$. So we obtain that $(*) K = \lim\limits_{N\to\infty} K_N(a) = (1-a)r^2z^2 \lim_{N\to\infty} q_N^*/w_N^*$. Here and in the rest of the proof, we suppress dependence on a when convenient; in particular denote $x = x_a$ and $\beta = \beta_a$.

We introduce a substitution variable θ as follows:

$$\beta := \sqrt{4x}\cos\theta; \quad \beta \pm \alpha = \sqrt{4x}(\cos\theta \pm i\sin\theta) = \sqrt{4x}e^{\pm i\theta}, \tag{16.74}$$

with $\Im\theta < 0$ for $|r| < 1, |y| < 1, z \neq 0$. The idea of the substitution (16.74) may be found in [8, p. 352]. By $(\beta+\alpha)^n - (\beta+\alpha)^n = (4x)^{n/2}e^{in\theta}(1+e^{-2in\theta})$, the formulae (16.13) may be rewritten, where by our convention for the sign of $\Im\theta$, $1 + e^{-2in\theta} = 1 + o(1)$, as $n \to \infty$. We then substitute these expressions into the formulae (16.41) and find that $q_n^*(a)/w_n^*(a)$ is given by:

$$\frac{(y^2 - q_0^*)q_n(x,\beta) + q_0^* w_n(x,\beta)}{(1-w_0^*)q_n(x,\beta) + w_0^* w_n(x,\beta)} = \frac{y^2(1-e^{-2in\theta}) - \sqrt{x}q_0^* e^{-i\theta}(1-e^{-2i(n-1)\theta})}{1 - e^{-2in\theta} - \sqrt{x}w_0^* e^{-i\theta}(1-e^{-2i(n-1)\theta})}.$$

Therefore, using $e^{-i\theta} = (\beta-\alpha)/\sqrt{4x}$, we obtain $\lim_{n\to\infty}\frac{q_n^*}{w_n^*} = \frac{y^2 - q_0^*(\beta-\alpha)/2}{1-w_0^*(\beta-\alpha)/2}$. Finally, we simplify this expression by multiplying both numerator and denominator by $1 - w_0^*(\beta+\alpha)/2$. The new denominator becomes $1 + x_a w_0^*(a)^2 - \beta_a w_0^*(a) = (1-a)^2r^2z^2/\tau_a^2$, by direct calculation. Therefore, by bringing the τ_a^2 of this last expression to the numerator, we obtain that $\lim_{n\to\infty}(1-a)^2r^2z^2q_n^*/w_n^* = \tau_a^2(y^2 - q_0^*(\beta-\alpha)/2)(1-w_0^*(\beta+\alpha)/2)$, or

$$[y^2 - \beta_a(q_0^* + y^2w_0^*)/2 + x_a w_0^* q_0^*]\tau_a^2 + \alpha_a[q_0^* - y^2w_0^*]\tau_a^2/2 = I + \alpha_a II,$$

after cancelation of terms $\pm q_0^* w_0^* \alpha\beta/4$. By direct calculation, we find that $I = 1 - \frac{1}{2}\beta_a$, and $II = -\frac{1}{2}$. Hence by $(*)$ the proof is complete.

We may view the excursion statistics in the case $a = b$ by the way they are weighted relative to one another. Indeed, a specific excursion path of $2n$ steps and $2k$ runs is weighted with the probability $\frac{1}{2}a^{2n-2k}(1-a)^{2k-1}$, for k peaks and $k-1$ valleys. In the unweighted case $a = \frac{1}{2}$, it is known that the joint distribution of

$(\mathbf{L}, \mathbf{R})$ is essentially the same as that of $(\mathbf{L}, \mathbf{L} - \mathbf{R})$ [see [16, A001263; symmetry of the Narayana numbers]].

Proof (*of Corollary* 16.2) We establish the joint generating function identity in the unweighted case via a direct calculation. Let r, u and z belong to the unit circle. By applying Corollary 16.3 with $a = \frac{1}{2}$, we obtain the joint generating function of runs, long runs, and steps by

$$K(\frac{1}{2})[ru, 1/u, z] = \frac{1}{16}\left(16 - 4z^2 + 4r^2z^2 + r^2z^4 - 2r^2uz^4 + r^2u^2z^4 - S\right),$$

with S given by $S = S_1S_2S_3S_4$, for:

$$\begin{aligned} S_1 &= \sqrt{4 + 2z + 2rz + rz^2 - ruz^2}, \\ S_2 &= \sqrt{4 + 2z - 2rz - rz^2 + ruz^2}, \\ S_3 &= \sqrt{4 - 2z + 2rz - rz^2 + ruz^2},\ S_4 = \sqrt{4 - 2z - 2rz + rz^2 - ruz^2}. \end{aligned}$$

On the other hand, with the very same main term S, we have

$$K(\frac{1}{2})[u/r, 1/u, rz] = \frac{1}{16}\left(16 + 4z^2 - 4r^2z^2 + r^2z^4 - 2r^2uz^4 + r^2u^2z^4 - S\right).$$

The two generating functions differ by

$$K(\frac{1}{2})[ru, 1/u, z] - K(\frac{1}{2})[u/r, 1/u, rz] = \frac{1}{2}z^2(r^2 - 1).$$

The difference is mirrored only in the event that $\mathbf{L} = 2$, when it happens that $\mathbf{R} = 2$ and $\mathbf{U} = 0$. Thus, (16.8) holds for $a = \frac{1}{2}$ and $n \geq 2$.

Perhaps the simplest way to obtain (16.8) for $a \neq \frac{1}{2}$ is to apply (16.8) for the case $a = \frac{1}{2}$. Consider an excursion path Γ with $\mathbf{L}(\Gamma) = 2n$ and $\mathbf{L}(\Gamma) - \mathbf{R}(\Gamma) = 2k$. Then $P_a(\Gamma) = \frac{1}{2}a^{2k}(1 - a)^{2n-2k-1}$. Here, $\mathbf{R}(\Gamma) - 1 = 2n - 2k - 1$ counts the number of turns in the path, so is the exponent of $(1 - a)$ under P_a. Alternatively, $\mathbf{L}(\Gamma) - \mathbf{R}(\Gamma)$ is the total length of long runs minus the number of long runs in Γ, and this gives the exponent of a in $P_a(\Gamma)$. If $2n \geq 4$, then by the first part of the proof there are exactly as many paths Γ with the joint information $\mathbf{L}(\Gamma) = 2n$, $\mathbf{L}(\Gamma) - \mathbf{R}(\Gamma) = 2k$, and $\mathbf{U}(\Gamma) = \ell$, as there are paths Γ' with $\mathbf{L}(\Gamma') = 2n$, $\mathbf{R}(\Gamma') = 2k$, and $\mathbf{U}(\Gamma') = \ell$. Therefore, since for any such path Γ', the probability assigned by the probability measure P_{1-a} yields $P_{1-a}(\Gamma') = \frac{1}{2}a^{2k-1}(1 - a)^{2n-2k}$, we have that $aP_{1-a}(\Gamma') = (1 - a)P_a(\Gamma)$, for all Γ with $\mathbf{L}(\Gamma) \geq 4$. Hence, (16.8) holds.

4 Proofs of Theorems 16.1 and 16.2

Proof (*of Theorem* 16.1) We fix $t \in \mathbb{R}$. All big oh terms in the proof will refer to the parameter $N \to \infty$ with implied constants depending only on a, b, and t. Since, by [13], for fixed $m > 0$ and $f \sim \eta N \to \infty$, $P(\mathbf{X}_j = 0 \text{ before } \mathbf{X}_j = f | \mathbf{X}_0 = m) \to 1$, we may assume that $\mathbf{X}_0 = 0$. Let

$$r_N := e^{-it(2-a-b)/((1-a)(1-b)N)}, \; y_N := e^{it/((1-a)(1-b)N)}, \; z_N := e^{it/N}. \quad (16.75)$$

Since $\{(1+\frac{1}{N})X_{N+1}\}$ converges in distribution if and only if $\{X_N\}$ does, by (16.2) it suffices to establish that $E\{e^{it(1+\frac{1}{N})X_{N+1}}\} = \hat{\varphi}(t)$, as $N \to \infty$. It is clear that $a(2-a)z_N h_a[r_N, y_N, z_N] \to 1$ as $N \to \infty$. Therefore, by (16.28), we must show that $\lim_{N\to\infty} g_{0,N}[r_N, y_N, z_N]$ equals the limit in (16.3). By Proposition 16.4(I)(1), we have a formula for $g_{0,N}$, and by Proposition 16.3, we have a formula for its denominator $\overline{w}_{0,N}$. The main work is in calculating an asymptotic expression for $\overline{w}_{0,N}[r_N, y_N, z_N]$.

We now make substitutions analogous to (16.74), one for each stratum:

$$\begin{aligned} &\cos(\theta_1) := \beta_a/\sqrt{4x_a}\,; \;\; \cos(\theta_2) := \beta_b/\sqrt{4x_b}, \\ &\beta_a \pm \alpha_a = \sqrt{4x_a}e^{\pm i\theta_1}, \\ &\beta_b \pm \alpha_b = \sqrt{4x_b}e^{\pm i\theta_2}, \end{aligned} \quad (16.76)$$

where all functions on the right sides of these expressions are composed with $[r_N, y_N, z_N]$ of (16.75). Here, we write $\sqrt{4x_a}$ as a shorthand for the expression $2az\tau_a$; see (16.38). Note that the coefficients in (16.2) have been chosen such that the first order term of the Taylor expansions about $t = 0$ of the substitutions $\cos\theta_j[r_N, y_N, z_N]$, $j = 1, 2$, do in fact vanish in the following:

$$\cos\theta_1 = 1 + \frac{1}{2}\frac{\sigma_1^2 t^2}{(1-b)^2N^2} + O\left(\frac{1}{N^3}\right); \quad \cos\theta_2 = 1 + \frac{1}{2}\frac{\sigma_2^2 t^2}{(1-a)^2N^2} + O\left(\frac{1}{N^3}\right), \quad (16.77)$$

where σ_1^2 and σ_2^2 are as defined in the statement of the theorem, and we obtain (16.77) by direct computation. Therefore, by (16.77), and by applying the Taylor expansion of $\arccos(u)$ about $u = 1$, we find that θ_1 and θ_2 are both of order $1/N$ as follows:

$$\theta_1 = i\frac{\sigma_1 t}{(1-b)N} + O\left(\frac{1}{N^3}\right); \quad \theta_2 = i\frac{\sigma_2 t}{(1-a)N} + O\left(\frac{1}{N^3}\right). \quad (16.78)$$

By Proposition 16.3,

$$\overline{w}_{0,N} = d_1(f)q^*_{N-f}(b) + d_2(f)w^*_{N-f}(b). \quad (16.79)$$

We focus first on the coefficients $d_j(f)$, which are written in terms of $w_f^*(a)$ and $w_{f+1}^*(a)$ by (16.53). By (16.13) and (16.41), suppressing dependence on a, $w_f^* = (1 - w_0^*)q_f + w_0^*(q_f - xq_{f-1}) = q_f(x, \beta) - w_0^* x q_{f-1}(x, \beta)$. Thus, by (16.13), and (16.76),

$$
\begin{aligned}
w_f^*(a) &= 2i\alpha_a^{-1}\,(az\tau_a)^f\,\left\{\sin f\theta_1 - \sqrt{x_a}\,w_0^*(a)\sin(f-1)\theta_1\right\}\\
w_{f+1}^*(a) &= 2i\alpha_a^{-1}\,(az\tau_a)^f\,\sqrt{x_a}\left\{\sin(f+1)\theta_1 - \sqrt{x_a}\,w_0^*(a)\sin f\theta_1\right\}; \qquad (16.80)
\end{aligned}
$$

with verification by direct algebra for $q_f(x_a, \beta_a) = 2i\alpha_a^{-1}\,(az\tau_a)^f\,\sin(f\theta_1)$, and with $\sqrt{x_a}$ to stand for a factor of $(az\tau_a)$. Next denote

$$
e_j = e_j(f) := \frac{d_j(f)}{\Lambda_1},\ j = 1, 2,\ \text{for}\ \Lambda_1 := 2i\alpha_a^{-1}\,(az\tau_a)^f = (\sin\theta_1)^{-1}\,(az\tau_a)^{f-1}, \qquad (16.81)
$$

since $\alpha_a = i\sqrt{4x_a}\sin\theta_1 = 2iaz\tau_a\sin\theta_1$. By (16.80)–(16.81) and direct algebra, through (16.53) we can write an expression for e_j as follows:

$$
\left(\mu_{j,1} - \mu_{j,2}x_a w_0^*(a)\right)\sin f\theta_1 + \sqrt{x_a}\left[\mu_{j,2}\sin(f+1)\theta_1 - \mu_{j,1}w_0^*(a)\sin(f-1)\theta_1\right]. \qquad (16.82)
$$

Next we apply the trigonometric identity for the sine of a sum or difference to $\sin(f+1)\theta_1$ and $\sin(f-1)\theta_1$ in (16.82). At this point, we also introduce some abbreviations to keep the notation a bit compact. Thus, write

$$
\mathbf{s}_1 := \sin f\theta_1;\quad \mathbf{c}_1 := \cos f\theta_1. \qquad (16.83)
$$

We rewrite (16.82), with abbreviation $w_0^* = w_0^*(a)$, by collecting terms with a factor $\sqrt{x_a}$. Thus for each $j = 1, 2$,

$$
\begin{aligned}
e_j =&(\mu_{j,1} - \mu_{j,2}x_a w_0^*)\mathbf{s}_1\\
&+ \sqrt{x_a}\left\{\mu_{j,2}(\mathbf{s}_1\cos\theta_1 + \mathbf{c}_1\sin\theta_1) - \mu_{j,1}w_0^*(\mathbf{s}_1\cos\theta_1 - \mathbf{c}_1\sin\theta_1)\right\}. \quad (16.84)
\end{aligned}
$$

We introduce a book-keeping notation for the coefficient t_j of the variable $\mathbf{x}_j$ in square brackets, within a linear expression $\sum_i t_i\mathbf{x}_i$ in parentheses: $[\mathbf{x}_j]\left(\sum_i t_i\mathbf{x}_i\right) = t_j$. Our method for e_j is to asymptotically expand $[\mathbf{s}_1](e_j)$ and $[\mathbf{c}_1\sin\theta_1](e_j)$ by (16.84). We will treat $\sin\theta_1$ separately from the asymptotic expansions of the other terms due to the convenient fact that, by (16.78), we have $\sin\theta_1 = \theta_1 + O(N^{-3})$, and this will suffice for our purposes. Note that by (16.78) and (16.83), and $f \sim \eta N$, $\mathbf{s}_1$ and $\mathbf{c}_1$ are both $O(1)$. Further, by direct calculation, $\mu_{i,j}$ are polynomial, and $q_0^*(a)$ and $w_0^*(a)$ only involve negative powers of τ_a, where $\tau_a[\mathbf{1}] = 1$. Thus, the Taylor expansions of $[\mathbf{s}_1](e_j)$ and $[\mathbf{c}_1\sin\theta_1](e_j)$ about $t = 0$ are well behaved.

We next find reduced expressions for the terms $q^*_{N-f}(b)$ and $w^*_{N-f}(b)$ of (16.79). The approach is as above, but now with b in place of a, $N-f$ in place of f, and using the second substitution θ_2 in (16.76). Similar to (16.81), we introduce

$$q^* := \frac{q^*_{N-f}(b)}{\Lambda_2}, \ w^* := \frac{w^*_{N-f}(b)}{\Lambda_2};$$
$$\Lambda_2 := 2i\alpha_b^{-1}(bz\tau_b)^{N-f} = (\sin\theta_2)^{-1}(bz\tau_b)^{N-f-1}. \tag{16.85}$$

Similar as for (16.80), by (16.13) and both lines of (16.41) applied in turn, and (16.85),

$$q^* = y^2\sin(N-f)\theta_2 - \sqrt{x_b}q_0^*(b)\sin(N-f-1)\theta_2,$$
$$w^* = \sin(N-f)\theta_2 - \sqrt{x_b}w_0^*(b)\sin(N-f-1)\theta_2. \tag{16.86}$$

Introduce abbreviations also for the second stratum sines and cosines:

$$\mathbf{s}_2 := \sin(N-f)\theta_2; \ \ \mathbf{c}_2 := \cos(N-f)\theta_2. \tag{16.87}$$

We illustrate the book-keeping method by expanding $\sin(N-f-1)\theta_2 = \mathbf{s}_2\cos\theta_2 - \mathbf{c}_2\sin\theta_2$ to obtain by (16.86),

$$[\mathbf{s}_2]\left(q^*\right) = y^2 - \sqrt{x_b}q_0^*(b)\cos\theta_2; \ \ [\mathbf{c}_2\sin\theta_2]\left(q^*\right) = \sqrt{x_b}q_0^*(b);$$
$$[\mathbf{s}_2]\left(w^*\right) = 1 - \sqrt{x_b}w_0^*(b)\cos\theta_2; \ \ [\mathbf{c}_2\sin\theta_2]\left(w^*\right) = \sqrt{x_b}w_0^*(b).$$

To handle the asymptotic expansions for the four terms on the right side of (16.79), we expand the coefficients of $\mathbf{s}_1$, $\mathbf{c}_1\sin\theta_1$, $\mathbf{s}_2$, and $\mathbf{c}_2\sin\theta_2$ by direct computation and thereby find

$$\frac{\overline{w}_{0,N}}{\Lambda_1\Lambda_2} = O(N^{-2}) + \left[\left(-(1-a)(1-b) + 2(1-ab)\frac{it}{N}\right)\mathbf{s}_1\right]\left[\left(1+\frac{2}{1-a}\frac{it}{N}\right)\mathbf{s}_2\right]$$
$$+\left[\left(1-a-\frac{a(b-a)}{1-b}\frac{it}{N}\right)\mathbf{s}_1 + a\mathbf{c}_1\sin\theta_1\right]$$
$$\times\left[\left(1-b-\frac{b(2-a-b)}{1-a}\frac{it}{N}\right)\mathbf{s}_2 + b\mathbf{c}_2\sin\theta_2\right]. \tag{16.88}$$

Since, by (16.78), $\sin\theta_1$ and $\sin\theta_2$ are of order $1/N$, observe that the two terms of order 1 on the right-hand side of (16.88) are of form $\pm(1-a)(1-b)$ and therefore cancel. Also, since $\sin\theta_j = \theta_j + O(N^{-3})$ for θ_j as given by (16.78), we substitute these relations into (16.88) and collect the order $1/N$ terms to find by direct asymptotics that:

$$\frac{\overline{w}_{0,N}}{\Lambda_1\Lambda_2} = \left\{a\sigma_1\mathbf{c}_1\mathbf{s}_2 + b\sigma_2\mathbf{s}_1\mathbf{c}_2 + (b-a)^2\mathbf{s}_1\mathbf{s}_2\right\}\frac{it}{N} + O(N^{-2}). \tag{16.89}$$

To render a partial check on the book-keeping procedure for (16.89), write out a formula for $e_j = e_j(a, b, r_N, y_N, z_N, \mathbf{s}_1, \mathbf{c}_1 \sin\theta_1)$ of (16.84) by leaving $\sin\theta_1$ as an auxiliary variable. Then $\sin\theta_j$ is replaced by the order $1/N$ term of (16.78), whereas $\cos\theta_j$ is defined exactly by (16.76). So, expand $e_1 q^* + e_2 w^*$ as

$$e_1\left([\mathbf{s}_2](q^*)\mathbf{s}_2 + [\mathbf{c}_2\sin\theta_2](q^*)\mathbf{c}_2\sin\theta_2\right) + e_2\left([\mathbf{s}_2](w^*)\mathbf{s}_2 + [\mathbf{c}_2\sin\theta_2](w^*)\mathbf{c}_2\sin\theta_2\right),$$

and apply a Taylor series about $t = 0$ to recover (16.89); see [15].

Now plug (16.89) into the formula for $g_{0,N}$ in Proposition 16.4(I)(1), apply Proposition 16.1(I)(1) to rewrite $\Pi_{0,N}$, and recall Λ_j in (16.81) and (16.85). So

$$g_{0,N} = \frac{\omega_a \tau(a,b) r z^2}{a(2-a)\tau_a}\left(\frac{\sin\theta_1\sin\theta_2[(N-f)a + fb - (N-1)ab]}{\left[a\sigma_1\mathbf{c}_1\mathbf{s}_2 + b\sigma_2\mathbf{s}_1\mathbf{c}_2 + (b-a)^2\mathbf{s}_1\mathbf{s}_2\right]\frac{it}{N} + O(N^{-2})}\right).$$

Finally, to find the limit as $N \to \infty$ of this last expression, we substitute (16.78) into the definitions (16.83) and (16.87), and again employ $\sin\theta_j \sim \theta_j$. We note: $\lim_{N\to\infty}\omega_a[a(2-a)]^{-1} r_N z_N^2 \tau(a,b)\tau_a^{-1} = 1$, since $\omega_a[\mathbf{1}] = a(2-a)$ and $\tau(a,b)[\mathbf{1}] = 1$. Since by assumption $f \sim \eta N$, we have $[(N-f)a + fb - (N-1)ab] \sim N[(1-\eta)a + \eta b - ab]$, and since by (16.78), $\theta_1\theta_2 \sim i^2\frac{\sigma_1\sigma_2}{(1-a)(1-b)}t^2 N^{-2}$, we obtain, as $N \to \infty$,

$$g_{0,N} \sim \frac{i^2 t^2}{N}\frac{\sigma_1\sigma_2}{(1-a)(1-b)}\frac{(1-\eta)a + \eta b - ab}{\left[a\sigma_1\mathbf{c}_1\mathbf{s}_2 + b\sigma_2\mathbf{s}_1\mathbf{c}_2 + (b-a)^2\mathbf{s}_1\mathbf{s}_2\right]\frac{it}{N}}.$$

Here, we use implicitly that $\sin(ix) = i\sinh(x)$ and $\cos(ix) = \cosh(x)$, so that by (16.78), (16.83), and (16.87), and by definition of κ_1 and κ_2, $\mathbf{s}_j \sim i\sinh(\kappa_j t)$, $j = 1, 2$, and $\mathbf{c}_j \sim \cosh(\kappa_j t)$, $j = 1, 2$. Thus, we obtain, $\lim_{N\to\infty} g_{0,N}[r, s_N, t_N] = \hat{\varphi}(t)$, for $\hat{\varphi}(t)$ given by (16.3).

Proof (*of Corollary* 16.1) We now assume that $a = b$ and consider the random variable $sY_{1,N} + tY_{2,N}$ defined by (16.4) in place of tX_N in the proof of Theorem 16.1. By the definition (16.4), we write

$$sY_{1,N} + tY_{2,N} = \frac{1}{N}\left(t\mathscr{L}'_N + \frac{(1-a)s - (2-a)t}{(1-a)}\mathscr{R}'_N + \frac{t-s}{(1-a)}\mathscr{V}'_N\right).$$

Accordingly, define

$$r_{s,t,N} := e^{i((1-a)s-(2-a)t)/((1-a)N)},\quad y_{s,t,N} := e^{i(t-s)/((1-a)N)},\quad z_{s,t,N} := e^{it/N}. \tag{16.90}$$

It suffices to prove that, for each fixed pair of real numbers $s, t \in \mathbb{R}$,

$$\lim_{N\to\infty} g_{0,N}(a,a)[r_{s,t,N}, y_{s,t,N}, z_{s,t,N}]$$

exists and is given by the right side of (16.5). Define $\theta = \theta_{s,t,N}$ via $\cos\theta = \beta_a/\sqrt{4x_a}$, where the functions β_a and x_a are composed with the complex exponential terms in (16.90). It follows by making a direct calculation that

$$\cos\theta = 1 + \frac{1}{2}\frac{(1-a)s^2 + at^2}{N^2} + O\left(\frac{1}{N^3}\right); \quad \theta = i\frac{\sqrt{(1-a)s^2 + at^2}}{N} + O\left(\frac{1}{N^3}\right). \tag{16.91}$$

Since the model is homogeneous, we need only apply the first line of (16.80) with $f := N$ to obtain

$$w_N^*(a) = (\sqrt{x_a}\sin\theta)^{-1}[az\tau_a]^N\left\{\sin N\theta - \sqrt{x_a}w_0^*(a)\sin(N-1)\theta\right\}. \tag{16.92}$$

Expand $\sin(N-1)\theta = \mathbf{s}\cos\theta - \mathbf{c}\sin\theta$, for $\mathbf{s} := \sin N\theta$ and $\mathbf{c} := \cos N\theta$. Put $\Lambda := (\sin\theta)^{-1}(az\tau_a)^{N-1}$. After direct calculation, we find

$$1 - \sqrt{x_a}w_0^*(a) = 1 - a + O(N^{-1}).$$

Therefore, by (16.92), we have

$$\frac{w_N^*(a)}{\Lambda} = \mathbf{s} - \sqrt{x_a}w_0^*(a)(\mathbf{s}\cos\theta - \mathbf{c}\sin\theta) = (1-a)\mathbf{s} + O\left(\frac{1}{N}\right)$$

Note that there is no cancelation of the order 1 term in this expression. Now plug $\frac{w_N^*(a)}{\Lambda}$ into (16.37) to obtain

$$g_{0,N} = \frac{\omega_a}{a(2-a)}rz\tau_a^{-1}\frac{(N-(N-1)a)\sin\theta}{(1-a)\mathbf{s} + O(N^{-1})}.$$

Finally apply the asymptotic expression for θ in (16.91) and let $N \to \infty$.

Proof (*of Theorem* 16.2) By the same reasoning given at the outset of the proof of Theorem 16.1, we may assume that $\mathbf{X}_0 = 0$. By the fact that the absolute value process starts afresh at the end of each excursion, we have that $1 + \mathscr{M}_N$ is a standard geometric random variable with success probability $P(\mathbf{H} \geq N)$. Thus

$$P(\mathscr{M}_N = \nu) = [P(\mathbf{H} < N)]^\nu\, P(\mathbf{H} \geq N), \quad \nu = 0, 1, 2, \ldots. \tag{16.93}$$

Let $\mathbf{L}_N$, $\mathbf{R}_N$, and $\mathbf{V}_N$, respectively, be random variables for the number of steps, runs, and short runs, in an excursion, given that the height of the excursion is at most $N-1$. Therefore, in distribution, we may write:

$$\mathscr{R}_N = \sum_{\nu=0}^{\mathscr{M}_N}\mathbf{R}^{(\nu)}, \quad \mathscr{V}_N = \sum_{\nu=0}^{\mathscr{M}_N}\mathbf{V}^{(\nu)}, \quad \mathscr{L}_N = \sum_{\nu=0}^{\mathscr{M}_N}\mathbf{L}^{(\nu)},$$

where $\mathbf{R}^{(1)}, \mathbf{R}^{(2)}, \ldots; \mathbf{V}^{(1)}, \mathbf{V}^{(2)}, \ldots;$ and $\mathbf{L}^{(1)}, \mathbf{L}^{(2)}, \ldots$, respectively, are sequences of independent copies of $\mathbf{R}_N$, $\mathbf{V}_N$, and $\mathbf{L}_N$. Since the random variables $\mathbf{R}_N$, $\mathbf{V}_N$, and $\mathbf{L}_N$ already have built into their definitions the condition $\{\mathbf{H} \le N-1\}$, the probability generating function $K_{N-1} = E\{r^{\mathbf{R}_N} y^{\mathbf{V}_N} z^{\mathbf{L}_N}\}$ is calculated by Theorem 16.3. Thus by (16.93), and by calculating a geometric sum there holds:

$$E\{r^{\mathscr{R}_N} y^{\mathscr{V}_N} z^{\mathscr{L}_N} u^{\mathscr{M}_N}\} = \sum_{\nu=0}^{\infty} P(\mathscr{M}_N = \nu)\left(uK_{N-1}\right)^{\nu} = \frac{P(\mathbf{H} \ge N)}{1 - uP(\mathbf{H} < N)K_{N-1}[r, y, z]}. \tag{16.94}$$

We define (r_N, y_N, z_N) by (16.75), and also set $u_N := e^{-ita(b-a)/[(1-a)(1-b)N]}$. By (16.9), it suffices to show that $\lim_{N\to\infty} E\{e^{it(1+1/N)\mathscr{X}_{N+1}}\} = \hat{\psi}(t)/\hat{\varphi}(t)$; see (16.98). We define θ_1 and θ_2 by (16.76), so that also (16.77)–(16.78) hold. By the statement of Theorem 16.3, we must replace the calculation of $\overline{w}_{0,N}$, starting with (16.79), with instead $\overline{w}_{1,N+1}$. However, by (16.53), (16.64), and (16.79), the difference in the two calculations is simply accounted for by replacing f by $f-1$ in the calculation of $\overline{w}_{0,N}$, because j in (16.64) for $\overline{w}_{1,N+1}$ is determined by $j = N+1-f = N-(f-1)$, so $\frac{1}{\Lambda_2}\overline{w}_{1,N+1} = d_1(f')q^* + d_2(f')w^*$ with $f' := f-1$ in place of f in both (16.53) and (16.85). This is reflected by the fact that, by Lemma 16.2, $\overline{w}_{1,N+1} = \overline{w}_{N,0}$. We must now also calculate $\overline{q}_N = d_{q,1}(f)q^*_{N-f+1}(b) + d_{q,2}(f)w^*_{N-f+1}(b)$ given by Lemma 16.5, with $d_{q,j}(f) = \mu_{j,1}q^*_{f-1}(a) + \mu_{j,2}q^*_f(a)$, $j = 1, 2$, defined by (16.63) in the proof of Lemma 16.5. In summary, $f' = f-1$ yields $(\dagger)\overline{q}_N = d_{q,1}(f'+1)q^*_{N-f'}(b) + d_{q,2}(f'+1)w^*_{N-f'}(b)$. Thus, because we simply replace f by $f-1$ in the required substitutions, and since $f \sim \eta N$, we will not change the name of f. With this understanding, we may use the calculation of $\overline{w}_{0,N}$ in (16.79)–(16.89) verbatim in place of the calculation of $\overline{w}_{1,N+1}$, and we will do this without changing the names of e_j, q^*, w^* and Λ_j; see (16.81) and (16.85). Further with this understanding, by $(\dagger)$, with f now recouping the role of f', and with q^* and w^* defined by (16.85), we have $\frac{1}{\Lambda_2}\overline{q}_N = d_{q,1}q^* + d_{q,2}w^*$ for

$$d_{q,j} := \mu_{j,1}q^*_f(a) + \mu_{j,2}q^*_{f+1}(a). \tag{16.95}$$

Here, by (16.13), (16.41) and (16.76), in analogy with (16.80), we have

$$q^*_f(a) = 2i\alpha_a^{-1}(az\tau_a)^f\left\{y^2\sin f\theta_1 - \sqrt{x_a}q^*_0(a)\sin(f-1)\theta_1\right\}$$
$$q^*_{f+1}(a) = 2i\alpha_a^{-1}(az\tau_a)^f\sqrt{x_a}\left\{y^2\sin(f+1)\theta_1 - \sqrt{x_a}q^*_0(a)\sin f\theta_1\right\}.$$

Denote $e_{q,j} := d_{q,j}/\Lambda_1$. Therefore, by (16.95), the definition of Λ_1 in (16.81), and these equations for $q^*_f(a)$ and $q^*_{f+1}(a)$,

$$e_{q,j} = \left(y^2\mu_{j,1} - \mu_{j,2}xq^*_0\right)\sin f\theta_1 + \sqrt{x}\left\{y^2\mu_{j,2}\sin(f+1)\theta_1 - \mu_{j,1}q^*_0\sin(f-1)\theta_1\right\}, \tag{16.96}$$

where $x = x_a$ and $q^*_0 = q^*_0(a)$. Rewrite (16.96) by applying the notations (16.83). Thus $e_{q,j}$ is written, with dependence on a suppressed, by

$$(y^2\mu_{j,1} - \mu_{j,2}xq_0^*)\mathbf{s}_1 + \sqrt{x}\{y^2\mu_{j,2}(\mathbf{s}_1\cos\theta_1 + \mathbf{c}_1\sin\theta_1) - \mu_{j,1}q_0^*(\mathbf{s}_1\cos\theta_1 - \mathbf{c}_1\sin\theta_1)\} \tag{16.97}$$

In summary, by (16.95), we have $\overline{q}_N/(\Lambda_1\Lambda_2) = e_{q,1}q^* + e_{q,2}w^*$, for $e_{q,j}$ in (16.97), and Λ_j defined by (16.81) and (16.85).

To guide the asymptotic expansions of (16.97), we rewrite (16.94) by substituting the last line of the proof of Theorem 16.3:

$$E\{e^{it(1+1/N)\mathscr{X}_{N+1}}\} = \frac{P(\mathbf{H} \geq N+1)\overline{w}_{1,N+1}}{\overline{w}_{1,N+1} - (1-a)u_N r_N^2 z_N^2 \overline{q}_N}. \tag{16.98}$$

It turns out that there is a cancelation in the order of the denominator of (16.98). That is, the leading order of each of $\overline{w}_{1,N+1}/(\Lambda_1\Lambda_2)$ and $\overline{q}_N/(\Lambda_1\Lambda_2)$ will be some order 1 trigonometric factor times it/N; in fact there holds $(1-a)\overline{q}_N/\overline{w}_{1,N+1} \sim 1$, as $N \to \infty$. Define

$$\Delta_N := \overline{w}_{1,N+1} - (1-a)u_N r_N^2 z_N^2 \overline{q}_N. \tag{16.99}$$

By direct calculation, we will establish that $\Delta_N/(\Lambda_1\Lambda_2) = O(N^{-2})$, and we find the exact coefficient of the order N^{-2} term.

For the asymptotics of (16.97) we may still treat $\sin\theta_1 = \theta_1 + O(N^{-3})$ by (16.78), but must render precisely the $O(N^{-2})$ term in $\cos\theta_1 = 1 + O(N^{-2})$ of (16.77). In an appendix to [15], we display the many terms of the book-keeping method for this problem. For the present, we simply exhibit the asymptotics of (16.99) obtained by machine computation with $\sin\theta_j$ substituted by the corresponding order $1/N$ term of (16.78):

$$\frac{\Delta_N}{\Lambda_1\Lambda_2} = \frac{1}{(1-a)(1-b)}\frac{t^2}{N^2}\{-ab\sigma_1\sigma_2\mathbf{c}_1\mathbf{c}_2 - a\sigma_1(a-b)^2\mathbf{c}_1\mathbf{s}_2 + a^2\sigma_1^2\mathbf{s}_1\mathbf{s}_2\} + O\left(\frac{1}{N^3}\right). \tag{16.100}$$

Finally, we compute the limit of the ratio (16.98) by the asymptotic relations (16.76), and by (16.89) and (16.100). Thus, because by (16.66) and Proposition 16.1 we have that $P(\mathbf{H} \geq N+1) \sim C_{a,b}N^{-1}$ for $C_{a,b} = ab/[(1-\eta)a + \eta b - ab]$, we find $E\{e^{it(1+1/N)\mathscr{X}_{N+1}}\}$ is asymptotic to

$$C_{a,b}N^{-1}\left\{[a\sigma_1\mathbf{c}_1\mathbf{s}_2 + b\sigma_2\mathbf{s}_1\mathbf{c}_2 + (b-a)^2\mathbf{s}_1\mathbf{s}_2]\frac{it}{N} + O\left(\frac{1}{N^2}\right)\right\}$$
$$\Big/\left\{\frac{1}{(1-a)(1-b)}[-ab\sigma_1\sigma_2\mathbf{c}_1\mathbf{c}_2 - a\sigma_1(a-b)^2\mathbf{c}_1\mathbf{s}_2 + a^2\sigma_1^2\mathbf{s}_1\mathbf{s}_2]\frac{t^2}{N^2} + O\left(\frac{1}{N^3}\right)\right\}.$$

As in the proof of Theorem 16.1, we have $\mathbf{c}_j \sim \cosh(\kappa_j t)$, and $\mathbf{s}_j \sim i\sinh(\kappa_j t)$, $j = 1, 2$. Therefore, with $\tilde{C}_{a,b} := (1-a)(1-b)C_{a,b}$, we obtain that $E\{e^{it(1+1/N)\mathscr{X}_{N+1}}\}$ has the following limit as $N \to \infty$, where we refer to (16.3) and statement of Theorem 16.2 for the definitions of $\hat{\varphi}(t)$ and $\hat{\psi}(t)$:

$$\lim_{N\to\infty} E\{e^{it(1+1/N)\mathscr{X}_N}\} = \frac{\tilde{C}_{a,b}}{t} \times \frac{(b\kappa_1\sigma_2 + a\kappa_2\sigma_1)t}{\hat{\varphi}(t)} \times \frac{\hat{\psi}(t)}{ab\sigma_1\sigma_2}.$$

We have $\tilde{C}_{a,b} = ab\sigma_1\sigma_2/(a\sigma_1\kappa_2 + b\sigma_2\kappa_1)$, so the proof is complete.

Corollary 16.4 *Assume $a = b$. Define*

$$Z_1 = \frac{1}{N}\left(\mathscr{R}_N - \frac{1}{(1-a)}\mathscr{V}_N + a\mathscr{M}_N\right);$$
$$Z_2 = \frac{1}{N}\left(\mathscr{L}_N - \frac{1}{(1-a)}\mathscr{R}_N + \frac{a}{1-a}\mathscr{M}_N\right) - Z_1.$$

Then, $\lim_{N\to\infty} E\{e^{i(sZ_1+tZ_2)}\} = \frac{\tanh(\sqrt{(1-a)s^2+at^2})}{\sqrt{(1-a)s^2+at^2}}$.

Proof One simplifies the lines of proof of Theorem 16.2. We leave details in an appendix to [15].

For illustration of Theorem 16.1, see Fig. 4.

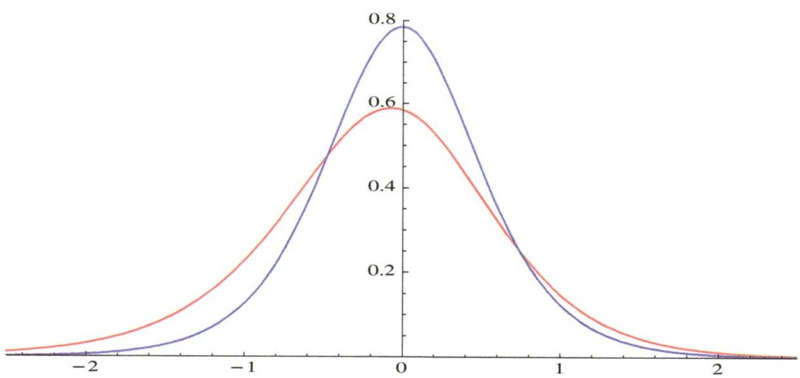

Fig. 4 The density $\varphi(x)$ whose transform $\hat{\varphi}(t) = \int_{-\infty}^{\infty} e^{itx}\varphi(x)\,dx$ is given by (16.6) for $a = \frac{1}{4}$, and the density $\frac{\pi}{4}\operatorname{sech}^2(\pi x/2)$, that is instead determined by $a = \frac{1}{2}$ and corresponds to simple random walk. Numerically, the mean of φ is $\int_{\infty}^{\infty} x\varphi(x)\,dx = -\frac{1}{4}$, and $\arg\max_x \varphi(x) = -0.131619$

Acknowledgements The author wishes to thank the referee who gave extensive suggestions that led to many improvements in the presentation. The companion document [15] would not have come into the public domain without the referee's helpful (and exuberant!) insight.

References

1. Bálint, P., Tóth, B., Tóth, P.: On the zero mass limit of tagged particle diffusion in the 1-d Rayleigh-gas. J. Stat. Phys. **127**, 657–675 (2007)
2. Banderier, C., Flajolet, P.: Basic analytic combinatorics of discrete lattice paths. Theor. Comput. Sci. **281**, 37–80 (2002)
3. Banderier, C., Nicodème, P.: Bounded discrete walks. In: Discrete Mathematics and Theoretical Computer Science, Proceedings, vol. AM, pp. 35–48 (2010)
4. Bousquet-Mélou, M.: Discrete excursions. Sém. Lothar. Combin. 57, Art. B57d, 23 pp. (2008)
5. Chihara, T.S.: An Introduction to Orthogonal Polynomials. Gordon and Breach, New York (1978)
6. de Bruijn, N.G., Knuth, D.E., Rice, S.O.: The average height of planted plane trees. In: Read, R.C. (ed.) Graph Theory and Computing, pp. 15–22. Academic, New York (1972)
7. Deutsch, E.: Dyck path enumeration. Discrete Math. **204**, 167–202 (1999)
8. Feller, W.: An Introduction to Probability Theory and Its Applications, vol. I, 3rd edn. Wiley, New York (1968)
9. Flajolet, P.: Combinatorial aspects of continued fractions. Discrete Math. **32**, 125–161 (1980)
10. Flajolet, P., Sedgewick, R.: Analytic Combinatorics. Cambridge University Press, Cambridge (2009)
11. Ismail, M.E.H.: Classical and Quantum Orthogonal Polynomials in One Variable. Cambridge University Press, Cambridge (2005)
12. Krattenthaler, C.: Lattice path enumeration. In: Bóna, M. (ed.) Handbook of Enumerative Combinatorics. CRC Press, Boca Raton (2015)
13. Mohan, C.: The gambler's ruin problem with correlation. Biometrika **42**, 486–493 (1955)
14. Morrow, G.J.: Laws relating runs and steps in gambler's ruin. Stoch. Proc. Appl. **125**, 2010–2025 (2015)
15. Morrow, G.J.: *Mathematica* calculations for laws relating runs, long runs, and steps in gambler's ruin, with persistence in two strata (2018). http://www.uccs.edu/gmorrow
16. On-Line Encyclopedia of Integer Sequences. http://oeis.org/
17. Poll, D.B., Kilpatrick, Z.P.: Persistent search in single and multiple confined domains: a velocity-jump process model. J. Stat. Mech. **2016**, 053201, 21 pp. (2016)
18. Reed, M., Simon, B.: Methods of Modern Mathematical Physics, vol. II: Fourier Analysis, Self-Adjointness. Academic, New York (1972)
19. Spitzer, F.: Principles of Random Walk. Van Nostrand, Princeton (1964)
20. Swamy, M.N.S.: Generalized Fibonacci and Lucas polynomials and their associated diagonal polynomials. Fibonacci Q. **37**, 213–222 (1999)
21. Szász, D., Tóth, B.: Persistent random walks in a one-dimensional random environment. J. Stat. Phys. **37**, 27–38 (1984)
22. Wolfram, S.: Mathematica. http://www.wolfram.com/mathematica/

Paired Patterns in Lattice Paths

Ran Pan and Jeffrey B. Remmel

Abstract Let $\mathscr{L}_n$ denote the set of all paths from $[0, 0]$ to $[n, n]$ which consist of either unit north steps N or unit east steps E or, equivalently, the set of all words $L \in \{E, N\}^*$ with n E's and n N's. Given $L \in \mathscr{L}_n$ and a subset A of $[n] = \{1, \ldots, n\}$, we let $ps_L(A)$ denote the word that results from L by removing the ith occurrence of E and the ith occurrence of N in L for all $i \in [n] - A$, reading from left to right. Then, we say that a paired pattern $P \in \mathscr{L}_k$ occurs in L if there is some $A \subseteq [n]$ of size k such that $ps_L(A) = P$. In this paper, we study the generating functions of paired pattern matching in $\mathscr{L}_n$.

Keywords Lattice paths · Words · Paired patterns · Generating function

2010 Mathematics Subject Classification Primary: 05A15 · Secondary: 05A19

1 Introduction

Let $\mathscr{L}_n$ denote the set of all paths from $[0, 0]$ to $[n, n]$ which consist of either unit north $[0, 1]$ steps or unit east $[1, 0]$ steps. The six paths in $\mathscr{L}_2$ are pictured at the top of Fig. 2. Clearly,

$$|\mathscr{L}_n| = \binom{2n}{n}.$$

We code elements in $\mathscr{L}_n$ as words over the alphabet $\{N, E\}$ with n N's and n E's. Given $L \in \mathscr{L}_n$ and a subset A of $[n] = \{1, \ldots, n\}$, we let $ps_L(A)$ denote the word that results from L by removing the ith occurrence of E and the ith occurrence of N in L for all $i \in [n] - A$, reading from left to right. For example, suppose

R. Pan (✉) · J. B. Remmel
Department of Mathematics, University of California, San Diego, La Jolla, CA 92093-0112, USA
e-mail: ran.pan.math@gmail.com

J. B. Remmel
e-mail: remmel@math.ucsd.edu

G. E. Andrews et al. (eds.), *Lattice Path Combinatorics and Applications*, Developments in Mathematics 58, https://doi.org/10.1007/978-3-030-11102-1_17

$L = NEEENN \in \mathscr{L}_3$, then $ps_L(\{1\}) = NE$, $ps_L(\{2\}) = EN$, $ps_L(\{3\}) = EN$, $ps_L(\{1, 2\}) = ps_L(\{1, 3\}) = NEEN$, and $ps_L(\{2, 3\}) = EENN$. We shall think of a word in $\{N, E\}$ with n N's and n E's as a *paired pattern* where the ith occurrence of E is paired with the ith occurrence of N, reading from left to right, for $i = 1, \ldots, n$.

Definition 17.1 Given a set of paired patterns $\Gamma \subseteq \mathscr{L}_k$ and word $L \in \mathscr{L}_n$, we say that

1. Γ **occurs** in L if there is an $A \subseteq [n]$ of size k such that $ps_L(A) \in \Gamma$,
2. There is a **Γ-match in L starting at the jth paired step** if $ps_L(\{j, j+1, j+2, \ldots, j+k-1\}) \in \Gamma$.
3. L **avoids** Γ if there is no Γ-matches in L.

Alternatively, we can code a path L as a $2 \times n$ array $T(L)$ where the bottom row of T consists of the positions of the east steps, reading from left to right, and the top row of T consists of the positions of the north steps, reading from left to right. We let $T(L)_{k,1}$ denote the element in the kth column of the bottom row of $T(L)$, and let $T(L)_{k,2}$ denote the element in the kth column of the top row. Given any $2 \times n$ array S filled with pairwise distinct positive integers, let the reduction of S, $\text{red}(S)$, denote the $2 \times n$ array which results from S by replacing the ith smallest integer in S by i. An example of the reduction operation red is pictured at the bottom of Fig. 1.

It is then easy to see that given $L \in \mathscr{L}$ and $A \subseteq [n]$, the array associated with $ps_L(A)$ corresponds to the array obtained by taking the columns in $T(L)$ corresponding to A and reducing. This process is pictured in Fig. 1. This given, we can restate our pattern matching conditions in terms of $2 \times n$ arrays. That is, the $\mathscr{T}_n$ denote the set of all $2 \times n$ arrays T filled with the numbers $1, 2 \ldots, 2n$ such that the rows of T are increasing reading from left to right. Given $T \in \mathscr{T}_n$ and $A \subseteq [n]$, we let $T[A]$ be the array that results by removing the columns corresponding to elements in $[n] - A$. For example, if $T = T(L)$ is the array pictured in Fig. 1, then $T[\{1, 4, 5\}]$ is pictured at the bottom left of Fig. 1.

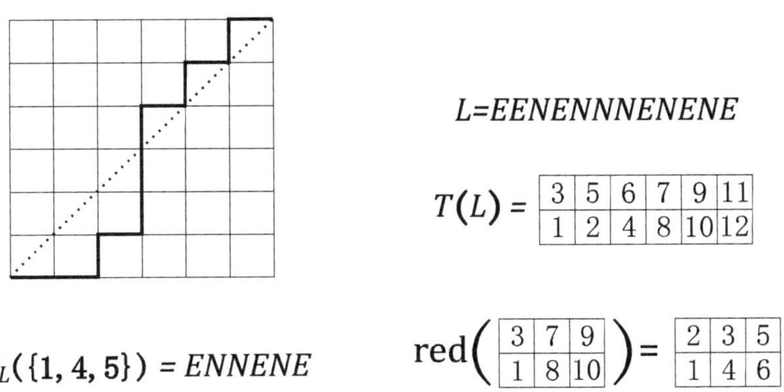

Fig. 1 Correspondence between paths and $2 \times n$ arrays

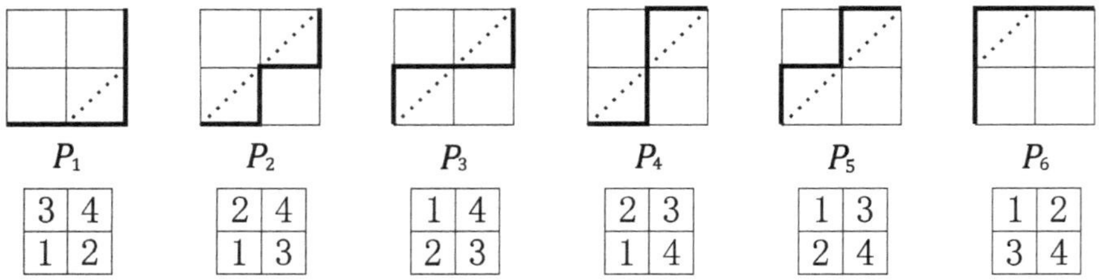

Fig. 2 $\mathscr{L}_2 = \{P_1, P_2, P_3, P_4, P_5, P_6.\}$

Then from the point of view of arrays in $\mathscr{T}_n$, our paired pattern matching conditions can be stated as follows.

Definition 17.2 Given a set of $2 \times k$ arrays $\Gamma \subseteq \mathscr{T}_k$ and a $2 \times n$ array $S \in \mathscr{T}_n$, we say that

1. Γ **occurs** in S if there is an $A \subseteq [n]$ of size k such that $\mathrm{red}(S[A]) \in \Gamma$.
2. There is a Γ**-match in** S **starting at column** j if $\mathrm{red}(S[\{j, j+1, j+2, \ldots, j+k-1\}]) \in \Gamma$.
3. S **avoids** Γ if there is no Γ-matches in S.

Note that from this point of view, Γ-matches correspond naturally to consecutive patterns matches in $2 \times n$ arrays. Results about consecutive patterns in arrays can be found in [4]. We let Γ-mch(L) denote the number of Γ-matches in L. If Γ-mch(F)=0, then we will say that L has no Γ-matches. If $\Gamma = \{P\}$ is a singleton, then we will write P-mch(L) for Γ-mch(L).

For example, there are six possible patterns of length four, as pictured in Fig. 2, namely $P_1 = EENN$, $P_2 = ENEN$, $P_3 = NEEN$, $P_4 = ENNE$, $P_5 = NENE$, $P_6 = NNEE$.

We note that paired patterns differ from classic consecutive patterns in words (e.g., [2, 10, 11]). Paired patterns actually describe relationships between paths and the diagonal $y = x$, the subdiagonal $y = x - 1$, and the superdiagonal $y = x + 1$. For our purposes, the set of Dyck paths $\mathscr{D}_n$ is the set of paths of $\mathscr{L}_n$ which stay weakly below the diagonal $y = x$. For example, in $\mathscr{L}_2$, the only two Dyck paths are P_1 and P_2. Actually, a path L is a Dyck path if and only if L has no $(\mathscr{L}_2 - \{P_1, P_2\})$-matches. More details and geometric interpretation of paired patterns can be found in Sect. 2.

By Theorems 17.2 and 17.3, we see that certain paired patterns are equivalent to returns (bouncings) and crossings of a path. These classical statistics have been studied in literature such as [3, 5–9, 13].

In this paper, we will focus on paired patterns of length 4 and pattern matching for subsets of these pattern. In other words, we would study generating functions of the form

$$F_{P_k}(x,t) := 1 + \sum_{n \geq 1} t^n \sum_{L \in \mathscr{L}_n} x^{P_k\text{-mch}(L)}, \tag{17.1}$$

where $k \in \{1, 2, 3, 4, 5, 6\}$, and

$$F_{\Delta}(\mathbf{x}, t) := 1 + \sum_{n \geq 1} t^n \sum_{L \in \mathscr{L}_n} \left(\prod_{j \in \Delta} x_j^{P_j\text{-mch}(L)} \right), \tag{17.2}$$

where Δ is a subset of $\{1, 2, 3, 4, 5, 6\}$.

Note there are two basic symmetries in our study of paired patterns. First, one can reflect a path $L \in \mathscr{L}_n$ about the diagonal $y = x$ which has the effect of interchanging E's with N's in the word of L or interchanging the rows in the diagram of $T(L)$ of L. Second, one can rotate the path by 180 degrees which has the effect of interchanging the E's and N's and then reversing the word of L. These symmetries immediately show that

$$\begin{aligned} F_{P_1}(x, t) &= F_{P_6}(x, t), \\ F_{P_2}(x, t) &= F_{P_5}(x, t), \text{ and} \\ F_{P_3}(x, t) &= F_{P_4}(x, t). \end{aligned}$$

Thus we need only to compute three generating functions of the form $F_{P_k}(x, t)$.

We can also give geometric interpretations to P_k-matches for each k. For example, we shall show that the number of P_1-matches in a path $L \in \mathscr{L}_n$ is the number of east steps that are below the subdiagonal $y = x - 1$. The formulas for the generating functions that we will derive then lead to many interesting bijective problems. For example, we will show that the total number of east steps that lie below the subdiagonal $y = x - 1$ over all paths $L \in \mathscr{L}_n$ equals the sum of the areas under all Dyck paths in $\mathscr{D}_n$.

The outline of this paper is as follows. In Sect. 2, we shall give the geometric interpretations of the number of P_k-matches in paths in $\mathscr{L}_n$. In Sect. 3, we shall derive closed formulas for the generating functions $F_{P_k}(x, t)$ for $k = 1, \ldots, 6$ and explore some of the consequences of such formulas. In Sect. 4, we derive a number of formulas for $F_{\Delta}(\mathbf{x}, t)$ for certain $\Delta \subseteq \{P_1, \ldots, P_6\}$. Finally, in Sect. 5, we discuss topics for future research such as finding bijections between paths with certain pattern matching condition and other known objects, extending the definition of paired patterns to Delannoy paths and finding generating functions $F_P(x, t)$ for paths P of length greater than 4.

2 The Geometric Interpretation of the Number of P_k-Matches

In this section, we shall give our geometric interpretations of P_k-matches for $k = 1, \ldots, 6$.

Theorem 17.1 *Let $L \in \mathscr{L}_n$. Then the number of P_1-matches in L is the number of east steps below the subdiagonal $y = x - 1$. Hence, by symmetry, the number of P_6-matches in L is the number of north steps above the superdiagonal $y = x + 1$.*

Proof Suppose that the ith east step in L occurs below the subdiagonal $y = x - 1$ and that this step corresponds to the jth letter in the word $w_1 \dots w_{2n}$ of L. Then it must be the case that the number of E's in $w_1 \dots w_j$ exceeds the number of N's in $w_1 \dots w_j$ by at least two. This means that when we restrict the diagram $T(L)$ to the letters $1, \dots, j$, then there are no elements in the $(i-1)^{th}$ and i^{th} columns of the bottom row. This means that $T(L)_{i-1,1} < T(L)_{i,1} < T(L)_{i-1,2} < T(L)_{i,2}$ so that $\mathrm{red}(T(L)[\{i-1, i\}])$ matches the array for P_1. Hence each such east step represents a P_1-match in L.

On the other hand suppose $\mathrm{red}(T(L)[\{i-1, i\}])$ matches the array for P_1. If $T(L)_{i,1} = j$, then in the word $w = w_1 \dots w_{2n}$ of L, the jth E, reading from right to left, must be preceded by at most $i - 2$ north steps which means that the east step corresponding to w_j is below the subdiagonal $y = x - 1$.

Given a path $L \in \mathscr{L}_n$, we let $\mathrm{bounce}^-(L)$ denote the number of points $[x, x]$ on L such that the preceding step is a north step N and the following step is an east step E. This means that the path bounces off the diagonal to the right. We let $\mathrm{bounce}^+(L)$ denote the number of points $[y, y]$ on L which is preceded by an east step E and followed by a north step N. This means that the path bounces off the diagonal to the left. For example, for the path L pictured in Fig. 3, the points [6, 6], [7, 7], and [8, 8] are points preceded by a north step and followed by an east step so that $\mathrm{bounce}^-(L) = 3$ and the point [4, 4] is preceded by an east step and followed by a north step so that $\mathrm{bounce}^+(L) = 1$.

Theorem 17.2 *Let $L \in \mathscr{L}_n$. Then the number of P_2-matches in L equals* $\mathrm{bounce}^-(L)$. *Hence, by symmetry, the number of P_5-matches in L equals* $\mathrm{bounce}^+(L)$.

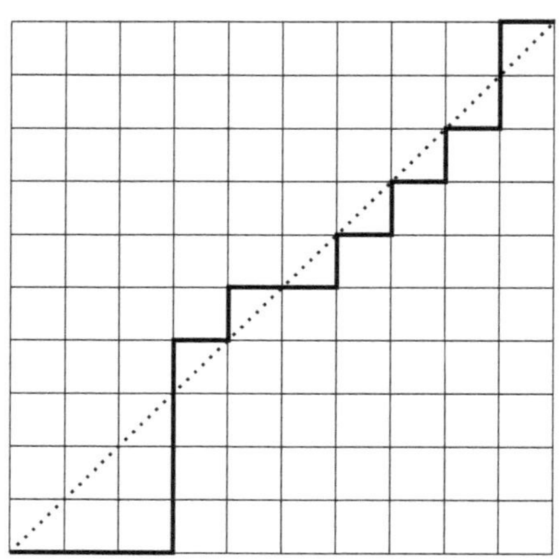

Fig. 3 $\mathrm{bounce}^+(L) = 1$, $\mathrm{bounce}^-(L) = 3$, $\mathrm{cross}^h(L) = 1$, and $\mathrm{cross}^v(L) = 2$

Proof Consider the diagram $T(L)$ of L. Then a P_2-match in L corresponds to a pair of columns $i-1$ and i such red$(T(L)[\{i-1,i\}])$ matches the array for P_2. This means that $T(L)_{i-1,1} < T(L)_{i-1,2} < T(L)_{i,1} < T(L)_{i,2}$. Now suppose that $T(L)_{i-1,2} = x$. It follows that all the elements in the columns to the right of x must be greater than x and all the elements in the columns to the left of x must be less than x. Since $x > T(L)_{i-1,1}$ it follows that $x = 2(i-1)$. Similarly, if $T(L)_{i,1} = y$, then all the elements in the columns to the right of y must be greater than y and all the elements in the columns to the left of y must be less than y. Since $y < T(L)_{i-1,1}$ it follows that $y = 2(i-1)+1$. This means that in the word of $w = w_1 \ldots w_{2n}$ of L, $w_{2(i-1)} = N$ and is preceded by i east steps and $i-1$ north steps so that point $[i,i]$ is on the path of L and is preceded by a north step and followed by an east step.

Vice versa, if $[i,i]$ is on the path of L and is preceded by a north step and followed by an east step, then it is easy to see that in the array $T(L)$ of L, we must have that $T(L)_{i-1,2} = 2(i-1)$ and $T(L)_{i,1} = 2(i-1)+1$ so that red$(T(L)[\{i-1,i\}])$ must match P_2.

Given a path $L \in \mathscr{L}_n$, we let cross$^h(L)$ denote the number of points $[x,x]$ on L such that the preceding step is an east step E and the following step is an east step E. This means that the path crosses the diagonal horizontally. We let cross$^v(L)$ denote the number of points $[y,y]$ on L which is preceded by a north step N and followed by a north step N. This means that the path crosses the diagonal vertically. For example, for the path L pictured in Fig. 3, there is a horizontal crossing of the diagonal at the point $[5,5]$ so that cross$^h(L) = 1$ and there are vertical crossings at the points $[4,4]$ and $[9,9]$ so that cross$^v(L) = 2$.

Theorem 17.3 *Let $L \in \mathscr{L}_n$. Then the number of P_3-matches in L equals* cross$^h(L)$. *Hence, by symmetry, the number of P_4-matches in L equals* cross$^v(L)$.

Proof Consider the diagram $T(L)$ of L. Then a P_3-match in L corresponds to a pair of columns $i-1$ and i such that red$(T(L)[\{i-1,i\}])$ matches the array for P_3. This means that $T(L)_{i-1,2} < T(L)_{i-1,1} < T(L)_{i,1} < T(L)_{i,2}$. Now suppose that $T(L)_{i-1,1} = x$. It follows all the elements in the columns to right of x must be greater than x and all elements in the columns to the left of x must be less than x. Since $x > T(L)_{i-1,2}$ it follows that $x = 2(i-1)$. Similarly, if $T(L)_{i,1} = y$. It follows all the elements in the columns to the right of y must be greater than y and all the elements in the columns to the left of y must be less than y. Since $y < T(L)_{i-1,1}$ it follows that $y = 2(i-1)+1$. This means that in the word of $w = w_1 \ldots w_{2n}$ of L, $w_{2(i-1)} = E$ and is preceded by $i-1$ east steps and i north steps so that point $[i,i]$ is on the path of L and is preceded by an east step and followed by an east step.

Vice versa, if $[i,i]$ is on the path of L and is preceded by an east step and followed by an east step, then it is easy to see that in the array $T(L)$ of L, we must have that $T(L)_{i-1,1} = 2(i-1)$ and $T(L)_{i,1} = 2(i-1)+1$ so that red$(T(L)[\{i-1,i\}]$ must match P_3.

3 Generating Functions

Let $F_i(x,t) = F_{P_i}(x,t)$ for $i = 1, \ldots, 6$. The goal of this section is to compute the generating functions $F_k(x,t)$ for $k = 1, \ldots, 6$.

To obtain a recurrence for Dyck paths, the usual way is to factorize Dyck paths based on where it returns to the diagonal for the first time. Application of this decomposition can be found in many papers focused on lattice path enumeration such as [2, 3, 11]. We shall show that similar ideas allow us to obtain recurrences for the number of P_k-matches.

3.1 *Pattern P_1*

For pattern P_1, consider the ordinary generating function $F_1(x,t)$ as follows,

$$F_1(x,t) := 1 + \sum_{n \geq 1} t^n \sum_{L \in \mathscr{L}_n} x^{P_1\text{-mch}(L)}. \tag{17.3}$$

We know for a path L, $P_1\text{-mch}(L)$ is equal to the number of east steps below subdiagonal $y = x - 1$. By our observation in the introduction, $F_1(x,t) = F_6(x,t)$.

We split the analysis of $P_1\text{-mch}(L)$ into two cases. Case 1 is when $P_1\text{-mch}(L) = 0$, that is, path L stays above $y = x - 1$. It is easy to see that the number of paths in $\mathscr{L}_n$ above $y = x - 1$ is $C_{n+1} = \frac{1}{n+2}\binom{2n+2}{n+1}$, the $(n+1)^{th}$ Catalan number. Case 2 is when $P_1\text{-mch}(L) > 0$, that is, path L has at least one east step below $y = x - 1$. Now consider the first time the path touches $y = x - 1$ and the first time after that where the path hits a point $[i, i]$ on the diagonal. It is easy to see that the two steps preceding the point $[i, i]$ must be north steps. An example of recurrence is pictured in Fig. 4, where two boxes are the two positions mentioned above and three diagonal dots stand for a whatever path follows the second box.

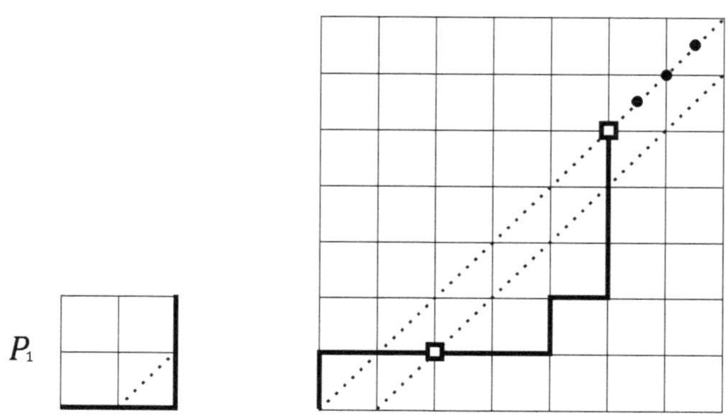

Fig. 4 An example of recurrence based on P_1

Suppose the position of the first box has coordinates $[j, j-1]$, $j \geq 1$, clearly there are C_j ways to choose steps before reaching $[j, j-1]$. Similarly, suppose the position of the second box has coordinates $[i+j, i+j]$, $i \geq 1$, clearly there are C_i ways to choose steps between $[j, j-1]$ and $[i+j, i+j]$.

Since the ordinary generating function for Catalan numbers is

$$C(x) = \sum_{n\geq 0} C_n x^n = \frac{1-\sqrt{1-4x}}{2x} = 1 + x + 2x^2 + 5x^3 + 14x^4 + 42x^5 + \cdots, \tag{17.4}$$

it follows that

$$\begin{aligned} F_1(x,t) &= \sum_{n\geq 0} C_{n+1}t^n + \sum_{i\geq 1}\sum_{j\geq 1} C_i C_j x^i t^{i+j} F_1(x,t) \\ &= \frac{C(t)-1}{t} + \sum_{i\geq 1} C_i (xt)^i \sum_{j\geq 1} C_j t^j F_1(x,t) \\ &= \frac{C(t)-1}{t} + \sum_{i\geq 1} C_i (xt)^i (C(t)-1) F_1(x,t) \\ &= \frac{C(t)-1}{t} + (C(xt)-1)(C(t)-1)F_1(x,t), \end{aligned}$$

and therefore by Eq. (17.4),

$$F_1(x,t) = \frac{C(t)-1}{t\,(1-(C(xt)-1)(C(t)-1))} = \frac{2x}{x\sqrt{1-4t}+\sqrt{1-4xt}+x-1}.$$

Some initial terms of $F_1(x,t)$ are

$$\begin{aligned} F_1(x,t) = 1 &+ 2t + (x+5)t^2 + (2x^2+4x+14)t^3 + (5x^3+9x^2+14x+42)t^4 \\ &+ (14x^4+24x^3+34x^2+48x+132)t^5 \\ &+ (42x^5+70x^4+95x^3+123x^2+165x+429)t^6 + \cdots. \end{aligned}$$

The number of paths in $\mathscr{L}_n$ avoiding P_1 is C_{n+1}, $(n+1)^{th}$ Catalan number. In general, the number of paths having exactly k P_1-matches has the generating function as follows,

$$\frac{1}{k!}\left.\frac{\partial^k F_1(x,t)}{\partial x^k}\right|_{x=0} \quad \text{or} \quad \frac{1}{k!}\left.\frac{\partial^k F_1(x,t)}{\partial x^k}\right|_{x\to 0}.$$

We evaluate the derivative at $x = 0$ or when $x = 0$ is a singularity of the derivative, we take the limit as x approaches zero. For example,

$$\left.\frac{\partial F_1(x,t)}{\partial x}\right|_{x\to 0} = \left(\frac{-1+\sqrt{1-4t}+2t}{2t}\right)^2$$
$$= t^2 + 4t^3 + 14t^4 + 48t^5 + 165t^6 + 572t^7 + 2002t^8 + \cdots.$$

The sequence $1, 4, 14, 48, 165, 572, 202, \ldots$ is sequence A002057 in the OEIS [14]. It have a number of combinatorial interpretations including the number of standard tableaux of shape $(n+2, n-1)$ and, with an offset of 4, the number of 123-avoiding permutations on $\{1, 2, \ldots, n\}$ for which the integer n is in the fourth spot. It follows from the hook length formula for the number of standard tableaux that the number of paths L in $\mathscr{L}_n$ with exactly one east below the subdiagonal $y = x - 1$ equals $4((2n-1)!)/((n-2)!(n+2)!)$ and is equal to the number of 123-avoiding permutations on $\{1, 2, \ldots, n+2\}$ for which the integer n is in the fourth spot.

Similarly, one can obtain the generating function for the number of paths having exactly two east steps below the subdiagonal as follows,

$$\left.\frac{1}{2!}\frac{\partial^2 F_1(x,t)}{\partial x^2}\right|_{x\to 0} = -\frac{\left(-1+\sqrt{1-4t}-2t\right)\left(-1+\sqrt{1-4t}+2t\right)^2}{8t^2}$$
$$= 2t^3 + 9t^4 + 34t^5 + 123t^6 + 440t^7 + 1573t^8 + 5642t^9 + \cdots.$$

The sequence $2, 9, 34, 123, 440, 1573, 5642, \ldots$ is sequence A120989 in the OEIS [14]. The nth term in this sequence counts the level of the first leaf in preorder of a binary tree, summed over all binary trees with $n-2$ edges. Thus the number of paths L in $\mathscr{L}_n$ with exactly two east steps below the subdiagonal $y = x - 1$ equals the sum of the level of the first leaf in preorder over all binary trees with $n-2$ edges. We leave open the problem of giving a bijective proof of this fact.

Next, we shall answer the following question, for a random path $L \in \mathscr{L}_n$, what is the expectation of P_1-mch(L), or in other words, on average how many east steps of L are below $y = x - 1$? Consider that

$$\left.\frac{\partial F_1(x,t)}{\partial x}\right|_{x=1} = -\frac{-1+\sqrt{1-4t}+2t}{2(1-4t)^{3/2}}$$
$$= t^2 + 8t^3 + 47t^4 + 244t^5 + 1186t^6 + 5536t^7 + \cdots. \quad (17.5)$$

For example, a random $L \in \mathscr{L}_7$, expectation of P_1-mch(L)

$$\mathbb{E}[P_1\text{-mch}(L) : L \in \mathscr{L}_7] = \frac{5536}{\binom{14}{7}} \approx 1.63,$$

which implies in average there are roughly 1.63 east steps below $y = x - 1$.

In general, by the OEIS, the coefficient of t^n in Eq. (17.5) has formula $\frac{1}{2}((n+1)\binom{2n}{n} - 4^n)$. Using Stirling's formula to approximate $n!$, we have

$$\mathbb{E}[P_1\text{-mch}(L) : L \in \mathscr{L}_n] = \frac{(n+1)\binom{2n}{n} - 4^n}{2\binom{2n}{n}} \sim \frac{n+1}{2} - \sqrt{\pi n}, \tag{17.6}$$

which implies when n is large, for a random path $L \in \mathscr{L}_n$, the expected number of east steps that lie below $y = x - 1$ is $\frac{n+1}{2} - \sqrt{\pi n}$.

The sequence $1, 8, 47, 244, 1186, 5536, \ldots$ from Eq. (17.5) is sequence A029760 and A139262 in the OEIS [14]. A029760 and A139262 count the total area under all the Dyck paths from $[0, 0]$ to $[n, n]$, the total number of inversions in all 132-avoiding permutations of length n and also total number of two-element anti-chains over all ordered trees on n edges. Again we leave open the problem of finding a bijective proof of these facts. We suspect that finding a bijective proof is a challenge because Dyck paths, 132-avoiding permutations, and ordered trees are all Catalan objects while lattice paths in $\mathscr{L}_n$ are not.

Next, by manipulating $F_1(x, t)$ we can also find the number of paths having an even number of east steps below the subdiagonal $y = x - 1$. The generating function is as follows,

$$\frac{1}{2}(F_1(1,t) + F_1(-1,t)) = 1 + 2t + 5t^2 + 16t^3 + 51t^4 + 180t^5 + 622t^6 + 2288t^7 + \cdots.$$

Similarly, the generating function for the number of paths having an odd number of east steps below the subdiagonal $y = x - 1$ is

$$\frac{1}{2}(F_1(1,t) - F_1(-1,t)) = t^2 + 4t^3 + 19t^4 + 72t^5 + 302t^6 + 1144t^7 + 4643t^8 + \cdots.$$

Neither of the series correspond to entries in the OEIS [14].

3.2 Pattern P_2

For pattern P_2, P_2-mch(L) counts the number of times L bounces off the diagonal $y = x$ to the right, in other words, $P_2\text{-mch}(L) = \text{bounce}^-(L)$. We shall study

$$F_2(x, t) := 1 + \sum_{n \geq 1} t^n \sum_{L \in \mathscr{L}_n} x^{P_2\text{-mch}(L)}. \tag{17.7}$$

As we observed in the introduction, $F_2(x, t) = F_5(x, t)$.

We shall consider two cases. Case 1 is the paths that start with an east step and Case 2 is the paths that start with a north step. We define

$$G_2(x, t) := \sum_{n \geq 1} t^n \sum_{L \in \mathscr{L}_n \text{ starting with } E} x^{P_2\text{-mch}(L)}$$

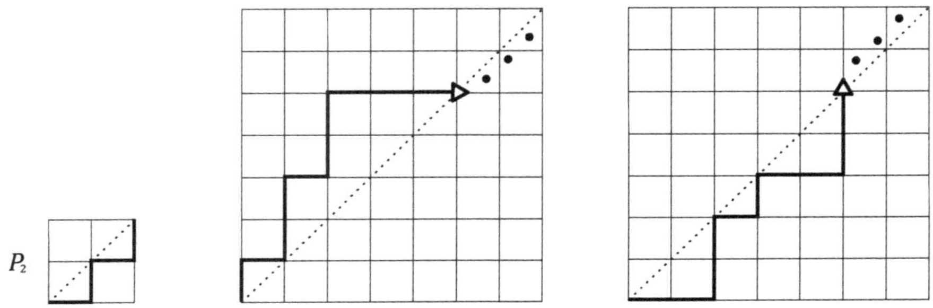

Fig. 5 An example of recurrence based on P_2

and

$$H_2(x,t) := \sum_{n\geq 1} t^n \sum_{L\in\mathcal{L}_n \text{ starting with } N} x^{P_2\text{-mch}(L)}.$$

Clearly, $F_2(x,t) = 1 + G_2(x,t) + H_2(x,t)$. For $H_2(x,t)$, we consider where is the first time a path starting with a north step crosses the diagonal $y = x$ horizontally. In the middle diagram of Fig. 5, the three dots stand for a path starting with 'E' or an empty path.

$$H_2(x,t) = \sum_{j\geq 1} C_j t^j \, (G_2(x,t)+1) = (C(t)-1)(G_2(x,t)+1). \tag{17.8}$$

Similarly, for $G_2(x,t)$, we consider where is the first time a path starting with an east step crosses the diagonal $y = x$ vertically. In the right diagram of Fig. 5, three dots stand for a path starting with 'N' or an empty path. Since we want to keep track of P_2-matches, here we need to introduce Catalan's triangle $C_{i,j}$, which is the number of Dyck paths in $\mathcal{L}_{2j}$ with i returns to the diagonal [2]. By [2], $C_{i,j}$ has generating function as follows,

$$C(x,t) = \sum_{i\geq 0}\sum_{j\geq 0} C_{i,j} x^i t^j = 1 + \frac{1-\sqrt{1-4t}}{(\sqrt{1-4t}-1)x+2} \tag{17.9}$$

Then

$$G_2(x,t) = \sum_{i\geq 0}\sum_{j\geq 1} C_{i,j} x^i t^j \, (H_2(x,t)+1) = \frac{1-\sqrt{1-4t}}{(\sqrt{1-4t}-1)x+2}(H_2(x,t)+1). \tag{17.10}$$

By Eq. (17.8),

$$G_2(x,t) = \frac{1-\sqrt{1-4t}}{(\sqrt{1-4t}-1)x+2}((C(t)-1)(H_2(x,t)+1)+1). \tag{17.11}$$

We can then solve $G_2(x,t)$ to obtain that

$$G_2(x,t) = \frac{\frac{1-\sqrt{1-4t}}{(\sqrt{1-4t}-1)x+2}C(t)}{1+\frac{1-\sqrt{1-4t}}{(\sqrt{1-4t}-1)x+2}-\frac{1-\sqrt{1-4t}}{(\sqrt{1-4t}-1)x+2}C(t)}$$
$$= \frac{(\sqrt{1-4t}-1)^2}{2(\sqrt{1-4t}(t(x-1)+1)-t(x-5)-1)}.$$

Then we have

$$\begin{aligned}F_2(x,t) &= 1+G_2(x,t)+H_2(x,t)\\ &= 1+G_2(x,t)+(C(t)-1)(G_2(x,t)+1)\\ &= (G_2(x,t)+1)C(t)\\ &= \frac{-(\sqrt{1-4t}-1)(\sqrt{1-4t}x-\sqrt{1-4t}-x+3)}{2(\sqrt{1-4t}xt-xt-\sqrt{1-4t}t+5t+\sqrt{1-4t}-1)}.\end{aligned}$$

A few initial terms of $F_2(x,t)$ are

$$F_2(x,t) = 1+2t+(x+5)t^2+(x^2+4x+15)t^3+(x^3+5x^2+16x+48)t^4 + (x^4+6x^3+23x^2+62x+160)t^5+\cdots.$$

$F_2(0,t)$ is the generating function for the number of paths that do not bounce off the diagonal to the right. One can compute that

$$\begin{aligned}F_2(0,t) &= \frac{2(t+\sqrt{1-4t}-1)}{(\sqrt{1-4t}-5)t-\sqrt{1-4t}+1}\\ &= 1+2t+5t^2+15t^3+48t^4+160t^5+548t^6+1914t^7+\cdots.\end{aligned}$$

The sequence $1, 2, 5, 15, 48, 160, 548, 1914, \ldots$ does not appear in the OEIS [14].

Similarly, we can compute the generating function of the number of paths that bounce at diagonal to right exactly one time. That is,

$$\begin{aligned}\left.\frac{\partial F_2(x,t)}{\partial x}\right|_{x=0} &= \left(\frac{-1+\sqrt{1-4t}+2t}{1-\sqrt{1-4t}+\left(-5+\sqrt{1-4t}\right)t}\right)^2\\ &= t^2+4t^3+16t^4+62t^5+238t^6+910t^7+\cdots.\end{aligned}$$

The sequence $1, 4, 16, 62, 238, 910, \ldots$ does not appear in the OEIS [14].

Also we could ask, for a random path $L \in \mathscr{L}_n$, what is the expectation of P_2-mch(L), or in other words, on average how many times do L bounce at $y = x$ to right? Consider that

$$\left.\frac{\partial F_2(x,t)}{\partial x}\right|_{x=1} = \left(\frac{-1+\sqrt{1-4t}+2t}{-1+\sqrt{1-4t}+4t}\right)^2$$
$$= t^2 + 6t^3 + 29t^4 + 130t^5 + 562t^6 + 2380t^7 + \cdots . \quad (17.12)$$

Coefficient of t^n in Eq. (17.12) agrees with sequence A008549 of the OEIS [14] which counts the total area of all the Dyck excursions of length $2n-2$. By OEIS [14], the coefficient of t^n is given by the formula $4^{n-1} - \binom{2n-1}{n-1}$. Using Stirling's formula to approximate $n!$, one finds that

$$\mathbb{E}[P_2\text{-mch}(L) : L \in \mathscr{L}_n] = \frac{\sum_{L\in\mathscr{L}_n} \text{bounce}^-(L)}{|\mathscr{L}_n|} = \frac{4^{n-1} - \binom{2n-1}{n-1}}{\binom{2n}{n}}$$
$$\sim \frac{\sqrt{\pi n}}{4} - \frac{1}{2} \approx 0.443\sqrt{n},$$

which implies when n is large, the expected number of times a random path $L \in \mathscr{L}_n$ bounces off the diagonal to the right is roughly $0.443\sqrt{n}$.

Next, by manipulating $F_2(x,t)$ we can also find the number of paths having even number of bounces off the diagonal to the right. The generating function is as follows,

$$\frac{1}{2}(F_2(1,t) + F_2(-1,t)) = 1 + 2t + 5t^2 + 16t^3 + 53t^4 + 184t^5 + 654t^6 + 2368t^7 + \cdots .$$

Similarly, the generating function for the number of paths having odd number of bounces off the diagonal to the right is

$$\frac{1}{2}(F_2(1,t) - F_2(-1,t)) = t^2 + 4t^3 + 17t^4 + 68t^5 + 270t^6 + 1064t^7 + 4181t^8 + \cdots .$$

Again, neither of the series correspond to sequences in the OEIS [14].

3.3 Pattern P_3

For pattern P_3, as discussed in Sect. 2, P_3-mch(L) counts the number of times L crosses the diagonal $y = x$ horizontally. We shall study

$$F_3(x,t) := 1 + \sum_{n\geq 1} t^n \sum_{L\in\mathscr{L}_n} x^{P_3\text{-mch}(L)}. \quad (17.13)$$

By our observation in the introduction $F_3(x,t) = F_4(x,t)$.

Similar to the discussion of P_2, we consider two cases. Case 1 is the paths that start with a north step and Case 2 is the paths that start with an east step. We define

$$G_3(x,t) := \sum_{n\geq 1} t^n \sum_{L\in\mathscr{L}_n \text{ starting with } E} x^{P_3\text{-mch}(L)}$$

and

$$H_3(x,t) := \sum_{n\geq 1} t^n \sum_{L\in\mathscr{L}_n \text{ starting with } N} x^{P_3\text{-mch}(L)}.$$

Clearly, $F_3(x,t) = 1 + G_3(x,t) + H_3(x,t)$. Essentially, the way we shall decompose the paths in this case is the same as how we decomposed the paths for pattern P_2. For paths starting with a north step, we consider where is the first the path crosses the diagonal $y = x$ from left to right and then it is followed by a path starting with an east step or an empty path. Then

$$H_3(x,t) = \sum_{j\geq 1} C_j t^j \left(xG_3(x,t) + 1\right) = (C(t)-1)(xG_3(x,t)+1) \tag{17.14}$$

Similarly, for paths starting with an east step, we consider where is the first time that the path crosses the diagonal vertically. Then

$$G_3(x,t) = \sum_{j\geq 1} C_j t^j \left(H_3(x,t) + 1\right) = (C(t)-1)(H_3(x,t)+1). \tag{17.15}$$

Then by Eq. (17.14),

$$H_3(x,t) = (C(t)-1)\left(x(C(t)-1)(H_3(x,t)+1)+1\right)$$

and we solve the formula above for $H_3(x,t)$

$$\begin{aligned} H_3(x,t) &= \frac{(1-C(t))(x(C(t)-1)+1)}{x(C(t)-1)^2-1} \\ &= -\frac{\left(2t+\sqrt{1-4t}-1\right)\left(2t(x-1)+(\sqrt{1-4t}-1)x\right)}{2\left(2t^2(x-1)+2\left(\sqrt{1-4t}-2\right)tx-\sqrt{1-4t}x+x\right)} \end{aligned}$$

Therefore,

$$\begin{aligned} F_3(x,t) &= 1 + G_3(x,t) + H_3(x,t) \\ &= 1 + C(t)H_3(x,t) + C(t) - H_3(x,t) - 1 + h_N(x,t) \\ &= (H_3(x,t)+1)C(t) \\ &= \frac{2}{(2t(x-1)+(\sqrt{1-4t}-1)x+\sqrt{1-4t}+1)} \end{aligned}$$

A few initial terms of $F_3(x,t)$ are

$$F_3(x,t) = 1 + 2t + (x+5)t^2 + (6x+14)t^3 + (x^2+27x+42)t^4 + (10x^2+110x+132)t^5 + \cdots.$$

Next, we shall find the generating function of the number of paths crossing the diagonal horizontally exactly once.

$$\begin{aligned}\left.\frac{\partial F_3(x,t)}{\partial x}\right|_{x=0} &= -\frac{2\left(-1+\sqrt{1-4t}+2t\right)}{\left(1+\sqrt{1-4t}-2t\right)^2}\\ &= t^2+6t^3+27t^4+110t^5+429t^6+1638t^7+\cdots,\end{aligned}$$

The sequence 1, 6, 27, 110, 429, 1638, ... is sequence A003517 on OEIS [14]. This sequence has several combinatorial interpretations such as the number of standard tableaux of shape $(n+3, n-2)$ and the number of permutations of $\{1,\ldots,n+1\}$ with exactly one increasing subsequence of length 3. It follows from the hook length formula for the number of standard tableaux that the number of paths L in $\mathcal{L}_n$ with exactly one horizontal crossing equal $6((2n+1)!)/(n-2)!(n+4)!)$.

Similarly, the number of paths L in $\mathcal{L}_n$ with exactly 2 horizontal crossings has the following generating function:

$$\begin{aligned}\left.\frac{1}{2!}\frac{\partial^2 F_3(x,t)}{\partial x^2}\right|_{x\to 0} &= \frac{4\left(-1+\sqrt{1-4t}+2t\right)}{\left(1+\sqrt{1-4t}-2t\right)^2}\\ &= t^4+10t^5+65t^6+350t^7+1700t^8+7752t^9+\cdots,\end{aligned}$$

The sequence 1, 10, 65, 350, 1700, ... is sequence A003519 on OEIS [14]. It counts the number of standard tableaux of shape $(n-5, n-4)$ from which it follows that the number of paths L in $\mathcal{L}_n$ with exactly 2 horizontal crossings equals $\frac{10}{n+6}\binom{2n+1}{n-4}$.

Also we could ask, for a random path $L \in \mathcal{L}_n$, what is the expectation of P_3-mch(L), or in other words, on average how many times does L cross $y=x$ from left to right? In this case, we have computed that

$$\begin{aligned}\frac{\partial}{\partial x}F_3(x,t)|_{x=1} &= \frac{-1=\sqrt{1-4t}+2t}{-2+8t}\\ &= t^2+6t^3+29t^4+130t^5+562t^6+2880t^7+9949t^8+\cdots\\ &= \frac{\partial}{\partial x}F_2(x,t)|_{x=1},\end{aligned}$$

which means the total number of P_3-matches in paths in $\mathcal{L}_n$ is equal to the total number of P_2-matches paths in $\mathcal{L}_n$.

Next we give a bijection that shows this fact. Since the total number of P_3-matches in paths in $\mathcal{L}_n$ is half of total $\{P_3, P_4\}$-matches in paths in $\mathcal{L}_n$ and the total number of

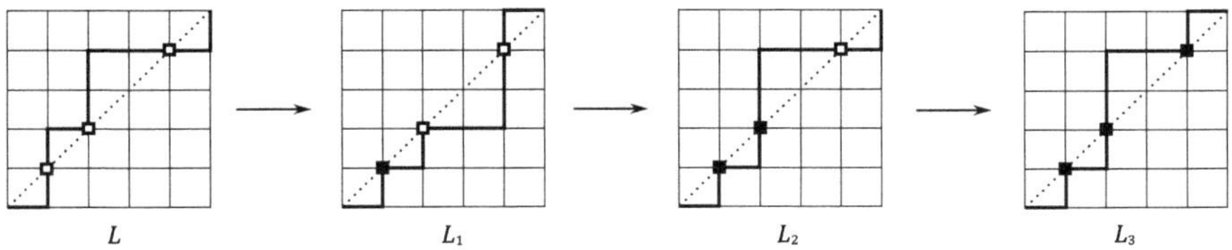

Fig. 6 L is mapped to L_3 by the bijection

P_2-matches in paths in $\mathscr{L}_n$ is half of total $\{P_2, P_5\}$-matches in paths in $\mathscr{L}_n$, we only need to show that the total number of $\{P_2, P_5\}$-matches in paths in $\mathscr{L}_n$ is equal to the total number of $\{P_3, P_4\}$-matches in paths in $\mathscr{L}_n$. In other words, we only need to show that the total number of times that all the paths in $\mathscr{L}_n$ bounce off the diagonal is equal to the total number of times that all the paths in $\mathscr{L}_n$ cross the diagonal.

By the reflection principle, we can construct a bijection between the set of paths in $\mathscr{L}_n$ crossing the diagonal k times, denoted by $\mathscr{C}_{n,k}$ and the set of paths in $\mathscr{L}_n$ bouncing off the diagonal k times, denoted by $\mathscr{B}_{n,k}$. The procedure of the bijection is as follows. For any path $L \in \mathscr{C}_{n,k}$, L crosses the diagonal k times and suppose L touches the diagonal j times at positions $\{p_1, p_2, \ldots, p_j\}$, $j \geq k$. First we retain the part between $[0, 0]$ and p_1 of the path and flip the path between p_1 and $[n, n]$ along the diagonal, then we can get a new path L_1. At the second step, we retain the part between $[0, 0]$ and p_2 of the path L_1 and flip the part between p_2 and $[n, n]$ along the diagonal, then we can get a new path L_2. We repeat the process above until we acquire L_j. L_j is a path in $\mathscr{B}_{n,k}$ because the procedure above transforms a crossing of L into a bouncing of L_j and a bouncing of L into a crossing of L_j. An example is pictured in Fig. 6. $L \in \mathscr{C}_{5,2}$ is mapped to $L_3 \in \mathscr{B}_{5,2}$ under the bijection.

Therefore,

$$\mathbb{E}[P_3\text{-mch}(L)] = \mathbb{E}[P_2\text{-mch}(L)] \sim \frac{\sqrt{n\pi}}{4} - \frac{1}{2} \approx 0.443\sqrt{n}.$$

Next, by manipulating $F_3(x, t)$ we can also find the number of paths having even number many horizontal crossings. The generating function is as follows,

$$\frac{1}{2}(F_3(1, t) + F_3(-1, t)) = 1 + 2t + 5t^2 + 14t^3 + 43t^4 + 142t^5 + 494t^6 + 1780t^7 + \cdots,$$

The sequence 1, 2, 5, 14, 43, 142, 494, ... is sequence A005317 in the OEIS [14] where no combinatorial interpretation is given. Thus we have given a combinatorial interpretation to this sequence.

Similarly, the generating function for the number of paths having odd number many horizontal crossings is

$$\frac{1}{2}(F_3(1, t) - F_3(-1, t)) = t^2 + 6t^3 + 27t^4 + 110t^5 + 430t^6 + 1652t^7 + 6307t^8 + \cdots,$$

in which coefficient of t^n also counts number of unordered pairs of distinct length-n binary words having the same number of 1's according to A108958 in the OEIS [14]. We leave open the problem of giving a bijective proof of this fact.

4 Multivariate Generating Functions

In this section, we shall study multivariate generating functions for Δ-matches for certain $\Delta \subseteq \{P_1, \ldots, P_6\}$. Our choices for the Δ that we consider are motivated by picking those pattern matching conditions which have the clearest geometric interpretations. Let

$$F_\Delta(\mathbf{x}, t) := 1 + \sum_{n \geq 1} t^n \sum_{L \in \mathscr{L}_n} \left(\prod_{j \in \Delta} x_j^{P_j\text{-mch}(L)} \right),$$

where Δ is a subset of $\{1, 2, 3, 4, 5, 6\}$. We start by looking at the two elements sets that have symmetry, namely $\Delta = \{1, 6\}$, $\Delta = \{2, 5\}$, and $\Delta = \{3, 4\}$.

4.1 P_1 and P_6

Pattern P_1 has one east step below $y = x - 1$ and P_6 has one east step above $y = x + 1$, as shown in Fig. 7. We know that for a path $L \in \mathscr{L}_n$, P_1-mch(L) and P_6-mch(L) are the numbers of east steps below $y = x - 1$ and above $y = x + 1$, respectively.

In this subsection, we shall consider the multivariate generating function

$$F_{1,6}(x_1, x_6, t) := 1 + \sum_{n \geq 1} t^n \sum_{L \in \mathscr{L}_n} x_1^{P_1\text{-mch}(L)} x_6^{P_6\text{-mch}(L)}.$$

We use essentially the same ideas as in Sect. 3.1 to decompose the paths in $\mathscr{L}_n$ to obtain recurrences that will allow us to compute $F_{1,6}(x_1, x_6, t)$. In this case, we take three cases into account. Case 1 is the paths that have no P_1-match or P_6-match. In addition, we can see paths avoiding P_1 and P_6 must stay between $y = x - 1$ and $y = x + 1$. It is easy to see that if the word of such a path is $u_1 \ldots u_{2n}$, then either

Fig. 7 Pattern P_1 and P_6

$u_{2i-1}u_{2i} = EN$ or $u_{2i-1}u_{2i} = NE$ for all i. Thus the number of paths in $\mathscr{L}_n$ bounded by $y = x - 1$ and $y = x + 1$ is 2^n. Case 2 are the paths L such that the first pattern matching of either P_1 or P_6 in path L is P_1 and Case 3 are the paths L such that the first pattern matching of either P_1 or P_6 in L is P_6. Then we have

$$\begin{aligned} F_{1,6}(x_1, x_6, t) &= \sum_{n\geq 0} 2^n t^n + \sum_{i\geq 1}\sum_{j\geq 1} \left(C_i 2^{j-1} x_1^i t^{i+j} + C_i C 2^{j-1} x_6^i t^{i+j}\right) F_{1,6}(x_1, x_6, t) \\ &= \frac{1}{1-2t} + \frac{t}{1-2t}(C(x_1 t) + C(x_6 t) - 2) F_{1,6}(x_1, x_6, t). \end{aligned}$$

Then solving above equation for $F_{1,6}$, we have

$$\begin{aligned} F_{1,6}(x_1, x_6, t) &= \frac{2x_1x_6}{\left(-1+\sqrt{1-4x_1 t}\right)x_6 + \left(-1+\sqrt{1-4x_6 t}\right)x_1 + 2x_1x_6} \\ &= 1 + 2t + (x_1 + x_6 + 4)t^2 + (2x_1^2 + 4x_1 + 2x_6^2 + 4x_6 + 8)t^3 \\ &\quad + (5x_1^3 + 9x_1^2 + 12x_1 + 5x_6^3 + 9x_6^2 + 12x_6 + 2x_1x_6 + 16)t^4 + \cdots. \end{aligned}$$

Clearly, $F_{1,6}(x, 1, t) = F_{1,6}(1, x, t) = F_1(x, t)$. Next, we discuss coefficients of $x_1 t^n$ and $x_1 x_6 t^n$ in $F_{1,6}(x_1, x_6, t)$ which count the number of paths in $\mathscr{L}_n$ having exactly one P_1 pattern and avoiding P_6 and the number of paths in $\mathscr{L}_n$ having exactly one P_1 and exactly one P_6. In general, the generating function for coefficients of $x_1^j x_6^k$ is

$$\frac{1}{j!k!}\frac{\partial^{j+k} F_{1,6}(x_1, x_6, t)}{\partial x_1^j \partial x_6^k}\Bigg|_{x_1=0, x_6=0}, \tag{17.16}$$

where if the derivative cannot be evaluated at zero, we take the limit.

By the symmetry of P_1 and P_6, the coefficient of $x_1 t^n$ in $F_{1,6}(x_1, x_6, t)$ equals the coefficient of $x_6 t^n$ in $F_{1,6}(x_1, x_6, t)$. By Eq. (17.16), the generating function for the coefficients of $x_1 t^n$ in $F_{1,6}(x_1, x_6, t)$ equals

$$\frac{t^2}{(1-2t)^2} = t^2 + 4t^3 + 12t^4 + 32t^5 + 80t^6 + 192t^7 + 448t^8 + \cdots.$$

The sequence 1, 4, 12, 32, 80, 192, ... is A001787 in the OEIS [14]. The nth term of this sequence is $n2^{n-1}$ which means that the number of paths $L \in \mathscr{L}_n$ with exactly one east step below the subdiagonal $y = x - 1$ and no east step above the superdiagonal $y = x + 1$ equals $(n-1)2^{n-2}$. This is easy to prove directly. That is, if L is such a path, the one east that occurs below the subdiagonal $y = x - 1$ must arise by starting at a point $[a, a]$ on the diagonal where $0 \leq a \leq n-2$ followed by a sequence $EENN$. If we remove this sequence from the word of L, we will end up with the word $u_1 \ldots u_{2n-4}$ of path $L' \in \mathscr{L}_{n-2}$ which has no east steps either below the subdiagonal $y = x - 1$ or above the superdiagonal $y = x + 1$. It is easy to see that in such a path L' either $u_{2i-1}u_{2i} = EN$ or $u_{2i-1}u_{2i} = NE$ for all i. Hence there are 2^{n-2} such paths L' so that the number of paths $L \in \mathscr{L}_n$ with exactly one east step below the subdiagonal $y = x - 1$ and no east step above the superdiagonal $y = x + 1$ equals $(n-1)2^{n-2}$.

The generating function of the coefficients of $x_1x_6t^n$ in $F_{1,6}(x_1, x_6, t)$ equals

$$\frac{2t^4}{(1-2t)^3} = 2t^4 + 12t^5 + 48t^6 + 160t^7 + 480t^8 + 1344t^9 + \cdots.$$

The sequence 2, 12, 48, 160, 480, ... is sequence A001815 in the OEIS [14]. We can show directly that the number of paths $L \in \mathcal{L}_n$ with exactly one east step below the subdiagonal $y = x - 1$ and exactly step above the superdiagonal $y = x + 1$ equals $(n-2)(n-3)2^{n-4}$. That is, if L is such a path, then the one east that occurs below the subdiagonal $y = x - 1$ must arise by starting at a point $[a, a]$ on the diagonal where $0 \leq a \leq n-2$ followed by a sequence $EENN$ and the one east that occurs above the subdiagonal $y = x + 1$ must arise by starting at a point $[b, b]$ on the diagonal where $0 \leq a \leq n-2$ followed by a sequence $NNEE$. We have $n-1$ choices for the point $[a, a]$. But these $n-1$ choices lead to different situations according to different values of a. If $a = 0$ or $a = n-2$, we have $n-3$ choices to choose a point $[b, b]$ on the diagonal followed by a sequence $NNEE$. If $0 < a < n-2$, there are $n-4$ choices to choose a point $[b, b]$ on the diagonal followed by a sequence $NNEE$. So the total ways to choose positions of one P_1-match and one P_6-match are equal to $2(n-3) + (n-3)(n-4) = (n-2)(n-3)$. We remove sequence $EENN$ and $NNEE$ from the word of L, we will end up with the word $u_1 \ldots u_{2n-8}$ of path $L' \in \mathcal{L}_{n-4}$ which has no east steps either below the subdiagonal $y = x - 1$ or above the superdiagonal $y = x + 1$. Hence there are 2^{n-4} such paths L' so that the number of path $L \in \mathcal{L}_n$ with exactly one east step below the subdiagonal $y = x - 1$ and exactly on east step above the superdiagonal $y = x + 1$ equals $(n-2)(n-3)2^{n-4}$.

If we are interested in counting lattice paths by the number of east steps below $y = x - 1$ or above $y = x + 1$, then we consider the following generating function:

$$F_{1,6}(x, x, t) = \frac{x}{-1 + x + \sqrt{1 - 4xt}}.$$

And also clearly,

$$\left.\frac{\partial F_{1,6}(x, x, t)}{\partial x}\right|_{x=1} = 2\left.\frac{\partial F_1(x, t)}{\partial x}\right|_{x=1}$$

because the symmetry of P_1 and P_6. Then by Eq. (17.6),

$$\mathbb{E}[\{P_1, P_6\}\text{-mch}(L) : L \in \mathcal{L}_n] = 2\mathbb{E}[P_1\text{-mch}(L) : L \in \mathcal{L}_n] \sim n + 1 - 2\sqrt{\pi n}.$$

Next, by manipulating $F_{1,6}(x_1, x_6, t)$ we can also find the number of paths having even number many east steps below $y = x - 1$ or above $y = x + 1$. The generating function equals (Fig. 8)

Fig. 8 Pattern P_2 and P_5

$$\frac{1}{2}\left(F_{1,6}(1,1,t)+F_{1,6}(-1,-1,t)\right)$$
$$= 1 + 2t + 4t^2 + 12t^3 + 36t^4 + 132t^5 + 456t^6 + 1752t^7 + \cdots . \tag{17.17}$$

Similarly, the generating function for the number of paths having odd number many east steps below the subdiagonal $y = x - 1$ or above $y = x + 1$ is

$$\frac{1}{2}\left(F_{1,6}(1,1,t)-F_{1,6}(-1,-1,t)\right)$$
$$= 2t^2 + 8t^3 + 34t^4 + 120t^5 + 468t^6 + 1680t^7 + 6530t^8 + \cdots . \tag{17.18}$$

Neither of the two sequences above is recorded in the OEIS [14].

4.2 P_2 and P_5

In this subsection, we shall study

$$F_{2,5}(x_2, x_5, t) := 1 + \sum_{n\geq 1} t^n \sum_{L\in\mathscr{L}_n} x_2^{P_2\text{-mch}(L)} x_5^{P_5\text{-mch}(L)}.$$

$F_{2,5}(x_2, x_5, t)$ is the generating function which keeps track of the number of lattice paths by the number of times it bounces off the diagonal to the right or to the left. By the symmetry induced by reflecting paths about the diagonal discussed in the introduction, it is easy to see that $F_{2,5}(x_2, x_5, t)$ is a symmetric function in x_2 and x_5. It is also clear that $F_{2,5}(x, x, t)$ is the generating function which counts number of times a lattice path in $\mathscr{L}_n$ bounces off the diagonal $y = x$.

First we define

$$G_{2,5}(x_2, x_5, t) := \sum_{n\geq 1} t^n \sum_{L\in\mathscr{L}_n \text{ starting with } E} x_2^{P_2\text{-mch}(L)} x_5^{P_5\text{-mch}(L)} t^n$$

and

$$H_{2,5}(x_2, x_5, t) := \sum_{n\geq 1} t^n \sum_{L\in\mathscr{L}_n \text{ starting with } N} x_2^{P_2\text{-mch}(L)} x_5^{P_5\text{-mch}(L)} t^n .$$

Clearly,

$$F_{2,5}(x_2, x_5, t) = 1 + G_{2,5}(x_2, x_5, t) + H_{2,5}(x_2, x_5, t).$$

Here we employ the decomposition of paths used in Sect. 3.2, then we have

$$\begin{aligned} G_{2,5}(x_2, x_5, t) &= \sum_{i\geq 0}\sum_{j\geq 1} C_{i,j} x_2^i t^j \left(H_{2,5}(x_2, x_5, t) + 1\right) \\ &= (C(x_2, t) - 1)(H_{2,5}(x_2, x_5, t) + 1) \end{aligned}$$

and

$$\begin{aligned} H_{2,5}(x_2, x_5, t) &= \sum_{i\geq 0}\sum_{j\geq 1} C_{i,j} x_5^i t^j \left(G_{2,5}(x_2, x_5, t) + 1\right) \\ &= (C(x_5, t) - 1)(G_{2,5}(x_2, x_5, t) + 1), \end{aligned}$$

where $C(x, t)$ is given as Eq. (17.9). Then

$$G_{2,5}(x_2, x_5, t) = (C(x_2, t) - 1)\left((C(x_5, t) - 1)\,(G_{2,5}(x_2, x_5, t) + 1) + 1\right).$$

Solving the above formula for $G_{2,5}$ we have,

$$\begin{aligned} G_{2,5}(x_2, x_5, t) &= -\frac{(C(x_2, t) - 1)C(x_5, t)}{C(x_2, t)(C(x_5, t) - 1) - C(x_5, t)} \\ &= -\frac{2(1 - x_5)t + (x_5 - 2)\left(1 - \sqrt{1 - 4t}\right)}{1 + \sqrt{1 - 4t} + x_2(x_5 - 1)\left(1 - \sqrt{1 - 4t}\right) - x_5 + \sqrt{1 - 4t}\,x_5 + 2(1 - x_2x_5)t}. \end{aligned}$$

Therefore,

$$\begin{aligned} F_{2,5}(x_2, x_5, t) &= 1 + G_{2,5}(x_2, x_5, t) + H_{2,5}(x_2, x_5, t) \\ &= 1 + G_{2,5}(x_2, x_5, t) + (C(x_5, t) - 1)(G_{2,5}(x_2, x_5, t) + 1) \\ &= C(x_5, t)(G_{2,5}(x_2, x_5, t) + 1) \\ &= \left(1 + \frac{1 - \sqrt{1 - 4t}}{2 - x_5\left(1 - \sqrt{1 - 4t}\right)}\right) \cdot \\ &\quad \times \left(1 - \frac{2(1 - x_5)t + (x_5 - 2)\left(1 - \sqrt{1 - 4t}\right)}{\begin{array}{r}\left(1 + \sqrt{1 - 4t} + x_2(x_5 - 1)\left(1 - \sqrt{1 - 4t}\right)\right. \\ \left. -x_5 + \sqrt{1 - 4t}\,x_5 + 2(1 - x_2x_5)t\right)\end{array}}\right). \end{aligned}$$

A few initial terms of $F_{2,5}(x_2, x_5, t)$ are

$$\begin{aligned} F_{2,5}(x_2, x_5, t) &= 1 + 2t + (x_2 + x_5 + 4)t^2 + (x_2^2 + 4x_2 + x_5^2 + 4x_5 + 10)t^3 \\ &\quad + (x_2^3 + 5x_2^2 + 14x_2 + x_5^3 + 5x_5^2 + 14x_5 + 2x_2x_5 + 28)t^4 \\ &\quad + (x_2^4 + 6x_2^3 + 21x_2^2 + 48x_2 + x_5^4 + 6x_5^3 + 21x_5^2 + 48x_5 \\ &\quad + 2x_2^2x_5 + 2x_2x_5^2 + 12x_2x_5 + 84)t^5 + \cdots . \end{aligned}$$

By Eq. (17.16), we can obtain the generating functions of the coefficients of x_2t^n in $F_{2,5}(x_2, x_5, t)$ which equals

$$\begin{aligned} \left.\frac{\partial F_{2,5}(x_2, 0, t)}{\partial x_2}\right|_{x_2=0} &= \frac{1 - \sqrt{1-4t} + 2t(-2 + \sqrt{1-4t} + t)}{2t^2} \\ &= t^2 + 4t^3 + 14t^4 + 48t^5 + 165t^6 + 572t^7 + 7072t^8 + \cdots \\ &= \left.\frac{\partial F_1(x, t)}{\partial x}\right|_{x \to 0} . \end{aligned}$$

This implies there exists a bijection between paths having exactly one P_2-match but no P_5-matches and paths having exactly one step below $y = x - 1$. We leave this as an open problem.

Similarly, we can get coefficients of $x_2x_5t^n$,

$$\left.\frac{\partial^2 F_{2,5}(x_2, x_5, t)}{\partial x_2 \partial x_5}\right|_{x_2=x_5=0} = 2t^4 + 12t^5 + 56t^6 + 236t^7 + 948t^8 + 3712t^9 + \cdots .$$

The sequence 2, 12, 56, 236, 948, ... is not in the OEIS [14].

It is also the case that $F_{2,5}(1, x, t) = F_5(x, t) = F_2(x, t) = F_{2,5}(x, 1, t)$ and

$$\begin{aligned} F_{2,5}(0, 0, t) &= 1 + 2t + 4t^2 + 10t^3 + 28t^4 + 84t^5 + 264t^6 + \cdots \\ &= 1 + 2C_1t + 2C_2t^2 + 2C_3t^3 + 2C_4t^4 + 2C_5t^5 + \cdots , \end{aligned}$$

where C_k is the kth Catalan number. $F_{2,5}(x, x, t)$ is the generating function over paths L in $\mathscr{L}_n$ by the number of times L bounces off the diagonal:

$$\begin{aligned} F_{2,5}(x, x, t) &= \frac{1 - \sqrt{1-4t} - t - x + x^2t}{-x + (1 + x^2)t} \\ &= 1 + 2t + 2(x + 2)t + 2(x + 2)t^2 + 2(x^2 + 4x + 5)t^3 \\ &\quad + 2(x^3 + 6x^2 + 14x + 14)t^4 + 2(x^4 + 8x^3 + 27x^2 + 48x + 42)t^5 + \cdots . \end{aligned}$$

We take the partial derivative of $F_{2,5}(x, x, t)$ with respect to x and evaluate at $x = 1$,

$$\begin{aligned}\left.\frac{\partial F_{2,5}(x,x,t)}{\partial x}\right|_{x=1} &= \frac{\sqrt{1-4t}}{-1+4t} - \frac{1-2t}{-1+4t}\\ &= \sum_{n\geq 2}\left(\frac{4^n}{2} - 2\binom{2n-1}{n-1}\right)t^n\\ &= 2t^2 + 12t^3 + 58t^4 + 260t^5 + 1124t^6 + 4760t^6 + 19898t^8 + \cdots\\ &= 2\left.\frac{\partial F_2(x,t)}{\partial x}\right|_{x=1}.\end{aligned}$$

It follows that

$$\mathbb{E}[\{P_2, P_5\}\text{-mch}(L) : L \in \mathcal{L}_n] = 2\mathbb{E}[P_2\text{-mch}(L) : L \in \mathcal{L}_n] \approx \frac{\sqrt{\pi n}}{2} - 1 \approx 0.886\sqrt{n}$$

gives the expected number of times a path in $\mathcal{L}_n$ bounces off the diagonal.

$\frac{1}{2}(F_{2,5}(1, 1, t) + F_{2,5}(-1, -1, t))$ is the generating function of the number of lattice paths in $\mathcal{L}_n$ that bounce off the diagonal an even number of times. We have computed that

$$\begin{aligned}&\frac{1}{2}\left(F_{2,5}(1,1,t) + F_{2,5}(-1,-1,t)\right)\\ &\quad= \frac{1-\sqrt{1-4t} + \left(-6+4\sqrt{1-4t}\right)t + 4t^2}{1-\sqrt{1-4t} + \left(-4+2\sqrt{1-4t}\right)t}\\ &\quad= 1 + 2\sum_{n\geq 1}\binom{2n-2}{n-1}t^n\\ &\quad= 1 + 2t + 4t^2 + 12t^3 + 40t^4 + 140t^5 + 504t^6 + \cdots.\end{aligned}$$

The sequence 2, 4, 12, 40, 140, ... is sequence A028329 in the OEIS [14]. It would be nice to have a direct combinatorial proof that the number of lattice paths in $\mathcal{L}_n$ that bounce off the diagonal an even number of times equals $2\binom{2n-2}{n-1}$.

$\frac{1}{2}(F_{2,5}(1, 1, t) - F_{2,5}(-1, -1, t))$ is the generating function of the number of lattice paths L in $\mathcal{L}_n$ that bounce off the diagonal an odd number of times. We have computed that

$$\begin{aligned}&\frac{1}{2}\left(F_{2,5}(1,1,t) - F_{2,5}(-1,-1,t)\right)\\ &\quad= 2\sum_{n\geq 2}\binom{2n-2}{n-2}t^n\\ &\quad= 2t^2 + 8t^3 + 30t^4 + 112t^5 + 420t^6 + 1584t^7 + \cdots.\end{aligned}$$

Fig. 9 Pattern P_3 and P_4

The sequence 2, 8, 30, 112, 420, 1584, ... is sequence A162551 in the OEIS [14]. It would be nice to have a direct combinatorial proof of that the number of lattice paths in $\mathscr{L}_n$ that bounce off the diagonal an odd number of times equal $2\binom{2n-2}{n-2}$ (Fig. 9).

4.3 P_3 and P_4

We define

$$F_{3,4}(x_3, x_4, t) := 1 + \sum_{n\geq 1} t^n \sum_{L\in\mathscr{L}_n} x_3^{P_3\text{-mch}(L)} x_4^{P_4\text{-mch}(L)}, \tag{17.19}$$

where x_3 is used to keep track of the number of horizontal crossings and x_4 is used to keep track of the number of vertical crossings. Clearly, $F_{3,4}(x_3, x_4, t)$ is symmetric in x_3 and x_4.

We also define

$$G_{3,4}(x_3, x_4, t) := \sum_{n\geq 1} t^n \sum_{L\in\mathscr{L}_n \text{ starting with } E} x_3^{P_3\text{-mch}(L)} x_4^{P_4\text{-mch}(L)}$$

and

$$H_{3,4}(x_3, x_4, t) := \sum_{n\geq 1} t^n \sum_{L\in\mathscr{L}_n \text{ starting with } N} x_3^{P_3\text{-mch}(L)} x_4^{P_4\text{-mch}(L)}.$$

We employ the same decomposition of paths used in Sect. 4.2 for P_3 and P_4. Then

$$H_{3,4}(x_3, x_4, t) = \sum_{j\geq 1} C_j t^j \left(x_3 G_{3,4}(x_3, x_4, t) + 1\right) = (C(t) - 1)(x_3 G_{3,4}(x_3, x_4, t) + 1)$$

and

$$G_{3,4}(x_3, x_4, t) = \sum_{j\geq 1} C_j t^j \left(x_4 H_{3,4}(x_3, x_4, t) + 1\right) = (C(t) - 1)(x_4 H_{3,4}(x_3, x_4, t) + 1).$$

Combining the two equations above, we can then solve for $G_{3,4}$ to obtain that

$$G_{3,4}(x_3, x_4, t) = \frac{(1 - C(t))((x_4C(t) - 1) + 1)}{x_3x_4(C(t) - 1)^2 - 1}$$
$$= -\frac{\left(-1 + \sqrt{1-4t} + 2t\right)\left(2t(1 - x_4) + \left(-1 + \sqrt{1-4t}\right)x_4\right)}{-2(-1 + \sqrt{1-4t})x_3x_4 + 4(-2 + \sqrt{1-4t})x_3x_4t + 4(x_3x_4 - 1)t^2}.$$

Then

$$\begin{aligned}F_{3,4}(x_3, x_4, t) &= 1 + G_{3,4}(x_3, x_4, t) + H_{3,4}(x_3, x_4, t)\\ &= 1 + G_{3,4}(x_3, x_4, t) + (C(t) - 1)(x_3G_{3,4}(x_3, x_4, t) + 1)\\ &= (x_3C(t) - x_3 + 1)\, G_{3,4}(x_3, x_4, t) + C(t)\\ &= \frac{1-\sqrt{1-4t}}{2t} - \left(1 - \frac{1-\sqrt{1-4t}}{2t}\right)\\ &\quad\times \frac{\left(1 - x_3 + \frac{1-\sqrt{1-4t}}{2t}x_3\right)\left(1 - x_4 + \frac{1-\sqrt{1-4t}}{2t}x_4\right)}{-1 + \left(-1 + \frac{1-\sqrt{1-4t}}{2t}\right)^2 x_3x_4}.\end{aligned}$$

A few initial terms of $F_{3,4}(x_3, x_4, t)$ are

$$\begin{aligned}F_{3,4}(x_3, x_4, t) &= 1 + 2t + (x_3 + x_4 + 4)t^2 + (4x_3 + 4x_4 + 2x_3x_4 + 10)t^3\\ &\quad + (14x_3 + 14x_4 + x_3^2x_4 + x_3x_4^2 + 12x_3x_4 + 28)t^4\\ &\quad + (48x_3 + 48x_4 + 2x_3^2x_4^2 + 8x_3^2x_4 + 8x_3x_4^2 + 54x_3x_4 + 84)t^5\\ &\quad + \cdots.\end{aligned}$$

By symmetry, $F_{3,4}(1, x, t) = F_4(x, t) = F_3(x, t) = F_{3,4}(x, 1, t)$. It is also clear that $F_{3,4}(0, 0, t) = F_{2,5}(0, 0, t) = 2C(t)$, where $C(t)$ is the generating function of Catalan numbers, since if a path in $\mathscr{L}_n$ has no vertical or horizontal crossings, then the path either stays on or below the diagonal or on and above the diagonal.

By Eq. (17.16), we see that coefficients of x_3t^n in $F_{3,4}(x_3, x_4, t)$ yield the generating function of the number of paths in $\mathscr{L}_n$ that have exactly one horizontal crossing and no vertical crossings. We have computed that

$$\begin{aligned}\left.\frac{\partial F_{3,4}(x_3, 0, t)}{\partial x_3}\right|_{x_3=0} &= \frac{1 - \sqrt{1-4t} + 2t(-2 + \sqrt{1-4t} + t)}{2t^2}\\ &= t^2 + 4t^3 + 14t^4 + 48t^5 + 165t^6 + 572t^7 + 7072t^8 + \cdots\\ &= \left.\frac{\partial F_1(x, t)}{\partial x}\right|_{x\to 0}\\ &= \left.\frac{\partial F_{2,5}(x_2, 0, t)}{\partial x_2}\right|_{x_2=0},\end{aligned}$$

which implies the number of paths in $\mathscr{L}_n$ having exactly one P_3-match and avoiding P_4 is equal to the number of paths in $\mathscr{L}_n$ having exactly one P_2-match and avoiding P_5. This can be verified by the bijection defined in Sect. 3.3. However, coefficients of $x_3x_4t^n$ in $F_{3,4}(x_3, x_4, t)$ are not equal to the coefficient of $x_2x_5t^n$ in $F_{2,5}(x_2, x_5, t)$. This is due to the fact that a path in $\mathscr{L}_n$ cannot cross the diagonal horizontally twice without crossing the diagonal vertically. We have computed that

$$\begin{aligned}\left.\frac{\partial^2 F_{3,4}(x_3, x_4, t)}{\partial x_3 \partial x_4}\right|_{x_3=x_4=0} &= 2\left.\frac{\partial F_3(x, t)}{\partial x}\right|_{x=0} \\ &= -\frac{8t^2\left(-1+\sqrt{1-4t}+2t\right)}{\left(\sqrt{1-4t}\left(1+\sqrt{1-4t}-2t\right)\right)^3} \\ &= 2t^2 + 12t^3 + 54t^4 + 220t^5 + 858t^6 + 3276t^7 + \cdots,\end{aligned}$$

The sequence 2, 12, 54, 220, 858, 3276, ... is Column 2 in A118920 and the exactly same interpretation is given by Emeric Deutsch in the OEIS [14].

For $F_{3,4}(x, x, t)$, we can show that $F_{3,4}(x, x, t) = F_{2,5}(x, x, t)$ by the bijection defined in Sect. 3.3, which gives us that for a random $L \in \mathscr{L}_n$, the expectation of the number of crossings has asymptotic approximation as follows,

$$\mathbb{E}[\{P_3, P_4\}\text{-mch}(L)] \sim \frac{\sqrt{\pi n}}{2} - 1 \approx 0.886\sqrt{n},$$

and also

$$\frac{1}{2}(F_{3,4}(1, 1, t) + F_{3,4}(-1, -1, t)) = \frac{1}{2}(F_{2,5}(1, 1, t) + F_{2,5}(-1, -1, t)).$$

4.4 P_2 and P_4

Due to space limitations, we shall consider only one more set of patterns of size 2, namely $\Delta = \{2, 4\}$ as shown in Fig. 10. First, we define

$$F_{2,4}(x_2, x_4, t) := 1 + \sum_{n\geq 1} t^n \sum_{L\in\mathscr{L}_n} x_2^{P_2\text{-mch}(L)} x_4^{P_4\text{-mch}(L)}.$$

$F_{2,4}(x_2, x_4, t)$ counts the number of lattice paths by the number of times it bounces off the diagonal to the right and by the the number of times it crosses the diagonal vertically. It follows that $F_{2,4}(x, x, t)$ is the generating function over lattice paths L in $\mathscr{L}_n$ by the number of times L touches the diagonal with a north step. By symmetry, $F_{3,5}(x, x, t)$ is also the generating function over lattice paths L in $\mathscr{L}_n$ by the number of times L touches the diagonal with an east step.

Fig. 10 Pattern P_2 and P_4

First we define

$$G_{2,4}(x_2, x_4, t) := \sum_{n\geq 1} t^n \sum_{L\in\mathscr{L}_n \text{ starting with } E} x_2^{P_2\text{-mch}(L)} x_4^{P_4\text{-mch}(L)}$$

and

$$H_{2,4}(x_2, x_4, t) := \sum_{n\geq 1} t^n \sum_{L\in\mathscr{L}_n \text{ starting with } N} x_2^{P_2\text{-mch}(L)} x_4^{P_4\text{-mch}(L)}.$$

Clearly,

$$F_{2,4}(x_2, x_4, t) = 1 + G_{2,4}(x_2, x_4, t) + H_{2,4}(x_2, x_4, t).$$

Employing the same decomposition that is used in Sect. 4.2, we have

$$\begin{aligned} G_{2,4}(x_2, x_4, t) &= \sum_{i\geq 0}\sum_{j\geq 1} C_{i,j} x_2^i t^j \left(x_4 H_{2,4}(x_2, x_4, t) + 1\right) \\ &= (C(x_2, t) - 1)\left(x_4 H_{2,4}(x_2, x_4, t) + 1\right) \end{aligned}$$

and

$$\begin{aligned} H_{2,4}(x_2, x_4, t) &= \sum_{j\geq 1} C_j t^j (G_{2,4}(x_2, x_4, t) + 1) \\ &= (C(t) - 1)(G_{2,4}(x_2, x_4, t) + 1). \end{aligned}$$

Combining the two equations above, we can solve them for $G_{2,4}$,

$$G_{2,4}(x_2, x_4, t) = -\frac{(C(x_2, t) - 1)(x_4(C(t) - 1) + 1)}{x_4(C(x_2, t) - 1)(C(t) - 1) - 1}.$$

Then

$$\begin{aligned} F_{2,4}(x_2, x_4, t) &= 1 + G_{2,4}(x_2, x_4, t) + H_{2,4}(x_2, x_4, t) \\ &= 1 + G_{2,4}(x_2, x_4, t) + (C(t) - 1)(G_{2,4}(x_2, x_4, t) + 1) \\ &= C(t)(G_{2,4}(x_2, x_4, t) + 1) \\ &= \frac{(x_2 - 2)\left(-1 + \sqrt{1-4t}\right) + 2(x_2 - 1)t}{x_4\left(-1 + \sqrt{1-4t}\right) + x_2\left(2 + \left(-1 + \sqrt{1-4t}\right) + 3x_4 - x_4\sqrt{1-4t}\right)t}. \end{aligned}$$

A few initial terms are

$$\begin{aligned}F_{2,4}(x_2,x_4,t) &= 1+2t+(x_2+x_4+4)t^2+(x_2^2+3x_2+5x_4+x_2x_4+10)t^3\\ &\quad+(x_2^3+4x_2^2+9x_2+x_4^2+19x_4+x_2^2x_4+7x_2x_4+28)t^4\\ &\quad+(x_2^4+5x_2^3+14x_2^2+28x_2+8x_4^2+68x_4+x_2^3x_4+2x_2x_4^2\\ &\quad+9x_2^2x_4+32x_2x_4+84)t^5+\cdots.\end{aligned}$$

By Eq. (17.16), the coefficient of x_2t^n in $F_{2,4}(x_2,x_4,t)$ is the number of paths in $\mathscr{L}_n$ which bounce off the diagonal to the right exactly one time but do not cross the diagonal vertically. We have computed that

$$\begin{aligned}\left.\frac{\partial F_{2,4}(x_2,0,t)}{\partial x_2}\right|_{x_2=0} &= -\frac{\left(-1+\sqrt{1-4t}\right)^3}{8t}\\ &= t^2+3t^3+9t^4+28t^5+90t^6+297t^7+1001t^8+\cdots.\end{aligned}$$

The sequence $1, 3, 9, 28, 90, 297, \ldots$ is sequence A000245 in the OEIS [14] which has several interpretations such as the number of permutations on $\{1, 2, \ldots, n+2\}$ that are 123-avoiding and for which the integer n is in the third spot, the number of lattice paths in $\mathscr{L}_{n-1}$ which touch but do not cross the $y = x - 1$ and the number of Dyck paths in $\mathscr{L}_n$ that start with 'EE.'

Similarly, the coefficient of x_4t^n in $F_{2,4}(x_2,x_4,t)$ is the number of paths in $\mathscr{L}_n$ which have exactly one vertical crossing but never bounce off the diagonal to the right. We have computed that

$$\begin{aligned}\left.\frac{\partial F_{2,4}(0,x_4,t)}{\partial x_4}\right|_{x_4=0} &= -\frac{\left(-3+\sqrt{1-4t}\right)\left(-1+\sqrt{1-4t}+2t\right)^2}{8t^2}\\ &= t^2+5t^3+19t^4+68t^5+240t^6+847t^7+3003t^8+\cdots.\end{aligned}$$

The sequence $1, 5, 19, 68, 240, \ldots$ is sequence A070857 in the OEIS [14] which has no combinatorial interpretation. Thus we have given a combinatorial interpretation to this sequence.

The coefficient of $x_2x_4t^n$ in $F_{2,4}(x_2,x_4,t)$ is the number of paths in $\mathscr{L}_n$ which bounce off the diagonal to the right exactly once and cross the diagonal vertically exactly one. The corresponding generating function equals

$$\begin{aligned}&\left.\frac{\partial^2 F_{2,4}(x_2,x_4,t)}{\partial x_2\partial x_4}\right|_{x_2=x_4=0}\\ &= -\frac{\left(-1+\sqrt{1-4t}\right)^3\left(-2+\sqrt{1-4t}\right)\left(-1+\sqrt{1-4t}+2t\right)}{16t^2}\\ &= t^3+7t^4+32t^5+129t^6+495t^7+1859t^8+\cdots,\end{aligned}$$

which has no matches in the OEIS [14].

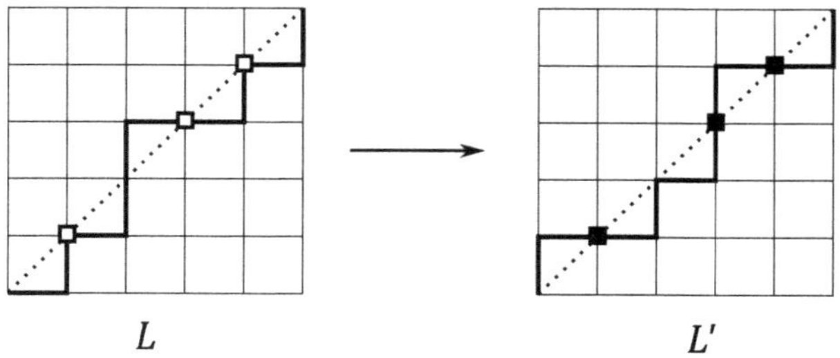

Fig. 11 L is mapped to L' by the bijection

As we mentioned, $F_{2,4}(x, x, t)$ is the generating function for the times of paths touching the diagonal $y = x$ with a north step,

$$\begin{aligned} F_{2,4}(x, x, t) &= \frac{1 - \sqrt{1-4t} - t - x + x^2 t}{-x + (1+x)^2 t} \\ &= 1 + 2t + (2x+4)t^2 + (2x^2 + 8x + 10)t^3 + (2x^3 + 12x^2 + 28x + 28)t^4 + \cdots \\ &= F_{2,5}(x, x, t) = F_{3,4}(x, x, t). \end{aligned}$$

This fact can be also shown by constructing a bijection. Let $\mathscr{C}_{n,k}$ denote the set of paths in $\mathscr{L}_n$ that cross the diagonal k times and $\mathscr{T}_{n,k}$ denote the set of paths in $\mathscr{L}_n$ that touch the diagonal with a north step k times.

Next, we shall construct a bijection between $\mathscr{T}_{n,k}$ and $\mathscr{C}_{n,k}$, which is similar to the bijection defined in Sect. 3.3. For any path $L \in \mathscr{T}_{n,k}$, assume L touches the diagonal j times and these positions are denoted by $\{p_1, p_2, \ldots, p_j\}$. We let $p_0 = [0, 0]$ and $p_{j+1} = [n, n]$. If p_i is a bouncing right position or p_i is a horizontal crossing position, we flip the part between p_{i-1} and p_i along the diagonal. Then we obtain a new path L'. In this bijection, we can see that the number of crossings of L is equal to the number of north-touchings of L', and the number of north-touchings of L is equal to the number of crossings of L'. For example, in Fig. 11, L is mapped to L' and $\{P_2, P_4\}$-mch$(L) = \{P_3, P_4\}$-mch$(L') = 3$ and $\{P_3, P_4\}$-mch$(L) = \{P_2, P_4\}$-mch$(L') = 2$.

Because $F_{2,4}(x, x, t) = F_{3,4}(x, x, t)$, for a random $L \in \mathscr{L}_n$,

$$\mathbb{E}[\{P_2, P_4\}\text{-mch}(L)] \sim \frac{\pi n}{2} - 1 \approx 0.886\sqrt{n}$$

4.5 P_2, P_3, P_4, and P_5

The last example of this section is a generating function of a subset of $\{1, \ldots, 6\}$ of size 4. That is, we shall study the generating function

$$F_{2,3,4,5}(x_2, x_3, x_4, x_5, t) := 1 + \sum_{n\geq 1} t^n \sum_{L\in\mathscr{L}_n} x_2^{P_2\text{-mch}(L)} x_3^{P_3\text{-mch}(L)} x_4^{P_4\text{-mch}(L)} x_5^{P_5\text{-mch}(L)}.$$

For convenience, in this subsection we use $F_{2,3,4,5}$ to denote $F_{2,3,4,5}(x_2, x_3, x_4, x_5, t)$, $G_{2,3,4,5}$ to denote $G_{2,3,4,5}(x_2, x_3, x_4, x_5, t)$, and $H_{2,3,4,5}$ to denote $H_{2,3,4,5}(x_2, x_3, x_4, x_5, t)$ where

$$G_{2,3,4,5} := 1 + \sum_{n\geq 1} t^n \sum_{L\in\mathscr{L}_n \text{ starting with } E} \prod_{k=2}^{5} x_k^{P_k\text{-mch}(L)}$$

and

$$H_{2,3,4,5} := 1 + \sum_{n\geq 1} t^n \sum_{L\in\mathscr{L}_n \text{ starting with } N} \prod_{k=2}^{5} x_k^{P_k\text{-mch}(L)}.$$

Similar to the recurrences used in previous subsections, we have

$$\begin{aligned} G_{2,3,4,5} &= \sum_{i\geq 0}\sum_{j\geq 1} C_{i,j} x_2^i t^j \left(x_4 H_{2,3,4,5} + 1\right) \\ &= (C(x_2, t) - 1)\left(x_4 H_{2,3,4,5} + 1\right) \end{aligned}$$

and

$$\begin{aligned} H_{2,3,4,5} &= \sum_{i\geq 0}\sum_{j\geq 1} C_{i,j} x_5^i t^j \left(x_3 G_{2,3,4,5} + 1\right) \\ &= (C(x_5, t) - 1)\left(x_3 G_{2,3,4,5} + 1\right) \end{aligned}$$

Combining the two equations above, we can solve them for $G_{2,3,4,5}$,

$$G_{2,3,4,5} = \frac{(C(x_2, t) - 1)(x_4(C(x_5, t) - 1) + 1)}{x_3 x_4 (C(x_2, t) - 1)(C(x_5, t) - 1) - 1}.$$

Then

$$\begin{aligned} F_{2,3,4,5} &= 1 + G_{2,3,4,5} + H_{2,3,4,5} \\ &= 1 + G_{2,3,4,5} + (C(x_5, t) - 1)\left(x_3 G_{2,3,4,5} + 1\right) \\ &= C(x_5, t)\left(x_3 G_{2,3,4,5} + 1\right) + (1 - x_3) G_{2,3,4,5} \\ &= \frac{P(x_2, x_3, x_4, x_5, t)}{Q(x_2, x_3, x_4, x_5, t)}, \end{aligned}$$

where

$$P(x_2,x_3,x_4,x_5,t) = \Big((-1+\sqrt{1-4t}+2t)x_3(-1+x_4)+x_4-\sqrt{1-4t}x_4 \\ -2tx_4+x_2(-(-1+\sqrt{1-4t})(-2+x_5)-2t(-1+x_5)\Big) \\ +2\sqrt{1-4t}x_5+2tx_5-2(-2+\sqrt{1-4t}+x_5)$$

and

$$Q(x_2,x_3,x_4,x_5,t) = 2+(-1+\sqrt{1-4t}+2t)x_3x_4+(-1+\sqrt{1-4t})x_5 \\ +x_2\left(-1+\sqrt{1-4t}-(-1+\sqrt{1-4t}+2t)x_5\right).$$

One can imagine that even a few initial terms of $F_{2,3,4,5}(x_2,x_3,x_4,x_5,t)$ are very long so that we will not list them here. However, it is easy to verify that the constant coefficient of t^n is just $2C_n$ because there are two sets of Dyck paths, namely the ones that stay on or above the diagonal and the ones that stay on or below the diagonal.

By manipulating $F_{2,3,4,5}(x_2,x_3,x_4,x_5,t)$, one is able to answer certain complicated enumerative problems, such as how many paths in $\mathcal{L}_n$ are there that cross the diagonal vertically exactly once and horizontally exactly twice, and bounce off the diagonal to the right once but not to the left. The answer to this question has the generating function as follows,

$$\frac{1}{2!}\left.\frac{\partial^4 F_{2,3,4,5}(x_2,x_3,x_4,0,t)}{\partial x_2 \partial x_3^2 \partial x_4}\right|_{x_2=x_3=x_4=0} \\ = \frac{\left(1-\sqrt{1-4t}\right)^5}{16} \\ = 2t^5+10t^6+40t^7+150t^8+550t^9+2002t^{10}+7280t^{11}+\cdots.$$

Amazingly, the sequence 2, 10, 40, 150550, 2002, ... is twice the sequence A000344 in the OEIS [14], which has interpretations such as the number of paths in $\mathcal{L}_{n-3}$ that touch but do not cross $y = x - 2$ and the number of standard tableaux of shape $(n-1, n-5)$. We leave open the problem of finding a bijective proofs of these facts.

Next, we consider the formula $F_{2,3,4,5}(x,x,x,x,t)$ which gives us the generating functions for the times of touching the diagonal,

$$F_{2,3,4,5}(x,x,x,x,t) = \frac{1+(x-1)\left(-1+\sqrt{1-4t}\right)}{1+\left(-1+\sqrt{1-4t}\right)x} \\ = 1+2t+(4x+2)t^2+(8x^2+8x+4)t^3+(16x^3+24x^2+20x+10)t^4 \\ +(32x^4+64x^3+72x^2+56x+28)t^5+\cdots.$$

Next, we want to ask for a random $L \in \mathscr{L}_n$ how many times in average that L touches the diagonal. Applying the same idea that we used in previous sections, we see that

$$\begin{aligned}\left.\frac{\partial F_{2,3,4,5}(x,x,x,x,t)}{\partial x}\right|_{x=1} &= \frac{\left(\sqrt{1-4t}-1\right)^2}{4t-1}\\ &= 4t^2+24t^3+116t^4+520t^5+2248t^6+9530t^7+\cdots\\ &= 4\left.\frac{\partial F_2(x,t)}{\partial x}\right|_{x=1}.\end{aligned}$$

So for a random $L \in \mathscr{L}_n$, the expectation of times L touches the diagonal is that

$$\mathbb{E}[\{P_2, P_3, P_4, P_5\}\text{-mch}(L)] = \frac{4^n - 4\binom{2n-1}{n-1}}{\binom{2n}{n}} \sim \sqrt{\pi n} - 2 \approx 1.772\sqrt{n}.$$

Similarly, we can also obtain the generating functions for the number of paths touching the diagonal an even number of times or an odd number of times. We have computed that

$$\begin{aligned}&\frac{1}{2}\left(F_{2,3,4,5}(1,1,1,1,t)+F_{2,3,4,5}(-1,-1,-1,-1,t)\right)\\ &= \frac{4t+\sqrt{1-4t}}{4t+2\sqrt{1-4t}-1}\\ &= 1+2t+2t^2+12t^3+34t^4+132t^5+468t^6+1752t^7+6530t^8+\cdots. \end{aligned} \tag{17.20}$$

and

$$\begin{aligned}&\frac{1}{2}\left(F_{2,3,4,5}(1,1,1,1,t)-F_{2,3,4,5}(-1,-1,-1,-1,t)\right)\\ &= -\frac{2\left(-1+\sqrt{1-4t}+2t\right)}{-1+2\sqrt{1-4t}+4t}\\ &= 4t^2+8t^3+36t^4+120t^5+456t^6+1680t^7+6340t^8+\cdots.\end{aligned}$$

Neither of the two series have matches in the OEIS [14].

By observing Eqs. (17.17) and (17.18), we find that coefficient of t^k in Eq. (17.20) is equal to

$$\begin{cases}\text{coefficient of } t^k \text{ in } \frac{1}{2}(F_{1,6}(1,1,t)-F_{1,6}(1,1,t)), \text{ if } k \text{ is even}\\ \text{coefficient of } t^k \text{ in } \frac{1}{2}(F_{1,6}(1,1,t)+F_{1,6}(1,1,t)), \text{ if } k \text{ is odd.}\end{cases}$$

This is because all the six patterns in $\mathscr{L}_2$ are mutually exclusive. For any path $L \in \mathscr{L}_k$, $\mathscr{L}_2$-mch$(L) = k-1$, which implies that

$$\{P_1, P_6\}\text{-mch}(L) + \{P_2, P_3, P_4, P_5\}\text{-mch}(L) = k - 1.$$

If k is odd, $\{P_1, P_6\}$-mch(L) and $\{P_2, P_3, P_4, P_5\}$-mch(L) have the same parity and otherwise, they do not.

5 Future Research

In this paper, we computed the generating functions $F_{P_k}(x, t)$ for $k = 1, \ldots 6$ and $F_\Delta(\mathbf{x}, t)$ for certain selected $\Delta \subseteq \{1, \ldots, n\}$. In a subsequent paper, we will systematically compute $F_\Delta(\mathbf{x}, t)$ for all sets of size two. There are only nine such generating functions up to symmetry and we have computed five of them since $F_{P_2,P_3}(x_2, x_3, t)$ is a specialization of $F_{2,3,4,5}$. The ones that we have not computed in the paper are represented by $F_{P_1,P_2}(x_1, x_2, t)$, $F_{P_1,P_3}(x_1, x_3, t)$, $F_{P_1,P_4}(x_1, x_4, t)$, and $F_{P_1,P_5}(x_1, x_5, t)$. As a special case for pattern P_1, P_2, $F_{1,2}(x, x, t)$ keeps track of the number of paths in $\mathscr{L}_n$ that have k steps below the diagonal. For any fixed k, the coefficient of $x^k t^n$ in $F_{1,2}(x, x, t)$ is also Catalan number C_n, shown by Chung and Feller [1]. We shall explore these generating functions in a subsequent paper where we will also add some additional parameters which keep track of both the area below the diagonal and the area above the diagonal in path in $\mathscr{L}_n$.

There are many interesting bijective problems that arise from our results. For example, in Sect. 3.1, we find that the total east steps below $y = x - 1$ of all the paths in $\mathscr{L}_n$ is equal to the total area under all Dyck paths in $\mathscr{L}_n$. We take $\mathscr{L}_3$ as an example, there are 6 paths having P_1-matches and there are 5 Dyck paths, pictured in Fig. 12. The total east steps below $y = x - 1$ are equal to $2 + 2 + 1 + 1 + 1 + 1 = 8$ and the total area under all the Dyck paths is also equal to $0 + 1 + 2 + 2 + 3 = 8$. Although how to design the bijection is unknown, it is interesting to see paired pattern matching does have connection to other statistics for lattice paths.

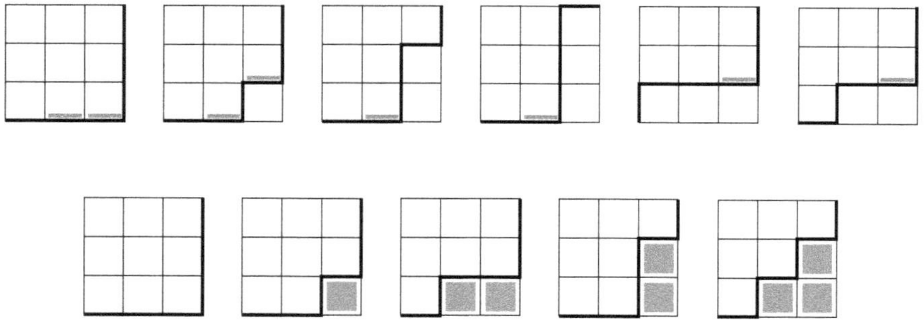

Fig. 12 Total number of east steps below $y = x - 1$ in $\mathscr{L}_n$ equals the total area below Dyck paths in $\mathscr{D}_n$, $n = 3$ as an example

Another direction for further research is to consider Delannoy paths. In this paper, we only consider paths consisting of north steps [0, 1] and east steps [1, 0]. Naturally, we can extend our definitions to Delannoy paths which are paths consisting of east steps [1, 0], north steps [0, 1], and northeast steps [1, 1] which start at [0, 0] and end at $[n, n]$. We denote the steps [1, 0], [0, 1], and [1, 1] by E, N, and D, respectively. The set of all the Delannoy paths from [0, 0] to $[n, n]$ is denoted by $\mathscr{S}_n$.

According to [12], a Schröder path is a path from [0, 0] to $[n, n]$ consisting of east steps [1, 0], north steps [0, 1], and northeast steps [1, 1] which never goes above the diagonal $y = x$. The number of Schröder paths from [0, 0] to $[n, n]$ is counted by large Schröder number D_n whose ordinary generating function equals

$$D(x) = \sum_{n\geq 0} D_n x^n = \frac{1 - x - \sqrt{1 - 6x + x^2}}{2x} = 1 + 2x + 6x^2 + 22x^3 + 90x^4 + 394x^5 + \cdots .$$

The nth little Schröder number $\tilde{D}(n)$ counts the number of Schröder paths from [0, 0] to $[n, n]$ without northeast steps on the diagonal $y = x$ whose ordinary generating function equals

$$\tilde{D}(x) = \sum_{n\geq 0} \tilde{D}_n x^n = \frac{1 + x - \sqrt{1 - 6x + x^2}}{4x} = 1 + x + 3x^2 + 11x^3 + 45x^4 + 197x^5 + \cdots .$$

Here, we adopt the same definition of paired pattern for Delannoy paths. For example, in Fig. 13, $L = EDNDDNNEDE \in \mathscr{S}_7$. $ps_L(1, 2) = ENNE = P_4$ and $ps_L(2, 3) = NNEE = P_6$, that is, P_4-mch$(L) = P_6$-mch$(L) = 1$. It matches our observation: L crosses the diagonal $y = x$ 'vertically' once and there is one east step above $y = x + 1$.

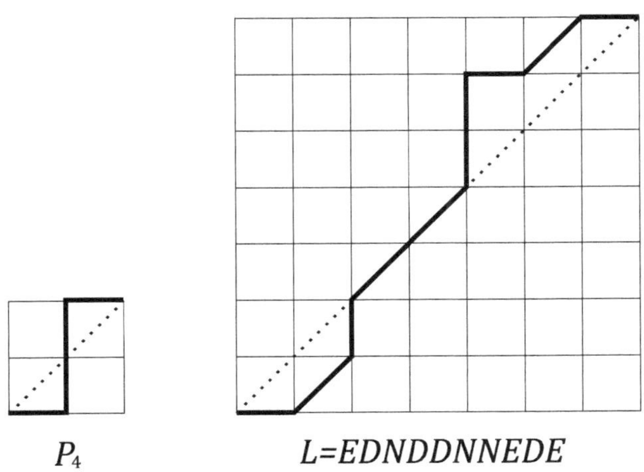

Fig. 13 $L = EDNDDNNEDE \in \mathscr{S}_7$

We take pattern P_4 as example, P_4-mch(L) is the number of times L crosses the diagonal $y = x$ vertically. We shall study the ordinary generating function

$$FS_4(x,t) := 1 + \sum_{n\geq 1} t^n \sum_{L\in\mathscr{S}_n} x^{P_4\text{-mch}(L)}.$$

We split the discussion into two cases. Case 1 is the paths in $\mathscr{S}_n$ that start with a north step and Case 2 is the path in $\mathscr{S}_n$ that start with an east step or a northeast step. We define

$$GS_4(x,t) := \sum_{n\geq 1} t^n \sum_{L\in\mathscr{S}_n \text{ starting with } E \text{ or } D} x^{P_4\text{-mch}(L)}$$

and

$$HS_4(x,t) := \sum_{n\geq 1} t^n \sum_{L\in\mathscr{S}_n \text{ starting with } N} x^{P_4\text{-mch}(L)}.$$

Clearly,

$$FS_4(x,t) = 1 + GS_4(x,t) + HS_4(x,t).$$

We obtain following formulas based on recursion on where is the first time the path starting with 'E' or 'D' crosses the diagonal $y = x$ from bottom to top.

$$GS_4(x,t) = \left(D(t) - \frac{1}{1-t}\right)(xHS_4(x,t) + 1) + \frac{t}{1-t}(HS_4(x,t) + 1)$$

Similarly, we consider where is the first time a path starting with a north step and having no northeast steps on the diagonal crosses the diagonal 'horizontally.'

$$HS_4(x,t) = \left(\tilde{D}(t) - 1\right)(GS_4(x,t) + 1)$$

Solving for $GS_4(x,t)$, we have

$$GS_4(x,t) = -\frac{(t-1)D(t)((\tilde{D}(t)-1)x+1) + (\tilde{D}(t)-1)x - t + 1}{(\tilde{D}(t)-1)x(D(t)(t-1)+1) - 2t + 1}$$

Then we have

$$\begin{aligned}
FS_4(x,t) &= 1 + GS_4(x,t) + HS_4(x,t) \\
&= 1 + GS_4(x,t) + \left(\tilde{D}(t) - 1\right)(GS_4(x,t) + 1) \\
&= \tilde{D}(t)(GS_4(x,t) + 1) \\
&= \frac{2}{3 + \sqrt{1-6t+t^2} - \frac{2(x-1)}{t-1} + t(x-1) - 3x + \sqrt{1-6t+t^2}x}.
\end{aligned}$$

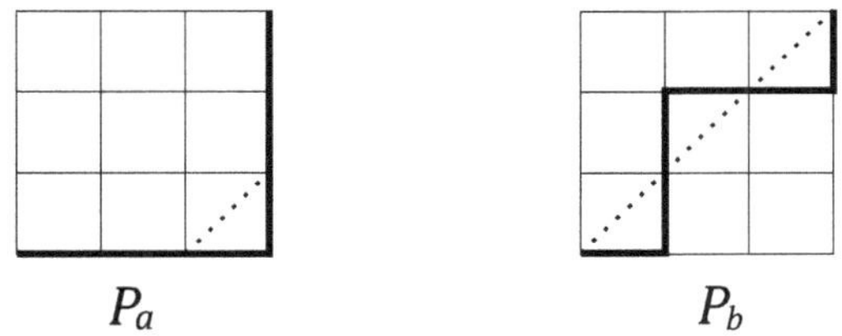

Fig. 14 Examples of two patterns in $\mathscr{L}_3$

A few initial terms of $FS_4(x,t)$ are

$$\begin{aligned} FS_4(x,t) =& 1 + 3t + (x+12)t^2 + (11x+52)t^3 + (x^2+84x+236)t^4 \\ &+ (19x^2+556x+1108)t^5 + (x^3+220x^2+3428x+5340)t^6 + \cdots . \end{aligned}$$

By setting $x = 0$ in $FS_4(x,t)$, we obtain the generating function of the number of Delannoy paths that do not cross the diagonal vertically,

$$\begin{aligned} FS_4(0,t) &= \frac{(t-1)\left(-1+3t+\sqrt{1-6t+t^2}\right)}{t^2\left(3-t+\sqrt{1-6t+t^2}\right)} \\ &= 1 + 3t + 12t^2 + 52t^3 + 236t^4 + 1108t^5 + 5340t^6 + \cdots . \end{aligned}$$

The sequence $1, 3, 12, 52, 236, \ldots$ does not appear in the OEIS [14].

Finally, one can study pattern matching for paired patterns in both lattice paths and Delannoy paths for patterns P of length ≥ 6. For example, based on Definitions 17.1 and 17.2 and Theorems 17.2 and 17.3, one can obtain geometric interpretations for the number of P-matches in a path L.

For example, consider the two patterns P_a and P_b are pictured in Fig. 14. Note that P_a has one east step below $y = x - 2$ and P_b has a vertical crossing immediately followed by a horizontal crossing. For any path $L \in \mathscr{L}_n$, P_a-mch(L) can be interpreted as the number of east steps of L below $y = x - 2$ and P_b-mch(L) can be interpreted as the number of such pairs of crossings of L.

References

1. Chung, Y., Feller, W.: On fluctuations in coin tossing. Proc. Natl. Acad. Sci. USA **35**, 605–608 (1949)
2. Deutsch, E.: Dyck path enumeration. Discrete Math. **204**, 167–202 (1999)
3. Goldman, J., Sundquist, T.: Lattice path enumeration by formal schema. Adv. Math. **13**, 216–251 (1992)
4. Harmse, J., Remmel, J.B.: Patterns in column strict fillings of rectangular arrays. Pure Math. Appl. **22**, 131–171 (2011)

5. Humphreys, K.: A history and a survey of lattice path enumeration. J. Stat. Plan. Inference **140**, 2237–2254 (2010)
6. Krattenthaler, C.: Lattice path enumeration. In: Bóna, M. (ed.) Handbook of Enumerative Combinatorics. Discrete Mathematics and Its Applications, pp. 589–678. CRC Press, Boca Raton (2015)
7. Manes, K., Sapounakis, A., Tasoulas, I., Tsikouras, P.: Counting strings at height j in Dyck paths. J. Stat. Plan. Inference **141**, 2100–2107 (2011)
8. Mansour, T.: Statistics on Dyck paths. J. Integer Seq. **9** (2006). Article 06.1.5
9. Mohanty, S.G.: Lattice Path Counting and Applications. Academic Press, New York (1979)
10. Sapounakis, A., Tsikouras, P., Tasoulas, I., Manes, K.: Strings of length 3 in grand-Dyck paths and the Chung–Feller property. Electron. J. Comb. **19**(2) (2011). Article #P2
11. Sapounakis, A., Tasoulas, I., Tsikouras, P.: Counting strings in Dyck paths. Discrete Math. **307**, 2909–2924 (2007)
12. Schröder, E.: Vier kombinatorische Probleme. Z. Math. Phys. **15**, 361–376 (1870)
13. Sen, K., Jain, J.L., Gupta, J.M.: Lattice path approach to transient solution of M/M/1 with $(0, k)$ control policy. J. Stat. Plan. Inference **34**, 259–267 (1993)
14. Sloane, N.J.A.: The on-line encyclopedia of integer sequences. http://oeis.org/

GPSR Compliance
The European Union's (EU) General Product Safety Regulation (GPSR) is a set of rules that requires consumer products to be safe and our obligations to ensure this.

If you have any concerns about our products, you can contact us on

ProductSafety@springernature.com

In case Publisher is established outside the EU, the EU authorized representative is:

Springer Nature Customer Service Center GmbH
Europaplatz 3
69115 Heidelberg, Germany

www.ingramcontent.com/pod-product-compliance
Ingram Content Group UK Ltd.
Pitfield, Milton Keynes, MK11 3LW, UK
UKHW021315130926
15127UKWH00002B/10
9783030111014